MICROCOSMUS.

MICROCOSMUS:

AN ESSAY CONCERNING MAN AND HIS RELATION TO THE WORLD.

BY

HERMANN LOTZE.

Translated from the German

BY

ELIZABETH HAMILTON AND E. E. CONSTANCE JONES

IN TWO VOLUMES.

VOL. II.

BOOKS FOR LIBRARIES PRESS
FREEPORT, NEW YORK

First Published 1885
Reprinted 1971

122653

NOTE.

THE whole of this volume has been translated by Miss
E. E. CONSTANCE JONES.

INTERNATIONAL STANDARD BOOK NUMBER:
0-8369-5989-2

LIBRARY OF CONGRESS CATALOG CARD NUMBER:
76-169769

PRINTED IN THE UNITED STATES OF AMERICA

TABLE OF CONTENTS.

——o——

BOOK VI.

THE MICROCOSMIC ORDER;[1] OR, THE COURSE OF HUMAN LIFE (*Der Weltlauf*).

CHAPTER I.

THE INFLUENCES OF EXTERNAL NATURE.

CHAPTER II.

THE NATURE OF MAN.

CHAPTER III.

MANNERS AND MORALS.

Cf. Book VI. ch. i. § 1, especially pages **4, 5.**

                        BOOK VII.

                         HISTORY.

                        CHAPTER I.

                   THE CREATION OF MAN.

CHAPTER II.

THE MEANING OF HISTORY.

CHAPTER III.

THE FORCES THAT WORK IN HISTORY.

CHAPTER IV.

EXTERNAL CONDITIONS OF DEVELOPMENT.

CHAPTER V.

THE DEVELOPMENT OF HISTORY.

BOOK VIII.

PROGRESS

CHAPTER I.

TRUTH AND SCIENCE (*das Wissen*).

CHAPTER II.

WORK AND HAPPINESS.

CHAPTER III.

BEAUTY AND ART.

CHAPTER IV.

THE RELIGIOUS LIFE.

CHAPTER V.

POLITICAL LIFE, AND SOCIETY.

BOOK IX.

THE UNITY OF THINGS.

CHAPTER I.

OF THE BEING OF THINGS.

CHAPTER II.

THE SPATIAL AND SUPERSENSUOUS WORLDS.

CHAPTER II.

THE REAL AND THE IDEAL.

CHAPTER IV.

THE PERSONALITY OF GOD.

CHAPTER V.

GOD AND THE WORLD.

---

[1] I have tried to make plain the antithesis in this Chapter between (1) *real, Reale, Realität,* and (2) wirklich, Wirkliche, Wirklichkeit, by writing *Real, Realness* in the text for (1). As this way of marking the difference did not occur to me until the Chapter was in print, the question of making previous Chapters correspond could not be considered.

BOOK VI.

THE MICROCOSMIC ORDER.

CHAPTER I.

THE INFLUENCES OF EXTERNAL NATURE.

THE INFLUENCES OF EXTERNAL NATURE.

History, and the Microcosmic Order—The Effects of Cosmic and Terrestrial
Influences upon the Human Soul—Parallelism between the Macrocosm and
the Microcosm—Natural Features of a Country, and Character of the
Inhabitants—Life with Nature—Relation of Man to Nature.

§ 1. BYGONE times which are beyond the reach of our
own recollection seem to imagination extremely
obscure. All the serious interests of life and all the trifling
and folly by which we ourselves are stirred, are so closely
bound up with clear and definite images of our surroundings,
that we feel perplexed and astray when we would picture to
ourselves the same varied wealth of existence in times divided
from the present by an infinite series of changes by which the
background and accessories of life have been transformed
We almost fancy that in those olden days the sun must have
shone with a different radiance, that the voices of Nature
must have spoken in different tones, and the world have lain
in twilight as contrasted with our present life of noontide
brightness. History indeed depicts for us on this sober background great deeds and mighty events, but is for the most
part silent concerning the small causes which combined
to produce them. How the heroes of classic times were
housed and clothed, what was their manner of speech, and
how they filled up the blank intervals of time between
their mighty deeds, is left for the most part to be determined
by our own wandering fancy. There are but few periods of
human history which have left us, in works of art, speaking
monuments from which, besides the glory of heroic deeds,
we also learn something of the stirrings of men's minds, the
philosophic views, the conflict and the joy from which sprang

those great results.   But however truly and naturally poetry may reproduce for us many of the features of everyday life, it must naturally leave many gaps, and it is most difficult for us to ascribe to the thoughts of such far-off personages in their treatment of common things that familiarity and supple ease upon the vividness and completeness of which our own sense of life principally depends.   Every delineation of long past times which we attempt seems to us true in proportion as it emphasizes particular points of importance, jumping from one to another—and this not only because, on account of our lack of historical knowledge, we are unable to clothe the skeleton of narrative with the flesh and blood by which its different parts were connected in reality, but also because it is extremely difficult for us to get rid of the notion that in those old days everything was said and done after a stilted fashion that would have suited the immobility of marble statues.   If in the writings of antiquity we come across some graceful trait instinct with life, some touch of unaffected fun, some vivid description of scenery sketched in a few careless strokes, how great, even now, is the concourse of wondering interpreters calling upon us to admire this classical revelation of genuine human nature !   As if we could have expected anything else—as if we might not have supposed that a cultivated people of antiquity would be susceptible to all the minor charms and beauties of life, and would have found expressions for their emotions as adequate as those which are familiar to the mouth of every modern booby !   No doubt the course of history has by degrees produced variations of colouring in human imagination, and greatly widened its scope by increasing knowledge of men's earthly abiding-place, by intercourse between nations, and by gradually enlarging acquaintance with the world of Ideas ; but not the whole of life is included in this forward movement ; there is a region of human existence in which, at all times, the same ends, motives, and customs recur without any alteration.   All the generations that have passed away have dreamed and observed, loved and hated, hoped and despaired, worked and played, just as we do, and those who

come after us will do the same. The same passions which
move us, the same intriguing calculations of greed and ambition, the same hidden motives, or the same unreserved devotion
of affection, which we praise or blame in one another—all
these have from the earliest times worked in human hearts.
And though external results exhibit various forms and dimensions according to the direction and degree of culture at any
given time, there is still no doubt that we are mistaken if,
putting faith in a foolish analogy, we imagine that we can
find among primitive men nothing but the inconsiderateness
and empty-headedness of children.

This it is that we mean by the *Microcosmic Order*—
the impulses ever fresh and ever the same, out of which
have sprung the many-hued blossoms of history, the eternal
cycle in which human fates revolve. It is indeed true that
this order may not be strictly a cycle, but that the apparent
recurrence may include some hidden progress. Still even we,
who live in times in which at any rate the outward splendour
of progress is unfolded more vividly than ever before our eyes,
even we may say to ourselves that the true value of our inner
life is but slowly if at all increased by all this. There arise
no fresh springs of enjoyment which had not flowed before, or
if indeed the springs are new, yet that which they distribute is
still but the old pleasure for which our nature is designed; our
cognition may be enlarged boundlessly, but the results almost
always lead us back to thoughts which men have had long
ago. It seems as though former ages had extracted from
different and perhaps poorer material those same treasures of
happy or exalted feeling which we with far greater expenditure
of scientific and technical power imagine we are discovering
anew. In the ordinary view all our labour is for the most
part only a more extensive preparation for life and not itself
a fuller life, though indeed we frankly confess that this is not
altogether true. Progressive culture is not unlike a majestic
waterfall which, seen from a distance, seems to promise great
things, and which yet when we look nearer does not appear
to shower upon the soil of life a greater amount of refreshing

and really fertilizing spray than was afforded for the refreshment and satisfaction of the quieter life of antiquity by the more modest stream of a less splendid civilisation.

We cannot renounce the hope that in this flux and reflux of human development there may be found a tendency towards some finite goal; but before we attempt to trace a plan of historic progress and training of the human race, we would linger for a while over the stationary aspect which is at first presented to us by the struggles and the destiny of men. The spectacle is one which may be regarded with very various feelings. We cannot without an emotion of melancholy see the same evil, the same passions, the same seeds of all wickedness recurring in every age; but, on the other hand, it is a consoling thought that every age has also had access to everything in which men's hearts can find real and essential happiness, and that every age in its own fashion—a fashion which satisfied it—had part in that higher world which has indeed become clearer to us, but is not on that account grasped more strongly by our minds. Our intention for the present is to seek in the nature of human intelligence, and in the ever-recurring conditions of man's life, the ready-made instruments with which Providence works in history; to seek out, that is, the natural order of the world, regarding which we may in a later chapter ask, To what end does the Supreme Will bend the course of its uniform progress?

§ 2. Such being our aim, attention is in the first place attracted to the conditions of external Nature under which we are placed, and their varied influence upon us, whether obvious or unobserved. In so far as these circumstances affect our corporeal life or provide us with means for the satisfaction of our wants, their action is on the whole plain, and in a more detailed consideration than we can here attempt, nothing more would be required than to establish in special cases the relative worth for civilisation of each one of these influences. But reflection is very commonly disposed to take a more profound view of the relation between man and Nature, and instead of measuring the gain or harm which we receive from the latter, or seeking to find the direction which it gives to our action, people prefer to speak of an immediate and more mysterious sympathy which binds man to Nature, and especially to his dwelling-place the earth. Indeed, they prefer to speak of the earth as not merely his dwelling-place, but compare the relation between him and it to the intimate relation subsisting between mother and child, or between a parasite and the organism which supports it; they speak of the powers and tendencies to development which are inherent in the earth as being repeated under more significant forms in the bodies of men; of every internal fluctuation of telluric life as finding an echo in changes of human organization; and say that what earth herself vainly struggles to express, receives a spiritualized manifestation in the constitution of conscious beings.

We have already remarked at length upon the great extent to which the character of organic beings inhabiting the surface of a planet, is determined by the special nature of the planet itself—in respect, that is, of the materials which compose it, and the conditions of mobility and capacity of combination which it prescribes to them. We have referred to a view according to which the connection between the earth and man is different from that just indicated; a view according to which not only is man forced by the nature of his material abiding-place to use particular means for the attainment of *his own* ends, not only is he provided by its continual influences with fresh material which the organism appropriates and elaborates after *its own* fashion, but moreover the whole of this human life is after all only a mystical repetition of the life of the earth, and of its internal tremors. This view seems to owe its convincing power to the strange inclination which men so often have to regard what is unintelligible and indemonstrable as having pre-eminent truth and profundity, especially in cases where the unintelligibility is such that a sort of mysterious awe may attach to it. There is no occasion to deny any one of the actual facts which are usually brought together with reference to the reciprocal relation which we are discussing, but we may be sure that it is only a capricious liking for obscurity which requires that they should be judged from this particular point of view.

How often do such perverse considerations begin with a reference to that alternation of sleep and waking in men, which is in sympathy with the day and night of the world, and to that emotion of dread when darkness sets in from which no one is altogether free! What in fact are night and day for the earth? Is it anything more than an arbitrary play of fancy to call the earth asleep because the noise ceases which we and the other animals are accustomed to make during the day-time? Or because there are no longer those oscillations of ether which by day make it light to *our* eyes, but affect the earth merely by causing a rise of temperature which extends only a few inches below the surface? What other activities are there which rest during the night? Or what dread and fear is there in Nature itself, with which we are in sympathy? It is in us that there is light or darkness, in us that there is serenity or fear, and neither the one nor the other results from our being affected by some pervading condition of the earth, but from the fact that alterations of outward circumstances, indifferent in themselves, are at one time favourable, at another time unfavourable to the requirements of our active nature. Such circumstances act upon the sensitive constitution of our mind, which feels not only how much but also in what way they aid or hinder us, and is able to connect all this with various trains of thought; and these circumstances, so acting, produce mental conditions which are our own property, and are not mere participations in a universal life such as Nature is certainly not capable of. How often, too, is it said that with the changing seasons of the year the bodies and souls of individuals suffer from sympathetic affections, and even that in the course of geologic ages the very nature of men rejoices and mourns with the youth and age of the earthly sphere itself; that convulsions of Nature correspond to all the revolutions of human history; that the temperament and national fancy of the inhabitants of any country are directly affected by the conformation of its land and the prevailing hue of its sky ! We would not deny that these statements have a certain basis of fact ; but it would be better to try and find in each individual case the means by which any natural circumstances have produced in organic life an impression or an echo of themselves. One gains little more than the weird charm of a ghost story by exaggerating with devout admiration what is incommensurable and irrational in these circumstances, instead of trying to remove it by close investigation.

One cannot think without serious regret of this perversion of thought which has, as it were, taken up the mantle of astrology. It has not merely delayed the commencement of more exact research, but has moreover introduced a general fashion of romancing about phænomena which is supposed, with but little show of reason, to involve some specially profound understanding of them. It would no doubt be interesting to investigate historical fluctuations in the bodily and mental condition of mankind in their relations to the physical alterations of the surface of the earth. The history of epidemics teaches us that every visitation of any pestilence encounters different receptivity and various modes of reaction in living bodies, and we can mark out considerable periods of time within which the human frame has a special predisposition to sickness of some one particular type. It is probable that the same combination of inner and outer conditions which causes this striking one-sided susceptibility produces also in persons who are in a healthy state some peculiar modification of general condition and tone. The higher mental interests of mankind might thus at different times be modified by various emotional conditions, sometimes by a relaxed apathetic state, sometimes by a state of great and anxious excitability, and it may possibly be the fact that the peculiar influences of outward Nature upon man in every age have left their traces in the productions of that age, in the colouring of its poetry, in the nature of its favourite superstitions, in the general direction of its intellectual powers. But the most necessary rule of such investigations would be not to try and find what is not there and

not to over-estimate these influences of Nature as compared with
the much more obvious influences which are to be sought in
the uninterrupted transmission by education of the same wants,
problems, interests, and sorrows from one generation to another,
and in the solidarity of social life.   The assertions of parallelism between natural and spiritual revolutions are for the
most part innocent of any such cautious procedure.   It is as
easy to understand how widespread and devastating disease is
developed on the direct path of immediately causal influence,
by a great social upheaval with all its train of unusual bodily
and mental exertion, privations, and wretched substitutes for
the ordinary means of subsistence, as it is to understand how,
conversely, striking natural events, earthquakes, inundations,
or epidemics have caused social movements to result from
physical necessities.   Accounts of plagues in times most
remote from one another unite with melancholy unanimity
in showing us how quickly all the moral obligations of
order, duty, and affection are dissolved under the influence
of terror, how excited and terrified imaginations become
incapable of any sober judgment, and the wildest superstition, alternating with the densest folly, rages unchecked.
Yet it can hardly be, that any great historical revolution has
arisen entirely from such a source; if it were so, the storm of
revolt would be allayed by the alleviation of the physical
distress which aroused it.   To seek here for any more
mysterious connection between cosmic and human life is
but to follow a will-o'-the-wisp.   It is easy to bring forward the facts that the downfall of Grecian civilisation in
the Peloponnesian war, the last struggles of the Roman empire,
the rise of Mohammedanism, the Crusades, the discovery of
America, and the Reformation were contemporaneous with devastating epidemics; but when these coincidences are adduced,
it is forgotten that at many other periods of less historical
importance, and in countries not included in the great stream
of historical development, similar plagues have sometimes
raged under like conditions, and sometimes, having first
arisen under unknown conditions, have spread by means of

the ordinary channels of communication to districts with the social circumstances of which they had originally nothing to do. Plague and yellow fever have continued their ravages in their native haunts down to recent times without any connection with great political events, and it is hardly credible that an outbreak of cholera in India should have been in necessary correspondence with contemporaneous revolutionary movements in Paris.

The same insecurity hovers about our views as to the influence of climate upon the character of a people. We cannot seriously decide offhand that the backward civilisation of negroes is due to the blazing sun that beats down upon their heads and makes it impossible for them to gaze upwards, and by heating their blood to fever point, inspires them with ungoverned passions. Even at the equator the sun is not in the zenith day and night; and when we think of the relaxing and yet exciting effect which our own greatest summer heat has upon ourselves, we forget that in consequence of the long acclimatization of the negro this effect may for him have become so modified as to be merely one of those pleasures of existence which are regarded as matters of course, and are certainly not self-evident barriers to the progress of development. The monotony of the tropical year in contrast to our changing seasons, has also been adduced as another hindrance in the way of advanced civilisation. There are undoubtedly present vital changes which we as certainly feel, changes produced by the transitions of the seasons that cause alterations in our bodily economy, but these changes are little known to us in detail. The mental effect of these natural circumstances is to be found rather in the facts which they present for our observation than in the impressions which our senses immediately receive from them. We learn abundantly from the songs of poets how significant for our emotional life are these great periodic alternations of decay and resurrection to life, with all the hopes and remembrances that attach to their different phases. Not only do we here see our own destiny symbolized in a thousand images appealing to the senses, but also a deeper feeling for the slowly passing phases of human life, and the characteristic advantage of each may certainly be connected with this clear marking-off of time into divisions. Such occasions for thoughtful and self-examining reflection no doubt occur less effectively where blossoms and fruit are always growing and blooming and ripening at the same time as the fresh shoots are budding forth; but even with ourselves the impression of human transitoriness is softened by the way in which the gaps left by death are unobtrusively filled up every moment by creatures newly born into the world. How different it would be if the human race, like the vegetable life of these climates, all together growing old, or blooming in fresh youth, were to die off completely in fixed periods and be replaced by a new growth! But in such a case, who could deduce the absence of historical recollection and historical progress among the black races from the absence of clearly-defined seasons of the year?

The character of the African continent, its isolation and inaccessibility, without bays or gulfs, has seemed to many to be mirrored in the mental constitution of the negro. We cannot deny the influence of this conformation of the land, though we may not hold that it consists in this inexplicable mirroring. It is to be found in the material hindrances to intercourse between nations presented by a wide extent of continent without a corresponding supply of navigable rivers, and the obstacles to a clear comprehension of their position and proximity to one another presented by the absence of any large gulfs and of numerous and well-distributed mountain ranges. In comparing views of scenery, we feel directly that in the simultaneous presentation of a wide extent of country there is something that does one good and seems to enlarge the soul, and that there is a keen pleasure in being able to comprehend in one view a multitude of different but connected objects, enclosed as it were in a firm network of relationships. The notion of being able to reach any place by a given amount of movement in a given direction can never be a substitute for the peculiar impression of clearness which we receive from actually seeing its position with reference to other places. The dweller in the wilderness has at any rate a boundless horizon spread before him; in the interior of a continent where there are no mountain-tops from which one may survey the country, which is otherwise impenetrable to the view on account of its luxuriant vegetation, permanent obscurity invests even adjoining districts, and fancy here could never look with such a far-seeing and penetrating glance into the comprehensive connectedness of human life, as it has done since ancient times from the favoured shores of the Mediterranean Sea. Many points could be found there upon the mountains, the level coast, and the sea itself, from which one could behold at once numerous countries and islands like a wreath of many-coloured flowers, and could watch the busy traffic which connected them all together. In every case where anything complex falls into well-defined groups distinctly marked off from one another, a clearer and more intelligible picture is presented than where immeasurable continuity offers no fixed points of support to the imagination; in this way it is that the alternation of land and water in the Mediterranean region has much facilitated geographical apprehension, and also at the same time aided a part of our knowledge concerning the relation of man to the universe. But we cannot regard these influences nor the hindrances offered to commerce by great unbroken extents of continent as being in themselves sufficient explanation of the backwardness of the negro races; we ourselves look at these latter circumstances chiefly as hindrances to the eager zeal of discoverers; but they could not present really formidable barriers to a steadily progressive, long-continued struggle of native tribes unless reinforced by other causes.

§ 3. If the other condition which must be added to the favour or disfavour of geographical situation in order to explain a small or great degree of progress be sought in the character of the country which is reflected sometimes usefully, sometimes detrimentally in the mental dispositions of its inhabitants, we get upon still more slippery ground, and the observations on this point which we fancy we make do certainly contain an extraordinary amount of æsthetic self-deception. We are justified in expecting that extreme coldness and severity of climate will produce dispositions deficient in quickness and activity, and that greater warmth and uniformity will cause a boundless development of all bodily and mental capacities ; and when our conclusions go no further than this, they are confirmed by a comparison of different nations and the countries which they inhabit. But when we go beyond this and think that we find men's special peculiarities of imagination, civilisation, and mode of thought to be in direct and perceptible harmony with the countries in which they dwell, we are led astray by the circumstance that a country and its inhabitants are ever presented to us in conjunction as making one picture, having therefore that appearance of intrinsic æsthetic connection which comes to be assumed by any fact which is continuously presented to us. The Dutchman in Holland seems to us to be suited only to his own flat, fertile, lowland home, the North American Indian we imagine to be the only fitting denizen of his forests and steppes ; but if we see Mynheer in the Sunda Islands, or the Anglo-Saxon pioneer in the far west of North America, we can hardly say that either the one or the other is in irreconcilable contradiction with his new surroundings, unlike as they are to those of his native place—unless indeed we look with an eye prejudiced by recollection. The same ground which the ancient Greeks once trod is now pressed by the foot of the Turk, and it seems to us that the one race matches the physical background just as naturally and harmoniously as the other. The physical nature of any country is a whole composed of very varied parts, and the nature of its inhabitants is equally complex. The comparison of two pictures both so many-sided and composite is sure to furnish him who is seeking to establish a relationship between them with some evidence in support of his view, if he has a capacity for skilful combination ; it will also furnish without much difficulty, to him who seeks them, points enough in which there exists an inexplicable contrast between the two. The creative power of Nature which produced in India the colossal elephant, produced there also a race of men by no means equally colossal, but on the contrary surprisingly feeble ; one might, however, fancy the cunning and incalculable fierceness of its beasts of prey to be repeated in the dispositions of the human inhabitants, as some have thought that they saw reproduced in them the slender grace of various native plants.

Often the only effect that magnificent scenery has upon the minds of the inhabitants, results from the hindrance which features of great natural beauty present to the ordinary occupations of life ; the dweller in the Alps owes to the character of his home an unusual development of bodily strength, and also of conscious worth fostered by the necessity of continual self-reliance, but he does not receive from it the freedom and breadth and fulness of spiritual interests which it seems to us would fitly correspond with the boundless horizon stretched out before him. The false notions which people so often have of the connection between a country and the dwellers in it, result from neglecting to investigate the actual means by which Nature really comes to operate in mental life. That any object has a definite form and position, is no sufficient reason for our necessarily perceiving that it has that form and position, or even for perceiving it at all ; our doing so depends upon whether its form and position and all its qualities are presented to our eye and our mind through the effects which it has upon us. And it is not enough that the vault of heaven should stretch above us in various degrees of blueness and purity and brightness, that we should be surrounded with bolder mountains and more luxuriant vegetation; in order to understand the educative influence of Nature upon us, we ought to know first what circumstances make a noticeable physical impression upon us ; secondly, for how much of the æsthetic worth of these spectacles we have the capacity of reception which is a condition of feeling this worth and of assimilating it for the needs of general development ; and lastly, how much of it all is lost upon us because it is obscured

in our consciousness by other influences which are responded
to by more pressing natural interests.   In so far as mental
character depends upon external Nature, it does not depend
upon what this Nature is, but upon how it affects the as yet
untutored minds of men who are habitually surrounded by it.
The effect of external Nature is not to be directly estimated
by considering the impression that it makes upon a mind that
is already educated, and that comes to it merely as a spectator
and not as dwelling with it and in the midst of it.

On the whole, one hears much said of those happier times
when there was more intimate communion between man and
Nature, and we wish that we could return to that transparently
simple existence and leave the clouds of sophistication in
which our modern life is wrapped.   This longing may be
justified if what it desires is social arrangements a little more
in accordance with the natural impulses of humanity, and free
from the excess of traditional trammels by which we are at
present hemmed in; but it is certainly wrong if it expects
that a fuller enjoyment of external Nature as contrasted with
social life, would produce more exalted happiness and a truer
development of humanity.   When a man exhausted by the
interminable distractions of his daily occupation hastens to
open the great book of Nature and to read therein, he scarcely
notices that which is the only redeeming touch of truth amid
all the pedantry and folly of the fancy picture to which we
have referred; the admission that Nature has a permanent
charm only for the mind accustomed to dwell on some
great connected system of interests, whether scientific or social,
or for the soul that having been thus exercised now finds in
external phænomena innumerable reminders of the experience
of his life, living solutions of his doubts, refutations of his
prejudices, confirmations of his hopes, and incitements to
further investigation.   It is the culture of the heart and the
understanding developed by the relations between man and
man which first makes us capable of receiving further culture
at the hand of Nature; a man who has always lived and who
continues to live alone with Nature, would be hardly more

stirred by her influences than the wild animals who live on amidst all this glorious beauty without being softened or ennobled by it.

One may be enthusiastic about the life which a hunter leads who wanders through the American forests and prairies alone with Nature; but the intelligent glance that can take in and enjoy the changing phænomena of his surroundings he owes to his early education and to the (perhaps long unheard) language of his own people, which calls up along with every fresh thought a thousand remembrances of the home and the civilisation which he has left; and the intellectual dower which he has received from these is just as indispensable to him as are the material aids of civilisation. How the young romance over adventurous wanderings, and think that they could plunge with full satisfaction into the lonely enjoyment of Nature! And they do not remark what a large share of their pleasure is due to the sociability of travel, and how little the continuous absence of friendly intercourse with men could be supplied even by the countless occasions of far-reaching trains of thought which Nature furnishes to the instructed mind. The continued view of some striking natural beauty, operates upon the mind, if we are alone, as a gradually increasing pressure, as an impulse which fails to find its object. This tension, half pleasant, half painful, is lessened but not quite removed by the consciousness that others share it with us; for it arises not only from the need for sympathy, but also from our feeling that we really do not know exactly what this beauty of Nature should prompt us to do. For it is in human nature to be prompted to some action by anything which interests it; it cannot remain long in a condition of passive enjoyment without feeling the inner restlessness of unsatisfied activity. But it is into such a position that Nature always forces us at first; all its visional splendour, however clearly it may be spread before us, is yet something of which in itself we can have no intimate comprehension. It is indisputable that the light, and the sunset glory, and the fresh green of spring, and the wonderful outlines of hill and valley, take our spirit captive with their charm; but all this glorious beauty is voiceless, nor do we know aright what we would have of it; we can never get nearer to any of these phænomena, and though the light should shine for ever, and the woods be ever green, our enjoyment of these living pictures would never be heightened or increased in significance if we did not supplement them by thoughts of our own. What, indeed, are they to us? The answer would be easy if we could embrace the sunset glow, or feed upon the green beauty of the woods; or if it were possible, in any way, to probe somewhat deeper, and with a more active exercise of our own powers, the " open secret" of Nature—open and yet so close—to sound this seeming depth, which on nearer inspection is ever seen to be for us a mere—and yet impenetrable—surface. Since however this cannot be, our interest in a riddle which seems insoluble dies out; we always indeed retain a capacity of being freshly roused by it, but it cannot occupy the mind continuously and alone. Suppose we have reared some plant with the greatest care and pains, when at last the blossoms appear, a sort of helplessness comes over us, as if we did not know what to do next; our interest is momentarily re-awakened when we show it to others; but to look at it for long together, makes us inclined to ask, What is the use of it? We should not wish to see the most charming prospect spread out before us for ever without alteration; there is not enough meaning in it; all these things suffice only to make a pleasing background for life itself; they are graces of existence which we lay aside and return to again. A day of lonely enjoyment of Nature, although enriched by all the intellectual delight that may be derived from solitary reading, secretly seems to us incomplete and half-wasted, unless a word with some fellow-creature crowns the day, reminding us of that community of human life in which we are included. I believe that such emotions occur in every one who observes himself, and they explain the profound sense of discord and the discomfort produced in us by the laboured attempts of a good deal of feeble poetry to entertain us by continual immersion in the mystery and romance of natural phænomena, whilst our heart is hungering not for mere symbols and analogies but for the full pulse of life itself, and thirsting for reality.

These are feelings which belong to civilised life. He who thinks that life is spoiled by such sentiments, and glorifies the primitive condition of mankind as if Nature had been then less impenetrable to human intelligence, indulges fancies which are extremely improbable. We find that the understanding of Nature among those who still have the advantage of living in closer contact with her, is not greater but less than that of those who come to her fresh from social life ; the former are just those to whom that which is useful and the handiwork of men seems decidedly more valuable than the poetry of Nature. And even in the present day we can see by reference to those socially undeveloped peoples who inhabit tracts of land as fertile and beautiful as Paradise, how little immediately educative power there is in the unelaborated influences of Nature. Isolated, and deprived of even the imperfectly organized community with their fellows which these tribes enjoy, men would only feel with still more force that enervating influence which is exercised by natural surroundings, however full of sensuous beauty, as long as they do not arouse either the keenness of scientific search, or that practical faculty of the mind which takes delight in laborious transformations of material objects. But in fact Nature does arouse both, when she creates wants and at the same time affords the means of gratifying them. It has been long maintained, and with truth, that higher development is hindered not only by the extreme disfavour of Nature but also by that excess of bounty which enables men to supply the needs of life without exertion on their part. Human culture began when men began to regard the earth as a fruitful field of labour ; but the beauty and ideal meaning of natural scenery has of itself produced no culture ; it has in fact only become intelligible in proportion as the school of work has trained human thought to form plans and to appreciate the worth of *success*, that is, the worth of the harmony established among disconnected beginnings by their joint contribution to one final result. Man learns to know and to estimate the great value of truth and of faithful law-abiding constancy on which one can depend, when he finds that the soil with unfailing regularity causes the seed entrusted to it to spring up and ripen, or when a successful result crowns some simple attempt in which, relying upon the teaching of his own experience, he seeks to make an artificial arrangement of natural powers serviceable to his own ends. By this time there has crept into his consciousness by imperceptible degrees the conviction of a connection between things which in a general way guarantees some conclusion to every beginning, some result to every experiment, to every like cause a like effect, to all events the possibility of ordered harmony, to every individual thing in the world a certainty of not being isolated or in vain, but of ever finding some way open by which its longing and its activity may be added to the sum of the universe, and in the end make its worth felt. Under whatever forms early mythologic fancy may have pictured the life of Nature, it was in truth a perception of the ordered mechanism of the external world which educated mankind, and it was the steady immutability of this mechanism which first impressed man's sense. He only learnt to understand the frank beauty of Nature in proportion as he became able on the one hand to rejoice in the pervading order of the universe, and on the other hand to feel the bitterness of temporary discord between it and his own individual wishes—becoming able, with the help of such experience, to find the meaning of natural phænomena.

§ 4. Our sceptical observations have up to this point been directed partly against the opinion that the peculiarities of the planet to which we belong reappear in the general features of the human mind, or that particular peoples present a kind of spiritualized reflection of the character of their native land; they are also partly directed against the belief that these mysterious influences of cosmic life further the development of humanity. In making these observations we are renewing a warfare, begun long ago, against the inclination to see in every individual department of reality merely an imitative echo or a prophetic indication of some other department, and in the whole great circle of phænomena nothing more than a continuous shadowing forth of the higher by the lower, and of the lower by the higher. The life of the soul does not appear to us as an image of the life of the body, does not seem to us to be bound to develop some inner activity as a counterpart of every individual function of the body; on the contrary, we hold that all which is material is but a system of means which the mind uses for other than material ends, and with the useful results of which it is concerned, without asking by what system of activities the body has secured this net produce of available stimulation. Again, man is not a mere copy of external Nature, but is a living product, unique in kind— receiving, indeed, innumerable impressions from Nature, yet not in order that he may reflect them back in the form in which they were received, but that he may, in accordance with his nature, be roused by them to reactions and developments, the explanatory cause of which lies in himself, and not in what is external. We are not here denying out and out any determination of man by Nature; we even admit that kind of dependence in accordance with which fluctuations of natural circumstances tinge our inner life with changing hues. We may and do admit that our organic feelings depend upon the weather, our moods upon light and air, the tone of our thought upon season and climate. But on the one hand, it is mere superstition to lay extravagant emphasis upon conditions so difficult to calculate, whilst clear and imperative motives of our reciprocal action are seen much more obviously in human passions and circumstances; on the other hand, that which is thus subject to the influences of Nature is only our moods, those vague states of mind which may indeed hinder or further an impulse to development which has originated elsewhere, but which could never of themselves have guided human progress in any definite direction.

When, however, from these considerations we turn to the question, By what definite ideas of action could Nature favour

the moral development of mankind ? the beginning of all
human culture seems still more wonderful. For it is clear
how fruitless must be any attempt to borrow from soulless
reality rules which have an unconstrained and natural relation
to our action with its totally different motives and aims. To
a mind already alive to the worth of law and order, the fact
of their universal prevalence is a point—and the only point—
in Nature which it can recognise as presenting some similarity
to the constitution of its own conscience, and as affording a
clear lesson for its own guidance ; but to attempt to model
the duties of creatures that have mind and the arrangements of
their social intercourse after the particular forms in which the
phænomena of the external world depend on one another, is one
of the most grievous and barren blunders of that sentimental
symbolism which we are opposing. What suits stars and
flowers need not on that account suit us ; the most we could
expect would be that the sure instincts which guide these
creatures nearest to us in the scale of creation might perhaps
furnish a true and unsophisticated indication of what Nature
requires of man, and whereto she has destined him. We
know the ideals with which this department of life can furnish
us. Beside the strength and grace of one animal we see the
sloth and stupidity of another, beside isolated moments of self-
sacrificing love and fidelity the treachery of the most blind
and inconsiderate selfishness, and in some creatures dainty
grace and timid beauty, combined with a cruelty that delights
in tormenting prey ; and the whole of this motley picture in a
perpetual ferment, one part cancelling another. What sort of
conviction of an intelligible connection of the world, and what
sort of a consciousness of our own duties could result from
such observations as the foregoing ? It is unquestionable
that he who takes the nature of brutes as his pattern will
attain a development, not of humanity but of bestiality. He
however who begins to distinguish between the indications of
universal validity which Nature affords us even in the life of
brutes and the impulses prompted by blind instinct, though he
refuses to recognise a higher law of conscience, has already

reached a stage of criticism at which any worth of natural impulses considered as furnishing a standard of right must disappear altogether. For he will not be able to deny that in his own nature also, many of these condemned impulses occur, and that too with all the force of importunate attraction, and he will then perceive that physical Nature cannot teach right or duty until its indications have been approved by the higher law which is in man himself, and until they have become part of the intelligible connection of a supersensuous rule of life.

CHAPTER II.

THE NATURE OF MAN.

Temperaments—The Meaning of Temperament—Differences of Temperament—
The Successive Stages of Human Life—Connection between the Vital
Feelings which have a Corporeal and those which have a Mental Origin
—Differences between the Sexes—General Mental Peculiarities of Women
—Heredity, and Original Difference of Endowment.

§ 1. IF from external Nature, the influence of which we
could neither deny, nor admit without qualification,
we turn back to ourselves, we find that the original peculiarity
of our own nature sets numerous limits to the development
of our individuality. In temperament, in innate capacities, in
those changes of the whole background of our mental life
which are inevitably caused by changes of age, in difference
of sex, in the varieties of susceptibility and impulse which
mark different nationalities, are to be found rules and limits
from which our development cannot escape. And from which,
indeed, in many respects it ought not to escape. The ideal
of humanity may find in these natural endowments more or
less hindrance to its realization ; but it is not of the essence
of this ideal to require a uniformity from which every tinge
of individuality has been expunged. It is only among brutes
that such conformity to the type is regarded as a perfection ;
among men it is more in accordance with the ideal that the
special nature of each individual should impart to his conduct
(of which the general outlines are the same for all men) its
characteristic tone and colour.

We are little acquainted with the circumstances upon which
these varieties of human endowment depend. They may be
for the most part conditioned by bodily constitution, or they
may result from the gradual summation of innumerable similar

impressions; whether it be that these continued influences
have become as it were to a certain extent fixed as tendencies
to development in the bodily constitution, or whether it be that
the mental development of our ancestors has been transmitted, as
innate capacity, to their descendants, after a spiritual fashion
which is still less comprehensible to us. However this may
be, the differences exist, and we cannot altogether neglect a
consideration of their consequences, though we may leave the
question of their origin undecided.

Varieties of *Temperament*, as of all other innate natural
capacities, appear to us to be most marked under conditions
of advanced civilisation. This may result from our imperfect
knowledge of the more simple forms of life, the distant view
making their uniformity seem greater than it is, or it may be
that only high culture affords scope for any great development
of the characteristic talents and dispositions of individuals.
Clear as these differences themselves may be in many cases,
the signification of the name—*temperament*—by which they
are distinguished, continues vague. The original meaning of
the word seems to indicate that we should understand temperament to signify general characteristics of the course of
mental life which do not of themselves exclusively predetermine
either a fixed degree or a fixed direction of culture, but which
certainly promote or hinder in various ways the development
of intelligence and of moral character. These we cannot
pronounce to be either altogether unconnected with, or indissolubly attached to, special varieties of bodily constitution
and predispositions to particular forms of disease. Under the
head of temperaments comes a consideration of the throng
of ideas which pass through consciousness together, the swiftness with which one succeeds another, and the force with
which thought works, either in one direction specially, or
several simultaneously, calling up a more or less numerous
and harmonious association from the ranks of previous impressions; of the fidelity with which previous perceptions are
retained, or the rapidity with which they melt into vague
general images; of the constancy with which an idea once

taken up with interest is held fast in the midst of numerous changes, or the ease with which sympathy and attention are diverted from their original object to a host of importunate accessories; of the general degree of feeling roused by impressions, and the permanence or transiency of this feeling; of the concentration of effort at certain points of enduring interest, or the inclination to jump from one occupation to another, and of the various strength of the impulse to express one's feelings in movements, words, and gesture. Differences of temperament are just like those differences in the movement of a current which are due to the original nature of its source and channel; according to the original density of the fluid, according to the direction of its fall and the nature of its bed, the various obstacles with which it meets cause it to be disturbed in some cases by deep, slow movements, in others by waves which merely fret its surface.

§ 2. If out of the innumerable varieties of individual temperament which we must recognise in experience, we would emphasize some striking forms in which the distinctive features we have noticed are grouped with most coherence, we shall naturally recur to the quaternion (*Vierzahl*) to which antiquity, combining groundless theory with sound observation, gave names which are still retained. But nothing would be gained by painting here over again these oft-presented pictures; we shall be better occupied in considering how, in the individual and in society, temperaments akin to these do to some extent naturally occur, and how to some extent they should occur in a regular course of development.

The health of the body depends a good deal upon its different parts not being so intimately connected as to cause every shock received by one to be communicated to the others. It is a sign of morbid weakness of nerve when the wholesome resistance to diffusion which prevents the spread of excitation is so far diminished, that every slight irritation affects the whole frame, and when disturbances of organic feeling which are by no means immoderate immediately call forth a variety of secondary sensations, produce convulsive movements, and accelerate secretions, or change their character. On the other hand, one might ask whether this general sensitiveness to stimulation is not the right state for a mind to be in prior to experience. Minds are not of course destined to remain permanently in such a state, but the task of educating oneself, and of gradually establishing one's own character, can only be satisfactorily carried out when it is unhindered by any original rigidity or sluggishness of constitution. Permanent excess of this general capacity of reciprocal excitement among all psychical states and general sensitiveness of the soul to all outward stimuli, distinguish that temperament which with a tinge of disapprobation we designate the *sanguine.* We think that to be easily disturbed and so pass easily from one mood to another, is natural and fitting in childhood, an age of which the proper business is to collect impressions by which it may build up its mental life without prejudice or special preference, and in fact it is generally where this volatility exists without lasting too long that a child develops most rapidly. The liveliness of the sanguine temperament seems to us to be also natural among uncivilised tribes, the differences of whose interests in life are generally too slight and shallow to call forth such a one-sided pursuit of definite ends as to weaken men's original receptivity for impressions of all kinds. Only it must be remembered that favourable conditions of external Nature are necessary for the simultaneous development of quickness of mind and joyous activity of body.

But while this temper of mind is advantageous at the outset of development, it presents many hindrances to the later development of intelligence, as well as of the emotional and moral nature. Great rapidity in the succession of ideas, which is made possible by the short-lived interest awaked by each one, is to a certain extent necessary for a child. This rapidity produces knowledge of a multitude of individual facts, and moreover, by means of the many-sidedness of ideas which supplement and correct one another, it prevents the establishment of narrow notions and attachment to ideas accidentally got and not of universal validity — faults which men are only too apt to fall into in later life in consequence of the monotony of their particular occupations.   But on the other hand, this rapidity of change hinders the fixation of that which has been acquired, and a sharp demarcation of the regions within which easily attained generalities are valid but beyond which they cease to be applicable.   It is further necessary for a child that feeling should be easily roused by slight impressions and unimportant perceptions, and also that the fluctuations of such feeling should be as rapid as the fluctuations of its various occasions.   It would be ill if in children laughter did not follow in the wake of tears, and if instead of their happy forgetfulness of sorrow, and even to a certain extent, of salutary punishment, a tenacious memory for all evil, for injustice, affronts, and pain, were to occasion moods of considerable duration during which their ready receptivity would be disturbed.   This characteristic again, which is an advantage in the beginning, becomes a disadvantage later on.   The quickness with which feeling that is continually on the *qui vive* responds to every momentary impression, together with the small amount of effort which the excitation is capable of calling forth, leads to the instability which must mark a course of conduct prompted by motives not derived from comprehensive reflection, or from the combined tendencies of a formed character, but borrowed hastily and fragmentarily from isolated and transient occasions.   Every human life starting with infinite possibilities of varied development, has the task of limiting itself to the finitude of some definite characteristic form which leaves a thousand early hopes unfulfilled, but by way of compensation evolves from the few impulses which it really develops a thousand wonderful and characteristic results, the rich variety of which could never have been suspected in the beginning.   The man whose sanguineness of temperament has outlived its natural term, gives us, not inappropriately, the impression of being a grown-up child, and the social charm which we readily grant to his general responsiveness and easy adaptation to all circumstances, does not make up for the want of trustworthiness, and does not rouse that interest which we take in every individuality that has actually worked out its natural potentialities to some definite reality.

To correct such faults without sacrificing what is attractive in such a temperament should be the aim of subsequent development. The mind ought to retain all its receptivity, for both great and small, and for the most various kinds of stimulation; but it should at the same time learn to discriminate between that which is of great and that which is of little worth, and to regulate the amount of responsive reaction according to the significance that each impression has for the interests of human life, which gradually stand out more and more clearly as forming a coherent whole. The natural course of development begins the accomplishment of this task, the *sentimental* temperament of youth displacing the sanguine temperament of childhood. I choose this name in order to avoid an inexactness which is involved in the ordinary designation of the *melancholic* temperament, an expression which makes us think of sadness and dejection of mind, and though this unhappy humour may cast its gloom over the whole of a man's mental life, in consequence of bodily disease or of long-continued misfortune and the memories which succeed it, yet it is not itself one of those general types of inner life to which the name of temperament can be properly applied. Indeed, the fact is that this humour, like every other, is compatible with any temperament, although one may be more conducive to it than another; while what we mean by the sentimental temperament is not one humour which out of the many that we may experience has become predominant, but a general propensity to give oneself up to humours, to as it were lay oneself out for them, and to entertain them in greater force and to a greater extent than occasion warrants. Children do not pick and choose among impressions those that they will attend to; their curiosity is easily excited by facts of any kind which can furnish them with ideas. If we sometimes find them disinclined to learn, we should remember how very uninteresting to them those objects must be in which we are only interested because of our knowledge of their significance. If we consider this, we shall admit that there is in the child a disinterested readiness to appropriate the most various material, and that the results of this during the early years of life far exceed what is acquired in any equal space of time in after life. It is natural that this undiscriminating receptivity should diminish, the more the task of thoroughly organizing the acquired material of knowledge comes into prominence. The youth therefore is more discriminating than the child in his reception of impressions; much seems to him indifferent or repulsive which the mental digestion of the child readily assimilated. But in proportion as there have not yet arisen definite objects in life in connection with which all particular experience may be steadily systematized, the interest of the soul will become centred in the emotional worth of impressions; it will withdraw from all which does not promise satisfaction to its inclination for this kind of excitement, or conversely will use every imaginable impression merely as the peg on which to hang a succession of feelings, treating its intellectual content with unsympathetic neglect. Thus is formed the sentimental temperament which naturally gives the tone to mental life during the period of youth; and if it does not outlast its due time much that is valuable and noble in our development is due to it. Being specially capable of appreciating the harmony or discord which belongs to the formal relations of impressions, it is given to the dreamy repetition of all that is rhythmical and in general of all æsthetic impressions; little inclined for real hard work, but driven by restlessness of feeling to imaginative activity, it seeks an outlet partly in artistic creation and partly in framing ideals of a better state of things than that which actually obtains. But while susceptible to the emotional worth of perceptions, it is at the same time disposed to theoretical vagueness, in consequence of not having a sufficiently firm grasp of the definite points between which those relations extend which are themselves of so much consequence. Thus it becomes unpractical, wishing indeed to reproduce by its own activity the moods which it values, but having no sympathy for the uninteresting details of appropriate means; and just as often it is unjust, resenting the indifference or opposition of others to its own æsthetic prejudices with a bitterness which excludes all fair judgment and all toleration of divergent culture.

It is a happy peculiarity of our nature that past suffering does not live as vividly in our memory as past joy; but any pain at the moment when it affects us, stirs the spirit more powerfully, and produces a greater mental turmoil of thoughts seeking for utterance. Sometimes a man does not for the moment know what to make of pleasure, and often he has to wait until time shall have revealed all the individual happy consequences which some present good fortune involves, finding only then a fitting expression for his joy. This explains how it is that men are disposed to seek dissonances, or to exaggerate them when they exist, in order that by doing so they may as it were gain mediately a clearer consciousness of the harmonies which are actual or possible, and the worth of which stands out the more clearly in the contrasted presence of impending danger. Therefore sensitive souls love the gentle melancholy which is spread like a grey background behind the rainbow glory of isolated moments of delight, and the old view was not altogether wrong in giving to the sentimental temperament the designation of *melancholy*, with which humour that temperament is in fact thus naturally connected.

The great defect which attaches to this temper of mind is the ease with which the development or establishment of a sense of duty may be hindered by excitability of feeling. However indispensable this temperament may be not only for artistic genius but also for the truly humane ordering of practical life, yet if it continues in isolated predominance it leads both in art and practice to mere skill, which amuses itself but acknowledges no obligation to serious work. We need not refer to that repulsive form of sentimentality which turns all the circumstances of life to account in no other way than as occasions of emotional excitement; we may also trace the ill effects of the sentimental temper both in science and in art, It is shown in the latter by its way of dealing with the isolated lyric movements of emotion which naturally arise in men and have received a pleasing formal expression either from some gifted individual or from the cultivated general mind; these it is incapable of grasping and bringing together into a coherent whole in such a way as to attain to higher truth. It is shown in the scientific region by the numerous examples of men who, with great natural gifts, can be content to spend their ingenuity in constantly devising some new dress for the knowledge they have already acquired, in giving it a finer point and more exquisite arrangement, without ever honestly doing their part towards the final solution of any problem. A good deal also of apparently earnest effort has to be set down to this less emotional form of sentimentality; but what is great in life and in science has always been the result of concentrated energy, which, without denying the worth of other impressions, yet passes them by on the other side, as it presses towards its own goal, busying itself all the more eagerly about the means of attaining to it, though these being indifferent in themselves, are despised by the excited temper of youth in its search after worthy ends.

The *choleric* temperament is plainly that which we must desire to see developed in the time of manhood, as the natural successor of the sentimental temperament; its too early appearance would be as contrary to the perfection of human development as its not appearing at all. The diminished susceptibility to excitement which is ascribed to this temperament, together with the great force and endurance of its reaction, when feeling has once been aroused, are doubtless often the effect of a moral steadiness of character, which having chosen definite ends refuses to be lured from its path by irrelevant attractions; or it may be that they are the effect of a narrow range of ideas produced by the monotony of life, and in many cases blunting the interest which would naturally be felt. But that obstinate perseverance in a path once entered upon, which hindrances only serve to spur on to

greater activity, often occurs even in children ; we are therefore fully warranted in designating this state of mind as a
particular temperament. Its essential features are to be
found in its unreceptiveness for incidental attractions which
lie out of the beaten track of its thought; in the narrow scope
afforded to new impressions, these sufficing only to call up
the recollections most closely associated with them in one
particular groove ; and lastly in the small degree of feeling
which can be aroused by any perceptions but those which fall
in with the prevailing current of feeling. But when interest
is once awakened, it affects with equal steadiness the train
of ideas and the efforts of the will; thus this is the pre-
eminently practical temperament, both on account of the
definiteness of the ends which imagination presents to it,
and also because its less exacting and less touchy temper
does not shrink from the employment of indifferent or irksome
means which, while destitute of intrinsic worth, are indispensable for the attainment of the desired end. But the frequent
confusion of this temperament with what we call simple
wilfulness shows that it has drawbacks which are closely
related to its advantages. In fact its practical efficacy
is often impaired by a gradually increasing narrowness of
mental life, which having chosen some one exclusive end,
not infrequently fixes with equal exclusiveness and obstinacy
upon some one definite kind of means, and even sometimes,
reflecting itself as it were, seems as a final stage, to reject all
reference to intrinsically worthy ends, and develops into that
conscious stubbornness which is the caricature of rigid consistency. It is not in such results that the progress of development
which we desire is to be found. Later life ought to inherit
a fair share of that passion for everything which has emotional
worth which is characteristic of the sentimental temperament,
as well as of the mobility and sensitiveness of the sanguine
temperament, and the group of characteristics which best
becomes the ideal of human excellence is not to be found in
the unsympathizing or contemptuous disposition which a
narrow-hearted devotion to definite ends exhibits towards

all which lies out of the track of its own particular effort.

I shall perhaps be regarded as the advocate of a strange thesis when I say that I regard the *phlegmatic* temperament as the natural temper of advanced age, and at the same time as an improvement on the choleric temperament with its prejudices and narrowness. A description of the different temperaments so naturally presents each one as an exaggeration of its special characteristics, that at the very name of *phlegmatic* we are accustomed to think of a sort of mental lethargy very far from suggestive of advance in human development—a state in which susceptibility to impressions, as well as any pleasure in responding to them, has been almost wholly lost. But in this representation vacuity of mind is confounded with a form of activity which may belong to a full as well as to an empty mind. A state of steady equanimity would be intolerable and repulsive in a soul whose capacities were as yet only partially unfolded, and whose best development yet remained to be won among the manifold changes and chances of life; but such calm is to be reverenced in a mind which has passed victorious through chance and change, and has learnt by wide experience, neither to be carried from one mood to another by every changing impression, nor to give exclusive and one-sided approval to some one particular form and direction of human effort. It is true, indeed, that as long as we understand by temperament only a natural disposition as contrasted with any acquired attitude of mind, the immovability of the *phlegmatic* temperament must seem to us the least pleasing of any human character. And yet even in this we are often unjust; we conclude too hastily that disinclination to bodily movement indicates an equal sluggishness of thought, that the absence of foolish outbursts of emotion and omission of useless expressions of feelings are due to coldness of heart. Hence we are often surprised to see such minds stirred up by a great and impressive stimulus to some energetic passion, producing vigorous and long-sustained efforts; such an occurrence we have often enough seen " writ large " in the history of races whose national temperament is decidedly phlegmatic. We learn from such cases that it is unjust to attribute the immovability and incapacity of mere stupidity to that solid-ness of mental life which is hardly affected by individual passing impressions, but slowly stores them up until the time arrives for some supreme effort—or at any rate if no occasion for action arises is not haunted by a mental unrest which prompts the search for such an opportunity. Like all rest, this equanimity of soul is a phænomenon that may have many significations, and its worth is in proportion to the amount of dormant power which it holds in suspense. We blame the unreceptiveness which remains unmoved because it is wanting in all intelligence and sympathy ; but we all seek that peace which is not immoderately excited by anything, because nothing is any longer wholly new to it ; which has experienced every kind of emotion, but has long ago learned to assign to every passionate impulse its proper value in the whole intricate chain of human interests, appealing to this from any accidental strength of feeling which may be due to the circumstances of the moment ; which finally has ceased to have any part in the heat and hurry of self-willed effort, because it has learnt that the vicissitudes of destiny are too great, and the field of human activity too circumscribed to admit of our attributing absolute and unconditioned worth to any single work or any single performance of ours. We hope for this frame of mind as the natural temperament of old age, but we certainly do not see that it is generally attained ; on the other hand, however, we find that by innate favour of spiritual organization, some few happy souls have all through life this fine balance of mental temper. They receive with pure-hearted and ever fresh interest, impressions of all degrees of importance ; they are not indifferent to any class of feelings, but on the other hand, none carries them away into the tangled paths of a one-sided and narrow humour ; with clear vision and patient hand, they quietly compass the means to some steadfastly pursued end, without the unsympathizing harshness which refuses to endure any interruption of its work, and without that contempt for other paths which is natural to him who knows none but his own. It is not of the great names ot history that we are thinking now, but of those gentle and blessed natures who pass noiselessly through life, seeming as it were the very embodiment of our ideal; those who have had a strongly marked effect upon the course of history, have much oftener been men whose minds were not thus finely balanced, and who owed their influence to the one-sided harshness with which they have succeeded in forcing their own views upon the world, undisturbed by any acute sense of the comparative worth of conflicting opinions.

Observation does not show us that more than a distant approach to this gradation of human development actually exists. In order to go through it completely, and to let each of the temperaments run its whole course in full and unmixed current, unusually favourable conditions both of natural disposition and of outward circumstances would be required. It is only when culture has advanced rather far that it can furnish the different periods of life with that variety of interests from which each particular phase of character can draw material for vigorous development; hence the monotony of a very simple mode of life would weaken the characteristic differences of temperament. But on the other hand, the multifarious complications of life may hinder regular development by events which press with such a weight upon the soul that completeness and spontaneity of further development becomes impossible. And finally, the more thorough-going has been the development of mind and character in any generation by a life of varied culture, the more are the natures of the next generation likely to diverge from one another, exhibiting characters of striking individuality, the course of development of which often differs strangely from that of the ordinary type. Then there are numerous diseases which have a powerful effect on temperament and humour, and numerous bodily disorders which, before they declare themselves as disease, appear in disturbances of organic feeling which, inexplicable even to him who suffers them, imperceptibly give a tone to the totality of his views and feelings. It would be extremely interesting if it were possible, to investigate the causes of these phænomena. But it is indeed impossible to discriminate in them between what has its origin in the region of mind, in the impenetrable windings of every individual development, and to some extent reacts upon the bodily organization, and what on the other hand is due to organic development and its disturbances, and has a share in influencing the growth of the inner life. Perhaps too much weight is sometimes attributed to the last factor, but still there is no doubt that it does have a very important effect. We see tardiness or precocity of bodily development accompanied by a like tardiness or precocity of the mental dispositions corresponding to these stages of physical growth ; and on the whole nothing is more natural than the assumption that the full tide of organic feeling receives at different times a different colouring in proportion as this or that organ or department of the bodily economy makes its influence more or less felt by innumerable constant excitations, singly imperceptible, which vary according to the rapidity or backwardness with which the organ or department in question develops its activity. But while the time is gone by for explaining such matters by reference to the black bile and the yellow bile, the time is not yet come when we may have recourse to exact observation for an explanation of the importance of different functions at different times, and for trustworthy information as to their influence on mental life.

How intimately permanent bodily conditions may be connected with permanent mental dispositions, is shown by observation of cases in which their reciprocal influence is temporary. It has been said, and not without truth, that we think differently when we are lying down and when we are standing up ; a constrained and cramped position of the body has a depressing effect upon the spirits ; again, we find it difficult to be devotional in a comfortable and careless attitude; rage is quieted by muscular repose—it is a dictate of prudence

to get a furious man to sit down in an easy-chair; and the
hand which smooths the wrinkles from one's brow, smooths
away trouble too.   It may be asked whether æsthetic and
moral judgments or our thoughts about future joy and sorrow
do not primarily receive their vividness and intensity from
accompanying sensations in which that which is of intrinsic
worth appears to us as harmonizing with the innermost conditions of our own individual existence.   There are plenty of
apathetic states in which these attendant feelings are wanting
—in which we may see as plainly as before the objective
excellence of one kind of conduct, and the blameworthiness of
another, and recognise the just claims of others on our love
and sympathy without being in the slightest degree capable of
conjuring up that glow of feeling which we know would be
appropriate to the occasion.   How often does the same thing
happen in our enjoyment of beauty!   Appreciation of it is
not mere abstract delight in harmonious relations, delight in
general is not a merely mental process, but something by
which our whole being seems to be exalted and carried away,
something which makes us breathe more fully and freely,
which quickens our pulse and gives elasticity to our muscles;
remorse for what is past is not the mere moral sentence of
condemnation which, pronounced by conscience, is simply
apprehended by the soul; the relaxedness of the limbs, the
oppression of the heart, perhaps in anger an actual spasmodic
contraction of the throat and rising of the gorge which prevent
our swallowing the morsel already in our mouth—these show
the sympathy of the bodily organization, and as it were
symbolize the attempt to get rid of some detested burden
under the pressure of which we suffer.   Even devotional
feeling is not a purely mental exaltation; but whilst it makes
us unconsciously forego the careless haste of our ordinary
gait, and causes our movements to be slower and more self-
restrained, and our attitude to take a peculiar stamp, not of
relaxedness, but of strength which voluntarily submits, there
flows back into consciousness from all these bodily effects an
echo of feeling strengthening the intellectual mood.   We can

understand what a difference it must make if the body return this echo imperfectly or with a tone altered by disease, and how in fact similar moods of some special individuals can never be quite comparable one with another. It is in the bloom of youth that we find this correspondence between mental life and its material vesture developed in the most attractive and perfect form ; in later life the gradual increase of obstacles and of friction causes the imperfections and incoherences in the connection between the two orders of affection to become more and more prominent. We can no longer read the whole soul in movement, gait, and carriage; ordinary daily actions are got through with unsympathetic dispatch, eating and drinking often with ugly and soulless eagerness; and it is always a sign of profound culture of the heart when the thoughts of a man advanced in years do not meet the sensuous warmth of any passing event with the uninterested and unsympathetic coldness of age.

§ 3. We feel afresh the want of trustworthy knowledge concerning the psychical importance of the bodily organs and their connections, now that we are come to that difficult part of our task, a consideration of the mental differences of the two sexes. I will not stay to compare the undulating outlines of the woman with the more angular build of the man ; it may be that there is foundation for the idea that the latter indicates the preponderance of some impulse towards characteristic individualization, and that the perhaps really greater bodily likeness among women is to be regarded as evidence of their greater mental conformity to some general type. Even here where the outward form is to others indicative of the inner life, I find myself able to lay little stress upon the merely symbolical significance of the bodily form ; it would be much more interesting to show, if one could, what particular organic feeling the body comes to have in consequence of its functions and of the particular proportions of its parts.

Of all this we know but little. The relations of the different parts of the skeleton and of the muscular system show that there is less power of work in the frame of the woman, the shoulders and chest are not adapted for lifting, carrying, and moving heavy weights and obstacles, nor are the hips and legs framed for swift running, or for walking firmly under a heavy burden ; the muscles seem less fitted to endure continuous strain, great as may be their capacity of work when they have frequent alternations of activity and rest.   These circumstances can hardly fail to influence organic feeling, a very important part of which always depends on a consciousness of the ease, elasticity, and peculiar security of our position, attitude, and mode of progression.   The fact that a man's body forms an oval with its greatest diameter through the shoulders, and a woman's body an oval which is widest across the hips, is in itself indifferent ; but it may be that on the man the preponderant weight of the upper part of his body may have the effect of a burden which demands to be carried forward swift and sure in opposition to all obstacles, while the woman, feeling more fettered, most naturally finds her sphere of work nearer home, and expects it to come to her thither from the dim distance.

This inferiority in strength is compensated by a greater capacity of adaptation to the most various circumstances.   The bodily wants of women are much less than those of men ; they eat and drink less, they breathe less air, and are said to be less easily suffocated ; with regard to hardships—at least those which are continuous and of gradual growth—and privations, they bear them to some extent more easily than men, and in some respects with less of ill effect than might have been expected from their degree of bodily strength. They endure great loss of blood and continuous pain better ; and even the greater irritability of their nervous system, on account of which many unimportant disturbances have a great effect, seems to favour the rapid and harmless dispersion of any shock that may be experienced.   Hence even under unfavourable circumstances, they often reach a great age, although the examples of extreme old age, lasting on far into a second century, are to be found almost exclusively among men.

They are naturally disinclined to very vehement sensuous gratifications, and often have only a sort of emotional aversion for disagreeable impressions in cases where a man would be almost overcome by absolute physical disgust; the work of restoring cleanliness is always in itself uncleanly. The same capacity of accommodation is shown in the various circumstances of life. It is an old and true remark that women can much more easily suit themselves to new conditions of life, to a different rank in society and changes of fortunes, whilst it is hardly possible for a man to efface the signs of his early training. Acquired habits also have a stronger hold on him, and when accustomed order is interrupted or the usual hour for work or food comes round empty-handed, his general comfort is much more greatly disturbed. With the above characteristics of women there is naturally combined a mixture of that liveliness proper to the sanguine temperament and that warmth of heart, belonging to the sentimental stage, the absence of which we regret in any woman, counting it an imperfection. In her, varieties of education hide much ; but even in the most extreme cases we shall hardly fail of finding a propensity (akin to inquisitiveness) to talk for the sake of talking, and some trace of pleasure in beautiful and harmonious arrangements.

But the question, How is the higher mental life of both sexes characteristically distinguished, with reference either to these natural features or to any others ? is one which it seems hardly possible to answer. The innumerable observations, partly ingenious and partly also at the same time true, to which this question has given rise, have seldom been concerned to distinguish between what is to be regarded as original disposition, and what as a remote result of the circumstances of life and educational routine which have affected the two sexes very differently, although in harmony with their natural dispositions. However often the attempt may be made to reduce to simple intelligible expression the multitude of these particular characteristics, which only a life's experience can teach, and only the plastic creations of poetry can reproduce, it will always be found that such attempts must be content to give merely an extremely colourless outline of that which in its boundless wealth of colouring furnishes the philosopher of common life with an inexhaustible field of interest.

I do not believe that the intellectual capacity of the sexes differs, except in so far as the special emotional interests of each have prescribed the course of their intellectual life. There is perhaps no subject which a woman's mind could not understand, but there are very many things in which women could never learn to be interested. Though it is often said that in knowledge a man is attracted by the universal, a woman by the particular, yet in very many cases we should find, that it is just the individualizing power of women which is inferior, and their delicate instinct for the universal which is superior; and besides, this division of the work of knowledge to which we have just referred is inconsistent with the current attribution of egoistic effort to the masculine will, and of subordination to universal rules to womanly self - suppression. There would perhaps be more truth in the opinion that the knowledge and will of men aim at *generality*, those of women at *completeness*. It is masculine philosophy to analyse striking phænomena and to find out from what complication of general conditions each of them inevitably and necessarily resulted, however much it may seem to be some arbitrary and chosen product of Nature; it is characteristic of women to hate analysis and to enjoy and admire the beauty and intrinsic worth of any whole that may be presented to them in finished completeness. All mechanical inventions have been made by men, and to men belongs delight in the mediate production of effects by the application of general forces according to general laws; while the actual manipulation belongs rather to women, and to them also the desire to find that the warmth of living feeling is being as it were transferred immediately to the product of their activity. Characteristic of masculine thought is the deep conviction that all which is greatest and most beautiful in the world has its mechanical conditions, and that no result which is premature and which evades this fixed order of realization can be permanent and stable; it is to this thought that is due the order by which life is organized, an order that is everywhere dependent on the principle of law, that is on the belief that the universally valid conditions of truth must be satisfied before there can be any question of a result that may be desirable in some particular case. On the other hand, the faith of women—which is both just in itself and as necessary as the other to the happiness of life—is that no general principle and no form can ever have an independent and unconditioned value, but that such value belongs exclusively to the living reality which may be founded on them; from this faith flow all the beauty and compensations of life, for it is a faith which is everywhere dependent on that principle of equity, which makes men feel bound to soften the harshness of law by unowed love and kindness; the misfortune is that this desire to show kindness is often in danger of hastily and unjustly breaking through forms of law which hinder the fulfilment of its intention.

All masculine effort depends upon profound reverence for general principles; a man's pride even and ambition are not satisfied by groundless homage, but he founds his claim upon the sum of generally recognisable superiority which he believes himself to possess; he feels that he is undoubtedly something more than a mere example of the universal, and he demands to be compared with others by means of some common standard. Just as devout is the sentiment of the feminine mind towards completeness; a woman no more desires to be considered as an example among others than the beauty of one flower requires to be compared with that of others according to some standard of comparison; and while a man cheerfully joins himself to others who are like-minded and cheerfully perishes with them for the sake of some general principle, a woman would rather be sought and loved as something fair and complete in herself, and for the sake of her own individuality, which is a thing that is not susceptible of comparison, nor explicable by reference to other individuals. For certainly in the feeling with which we regard such a whole, love in the strict sense is more prominent than esteem, but it is pre-eminently esteem and not love which a man requires in the feeling with which he is regarded ; he is not merely willing that his worth should be measured by a common standard, but he demands that it should be so measured.   No one, of course, will so misunderstand this contrast as to imagine that we mean that a woman's nature has, like the unanalysable fragrance of a flower, no pretensions to call forth the sentiment of esteem, which is in fact aroused in a very high degree by particular virtues which appear in women, and which are susceptible of comparison.

We only need to look about us as we go through life in order to find a thousand traits which bear witness to this general dissimilarity.   The business communications of men are brief, those of women are wordy, and generally abound in repetition ; it is plain that they have little faith in the trustworthiness of a promise which is guaranteed merely by the general obliga-tion to truth and good faith, and is not clenched by a variety of small considerations drawn from a comprehensive survey of the case in hand.   Men lay less stress upon the harmonious arrangement of their spatial surroundings, except in as far as these secure the immediate and ready applicability of means to desired ends ; but they value punctuality as regards time, which is in a much higher degree a mechanical condition of all success ; women have the happy knack of arranging a multitude of objects in space in such a way as to produce a pleasant effect on the whole without any rigid adherence to system ; but they show less management with regard to time, which is something that cannot be seen.   When men and women speak of regard to form they generally mean very different things ; the womanly nature is concerned to round off into a graceful and consistent whole the final product of any activity ; her skill lies in knowing what is appropriate to the case in hand, which very often is exactly that which the man's judgment disapproves ; for the forms which he would

THE NATURE OF MAN.

choose to have observed are general rules of orderly procedure, which must be carried out even at the cost of producing some isolated discords. With the above is closely connected the well-known unjudicial character of women. It does not consist in an incapacity to sacrifice individual claims, for nothing could exceed the cheerfulness with which women make such sacrifices, as soon as they have actually set before them that good of others for the sake of which the sacrifice is to be made. But they feel aggrieved because very often the law in considering any given case does not regard it as a whole, but brings it under some general definition in virtue of some special characteristics, the selection of which seems to the woman's mind to be arbitrary; the definition itself seeming to be not less arbitrary, because, being a general rule of procedure, the ultimate good which it seems to secure is not directly presented. A man does not rebel against undertaking things of which he cannot see the result, if the carrying out of some general principle is concerned; women require to have the future results set plainly before their eyes, they want to anticipate beforehand the final form of the whole, to know in what shape the unrest of action will be embodied in the end. This disposition, this happy faith that there is some answer to every puzzle, some mode of reconciling every conflict, some way of gathering up in the end the loose threads of broken effort, has unquestionably an injurious as well as beneficial effect on the masculine mind, women being able to produce this effect in consequence of the share which they have in education. That consideration of possible results which holds men back from action at moments when inevitable duty is in question, is generally due to maternal influence.

A man generally regards his property as what it really is, as a collection of usable and divisible means to various ends, and his liberality is not disturbed by the idea of breaking into some imaginary completeness which attaches to it as a whole; when women are extravagant, their extravagance generally consists in making purchases for which they will not themselves pay the money. On the other hand, property which they

have once acquired and actually have in possession often seems
to them a kind of sacred deposit, all the parts of which belong
to one another, and which it would therefore be wrong to disturb.
What draws down upon their management the suspicion of a
leaning to avarice, is not exactly an unwillingness to impart
to others, but certainly, to some extent, that reverence for the
intrinsic coherence of things, which is expressed equally in
their horror of disturbing some treasured remembrance, breaking up some possession with which the whole of life seems to
be entwined, and in their mysterious satisfaction in exacting
" good measure."

Finally, I would venture the assertion that to the soul of
a woman truth does not mean the same thing that it does to
a man's mind.   For women everything is true which is justified
by a capacity of fitting in harmoniously and significantly into
the rest of the world considered as a whole, with all its system
of relations; they do not care so much about its being at the
same time a reality.   Hence they are inclined not to lying but
to making a fair show, and if something presents the appearance they desire in some connection which they regard as
important, they care little whether or not it would prove on
investigation to be something which has any right to present
that appearance.   To wish to seem what one is not, is indeed
a failing common to all humanity, but a man is accustomed to
require, at any rate in the goods which he possesses, solidity
and genuineness; among women, on the contrary, there is a
widespread predilection for shams.   Having such leanings, they
are not given to scientific labours, and their mode of thought
is artistic and intuitive.   As a poet does not create characters
by analysis and calculation, but is assured that they are true
to nature if he can himself in his own mind follow their whole
action with natural and spontaneous sympathy, so women love
to put themselves in imagination in the place of things, and
as soon as they have succeeded in getting some idea of what
it is like to exist and move and develop in the way in which
any given thing exists, moves, and develops, they think they
understand it thoroughly.   That the possibility of things being

and happening as they do involves a scientific riddle, is something which it is hard to make a woman understand. It is easy to see the connection between all this and some of the great goods of life, for instance, firmness of religious belief, and calm assurance of moral feeling; but we also find this preponderance of living tact over scientific analysis in many small and inconspicuous traits. Women employ a thousand delicate technical devices in their daily work; but they can with difficulty describe, they can only show, that which they have skilfully accomplished. Analytic reflection upon their own movements is so little familiar to them that one may affirm, without fear of being very far wrong, that such expressions as, *to the right, to the left, across, reverse,* express in the language of women, not any mathematical relations, but certain particular feelings which one has when in working one makes movements in these directions.

§ 4. But I am in danger of trying to exhaust that which is inexhaustible; and I am the more bound to avoid this because a consideration of life in the concrete shows us everywhere the part taken by both sexes in the whole constitution of life and its enjoyments. Still a brief indication of the limits imposed on each sex by its own special nature, may guide us in a special consideration of the divergent developments which we see arise from the national character of different tribes and races. We often find among the people of one nation that many mental as well as bodily peculiarities are transmitted with great persistence from parent to child for several generations, especially talents for those arts which are concerned with combinations of many elements that can be intuitively apprehended. Examples of the inheritance of mathematical, musical, artistic, and technical capacity are not rare, and with these are connected primarily the transmission of similar temperaments, in which we have already recognised general formal peculiarities of mental life. Parents are often astonished at seeing reproduced in their children the same faults and the same little tricks of which they are conscious in themselves; in civilised life, indeed, where persons of the most differently developed characters unite to form new families, the very reverse of this is often seen, or at least we cannot trace the nature of the child to any mixture of the qualities of the parents; but on the other hand, among un-civilised tribes not only is a considerable constancy of such transmission to be expected, but the expectation is confirmed by actual experience.

We may try to derive all the national differences of civilised people from the influence of the peculiar conditions of their civilisation, which are to a great extent dependent on geographical position and the vicissitudes of history, this influence being of overpowering importance, and pervading the whole of life; but we cannot by so doing remove the general impression received from observation, that nowadays at least every new-born life comes into the world with some innate and inevitable national stamp, quite independent of its later contact with the civilisation of its nation.   It would be useless to try and explain this phænomenon, as observers differ so much in their opinion of the extent to which it occurs.   We cannot decide finally whether all races are capable of an equal degree of civilisation, but we find that the most favoured nations share in the development of humanity in unequal measure, and in ways peculiar to themselves, and we see that individuals of the same nations are very differently endowed with mental energy and activity; finally, we have every reason to believe that the savagery in which we find the coloured races of men is by no means a condition abso-lutely inseparable from their nature.   Our primary deduction from these considerations is the conviction that to attempt to deny all original difference of endowment is a superfluous undertaking, for when we have denied it of the great divisions of the human race, it infallibly recurs in individuals, and such a connate limit can be no more oppressive for the former than for the latter.   The only question is, whether all races of men have in common those capacities which are necessary to lead them to a participation in the moral inheritance of mankind, and to unite them in human fellowship.   It is not now our intention to give an answer to this question, which belongs to the Philosophy of History, but we shall find a preparation for the answer in a consideration of the general way in which the inborn nature of men is stirred up by the educative influences of Nature and of social relations to the production of all that is most essential to life.

CHAPTER III.

MANNERS AND MORALS.

Conscience and Moral Taste—Untrustworthiness of Natural Disposition—Food
—Cannibalism—Cruelty and Bloodthirstiness—Cleanness of Body and of
Mind—Modesty—Disparagement and Exaltation of Nature—Realism of
Individual Perfection, and Idealism of Work—Social Customs.

§ 1.　WHEN we sought in the human mind for the germ
of moral development, we did not seem to find
there any complete revelation directly enabling it to bring the
relations of life, or even those parts of human conduct which
are of most universal concern, into harmony with undoubted
precepts of moral order.　Even in an educated conscience, a
lively conviction of the worth of an ideal by no means guarantees
the simultaneous presence either of that sensitiveness of
judgment which is necessary for discriminating instances of
its genuine realization from spurious imitations, or of that
creative imagination which can apply the well-known general
type to particular cases without distortion or misapprehension.
Many a man whose soul was deeply stirred by thoughts of
the supremely good and beautiful, but who found in his own
age no artistically perfect expression of his ideal, has fancied
that he saw it realized in forms, the sorry poverty of which
calls forth the astonishment of a later and more developed
age.　Forced to satisfy its longing with something which it
has, the mind easily over-estimates those meagre outlines
which it invests with the life and colour of its own feeling;
and thus accustomed to take the will for the deed, it
becomes unreceptive, timid, and perverse towards that fuller
beauty which reality presents, and which if it only were
intelligible would much more effectually satisfy the soul's
needs.　This has been very much the case with moral de-

velopment. We may, indeed, certainly ascribe to the human mind the possession of innate general ideas of Right, of what ought to be; but the moral skill which enables us to find, in every individual case, the special form in which this Right should be realized, is decidedly a product of progressive civilisation, and happy traits of natural disposition are not a full and sufficient but only an extremely imperfect and fragmentary substitute for it.

This will appear to be self-evident with regard to all those more important human institutions, such as the State, or the organization of civil society, which, in as far as they are the intentional product of human skill, can only be founded on a knowledge of the thousand-fold relations which bind the members of a society both to one another and to the conditions of external life which they have in common; and this knowledge can only be attained and gradually perfected by the actual experience of life. But where man is related to his fellow in a way that does not involve any of these complicated relationships, or where he dwells alone face to face with external Nature, one might suppose that his conduct would be guided more unambiguously by the innate voice of Conscience, prescribing to him not only fitting ethical sentiments but also the manners and morals corresponding to these as their natural expression. However, a comparison of the different modes of human life teaches us the very contrary. What it is fitting a man should do or leave undone, in what way it is becoming that he should order his surroundings and his social behaviour, what he should esteem and what he should avoid, and what things are without claims upon him, and of no importance to him—finally how he ought to dispose all his conduct and every detail of his action, so that his life may be a harmonious whole—all this must be learnt in a long course of development, and never can be fully learnt. The innate goodness of mankind is very far indeed from leading directly to such a development of morality.

Many a simple custom of peoples who are yet uncivilised may well compare favourably with the distorted growths of our civilisation; the unsophisticated manifestation of isolated traits of natural nobility may well have a charm for us; but around these bright spots the shadows lie all the deeper, and the general character of this life of Nature, and of every people that is in a state of Nature, exhibits the instability, the incoherence, and the incalculable inconsistency with which, side by side with attractive manifestations of particular moral feelings, inhuman crime and the most astounding perversity of conduct flourish in rank luxuriance.   We are struck by some advantages of a state of Nature which are for the most part, though never necessarily, sacrificed by civilisation for the sake of higher ends, and we long to return to the simplicity of such a life—forgetting that it is civilisation itself which has sharpened our appreciation for it as presenting a pleasing contrast to the conditions that are evil in our own state, and that with the charm of such an existence there is associated a poverty which neither knows nor can produce a large proportion of the best goods of life.   In such moods we are but too apt to lose courage, and it is this which so often makes us turn back from the complication of great and not altogether successful undertakings to refresh ourselves with the complete success of more insignificant works, rather than push forward with a good courage notwithstanding.   A little flock is soon counted; and he who shrinks from venturing on the open sea and steering his course among the thousand conflicting claims of a civilised life which, as regards all mental interests, is stirred to its very depths, can easily construct an idyl on which the eye may dwell with momentary satisfaction, but only to turn away from it wearied after a very brief space.   A fine climate, inherited excellence of bodily organization, and absence of hard work, develop among men, as among beasts, the greatest beauty and suppleness of form, and a natural gracefulness of carriage, independent of any deep spiritual life; kindliness and good nature which we would gladly count among innate human qualities are very likely to brighten life and beautify it by traits of social refinement in cases where simple relations exist which give no occasion to lasting and deep-rooted conflict; but untutored spirits are not accustomed to take a comprehensive view of human life; they know not its significance and the aims which are set before it, and hence they find only too many barren spots in life, too many moral difficulties which receive no decided answer, too many practical questions which may, it seems, be answered indifferently this way or that—and which consequently are frequently decided in accordance with the impulse due to temperament and external circumstances, leading often to an extreme of inhumanity and a barbarism which are in the most violent contradiction to the amiable traits that promised so much.

This moral untrustworthiness is by no means peculiar to uncivilised peoples in their natural condition. Even in our own highly civilised state, many an evil disposition is kept under only by the unremitting pressure exercised on all sides by the authority of systematised social forces; and not only so, but narrowness of moral insight, want of a delicate perception of the way in which the moral ideal should include and animate even the simplest relations of life, and all the rudeness of mere selfish subjectivism might appear at any moment, even among us, with most confusing effect if past centuries had not preserved and matured mighty spiritual forces of objective validity which they have handed down to us in the treasures of science, art, law, and religion. It is these which help the nobler minds to recognise that close connection between all the most sacred spiritual possessions of men which the individual could not discover unaided, whilst they keep baser natures within bounds as a system of institutions which, though uncomprehended, happen to have the authority. And finally, at no time can we say either that this vast fabric of human civilisation is completed, or that all its parts are at the same stage of advancement. In all societies there are departments of life which, though susceptible of thorough moral cultivation, are yet given over to individual caprice arising from temperament, as though they were subject to no law or rule; on the other hand, there are customs, really indifferent in themselves, which have become established as having the force of absolutely binding commands, much to the detriment of progress.   Finally, our morality as a whole suffers from a deficiency which it never will, and indeed never ought to, surmount wholly; a deficiency, namely, of perfectly clear theoretic insight into the grounds of the binding validity of its demands—such an insight as would be capable of making faith in the dignity of moral institutions independent of any change of mood, and hence out of the reach of that scepticism which passion and the sharp troubles of our earthly lot only too easily arouse.

In saying that this deficiency ought not to be wholly surmounted, what we mean is that it would not be advantageous for moral development if the binding truth of all particular moral commands, and the indissoluble connection between them, were presented to individual minds with the theoretical certainty of an arithmetical proof, and if it were not left for every soul to fight its way through the battle of life, by living, believing action and effort, to this clearness of comprehensive moral intuition.   As a possibility of doing ill is everywhere a condition of the realization of what is good, so this peculiarity of moral cultivation makes possible both original divergence to barbarism and a relapse into it.   The dignity of any moral custom or ceremony can very seldom be convincingly shown when it is regarded in isolation and not in its connection with the whole spiritual significance of human life; having a thousand roots entwined in this, it is generally wholly incapable of a concise syllogistic proof that does not, in its turn, require to have its own presuppositions supported by an infinite series of proof.   Just on this account every moral command is exposed to the destructive sophistry which, taking anything that appears an abomination to our civilised ideas, can so separate it from its relations with the whole of life as to make it seem merely an innocent matter of fact.   And not only so, but we also learn how impossible it is for the untutored reflection of a so-called state of Nature to avoid developing what is crooked and barbarous, side by side with those elements of personal merit to which a good disposition prompts.

§ 2. It may not be uninteresting to recall some instances both of the dawning Moral Taste which led men gradually to seek emancipation from the guidance of mere natural instinct, and also of the mistakes to which reflection was exposed in this progress. If we begin with a consideration of the bodily wants which first roused men to barter, and to the adoption of some simple rules of life, we observe that no people have ever had any moral scruple with regard to the consumption of vegetable food. The whole course of vegetable life is so unlike our own that the ripening fruits seem expressly fitted for our use as mere means, equally removed from the unserviceable toughness of inorganic material, and from that animal life which checks the longing of appetite by a kind of natural repulsion. The pious anchorite, feeding on roots and fruit, or at the outside on honey—the product indeed of animal activity, but itself inanimate—and the tribes who, in primitive innocence, support existence on the produce of the bread-fruit tree and the date-palm, are pictures which are harmonious in themselves, and with which our fancy is familiar. But dawning civilisation soon grows ashamed of such an unsophisticated use of Nature's raw products; it seems not altogether becoming to live so directly from hand to mouth, and the fruits of the trees and of the fields come to be at least gathered together and stored up, before they are wanted for use. It is as though the mere lapse of time between the moment when Nature matures them, and the moment when we enjoy them, had loosened their connection with the outer world, or as though they had become more assimilated to our own nature through being in our possession for a time. But it is seldom that we stop here. The inventions of cookery may indeed be chiefly intended to enhance a pleasure of sense, but we may certainly find another and less obvious motive of culinary activity in the obscure impulse which urges us to disturb the form given by Nature's own hand, to alter the raw condition of nutritive

material, and to give to this before we use it, as far as
possible, the character of a product of our own fancy.   It
would be a mistake to object in answer to this that when
we escape from the ceremonious propriety of our civilised life,
we delight to climb the trees and eat the fruit as we pluck
it from the bough; it is just because our sense of civilised
existence is so strong that we take pleasure in divesting ourselves for a moment of that which we can always resume at
will, and in dwelling for a moment with satisfaction on the
consciousness that our life is a life of sense, and in close
connection with Nature.   The truth of this will be readily
seen if we imagine how odd it would look for man, the
thinking creature, to go out daily at meal-time into the fields
to devour a turnip on the spot, just as he had pulled it out
of the ground.

But nearly every dawning civilisation has had scruples
concerning the lawfulness and propriety of eating animal
food.   Man has such a deep horror of consuming the dead
bodies of those animals which have died a natural death,
that he has always preferred to undertake the intentional
killing of beasts, this destruction being to a great extent
made less repulsive to him by the excitement of having to
defend himself against their attacks.   But in the choice of
what we use for food an unquestionably moral taste has
gradually prescribed limits, the worth and significance of
which it would be hard to reduce to definite notions.   By
civilised peoples it is almost exclusively vertebrate animals
that are used for food, and even among these amphibia at any
rate have never been generally used; among the invertebrate
animals, on the other hand, we can mention a few, and but
a very few, such as the oyster and the crab, which people
venture to consume in their natural state, whilst some others,
as snails, are only endured as disguised ingredients of prepared dishes.   It may be easy for the doctrinaire mind to
prove that at bottom meat is flesh, if indeed it does not
succeed in establishing the still more remarkable discovery
that the range of our natural appetite is coincident with that

of albuminous material in the animal kingdom, and that it ceases when we come to the lower orders of animals where these materials are replaced by others of different composition and more heterogeneous to us; but spite of all reasoning, the natural taste of civilised men adheres obstinately to the opinion that animals do certainly differ from one another in being some clean and others unclean. To eat insects and worms, leeches, maggots, and vermin, will always be regarded as a mark of hideous barbarism, however great their nutritive value may prove to be.

It is partly the shapelessness of these living objects which disgusts us, partly the numerous disagreeable qualities attaching to their exterior—as, for instance, slimy coldness—partly the strangeness of their appearance, and even their small size: for though we may take animal food, eating of meat which comes before us in pieces of considerable size, there seems something repulsive in the idea of consuming whole organisms with all their vital apparatus, something revolting in swallowing an object that comprises in itself the variety of a complete though minute anatomy, that we cannot disjoint. We thus seem impelled by a natural instinct to the consumption of creatures which are of a higher order, and whose organization is more akin to our own.

How dangerous this indication may be in itself does not need to be specially emphasized; it is plain that logically followed out it leads to Cannibalism. And, indeed, it is hardly to be doubted that men in a paradisiacal state of Nature have often enough in all innocence followed it out to this result, seeing no evil in it—indeed, even when the dawn of reflection had broken, they were by no means at a loss for pretexts which should invest with the semblance of tender consideration a custom we regard as the very extreme of inhuman barbarism. What could be a more appropriate fate for the organic remains of beloved persons than to be converted forthwith into the living flesh and blood of their descendants, instead of being consigned to the horrors of corruption? A man may be absorbed in tender recollection of the friend whom he has eaten, as he plays with the bleached knuckle-bones of the dead, and he may listen in amaze to the horror expressed by a civilised stranger at such proceedings. It may be objected that even cannibalism revolts from devouring the bodies of those who have died a natural death, and that therefore as a matter of fact, the feast of a cannibal must always be preceded by murder. But what is there that could effectually restrain men who are in a state of Nature from killing their enemies, or even neighbours to whom they are otherwise quite amicably disposed? We should remember how fond we ourselves grow of the domestic animals which we feed for human consumption, and how, without feeling any particular moral contradiction, we give them a final caress the evening before they are to be slaughtered. So much that is contradictory finds room in our minds, that we ought hardly to feel boundless astonishment at hearing of wild tribes who invite their parents, when becoming aged, to let themselves be killed and eaten, and when we find that the soft, natural grace and friendly deportment of the South Sea islanders hides a craving for human flesh.

If one thinks how easy it would be for an ingenious mind to bring forward whole series of reasons, plausible and hard to be refuted, in justification of such atrocious customs, one sees the more what a vast moral effect civilisation produces by merely holding fast the opposite conviction, and by its unhesitating and energetic refusal of such sophistry. The real positive grounds of this civilised conviction will probably not be alleged ordinarily, for they do not lie on the surface of our civilisation as isolated maxims which can easily be collected thence, but are bound up with the very foundation of our whole philosophic view. The deeper our insight into human destiny becomes, the more sacred does every individual human being seem to us, and the more unconditionally do we refuse to attempt to take the measure of his relative worth, with a view to determining whether he has already accomplished his task and tasted his share of happiness, and may now be treated as mere matter devoid of rights, which we may, if we choose consign to destruction—finally, the more intolerable becomes the thought that the body, which, as the vesture of a human soul, belonged to that soul in an unique sense, should be disintegrated in any other way than by those natural forces to which it owed its formation, or that its substance should be used by others as a mere means for the support of animal life. The spirit of civilisation has set upon human personality that seal of inviolability which the perversity of a state of Nature sometimes sets upon external objects; and wherever our conduct is not actuated by this sentiment, wherever Law and Society still treat individuals as though they were *things*, there our civilisation is marred by a remnant of barbarism, and there we have not yet succeeded in vanquishing the principle of barbarism altogether.

Even to have vanquished it in essentials has not been easy, and a glance at very various periods of history is sufficient to convince us that the task is not yet completed—that a considerable degree of so-called civilisation is not incompatible with a sanguinary background of cruelty, sometimes proceeding from natural savagery, sometimes from cold-blooded bigotry We very often see in children a disposition to torment animals; and it is said that the North American Indian never passes a bird's nest without destroying it. Among barbarous tribes it is often found that not only the physical courage which they have in common with the beasts, but also many a trait of weak voluptuousness is combined with deliberate cruelty; and if thirst for blood is not a prime characteristic of human nature, neither is there implanted in it anything like such a horror of bloodshed as many an optimist thinks. In the early stages of almost all civilisations we find the custom of avenging blood by blood; and the fact that we meet with it as a custom, as an established duty, shows that this wild impulse of revenge was passed on from a state of barbarism to ordered societies, which were incapable of repressing it. The East Indian Thugs and the Assassins I will merely mention, for we very easily credit mystical fanaticism with utterly obscuring human feeling,

even in the midst of civilisation which is in other respects
far advanced.   But in the most enlightened age of Greece
and Rome we find the exposing of weakly children recom-
mended in the most open way, in an ideal constitutional construction ; and we find, in practice, the abomination of a system
of slavery, that could not claim even such justification as may
be found for the white slave-owner of the present day, in the
contempt that he feels for the black-skinned race which he
reckons as belonging to the inferior animals—a system in
which, on the contrary, men were enslaved by others of their
own race, and in which there was much more cold, systematic
cruelty than in modern slavery, and hardly less of passionate
savagery.   And all this in a Golden Age of art, and amid the
glory of one of the great kingdoms of the world.

But we do not need to go back to distant centuries for
instances of what lies at our very doors.   I am not alluding
to the evils inseparable from war—war which springs up
again afresh in every age, and which it is idle to hope that
we can charm away with the olive branch of peace.   When
advanced civilisation turns to this last resource, it is not
because any delight in outrage stirs it to the temporary
unchaining of murderous forces, but because it recognises
that the complication of the situation is too great to be
solved by existing human wisdom.   No one denies that,
spite of this recognition, the solution would often be really
very easy to find ; but the very fact that the right view does
not obtain general acceptance and realization, is one of the
inevitable deficiencies of every civilisation which has recourse
to the *ultimatum* of war.   So men betake themselves to
the extreme remedy of momentarily suspending those laws
of humanity by which we are ordinarily bound, and of
referring to force the decision which has been sought in
vain from wisdom ; yet still the suspension is only partial,
and men always regard as sacred, at least those forms of
intercourse which serve to facilitate the return at any
moment from a state of violence to peaceable relations.
Therefore, however lamentable it may be to see this appeal of

civilisation to force recurring again and again, we find even in
the appeal itself a reference to that good to which men hope it
will help them to return; but there are not wanting proofs
of a continued influence of barbarous philosophy in suggestions which are made unhesitatingly even in our own time;
in incitements to wars of extermination, in exhortations to
assassination, in instigations to go beyond legitimate self-
defence and the re-establishment of justice, to deeds of
immoderate and bloody revenge.

§ 3. Let us, however, turn back to those simple phænomena in which dawning civilisation betrays a gradual
heightening of the human sense of self-esteem. To keep one's
own body free from all accretions of extraneous matter is
an impulse of cleanliness which is everywhere a sign either
of the beginning of culture, or of a happy natural constitution that promises to favour the establishment of culture.
On the whole, we can hardly maintain that cleanliness is
natural to men in a higher degree than to the beasts; it
springs up spontaneously among people who are invited by
the proximity of the ocean to frequent indulgence in the
pleasure of the bath; but where this favouring condition is
absent, we find not only that barbarous nations are extremely
uncleanly, but that even among those who have pretensions to
belong to the civilised world, uncleanliness is quite compatible
on the one hand with effeminate good nature, and on the
other hand with active æsthetic taste for beauty in outward
form and movement. Uncleanliness is unendurable only to
those civilised nations who strive after order and consistency in their inner life, in their whole system of thought,
in their feelings and endeavours. Gifts of genius, as well
as benevolence of disposition, have in every respect an extraordinary compatibility with uncleanliness and disorder; on
the other hand, nations which are not so remarkable for these
endowments, but which produce more perfect characters,
will be inclined to the same nicety and systematic precision
with regard to their own persons which they introduce into
their occupations and surroundings.

I am not, I think, having recourse to a far-fetched analogy when I couple with this outward virtue the inner virtue of truthfulness. The worth of truth, and the total impossibility of carrying on human intercourse under a system of barefaced lying, is so strongly felt, even by men in the most barbarous state, that lying has always been regarded as the root of all evil, at least in certain circumstances in which men reckon upon truth. But the impulse to speak truth is not directly bound up with the recognition of its worth, and it is only in civilised society that the liar appears to himself worthy of condemnation, whilst the life of barbarians is in many respects founded on craft and carefully cultivated hypocrisy. We may remark that to a man of morally cultivated mind it is peculiarly hard to tell an isolated lie on the spur of the moment for some temporary and isolated end; he feels that it disturbs too conspicuously his consistency as an individual, and is conscious of being untrue to himself; it is much easier to make consistent lying the maxim of his conduct; in that case he can still be conscious of having a coherent individuality, not destitute of all method and order. The same thing is seen in the case of other moral relations; men hesitate to infringe one isolated law of social order the more if they still recognise the others, and by this recognition condemn their own deed; it is somewhat easier to set oneself in opposition to social order altogether, and to wage war against the world, like some monster cut adrift from it. In such a course there may yet be expressed — though misguided to the last degree — the impulse of an individualizing personality to establish the basis of its own conduct not in dependence on foreign conditions, but in systematic complete harmony with itself.

Around these rare cases of conscious grand systematic untruth clusters the incredible amount of petty incoherent falsity, which in the most varied forms pervades all strata of civilised society, and which seems to me much less akin to lying in the ordinary sense than to that impurity and untrustworthiness of the inner life which appear, only im perfectly veiled by fair appearance, as the general rule among barbarous men. To a character of thorough moral development every entangled complication of circumstances, every uncertainty regarding claims which it is entitled to make or called upon to satisfy, every doubt about its relations to others is as odious as bodily impurity. We need only compare with this the prevailing inclinations of the lower classes, in order to see those moral deficiencies which it is so hard for imperfect civilisation to avoid; the difficulty of extracting from them a definite, decided promise, their constant disposition to leave everything they can in a state of fluctuating uncertain indecision, their inaccessibility to the notion that one's word once given is of binding obligation, and—in wider circles—the propensity to cling to doubtful and untenable relations, the hope that if one never takes a decided step one will be able in the hurly-burly of events to snatch some advantage, of which one has at present no clear notion—in short, inexhaustible patience with all sorts of confusion, and a delight in wriggling on, with the help of procrastination, waiting about, half-admissions and retractions, and general uncertainty, through the course of events which to men thus inclined seems itself equally uncertain. Among the more intelligent upper classes the same deficiency recurs, but under other forms, or under the same forms, but in different connections; among them, as among those whose conditions of life are less favoured, the noble spirits are but few, but there are some of these in all ranks of life—souls who, with an unwearied impulse towards truth, renounce all those pretexts with which the slothful of heart seek to excuse this mental instability, and who, moved by the enthusiasm and force of moral conviction, not only desire to make their whole duty clear before their eyes at every step of this changing life, but also obey with unhesitating decision every clear call to action.

Unexpected perfidy and perfectly sudden and inexplicable changes of mood have always been the first warnings which have roused mistrust towards the deceptive friendliness of

barbarous men.   It is the nature of a beast to act in accordance
with the passion of the moment, but in a man the passionate
motives to action due to momentary feeling should be
moderated by the counterbalancing force of the other moral
motives which the memory of past experience has stored up.
Children and barbarous peoples lack this retarding or regulating flywheel that can hinder, as in machines, the precipitate
course of springs once set in motion, and we can as little
rely upon their moods as upon the course of the weather.
To realize this one must look into one's own mind.   How
easily is one inspired with momentary enthusiasm by some
noble thought, or the idea of performing some magnanimous
deed!   But this excitement is followed by a state of nervous
exhaustion, or, to state the case more simply and honestly, by
laziness; there wake up all sorts of little likes and dislikes
which were hushed at first; and at last, although the work
may, as we had pictured it to ourselves, be indeed a noble
one, yet all the same we find that we can get on without it,
and besides who would thank us for our pains if we were
to trouble about it ?   Here we see that moral weakness
which so lightly dons the cloak of heroism, but has not the
enduring strength necessary for holding fast the ideals of
youth, and then coolly, as though it had long ago weighed
the whole matter, rejects as an idle dream that which it was
too lazy to convert from a dream to reality.   In a mind
which has not been furnished either by education or by rich
experience with power sufficient to withstand this sloth, the
obscurer of all that is good, but which retains unimpaired the
capacity of appreciating every passing advantage or disadvantage, the sloth will be almost necessarily intensified to
falsity.   Any fancy that crosses the mind, any unfamiliar
association of Ideas, rouses mistrust, and disturbs the equilibrium of these poverty-stricken souls, for whom all steady
social intercourse makes shipwreck on the rock of their own
incapacity to calculate and guide the course of their inner
life——a course which is not amenable to any standard of reason-
ableness, of principle, and of self-government.   We find that

this running wild of the course of thought and of changes of mood is not confined to men who are in a state of barbarism, any more than other moral deficiencies are; on the contrary, it is found among all nations except those which by a long course of development embracing equally all departments of human life have become the very repositories of human culture; and alas! the genius of civilisation—quieter, self-centred, hemmed in by a thousand self-imposed limits—is but too often imposed upon by this as yet unexhausted "natural force." For we find ready to our hand this and other flattering names for such untamed and untutored wildness, which bribes our æsthetic judgment sometimes with the heroic noise of boundless passion that must have its way, regardless of consequences, sometimes with the different charm of something unique, incommensurable, supernatural. We too easily forget that much which looks extremely well in a picture and has a striking effect in poetry, would make us heartily ashamed of our prepossession if we were to see it, not at a single favourable moment but in the ordinary course of life, in connection with all its manifold results. The charm of what is strange and full of characteristic expression and one-sided originality, is so great that it leads every one to be sometimes unjust towards that consistent, thoughtful, steadfast order of civilised life which though less warm in colouring is ineffably more worthy.

§ 4. We now turn back once more to the most fundamental relations between Nature and Man; to the great mystery which joins our spiritual life to our bodily form, and mental excitations to external gesture and movement, which binds up the continuance of our personal life with the continuous activity of the physical machine of our body—that body which we so cherish as long as it serves us, and which we regard with such strange horror as soon as life has departed from it; by which, finally, our existence altogether is made dependent on the inexplicable secret of bodily reproduction. The more deeply conscious the soul is of itself and of its destiny, the more obnoxious to its self-esteem is the

direct unity presented by the combination of the inner life
with the marvellous material organism, the soul being inevitably forced to sympathize intensively, by pain and pleasure,
with all the excitations of the body, and to trust to it for the
expression, the accomplishment, and even the very quickening
of its endeavours.   For in truth the soul can enjoy the full
warm life which alone can satisfy it, only if the supersensuous
play of its states and activities is supplemented, as by a
sensuous echo, by the sum of all those feelings which seem to
make known to us the strength and elasticity, the tension or
relaxation, the rest or the sympathetic stirring of desire
which affect our material part.   Our spiritual nature is everywhere ashamed at finding itself in indissoluble connection
with the world of sense—at the consciousness that while its
own aims have intrinsic worth and are incommensurable with
material processes, we are yet bound by the mechanical order
of Nature, and that of our whole destiny no part could be
realized without those natural impulses by which our
endeavours are provided with tangible objects and means of
attainment : it is the dim consciousness of all this which in
the dawn of moral culture has produced those various developments of the sense of shame by which the human race is
everywhere prompted to veil the physical basis of its spiritual
existence, especially when this physical basis furnishes the
pre-eminently sensuous means by which we must reach the
most precious and spiritual treasures of life and love.

I will not attempt to decide what is the significance of
those traces of a sense of Modesty which appear even among
beasts, or to what extent it may be an innate natural feature
of the human race.   Observation of barbarous peoples reveals
to us sometimes a considerate delicacy and purity of manners,
but much more often a bestial absence of restraint in the
satisfaction of all physical wants ; and we are left in doubt
which of the two we should regard as original and which as
the result of dawning civilisation or of almost total relapse
into savagery, or, indeed, whether we should not refer differences in this respect to peculiarities which are not shared

by all mankind.   However, beside the moral sentiment on
these subjects which has on the whole become established
among all civilised peoples, civilised reflection and sophistry
have produced two one-sided but mutually-opposed views :
on the one hand, the exaggerated contempt with which a
fanatical spiritualism looks down upon all Nature as something in itself unclean, shameful, and degrading, and to be
resisted by every weapon of a gloomy asceticism; on the
other hand, the cool assumption that everything which is
natural is pure.   Neither the former opinion with its hatred
of Nature, nor the latter with its easy complaisance, has
succeeded in guiding the moral feelings of civilised humanity
on the whole; but both have had an important practical effect
on the temper of different times, and both have in many
ways obscured theoretic belief concerning the grounds of
such moral feeling and the demands made by it.

With regard to those deep and sacred joys of life which
we can reach only through the middle term of sense, it is
not a genuinely human feeling of modesty which leads us to
despise and reject them merely on account of this medium, to
which they are joined in the order of Nature; on the other
hand, in that intentional prying into this mysterious connection
which vainly seeks to justify itself by the pretence of serving
science, there is an unconscious immodesty; and not here
only, but also in analysing, for the confirmation of christian
humility, all the foulness and corruption on which rest the
beauty and proud gladness of our life—in brief, in the disposition to hunt after that which is impure and sinful, of
which there will be the more to be found in proportion as
the imagination which seeks it is the more corrupt.   The man
of genuine moral feeling sees primarily that which is pure and
noble and divine in things; the indissoluble connection of all
this with the world of sense seems to him to be entailed by
his own finiteness, but to have no power to destroy his faith
in the worth of those blessings which are only accessible
through the medium of sense.

But on the other hand, the principle that all which is

natural is to be regarded as pure, leads to a mode of thought and action which is rejected with equal decision by cultivated moral feeling.   It can naturally be no reproach to a finite creature to be subject to the wants entailed by his bodily organization, and to say merely this would be to show but cheap wisdom.   But that in our consideration of human life as a whole, we should regard these calls of Nature as entitled to put in an appearance without check or reserve, and to be reckoned in their primitive simplicity as among the phæno-mena of moral development—this is a notion which we must in all cases reject as a mark of inhuman barbarism.   It is difficult to say whether the claims of moral culture in life and in art are more deeply sinned against by the impassioned voluptuousness which breaks through many a moral barrier, and misuses poetry as a means to its own glorification, or by the cold unemotional temper, which—taking a pride in being beyond the reach of temptation and knowing nothing of what is seductive but only of what is unclean—seeks this last, and with naked plainness describes or practises it as being, or be-longing to, " human nature."   If voluptuousness leads sooner to the transgression of moral limits, yet at least there is in it the remembrance of a natural charm to which the human impulse is subordinate ; but in that realism, coarse and scornful by turns, which takes pleasure in emphasizing the inevitable earthly element in all that is fair and noble, and in recognis-ing with deliberate expressness the impurity which our nature cannot shake off—in this there is a corruptness of imagination which far more completely, though perhaps less quickly, blunts all moral sensitiveness.   Beside two such monstrous growths the principle of the purity of Nature will certainly for the most part lead to a middle path ; it will allow the general practical necessity of modest decency, but will blame as exaggerated sentimentality the wish to ignore those natural facts which it is in truth impossible to deny.   The conduct of grown-up men and women tends for the most part to be in agreement with this view, which with simple straight-forwardness inclines to call everything by its real name.

Unless people are guarded by a noble refinement of mind, the older they grow the less reticent do they become with regard to their physical nature—increasing bodily infirmities incessantly call attention to the functions of animal life, and give occasion to seek medical counsel and help ; and thus is gradually shattered the proud, shy modesty of the individual spirit, the attachment of which to its disintegrating envelope begins to be loosened. If in contrast to this we recall the indignation of some young and lofty soul, when in ordinary life in the intercourse of elder persons it hears others treat and discuss and bring before it with idle indifference circumstances which it feels impelled to conceal even from itself, we shall be constrained to admit that even the well-meaning moderate view of steady-going folk involves a sensible retrogression in moral refinement, and that of all kinds of enlightenment none is more hazardous than that which conflicts with the prepossessions of modesty.

We are in the habit of expecting this feeling to be most active in the intercourse of the two sexes, and in fact the forms by which such intercourse is regulated are all the more essential marks of high moral culture because definite forms are so little prescribed in this department of ordered life by mere natural circumstances. The only kind of marriage which would everywhere seem unnatural is that between parents and children, and this on account of the disparity of age ; but Nature enters no protest against marriage between brothers and sisters, and presents as many analogies in favour of polygamy as in favour of monogamy; indeed, mere Nature provides us with no reasons why we should substitute a life-long union for a temporary connection formed for the gratification of desire. All the limits which the human race has set to its desires of this kind are the product of a gradually awakening moral sense ; the attempt to find for them a natural foundation which does not exist, does not make them any the more sacred or intelligible. For we are neither justified in following the dictates of Nature merely as such, nor bound in duty to do so; it is only when we act contrary to those commands of

Nature, on obedience to which all successful action depends, that our procedure is vain and criminal ; but with regard to those things which she leaves to our option, the moral nature has to make a nicer choice, a choice which can only be justified by its ideal end.   There is no other particular of ordered life in reference to which there has been a more strange divergence in the variety of custom, and this variety is to be explained by a consideration of the different degrees of clearness with which the worth of human personality and of the individual soul was presented to the imagination of different ages and nations.   To some nations of antiquity, marriage between brothers and sisters seemed admissible; to us it seems so incomprehensible that its inevitable necessity, in case of the human race having sprung from a single pair, has been thought a sufficient argument in disproof of this view of our origin.   But to think this is clearly wrong ; for it is certainly an error to imagine that the sinfulness of such a connection is immediately declared by the voice of Nature. On the contrary, the voice which declares it is that of the most highly developed moral insight, which impresses upon men a horror of mingling two human relationships, of which each can be experienced in the whole fulness and beauty of its ethical significance only if it is kept uncontaminated, by isolation from the other.   This monition could have had no weight for those primitive brothers and sisters who were as yet all the world to each other.

As we associated purity of the inner life with bodily cleanliness, we would also assign to modesty a wider range than is generally considered to belong to it.   As it is certainly a mark of defective civilisation to neglect the development of the bodily frame and its capacities, so is it little in agreement with genuinely moral feeling to make one's bodily presence conspicuous and to wish to be esteemed on account of it.   The more highly civilised nations and the more cultivated classes of society consider as most essential to a fitting dignity of demeanour that correctness of external appearance which neither can be found fault with, nor attempts to show off any

personal advantages, and which is thus best adapted to prevent
any undue attention from being excited by one's personal
appearance.   On the other hand, it betrays a lower degree of
culture to show off physical strength and skill, except in work
in which they find appropriate employment, and to wish to
do one's work in the world by means of a noisy display of
one's bodily gifts.

In respect of this, nations and individuals are divided into
two distinct groups, the peculiarities of which pervade and
give a tone to all departments of culture.   There is one
disposition which—to employ here one of the most repulsive
phrases which modern times have invented—considers that the
business of life is *to develop oneself* (*sich darzuleben*); there
is another which, forgetting and neglecting self, tries to find a
reflex of its own Ideas in any finished work, any labour, any
external order ; each has for the other an antipathy which only
gives place to mutual admiration when they look at one another
from a distance, and the one sees its own deficiencies supplied
by the other's peculiarities.   We will, however, not conceal the
fact that in the interests of human culture we are decidedly in
favour of the last, notwithstanding all its shortcomings.   A
deep-rooted aversion to take in hand any hard instrument not
easy to manage, and to do a spell of honest work, is in the case
of men of the disposition which we first noticed, ordinarily
joined with an inclination to make a boundless fuss about
their own appearance and about all those physical powers
which the bodily organization graciously and gratuitously puts
at the disposal of the fancy.   Continual inquisitive activity
of the senses and quick receptivity makes such men good
observers while they do observe ; but their attention being
easily distracted, for the most part they grasp only the superficial harmonies or discords of external form, only what is
graceful or ludicrous.   They likewise feel an unceasing need
of manifesting their inner life with all its emotions, however
transitory and insignificant; and this on the one hand leads
them to be always making a show and trying to give a
picturesque and heroic air to their finery or their rags, and

on the other hand tends to bring their minds, even in solitude, into a dramatic frame, in which they take a secret pride and pleasure. Little inclined to real exertion, they make the most perfect theatrical use of their bodily gifts ; they are eloquent, and in their language indulge in far more of high-sounding and diffuse description, colouring, and ornamentation than there is any occasion for ; they are given to song and noise, and add to all this the luxury of expressive gesticulation. It is chiefly the southern countries of the temperate zone which by their fineness of climate have produced in their inhabitants both a bodily organization which combines beauty and strength, and also a keen satisfaction in the endowments and capacities of this corporeal frame, and in addition to these the passion and vividness with which they feel to the full the joy or admiration, the love or hatred, the devotion or despair which any situation may call forth. If we add to this the approving definition of their nature, long ago adopted by philosophic reflection, and say that in them and in their culture we see attained the highest development of the living human form, we think we shall have sufficiently indicated the short-coming which is attributed to them by men of the opposite disposition.

For to cultivate oneself, and to make oneself into a perfect human being, may easily seem to be the essential scope of all human tasks; but nevertheless we must admit a deficiency in this mode of thought, which aims solely at moulding its own being into a beautiful flexible whole, doing this partly with a kind of natural instinct, partly with doctrinaire self-conscious-ness — a deficiency, namely, in that submission and self-sacrifice which make one element of morality. And this remark does not apply merely to that so-called healthy natural sensuousness which, glorying in the endowments of the physical organization, does in truth accomplish no more than the production of a first-rate specimen of the species *man*, looked at from the point of view of natural history ; we must also blame, as a more refined kind of Egoism, the deceptive self-culture which does indeed always seek that which is good and noble, but only in order to adorn with all the ornaments of virtue that specially cherished central point which we call our Ego. All the duties imposed upon itself by a mind of this temper seem to it to be duties to itself alone; the dignity of its own personality is the end to which every effort of life is devoted.

It cannot be said that the other mode of thought which we contrasted with this does not accomplish the same results, but the consciousness of personal dignity comes to it rather as an accidental gain, because it does not aim primarily at this end, but, forgetting and denying self, works for the general realization of what is good in all the world. Indeed, it would be more in accordance with truth to say that what it gains is not the consciousness of personal dignity, but the habit of feeling and acting in accordance with this; and also it attributes less value to the efficacy of external expression, which will naturally belong in greater measure to him who regards himself as a work of art to be polished to the utmost pitch of perfection. To be of use in the world, and to do one's work in life by labouring for the general good, is the comparatively prosaic motto of men of this character; and their own personality is regarded as but one among many—the many who are to share in the general benefit and rejoice at it. Wherever at particular periods, or in particular nations, this mode of thought has preponderated, there has arisen delight in work of a kind that not only is advantageous to the community, but also affords in its products an objective reflection of individual personalities—products in the characteristic forms of which the worker sees embodied the worth of his being and his own creative fancy. Not himself, but what he has made, not his person, the product of cosmic forces, but that reflection of his own being in his surroundings which his bodily and mental labour and self-sacrifice have called forth—these it is which such a man regards as what entitle him to a place in the world, and in proportion as this feeling grows, there increases also his aversion to any ostentatious display of a personal strength and beauty which are the gift of Nature. To speak louder than is necessary, seems to him an uneducated display of vocal resources; to be more excited than the importance of the occasion justifies, appears to be a foolish yielding to the sheer power of the external stimulus; he regards as unendurable all liveliness of gesture, all pantomime and movements of the hands which accompany simple verbal expression as a mere luxury of bustle, wholly useless and ineffectual; and the objectless and overflowing manifestation of mental moods is as repugnant to him as to be everlastingly thinking how to pose effectively. It is easy to see how these contrasts of external demeanour are connected with points favourable and unfavourable to mental life, and how this disposition of mind, if it becomes still more self-contained, threatens the beauty of life and art with a self-absorption, a closeness, and a reserve which are in truth little in accordance with its original self-forgetting and self-sacrificing bent.

I must renounce the attempt to investigate, within the narrow limits of these observations, the other innumerable peculiarities of moral feeling which are expressed in the forms of daily intercourse among men, and the development of which is due partly to the special circumstances of life, and partly to the original disposition of particular nations. We may remark in general that as culture advances, expressly established rules of etiquette become more numerous, not only for the regulation of the conduct of inferiors towards superiors, but also to prevent personal dignity from being wounded in the ordinary intercourse of life by natural passion and curiosity, or to secure the performance of binding duties against which sloth and selfishness rebel, by the sanctity of inviolable custom, regulating even the minutest details. The less scope is allowed to arbitrary choice in determining the mode of any performance, the more imperative does the performance itself seem. (In saying this, we would by no means deny that the refinement and politeness of manners, hospitality, and other virtues which we find exercised in states of rudimentary culture, may not be partly founded on natural good-heartedness.) The further progress of civilisation generally breaks these trammels of conduct for good and also for ill. In modern life even in the cases in which etiquette is most thought of, generally speaking it either has a legal or political significance, which is of use not in personal intercourse, but as a symbol of that objective order which transcends all mere subjectivism—or, if it is really a form of intercourse, it is seldom of such rigidity that a cultured person would not be able to substitute for the ordinary form some other of similar significance. Here also culture drops the use of fixed and specialized precepts, and trusts more to that unconstrained moral feeling to the predominance of which it is due that the social intercourse of civilised peoples is superior to the ceremonious meetings of less developed nations. But we must equally admit that with the removal of this curb, social intercourse among the more uneducated classes is freed from all check; clumsy curiosity, intrusive indiscretion of every kind, and the absence of all respect for the inner life of another, make the intercourse of these classes far less dignified than the reserve with which the hospitality of simpler peoples receives the wanderer and provides for his wants without inquiring too precipitately how he is called, whence he comes, and whither he goes. It is becoming more and more rare to find societies in which customs handed down from antiquity with all their traditional circumstantiality and detail, still give to social intercourse a cast of grave and considerate cere-moniousness.

CHAPTER IV.

THE ORDER OF EXTERNAL LIFE.

Nature and Culture—Home (*die Heimat*)—The Life of Hunters—Of Shepherds
—Permanent Occupation of Land, and Agriculture—Home (*das Haus*)—
Family Life — Society — Division of Labour — Callings of Individuals —
Simple and Complex Structure of Society—Civilisation—History.

§ 1. WHO is there that amid the thousand cares and perplexities of life has not sometimes asked with a sigh, To what purpose is all this pain and struggle? To what purpose all the conventionalities which at one moment oblige us to useless exertion, and at another impose upon us constraints which are equally irksome? To what purpose all this haste to be rich, since our very organization prevents us from getting enjoyment, except in imagination, from the abundance of overflowing wealth? To what purpose is our sensitive regard to honour when the estimation which others have for us adds, directly at least, so little to our happiness? Why should we not restrict ourselves to the simple, natural wants of existence, and give up struggling after all those things which are but means to other objects more or less remote—objects which themselves, when looked at closely, are of only imaginary worth? In such moods it seems to us that Diogenes in his tub had found the true secret of practical wisdom, and that all the complex culture which surrounds us would do well to abolish itself, and no longer to hinder by the useless constraints of innumerable artificialities, the satisfaction of the few wants inseparable from human nature.

And yet it was in vain that Diogenes protested against the civilisation of his age; and all those individuals who since his time have turned their backs upon human culture have only been able to make their solitude endurable to themselves by knowledge, thought, and reflection which they owe to the very culture which they despise. Opposition to the complexities and details of civilisation has a charm only as long as it remains mere opposition; if mankind by sudden consent were to return to the simplicity of the most natural conditions, without doubt the same mental forces which had brought about this resolution would forthwith be as busy as before in reproducing in turn all the rejected superfluities of civilisation. We may frankly admit that there is very much in the complexity of our present mode of life which is in itself idle and unmeaning, and that, if we were free from certain wants, we should do more wisely and be more happy. But the truth is that we already have these wants, and the mere knowledge that they are not inseparably joined to human nature as a whole does not in the least alter the fact that they are so much the more firmly bound up with that definite type of human nature which is special to us, and which we owe to historical development and to education. We, as we are, should suffer from their non-satisfaction, and the same degree of happiness which men in the natural state could obtain by the use of scanty means is only possible for us through the simultaneous fulfilment of many conditions, or through the conscious and voluntary renunciation of many individual satisfactions. But, on the other hand, a voluntary oblivion of that towards which our hearts are yearning is not in our power; it is only great historical changes of fortune that may sometimes obscure a nation's remembrance of all the complex variety of its demands upon life, and make it capable of being satisfied with the simple and elementary enjoyments of returning barbarism.

Have we, however, a right to speak thus, and to prefer such culture to such barbarism? Seeing that in advanced culture satisfaction is dependent on so many conditions, and that it must involve so much self-denial, is not this condition of culture unhappier than that more natural life, which with greater ease and security reaches its state of equilibrium, and

seems to be exposed only to the inevitable ills of the course
of Nature ?   These are questions, however, which we can
easily answer.   For the more vividly we represent to ourselves
the simplicity of a state of Nature, the more clear does it
become not only that it could never suffice to satisfy our
souls, but also that those living impulses in us which stand in
the way of such satisfaction, have, with all their train of
unrest and failure, an unconditional right to be preferred to
that contented poverty of mental existence which only seems
to us now and then desirable as a break in our own more
agitated life.   The happiness to which the human soul is
destined by no means consists in the mere absence of all
disturbances which could hinder those impulses which proceed
most directly from Nature, or in the maintenance of favourable
conditions, securing to them an uninterrupted and uniform
satisfaction ; the course of civilisation is not merely a succession of compensatory efforts capable of re-establishing, under
less favourable conditions and by the use of more powerful
means, a lost equilibrium and a degree of happiness previously
possible.   On the contrary, by the opposition which the
natural course of things offers to a too easy satisfaction of
natural impulses ; by the labour to which man is compelled,
and in the prosecution of which he acquires knowledge of,
and power over, things in the most various relations; finally,
by misfortune itself and the manifold painful efforts which he
has to make under the pressure of the gradually multiplying
relations of life : by all this there is both opened before him
a wider horizon of varied enjoyment, and also there become
clear to him for the first time the inexhaustible significance
of moral Ideas which seem to receive an accession of intrinsic
worth with every new relation to which their regulating
and organizing influence is extended.   In the longing for a
return to a simpler life there is involved a temporary over-
estimation of merely physical wellbeing, and we soon bethink
us that a cultured mind possesses far more springs of happiness, the origin of which we cannot trace.   Perhaps we should
not seriously wish to be without even the suffering entailed by

self-denial. And then there is pain, the bitterness of which is only intelligible by reference to the refined relations of social life, and to the consciousness of combined victory and reconciliation springing from practised ethical insight—pain which gives rise to innumerable feelings not easily expressed, and pervading our whole life like a precious fragrance that we would on no account consent to renounce. Men are much inclined to delude themselves with the hope of combining two incompatible advantages, *i.e.* the simplicity of existence in a state of Nature, and the feelings with which we ourselves regard the external world—we who have been moulded by the influences of science, art, and religion. For we would certainly wish to take with us these feelings when we return to a state of Nature ; but we should remember that they are products of a culture which is unthinkable without all that intricate mechanism, the noise and inflexibility of which sometimes disturb us. We can choose only the one or the other ; either the simple monotonous harmony of an uneventful life according to Nature, or the full, articulated melody of civilisation, gradually unfolding through many a discord ; and no one can doubt that the latter presents the higher beauty, and that civilisation is not a mere roundabout means of attaining under altered conditions the same degree of enjoyment as was tasted in a state of Nature, but that it must, on the contrary, be regarded as a power which for the first time unfolds before us, in all the glory of the perfect flower, the full worth and joy of every moral relation.

On this subject I have now but a few plain remarks to add. I will not here go into the question of the first origin of civilisation, nor endeavour to point out either what definite causes (in the minds of individuals and nations or in external circumstances) aroused and guided the spirit of progress, or what obstacles were put in the way of general or special development either by conditions of life or, more obscurely, by national character ; these things will for the most part remain always unknown to us, and as much as we can hope to make clear, we defer to a later historical consideration. It is just as little my intention to institute here a comparison between the different epochs of civilisation through which mankind have hitherto passed, although such an attempt might admonish us to desirable caution in many respects. For this attempt would in the first place take us back to an observation which we have already made, to the effect that a clear advance in knowledge and power, and in all the external trappings of life, may take place without a simultaneous increase of those things that are good in themselves, for the sake of which all the labour of civilisation is employed. With the advance of civilisation and of its power over the external world there arise everywhere new relations and new sources of enjoyment, but the alteration of social conditions which is bound up with these other changes, unavoidably demolishes many a form of existence handed down from antiquity, to the joy and worth of which only poetry and not real life will ever again find access. Whether this is to be regretted, or whether on the whole in our destiny the good only makes way for the better, is a question the answer to which we can seek only in considering the history of the human race. But the worth of culture in general, as compared with that natural condition which we sometimes describe as a state of innocence and sometimes as barbarism, is not here called in question. And although a sharp line of demarcation dividing the two would only be possible if we could contrast a perfect humanity, hitherto unrealized, with complete brutishness, yet we may emphasize some individual features of social order, on the presence of which the excellence of any culture must depend, and on the more or less completely organized combination of which to a coherent structure is grounded the superiority of one stage of culture over another.

A man wants, in the first place, a home, and possessions, and a sphere of work, so that he may feel he has some definite place assigned to him in the ordered universe; he further wants not merely occasional contact with his fellows, but a lasting community of life with some one person at least, so that he may secure understanding and sympathy for his own nature and individuality. The family circle, too, requires that beyond its own narrow limits there should stretch a wider social background, by the common opinion, custom, and law of which its own life and effort are regulated, to which it belongs, and by which it is supported and judged; finally, it is in all cases inevitable that the mind of this society should connect its own common life and the existence of every individual with the future and the past by some theory of the earth's history, and should link all terrestrial existence to some still more comprehensive theory of the universe by a common religious belief.

§ 2. Not even beasts rove about altogether homeless over the surface of the earth; even where a wide extent of country everywhere offers them equal means of subsistence, they restrict their wanderings to a limited region, beyond which they are driven only by force or unaccustomed circumstances, and not by their own impulse. It is as though each living soul could only taste rest and happiness when, instead of feeling lost amid the restlessly changing multiplicity of new impressions, it can make the unvarying representation of its own familiar surroundings the centre around which are grouped, in diminishing degrees of clearness, the more distant variety of the outside world. Man's love of adventure, which would otherwise lead him to transgress more easily than beasts these self-imposed limits, is counterbalanced by another and more profound impulse, that of the spirit of acquisition which makes him wish that the results of his activity should not disappear with the crowd of changing objects on which it is expended, but should gradually accumulate in lasting monuments of his labour, and present in visible and connected form the gain acquired by his life's work.

Natural circumstances favour or hinder this inclination in various degrees. Where men as yet without fixed habitation are forced by the great abundance of animal life and the necessity of defending themselves from the attacks of wild beasts to take at first to the hunter's life, the dawn of higher civilisation meets rather with delays and hindrances than with rapid furtherance. The necessity of following wandering game, substitutes for the idea of a settled abode the wider and vaguer idea of a hunting-ground, and the ease with which the captured prey, after very slight preparation, can be applied to satisfy natural wants, as well as the way in which, in this kind of life, all the fruit of men's efforts is consumed as it were from hand to mouth, without leaving a trace behind, is not conducive to any thought of collecting the results of one's labour so as to make some lasting and coherent monument, or any thought of so arranging life as to connect into some scheme of development men's fitful attempts to trade and accumulate. Cunning patience and passionate fury of attack are the two capacities which this life demands and exercises, in alternation; both are but little calculated to promote higher human civilisation. Only the calm with which the North American Indian listens without interruption to the speech of another, and the passive courage which he shows under suffering, are useful elements, which, from the necessity of quietly enduring countless hardships and mishaps, have been cultivated in the school of this wild life, in which the hunter is early taught to watch with silent self-restraint every movement of the jaguar or buffalo, so as not to betray himself too soon by any disturbance of them. If there were not other ineradicable impulses of human nature impelling individual men to some combination among themselves, there would be little in the character of this mode of life that could lead the homeless hunter to social union and the development of human intercourse; the occupations of all are too uniform for any one to expect that any other should specially complement his own knowledge and capacity.

The pastoral mode of life brings with it conditions somewhat more favourable to development. It cannot altogether dispense with courage and activity, which are needed for the protection of the flocks and herds; but it is based not on destruction but on cultivation of animal life; and this life calls out alongside of a patience which is not sneaking and cowardly but calm and persevering, much forethought and providence, and leads to a growing variety of wants, and hence to the beginning of a division of labour in a small society of members all helping one another. In place of the sudden alternations between wholly inactive leisure and exhausting effort which are usual with those who lead a hunter's life, there is established a steady succession of occupations each of which reckons upon the rest, and which reciprocally make each other possible; social life takes the place of isolation, and the position in which different persons stand with regard to the property (whether held in common or by individuals) with the management of which all are concerned, calls forth of itself simple differences of social importance. With the possession of this moveable property arise the first elements of two notions which are foreign to the hunter's life, namely rural economy and society. Settlements of some kind, which although not necessarily permanent are yet of some duration, are indispensable; and if the custom of feeding the flocks by letting them graze on natural pastures necessitates a periodical change of abode, still a return to familiar grounds is always preferred to uncertain wanderings into distant localities. Thus life becomes more and more bound up with the region of country which now (for the first time) begins to be a *home*, with the fountains, hills, and woods of which there begins to be linked an ordered remembrance of past events, and which no longer is the mere scene of adventures that have been gone through, but supplies to coherent labour that background and basis of orientation which imagination always requires. But pastoral life in itself does not everywhere produce those fair first fruits of civilisation which we rejoice to see in some examples of it. Partly the nature and capacity of the domesticated animals, the kind of tendance they require, and the degree of their attachment to mankind, partly climatic and social conditions, and finally the incalculable peculiarity of national character modify greatly the degree of development. The pastoral tribes of the polar regions, pressed by the disfavour of Nature, and cut off from contact with a different and more advanced civilisation by wide reaches of country, present a poverty-stricken picture beside the life of the Semitic patriarchs, in the simple grandeur of which we find distinct traces of commerce and of pretty considerable contact at many points with the culture of stationary tribes.   It is not only that the barter of an infant commerce provides the shepherds with products of foreign industry with which they may adorn their life and make it easier—the mere knowledge that beyond their immediate horizon there stirs other human life with other forms and customs, must lift their apprehension above that monotony which with more isolated tribes arises from want of the idea of human society.   For indeed this idea is absent even now in cases where a larger association of families repeats each the same mode of life, the same occupation, and the same petty domestic organization.

Even antiquity knew that the real beginning of higher human civilisation, was in all cases to be found in the change from nomad life to permanent settlement, and knew it with a fresher and nearer feeling than is possible for us.   This change was a necessary result of the need for procuring means of subsistence from vegetable life, a more fruitful and certain source of supply than the animal world.   It is only luxuriant tropical lands that yield such a vegetable supply to a large extent, without any human labour; and in just those regions man would have remained most completely a parasite of his bread-fruit bearing land, if—among populations that were growing numerous and pressing one another on all sides —the impulse to social enjoyments, and many a sensuous desire, flaming up irrepressibly, had not either given rise to some regulation of this communal life, or at least by violent interruptions of such regulation, infused into existence an element of passion.   Where these food-bearing plants are scarcer, the spots where they abound mark the abodes of men, who settle down at the foot of the trees, but systematic civilisation is first developed where Nature has made work a necessary forerunner of enjoyment.   The benefits which the vegetable kingdom has bestowed upon man in the banana, the bread-fruit tree, the date-palm and the cocoa-nut tree, are

certainly not accepted by him without any thought whatever; and the imagination of the people who live upon the
products of these plants is sensitive enough to link with
their striking images, in grateful veneration, the dawning
poetical reflection of their simple life.   But far superior to
these incitements is the educative power everywhere exercised
by the various occupations involved in the cultivation of
cereals.   It is his own strength and effort which the tiller of
the ground must employ for the satisfaction of his wants;
Nature and the soil, with which he deals, neither offer their
gifts gratuitously, nor can they be swindled out of them, but
they yield them to unceasing and exact industry.   The
necessary attention to a number of small conditions which
all help to secure the result; the indispensableness of a
definite succession of occupations which cannot be altered by
caprice nor avoided by thoughtless presumption; patience not
only in struggling with the weather and the seasons, but
also in waiting for the slow maturing of the produce which
cannot be accelerated by any greedy haste; and finally the
spectacle of the uniformity with which in general the work
of natural forces proceeds—all these things teach the mind
to feel itself taken up by and involved in a trustworthy,
consistent, and complicated system of natural order; and they
will not fail to produce even in the most poorly endowed
mind a consciousness of the necessity of complete, connected, and systematic means to secure the success of
any work, and to show how little a life that proceeds as it
were upon the spur of the moment can reckon upon satisfaction and success.

The growing labours of agriculture involve the establishment of permanent settlements, and man now enters for the
first time into a relation of manifold opposition to Nature, on
which all further progress in civilisation depends.   For in
fact the powerful tie binding man to the soil, which first
strikes one in considering the stationary state, is not the
predominant element in this relation, and the nomad who
wanders hither and thither has little reason to look down with

scorn upon this tie; on the contrary, he is himself in a state
of much greater dependence upon Nature, to the scenery of
which he seems to belong almost as much as the flock which
he guides or the game which he pursues.   It is within the
four walls of home that a man begins to enjoy, in his leisure
time and secluded from all outward influences, the quiet concentration of family life, and to prepare the mechanical means
with which to make fresh excursions into the surrounding
world, and to secure and work up its products: these walls of
home are a much more powerful means towards freeing him
from dependence on the external world than is the fugitive
haste of the nomad, who restlessly changes one place for
another without finding access to any inner world, except in
the quiet interior of his tent in the intervals between his
journeys.   The walls of home enclose a new realm of human
thought and effort; within them rising generations find a
fenced and guarded region of existence, filled with memorials
of their forefathers, with whose banished forms the life of the
present is now for the first time in conscious and unbroken
community—the work which they have left behind being
added to, altered, and carried on by each generation, which
thus makes its own contribution to what went before.   But it
would be wholly unnecessary to describe the thoughts and
feelings which arise in every one at the name of *home*, and
which are repeated in all their freshness and fulness whenever there is founded any permanent settlement, intended to
become the scene, for an indefinite time, of a succession of
human joys, sorrows, hopes, and remembrances, all inextricably
bound up with one another.   Suffice it to say that in the
dawn of civilisation the contrast between Nature and the
world of mind appears first, and in its most expressive manifestation, as the contrast between domestic life and the un-
boundedness of the external world.

Even in our present life, in which the intricate connection
of mental interests obscures in many ways our relation to
Nature, we may easily observe what an important influence is
exercised upon our minds by the visible marks of our efforts

in external works. The artificer who frames a work of his own hands, and whose joy in it is not diminished by any existing social deficiencies, retains almost always a more even and contented humour than the inquirer who lives in a super-sensuous world of merely intellectual interests. It is true that the latter may be compensated for many a long struggle in the moment when the result of these takes form in artistic completeness; but it is seldom that this result is as certain and complete in itself as the external work which, with all its excellences and defects, is set before our eyes in visible shape and can be fully estimated, and which as it grew beneath our hands gave us, at every step, practical insight into the means of overcoming individual difficulties. So that all the more in the dawn of civilisation (as in the beginning of every individual life) must there be a joyous celebration of the awakening of self-regard as soon as self beholds the first-fruits of its inner thought and effort embodied in the form of a finished work of its own creation. Every tool or utensil that a man has con-structed bears for him the stamp of some thought of his own, and it represents to him at the same time, the future service which it will render, and the power with which his own mind is now armed, for influencing the external world—that mind having now a stronger and a wider grasp than when it had only the aid furnished by his own bodily organization. This pro-found need of seeing our own life reflected in surroundings which have been transformed by ourselves, governs us always. Not only must house and home present to us the traces of past activity, and the instruments of that which is to come; but even where more spiritual interests are concerned, to which no spatial phænomenon can adequately correspond, we like to be able at least to point out some definite spot as the centre from which any particular human activity is used to radiate. It is true that God is near us everywhere, but every civilisation in its earliest dawn founds local and permanent sanctuaries and altars, and men will only adopt, as their special place of prayer, those spots which they feel have been made sacred by the prayers of their forefathers and the common devotion of their contemporaries.   It is not merely the pressing necessity of maintaining life which leads to the establishment of permanent settlements; but when a man gets a home, he seems to take as it were spiritual possession also of his whole surroundings, or perhaps we might better say that it is then that spiritual life receives local manifestation.

§ 3. With the establishment of a steady centre and circle of work a prosperous development of other moral relations first becomes practicable.   It is hardly possible that in the wild life which hunters lead the intercourse of the two sexes should attain a higher significance than that which as a matter of fact it actually does reach.   Constant participation in the efforts of the man is by Nature made impossible for the woman, and if it were possible it would still be a partnership which would afford to the diverse mental natures of both very little opportunity for the development of their special characteristics.   Under such conditions masculine strength cannot find in the woman's mind any essential complement of its own insufficiency, because the life is so poor, and furnishes so few circumstances which are of emotional value, and in which both have a common interest; moreover, in consequence of the lack of property to be looked after, there is too little community of labour and of solicitude.   The other family relations also suffer from this absence of a common aim in life.   Among beasts we see the young lives environed by a parental love which is capable of self-sacrifice, but which suddenly cools when the need of help in the young diminishes; and just in the same way men in a state of Nature afford striking examples of the self-sacrifice of parents for their children, but we also see how easily, with them, this connection is dissolved, when the children have attained bodily maturity.   In fact where one generation never takes up and continues the work of that which preceded it, but each one, as though isolated and beginning afresh for itself, turns to universal Nature, in order to obtain the satisfaction of its wants in traditional modes, it is plain that there cannot be that intimate communion of souls having common interests in life and yet individually different characters and different imaginative bents, and that community and at the same time that conflict of wishes, hopes, and fears, by which in the civilised world there is developed from the natural bonds of kinship a moral community of hearts. It has been often observed how easily and painlessly the North American Indian can bid his parents a last good-bye ; and by man in his natural state the relation between brother and sister is felt to have even less—much less—significance and beauty than that between parent and child.

I might go on to pastoral life, and extol in it the higher meaning which men now feel in family ties—the freer condition of women, who from being the slaves of men have been raised to be their companions—the pleasure which is taken in carrying on genealogical tables, by the unbroken coherence of which each individual member of a society which has grown up by degrees is assured of his connection with ancestors whose names have been made illustrious by well-preserved traditions of glorious events and deeds. But the fact is that these fair beginnings of culture are found only among a few favoured races, and especially in that Semitic past which we are accustomed to regard as a mirror of the purest and most primitive human development. They are found much attenu-- ated and accompanied by far less depth of feeling in the warlike shepherd tribes which still enliven the wildernesses and steppes of the old world, and they almost disappear in the unpoetic savagery of the polar races. A more comprehensive ethnographical comparison than we can here attempt would make it clearer to us that the degree of cultivation attained is by no means wholly dependent on the particular modes of life which we are here considering ; and on the other hand would show how strikingly the unexplained differences of mental endowment which distinguish individual races of men lead to divergence in their course of development, under conditions which are in all other respects similar. More than this, much which we should be inclined to regard as the almost immediate effect of a mode of life determined by external circumstances, is perhaps the echo of some extinct civilisation, or a reflection from some other civilisation existing elsewhere, into fruitful relations with which the historical course of events has brought some tribe which has apparently developed in isolation.   Historical consideration may distinguish if it can the separate influences of these coefficient factors; but if we are merely concerned to estimate the ethical importance of modes of life on which modern civilisation is built, we shall not doubt that permanent settlements, and the sphere of work which first establishes itself in house and home, form the firm basis of consolidated family life, and indirectly through this of wider social order also.   It is not, indeed, possible in the nature of events, nor is it an imperative necessity of human nature, that clans gradually increasing in number should permanently continue to inhabit that native land of their forefathers in which they themselves were born, or that the bonds of relationship which link a numerous posterity both to one another and to their ancestors, should be held in distinct and present remembrance to degrees of indefinite remoteness.   Grandparents and grandchildren are held together by a strong natural bond, but when we get beyond the third generation (and similarly with the wide extension of kinship by marriage) these feelings of blood-relationship cool down rapidly into the mere general interest which men take in their fellow-men or fellow-countrymen.   This does not, however, destroy the charm that we shall always find in being able to look back through centuries of successive generations of which we know ourselves to be the latest representatives; but as such tradition is only made possible by the existence of cultured feelings of considerable strength, so its value must consist either in the consciousness of some transmitted histori-cal work which has to be carried on, or in reflection on the connection of human destinies which may here be followed clearly along a single continuous chain, whilst universal history in its consideration of the whole human race, loses sight of individual threads.   It is but few who can take such a retrospect, and to whom is granted the happiness of lingering in an old ancestral home and among memorials of their forefathers; for most, their parents' temporary home takes the place of an inherited estate. But even to such a paternal home fancy gladly looks back from amid the storms of later life; and after the dispersion of the family, when the difficulty which its members find in keeping up an acquaintance with each other's various pursuits and courses in life has weakened the feeling of connection between them, the yearning with which they look back to the past and deeply-felt happiness of domestic union bears witness to the worth which a settled establishment of families possesses even in our own civilised condition. This dispersion itself is, however, made less painful by the ever-increasing importance of society, which, in proportion as its internal structure becomes more elaborate and complete, gives rise to an increased number of other ethical relations between individuals—relations which are of as great worth as the ties of kinship, and, in some cases, of still greater. But it hardly needs showing that the moral strength of these social relations is itself rooted in the soil of domestic family life, and that every career, though its orbit may be apparently eccentric, really revolves about this centre, and derives its human worth from the fact that it had its origin in that life, and will find in it its consummation, or that at any rate it works for a community which is founded upon such life.

§ 4. If the natural course of things did not, setting out from a single original pair, produce a growing society, or if it did not, in the present condition of the world, place every one at the beginning of his life in the midst of an already existing *society*, each individual pair would have to long in vain for the help which such a living background of life can afford towards the full development of humanity and the satisfaction of all the wants of men's souls. I do not doubt that the smallest cottage is large enough for happy lovers; but we may be certain that without the remembrance of a society, the cultivating influence of which they experienced before their isolation, and without any return to this living circle, the happiness of their love would not be essentially greater than that which falls to the lot of the forest Indians, who, going about in melancholy couples, ungregarious and dumb, search for and partake together the means of satisfying their wants. The drama of life is too tame when it is played by only two persons; they want, at least, the chorus to keep them in mind of the inexhaustible fulness of human interests, of which only a small portion can be brought into consciousness by their own relations to one another.   Men and women cannot be satisfied by the solitary companionship of one other human being; they wish to observe his attitude to some third person, and to know that he also observes theirs; finally, they wish that the reciprocal influence of themselves and their companion should be seen and recognised by other intelligent beings; for to enjoy without other people's knowing anything about it is not much better than to be non-existent.   This need of others' recognition runs through our whole life; even the most modest love does not wish to hide its joy for ever, he who has a friend desires to show his pride in him before the world, and the praise which we receive from another does not please us so much as the consciousness of being honoured by it in the eyes of some third person; all artistic effort demands recognition, and the most unselfishly devoted scientific labour carried on in self-absorbed isolation from the world of contemporaries secretly reckons upon the generations to come and their appreciation. Finally, it is not without cause that men's favourite topic of conversation in all ages has been their fellow-men; for it is a fact that everything else in heaven and earth is of less immediate interest than the doings of men, in observing, investigating, praising, and blaming which we can best become conscious of our own advantages, deficiencies, efforts, and ends.

Now, as long as the mode of life of any considerable society causes complete uniformity of the aims and occupations of all, this mutual interest and sympathy cannot unfold its whole educative force.   It is fixed settlements and the many occupations made necessary by agriculture which first lead to a growing variety of callings in life, and the whole nature of a man is pervaded and influenced by the particular spirit of his calling, without its suppressing those human qualities common to all. In this there is a double advantage. On the one hand, any life-work which is chosen to the exclusion of every other, not only requires a thorough acquaintance with the objects about which it is concerned, and produces great habitual exactness and systematic technical consistency in the treatment of them, but it also introduces the worker to a manageable and coherent circle of thought, within which universal truths stand out with the more convincing force in proportion as the examples which illustrate them intelligibly and clearly are more special to, and as it were inherent in, the particular occupation at which the worker is employed day by day. In order to appreciate the truth of this, we need only recollect the store of proverbs and proverbial sayings in which all nations are accustomed to treasure up the practical wisdom of experience ; the most expressive of them show that the general truth which they contain has been abstracted, within the sphere of some definite calling, from particular examples occurring there, and there alone. On the other hand, every calling gives a special cast to the mind, a particular bent to the imagination, distinctive standpoints and modes of criticism to philosophic views—and it gives to the emotions and to the whole mental attitude of a man a harmonious and distinctive stamp; consequently every one is now an object of greater interest to others. When we are absorbed in the study of a character thus strange to us and so different from our own, beside the innumerable individual traits which arouse our sympathy, that which is common to human nature stands out so much the clearer, and our moral horizon becomes enlarged when we cease to think that we are justified in regarding our own special fashion of existence as the only one that is conceivable, or the only one that is praiseworthy. But as the opening of the Odyssey emphasizes what our modern passion for travel confirms—namely, the value of learning to know the cities and the modes of thought of many men—this aspect of the educative influence of society needs no further proofs. We will, on the contrary, glance at the dangers entailed by the ever-increasing variety of ways of life and the acquaintance of each with the others.

We need only refer in the briefest way to that narrowness of thought and bluntness of sensibility for essentially human interests which may be caused by restriction to some monotonous groove of occupation. But the coexistence and neighbourhood of different modes of life has disadvantages too, as well as advantages. The more uniform the occupations of a large society are, the more easily is there formed, as a standard for all actions, a fixed rule of custom, from which nothing is exempt; as long as this remains unshaken, it reduces the individual to little more than a mere sample of some typical national civilisation, at the same time, however, securing him from the misery of doubt and of moral instability. But where civilisation has produced greater division of labour and greater variety of life, especially where, in consequence of historical conflicts, there is a mixture of the kindred civilisations of different peoples, a confusing multiplicity of possible modes of existence is presented to the mind; the influence of this is, on the one hand, powerful in raising the intellect above the narrowness of transmitted prejudices; but on the other hand, it is equally powerful in disturbing the stability and security of all moral restraints. For this reason the numerous amalgamations of different nations which have happened in the course of history are from some points of view the most interesting epochs of human development. When any established and harmonious civilisation has been broken up, the imagination of men is given back to unrule; and yet strongly stirred by the influences of the past, it moves among the ruins full of haunting thoughts, loosed from all constraint, eagerly investigating in every direction, and inclined, from the lack of mental equilibrium, to splendid extravagances. Such times may, indeed, bring forth products in which there is more richness and variety and more of the fire of genius than there is generally even in the prime of any civilisation which has attained stable equilibrium, and is faithful to its ideal; but we must also remember that such times are fated to sink down into a state which is a mixture of genuine barbarism and isolated unnatural moral exaggerations. We see this morally dissolving force in the present day in all those abodes of men in which there is continual contact between strongly contrasted civilisations. It was long ago remarked, and with justice, that in the East weak minds must be very confusingly affected by the sight of so many different races who, some white and some black, some proud of their freedom, others servile slaves, pray in one place to these gods and in another place to those; who in some cases are faithful to the marriage tie, in others enjoy the pleasures of polygamy. Everything seems to be permitted—all seem happy in their own fashion, and there falls no bolt from heaven to pronounce judgment amid this chaos of opinions.

To pass from a national habit of life to a more self-conscious condition of humanity, civilisation must run this risk of scepticism, and history continually renews its efforts to increase the reciprocal influence of different divisions of the human race. It is seldom that individuals or nations are induced to wander far by want of the simplest and most natural means of subsistence; they are led to do so more often by a restless adventurous impulse, most often of all by a desire for objects of which the direct worth for human nature is but unimportant, and which partly charm the senses and the love of novelty, and partly acquire through habit, as civilisation advances, the character of imperative necessities of life. We find that even in ancient times poets and moralists spoke of the insatiableness of men which, urged by a thousand artificial wants, transcends all natural limits, and brings into a life that might pass simply and peaceably the danger and unrest of far-reaching undertakings. How much might be added to such complaints in these days! For now there is no department of Nature which does not attract men to infinite labour by its productions. In the mineral world, gold and precious stones, iron, brimstone, and coal have tempted them, and have led to the discovery of new countries and to a development of industry to which are due the birth and extended influence

of innumerable other human activities.   The vegetable kingdom by its edible products early gave an impulse to commerce,
but the interested spirit of enterprise has been called out in
much larger measure by sugar, coffee, tea, and the numerous
spices which people could do without as long as they had not
had them.   Finally, in the animal kingdom, the whale and the
furs of the Arctic quadrupeds have attracted courageous and
enterprising spirits to the inhospitable polar regions, and the
web spun by the insignificant silk-worm early led to commerce between civilised nations.   The boundless influence
exercised by all these circumstances on the development of
human capacities is too well known in our own time to need
more than a passing mention.   A life which could have been
contented with the satisfaction of its primary natural wants,
would have found little stimulus to further development;
while, on the other hand, luxuries that men might have done
without have caused all physical and mental powers to be
exerted to the utmost, and as there has been a continuous
increase in the degree of exertion necessary to ensure the hope
of success, science has grown great in this ministry, and in it
the constructive imagination of men has found inexhaustible
occupation, and moral courage has encountered innumerable
opportunities of proving its worth in new and peculiar
circumstances.

§ 5. We have so far considered culture only with reference
to the good things of life which it produces and offers to
individuals; the further it advances the more does it require
likewise fixed external rules of individual conduct, and a
definite system of administration securing the greatest amount
of general satisfaction that is rendered possible by the existing
or attainable means of enjoyment.   A society, with the
customs and rules which have grown up naturally, becomes
transformed into a *State*, which has to take the living moral
Ideas existing in the mind of the society and, scientifically
and with conscious calculation, to work them as governing
principles into the details of present circumstances; likewise
to present to the mind of each, as a systematic whole, with the

clear stamp of objective reality, that spiritual organism of which he is a member. This is not the place for describing an ideal of political order, a task to which we shall not return till the end of our considerations; but we must here briefly notice the necessity of following such an ideal with more or less success, and the inevitable relations with it into which each living individual must enter in the natural course of things. We shall find that there are two struggles perpetually going on, one between subjective self-will in general and the obligation of an objective order, the other between the wants of the individual and the mechanism of ordered political life by which these wants are not all satisfied.

Coeval with all the political organizations of the world are the hardships inflicted by their institutions on individual members of the community; hardships which are blame-worthy in all cases where merit and struggling capacity are by law denied room for development and the opportunity of winning a congenial position in life, excusable in cases where the political organization, while making all careers accessible to all, does not at the same time remove those hindrances to entering upon them which proceed partly from external circumstances, partly from human nature and its weaknesses and evil inclinations. We shall have special occasion, at a later stage, to consider these partly evitable and partly inevitable deficiencies of human arrangements; we only refer to them here in as far as they may awaken doubts of the general beneficence of civilisation, and excite the desire for a return to the simplicity of a state of Nature. There can of course be no question that the ever-increasing refinement of life does not benefit all in equal measure, that a full enjoyment of the physical and mental advantages of civilisation is the lot of only a favoured few, and that on the other hand in all ages a large fraction of mankind remains far below the level of attainable culture and far removed from its enjoyments. But this only makes it all the more erroneous to imagine that while culture raises the more favoured ones, it inevitably diminishes the measure of enjoyment of all the rest to a less amount than might be possible for them if all the trammels of a complex social order were to fall away. A sufferer, wounded in spirit, forgets in his pain very many benefits which he owes to this order, and which because they do not assume the tangible form of some private possession are as easily overlooked as the presence of the atmosphere that un-obtrusively surrounds us and makes respiration possible; he forgets the security of his person, the legal protection which is accorded to his claims, the possibilities of culture which are open to him, the use which he makes (and indeed the very existence) of various ready-made paths in which he may endeavour to employ his powers in a way advantageous to him-self. He forgets that all this, as well as his very knowledge of most of the good things which are denied him, is only made possible by the civilisation which he blames, and that on the other hand the simple state of Nature for which he yearns could not secure to a numerous population in most climates anything like the same satisfaction of its wants as civilisa-tion affords—indeed, there are but very few climates where it could even do this approximately and for a time. The evils of poverty and misery which we so often see in close proximity and saddening contrast to the growing splendour of wealth, should no doubt stir up earnest efforts for the improvement of social arrangements, but they do not invali-date the assertion that every man who is a member—though only in a subordinate and unfavourable position—of a civilised society, has, unless hindered by his own fault, not only participation in an infinitely richer mental life than would have been accessible to him as a result of his own isolated strength, but also possesses greater possibilities of material wellbeing.

To the consideration of the other conflict—that which we mentioned first—we will also devote just a few words. The pressure which is imposed on the individual in the interests of universal order, the limits which it sets to his humours, fancies, and passions, naturally causes in him a counter-current of effort, and he seeks either to escape from this condition of constraint, or—where that is not possible—to abolish or change the order itself which is the cause of the constraint. Society will feel justified in attempting to do the last when it suffers as a whole from those of its institutions which have become unsuitable. And to wish to maintain an established order in opposition to the needs of the whole community, for the satisfaction of which it exists, is but a mere empty devotion to forms. This established order, however, may not only stand opposed to the individual as authoritatively restraining his personal desires, it can also, in a moral point of view, not be so completely subject to the arbitrary will of the community as if it had been the result of arbitrary convention. The statutes of a society which has come together of its own accord may indeed be regarded as bind-ing by conscientious members, but no one regards them as sacred; indeed, their being looked upon as binding, and the observance of fidelity and faith with regard to them, seem to me to be possible only in a civilised society which has pre-viously become accustomed to reverence a binding moral order which is independent of its own arbitrary will. A great political community is thus, to a large extent, every-where a work of Nature, or rather not of mere Nature, but of a Moral Order which is independent of the individual, and the commands of which occur to men when they are living together in a life of social communion. It rests on the one hand on a pious regard for the work which our forefathers have begun, whether it is human labour or the development of humanity; on the other hand, on provident love for our descendants, since we wish to preserve for them that which we have inherited and to transmit it to them with interest. A humanity which aimed at forgetting completely both past and future and at making all the arrangements of life sub-ordinate merely to present satisfaction, would be distinguished from the beasts by nothing except a better choice of means. Therefore, although there is no question that the mechanism of civilised order exists for the sake of society, and not society for the sake of it, yet society is not to be considered as the mere sum of all the individuals of which at any given moment it is composed.   Even in the improbable case of all the members without a single exception agreeing with one another, yet even then these coexisting members would not constitute the community which from a moral point of view is entitled to decide with supreme authority on all the forms of its own constitution ; we must reckon as indispensable members of such a community both the generations that are past and those which are still hidden in the lap of the future.   A man cannot be truly called a citizen of a state or of the world, unless he feels himself included in this unbroken chain of the temporal development of humanity, endowed with innumerable benefits won for him by past generations, and hence bound body and soul to this historical whole, without which his own existence would be unthinkable, and whose unfinished work he is called upon to develop further by his own activity and intelligence.   Something of this feeling has stirred men in all ages, but a consideration of history will teach us how seldom the one-sided attachment to what is old and the blind and passionate love of innovation have consciously joined for the carrying on of this work of true development, and how much oftener it has been left for the unconscious and pressing necessity of circumstances to work out by degrees the progress that human wills had refused or had in vain attempted to carry on.

CHAPTER V.

THE INNER LIFE.

Doubts concerning the Ends and Aims of Human Life—Man as a Transitory
Natural Product—Spontaneous Judgments, and Reflections upon them—
Connections with the Supersensuous World—Superstition—Religiousness
—Unsteadiness and Incoherence of Human Effort.

§ 1. THE more complex and multiform the external order
of life is, the more pressing becomes the question,
What is the kernel of this hull, and what is the clear gain
which men are to purchase at the cost of their life's labour ?
It is not asked only by those whose unfavourable position in
the midst of a complex civilisation forces them to a long
struggle for existence and to a continuous series of efforts
in which every success only brings an immediate necessity for
fresh labour, and hardly affords the hope, even in the far
distance, of at last reaching a secure position. It is asked
just as often by those who enjoy all the good things of life
without having to take any trouble about winning and
establishing their footing in society; to them, too, it often
seems as though there were no objects and aims of existence
except such as men arbitrarily choose to set before themselves — as though nothing could stir the soul except the
passion of a struggle for something yet unattained, whilst
any good that one has succeeded in winning seems to melt
into thin air, and the tension of effort being relaxed there
remain in its place a tedium and lassitude which seem to
seek in vain for some new object that will not lose all charm
in the moment of attainment. There are indeed some lots
more favoured—lots in which spells of hard work and joyous
holidays, labour and compensating enjoyment, are fairly mixed ;
but even from the peaceful content of such lives men are rudely

roused by the doubts which are stirred in them, both by the injustice of Nature and by a consideration of history. The comfort that can be derived from a comprehensive consideration of human destinies and the traces of divine guidance in the history of mankind, is not within the reach of the majority; within the range that is accessible to them it is, generally speaking, only a soul that has already attained peace that can grasp the wide harmonies in which all lesser discords are lost. It is not always in the power of honest endeavour to struggle upwards to a satisfactory position in life, and even if we were to allow that no misfortune happened without some error on the part of him who suffers from it, still this admission would soften but little the bitterness which we feel at seeing incomparably greater faults repaid by the undeserved favour of circumstances. And then, finally, how many hopes are dashed to the ground by sickness and by death! How many souls appear on the stage of this earthly life only to quit it again forthwith without end or aim—without bringing forth any fruit of development in their brief existence! And if we take a survey of the fates of human beings as far as our own experience goes, what do we see but a perpetual repetition of the same labours and sorrows, the same misunderstandings and perversities, differing only in external accessories, and everywhere brightened only by the same isolated lights of transitory enjoyment? How great is the number of the hours and days which are spent in works and labours which we should never have undertaken except in the hope of a result which would more than counterbalance them; and how few are the moments in which it seems to us that we have really lived, and not been merely busied with preparations for living! There is scarcely a soul which is altogether free from reflections of this sort, although—very fortunately for mankind—they are in most men extremely transitory, being displaced by cheerfulness of temper or deliberately put aside—and the heart is thus enabled to surrender itself to the attraction of all the little charms of life, and to be satisfied with them for the moment. It is even as the old saying declares, We know not whence we come nor whither we go ; the wonder is that we can be as light-hearted as we are.

The short survey of human existence which we have been attempting as a preparation for the consideration of its historical development, can hardly aim at concluding with an answer to these pregnant questions. But the very fact that such questions are raised, and that men, with hope and doubt, with faith and vivid fancies, look for ends and aims of their existence, that they feel themselves to be in constant connection with a supersensuous world and by their very efforts to suppress the feeling only bear witness to its obstinate vitality—all these reflections and emotions (as well as the external order of society and even more emphatically than it) must be reckoned among the decisive facts which raise humanity far above any psychical development of which the inferior animals are capable. It is true that among some species of animals, the reciprocal action of their psychical mechanism and physical organization leads to an established order of social life ; but whilst in these animal polities a predetermined order, fixed in every detail, combines the actions of all the members to an ever uniform whole, it is among mankind alone that with the question, What are we, and to what are we destined ? there first breaks the dawn of a genuine inner life, for the development and enriching of which all our expenditure of external activity seems designed. The views of life which attempts at answering these questions have produced in the human mind at all periods, will form our topic for the rest of this chapter.

§ 2. There are scarcely any theoretic convictions that are more severely tried by comparison with experience than the opinions which we frame concerning our own human nature and destiny. In the quiet presence-chamber of speculative thought, it is what is good and noble and significant in human life that stands out as if it were the whole, and all the dross being refined away, the image of man is insensibly glorified into an ideal form which not only fits harmoniously into its place in the intelligible whole of universal order, but merits a place so prominent that it seems hardly possible to describe worthily the significance of its destiny and the profound import-ance of its position in the world.   This reverent conception of humanity receives a rough shock where we come into contact with its average individual representatives.   We do indeed find everywhere the general physical and mental capacities with which man is endowed for the accomplish-ment of his high destiny, but so little are these capacities consecrated to the service of that destiny, that love of the race and contempt for the individual are but too often found to be compatible.   The last may perhaps be modified by a fair consideration of those seeds of good which we may always find even in perverted human nature; but the impression which we receive on the whole from these everyday experiences should make us critical of that over-estimation of human worth which has become so familiar to our anthropological reflec-tion, and which in truth corresponds but ill with the far more modest judgment which men mete out to themselves in their unsophisticated daily thought.   In the same way that terrestrial Nature has been regarded as the only phænomenal world in which the wealth of the Creative Substance has been manifested, has it been quite common for philosophy to regard man as the isolated apex of this phænomenal world, and to imagine that there was nothing between him and God except a yawn-ing chasm, the blank emptiness of which could offer no great hindrance to our leaping across it.   He who will only trust to the most direct experience, a kind of experience which presents us with nothing that is supersensuous and shows man as supreme in the world of sense, is right from a certain point of view.   But he who once permits his imagination to stray beyond the boundaries of the sensible world is wrong if he does not at the same time admit the possible boundlessness of the supersensuous realm, but tries instead to put that which is highest in the known world of sense into the position of next neighbour to the keystone of the universe. It is not our business to fill up that wide expanse with dreams more or less daring and more or less uncertain; but

we must say that we regard as worthless any theory which—
vainly imagining that it has, by some dialectic method, possessed
itself of the equation to the curve which represents the law of
universal development—thinks to demonstrate that the human
mind is the crown and end of that which can know no
end—that human life and existence are the last link in the
great chain of self-developing Infinity.   Let us give up the
presumptuous attempt to extract from such supposed certainty
of the high position which we occupy in the scale of creation,
the secret of our being, of our hopes and our destiny ; let us
rather set out with the admission that we are a feeble folk,
often wearing out our hearts with doubt, bare of counsel and
of aid, and feeling nothing so keenly as the uncertainty of our
origin, of our fate, and of our aims.

The same exalted and solemn light in which the concept of
humanity appears to the eye of speculation, illumines with
still more striking brightness the calm figures of primitive men
as tradition shows them to us at the beginning of history,
wandering over the still youthful earth within the precincts
of Paradise or in patriarchal simplicity.   How quickly the
glory of this picture too is changed when we glance at the
countless swarm to which mankind have multiplied since
then !   In this noise and hurly-burly of most prosaic reality,
how hard it is for the imagination to retain the impression
which is so naturally produced by the contemplation of that
little community of the early world which we know so
well, and the poetic largeness of its simple modes of life !
We are only expressing a feeling which must be familiar to
all when we recall the humiliating and confusing effect
exercised upon us by a concrete consideration of the un-
measurable multitude of mankind, amid the throng of whom
our own individuality seems to be swallowed up.   It is not
perhaps the entirely solitary man who feels that God is
close to him, and that he is guarded and sheltered by direct
divine interposition, but it is likely that this happiness will
be experienced by one who, while involved in the sacred community of family life, feels that all the significant relations of

soul to soul which grow out of this community are interwoven with his own inner life, and is not disturbed by any thought of the thousandfold repetition in every corner of the globe which makes this significant harmony of existence seem a mere ordinary everyday occurrence in the course of events.   As our hearts are not large enough to embrace all with equally active affection, so do we shun the idea of sharing with a countless number of other persons our own relation to the Infinite, and it seems to us that the strength of the tie, and indeed our very assurance of its reality, decrease in proportion to the increase of the numbers to which it is extended.   The more mankind emerges from the retirement of patriarchal life, and becomes conscious of the inexhaustible fertility with which, from time immemorial, the earth has produced one race of men after another, differing greatly in external form and mental endowments, and yet all alike in essentials, indeed all in the mode and conditions of their life resembling to some extent those races of beasts which in still greater multitudes inhabit the most remote corners of the earth, and which arise and pass away in shoals—the more vividly all this is present to consciousness, the less ready will men be to enter on a consideration of the worth of their own existence, and their mind will be gradually possessed by a belief that mankind is but one of the transitory phænomena which an eternal primitive force, revelling in the work of alternate creation and destruction, brings forth, only that it may vanish in its turn.

In saying this I do not intend to suggest that at any period of history this view has been predominant among men, although it might in fact be recognised as giving the keynote of thought at various epochs.   I would rather point it out as a view that may be met with in all ages, never perhaps as an unquestioned faith, but rather as a widespread feeling that casts its shadow effectively enough over all human effort.   Indeed, this mean opinion of themselves which men hold, appears in a twofold aspect.   In the first place, it appears without being sharpened and developed by far - reaching reflection, as a direct consciousness of their own lowness and commonness, in the vast number of those who, confined by the disfavour of circumstances to a narrow circle of thought and compelled to a daily struggle with petty hindrances, can only be said to endure life as a burden imposed upon them. Familiar with the aspect of misery, they know how men are ignominiously reaped down in shoals by the course of Nature, whilst to him who is more happily placed the infrequent spectacle of dissolution has at least the comforting and elevating solemnity of an event which is out of the common. All the dark shadows of life, all the hardships inflicted by the ordinary course of events, stand out in naked prominence in their daily experience, and produce that passive resignation with which in all ages the bulk of the human race endures life and death. They do not live their life but they tolerate it from its beginning to its end, having no comprehensive aims, and only intent upon warding off in detail immediate ill, and winning in detail proximate small advantages ; in the same way they tolerate death as a necessity which it would be hardly worth while to escape for the sake of continuing such a life as theirs; for although they may remember some isolated enjoyments, they would hardly find that life held for them any great and permanent treasure of delight which they would feel impelled to try and secure from destruction. The same power that helps us over so many dark and fathomless chasms in life softens also the gloomy colouring of this mood of thought—I allude to the thoughtless forgetfulness with which the human soul entertains in close conjunction the most diverse opinions, never bringing them into clear contrast—a thoughtlessness which enables us to give ourselves up fully and entirely to the passing pleasure of the moment, although we entertain such a poor opinion of the worth of our life on the whole.

That which we have here been considering as spontaneous feeling, and an ordinary accompaniment of existence, reappears refined by reflection and intensified to explicit belief in countless varied forms of theoretic conviction which for the present

we will not attempt to investigate further.   There can have
been no period in which there did not exist views according
to which human life was regarded as a passing wave thrown
up by an unknown ocean in its continuous movement; but all
these views, with the slight worth which they attribute to the
individual as a mere mortal and vanishing phænomenon, have
only exercised a noticeable influence upon life itself in cases
where they have been the living outcome of that natural turn
of mind which we have been describing, the causes and consequences of which were by these views brought into clear
consciousness.   But where this spontaneous feeling has been
other and better, where minds have been stirred by the large
interests of culture and civilisation, and have been admitted,
by favourable circumstances of nurture and education, to a
living participation in those interests—in all such cases
living life has been stronger than the pantheistic and materialistic views developed in opposition to it by reflection or
scholasticism, and men have in reality lived and felt and
striven after another fashion than that set down in their own
theories concerning themselves.

I know that this will be denied, and that it will be maintained that all moral greatness and purity of life can be logically
combined with a faith that does in fact in perfect honesty
deny the existence of a supernatural order of things, our
connection therewith, and the continuance of our existence
beyond the limits of earthly life.   I admit the fact of this
combination, but not its logical consistency; on the contrary,
it is that very inconsistency of our nature which so often
saves us from being perverted by our theoretic errors, which
makes it possible for us to combine action accordant with a
sense of the dignity of humanity with views the logical effect
of which would be to annihilate that dignity, and this in a
fashion which, as it seems to me, is wholly contradictory.   It
is asserted that the obligation of the Moral Law is not altered if
we regard all mental life as merely so much mechanical action
of matter and its accidental combinations, having no higher end
than to persist, and to fluctuate hither and thither for as long

a period as is made necessary by the collocation of the material particles ; in this assertion, however, there certainly is, not a logical connection of thought, but a forcible moral resolve that has determined to hold fast by a reverence for morality, spite of the materialistic theory with which it is incompatible. It will perhaps be attempted to substitute, for a supersensuous mysterious world which is to us the source of the obligation of moral commands, the Dignity of Man and a Self-respect which isolates him from dependence on any superior, yet enjoins him to rule and keep in check the lower nature in himself. I doubt, however, if a view which recognises only a mechanical course of Nature can logically do anything with such ideas as those of reverence and so forth but reckon them among the morbid productions of imagination to which nothing real corresponds, and of which it has already learnt to reject so many. I doubt further whether a view which regards the individual as merely a passing phase in the spontaneous activity of an Infinite Substance, could have any logical reason for attributing to such a nonentity any obligation to maintain a dignity belonging to it in its individual and transitory character — a dignity which it should or could maintain by its own spontaneous activity —whether such dignity ought not much rather to have its presence or absence laid to the account of the Infinite Substance itself. The logical outcome of all such views can only be to let ourselves go as Nature prompts, and to use that mysterious sparkle of independent substantiality which shines within us, with what wisdom we may, for the attaining and enhancing of physical wellbeing. Thus moral commands could only be accepted as maxims of action on account of the secondary consideration that they are useful on the whole.

Meanwhile it is not possible, nor is it our intention, to discuss in this place the question whether these different views of the supersensuous world are intrinsically right or wrong; our intention has merely been to refer to them in as far as they are to be reckoned among the ordinary factors of human development. And here we must repeat that we doubt whether any one of these views which regard human beings as altogether dependent and transitory has ever become a really pervading sentiment of the whole nature, in spontaneous thought and action, as well as in reflection.   When an ancient poet, having scouted all ideas of deities and retribution after death as useless terrors by which the smooth and peaceful course of our natural pleasure in life is disturbed, turns upon us and inveighs against the fear of death, and asks, Do we, insatiable, desire to go on feasting for ever, and never to retire with dignity, as satisfied guests, from the banquet of life ? the effect produced is no doubt striking.   But in asking this, does he not forget that monitions to moderation and dignity must fall very flat on the ear of him who knows that in an hour he will cease to be ?   Or, in using this simile, which is quite out of keeping with his general tenor, is he not perchance secretly influenced by the truer thought that this life is indeed a banquet from which, as guests who have had enough, we must depart; but that we, not so transitory, depart from it only to enter another state of existence in which there will remain to us the memory of what we have before enjoyed ?   And on the other side, what poetic and glowing expression has often been given to pantheistic views !   But whilst they extol with devotional rapture the absorption of the individual in the universal, is not that which they are glorifying just the abiding and enduring joy which the mortal experiences in its reunion with the eternal ?   And do they not hereby assert the immortality of that mortal, which, though destined to extinction, is only destined to such an extinction as signifies its eternal preservation in some form or other ? This thought, which pantheistic poetry cannot escape, is one which cannot be got rid of either by the most prosaic reasoning or the most commonplace views.   People may seem to be as thoroughly convinced as you will of their own impending annihilation, and may speak of the disappearance of personal existence in the lap of universal Nature, and one may indeed imagine that that which used to happen may cease to happen, but one can never imagine that anything which has once existed can cease to be. And however much people may attempt to persuade themselves· that the self-conscious Ego is in fact only an event, a vanishing passage between atoms variously moved, still the immediate consciousness of our personal reality will always remain invincible to these attempts, and we can never think of ourselves as melting away in the great receptacle of universal Nature without thinking too that we shall still be preserved and go on existing in it in our dissolved condition.

I must repeat that I am not setting up these modes of thought as true, but am describing them as facts of our unsophisticated consciousness ; they may be right or wrong, but at any rate they are what we go through life with ; our reflections are never quite free from a presentiment of something supersensuous. On the other hand, we are not in a position to raise these presentiments to a condition of unquestioned authority, except by a summary act of faith ; it is the natural condition of man to fluctuate between the consciousness of an eternal destiny and the ever-recurring dread of being a mere indifferent and perishing production of the general course of Nature, both feelings being toned down by thoughtless light-heartedness. And even that apathetic mood of the majority which I have described is broken by suggestions of such pre-sentiments, and the monitions of conscience make it plain to them now and again that they are not altogether like the grass of the field and the perishing productions of the vegetable world ; and conversely the security of the most earnest conviction of the eternal significance of man's spirit is shaken by the unmistakeable and peremptory clearness with which the course of Nature seems to declare that no other fate can await the living mind than the fate of sharing in that destruction which befalls the living form, and of disappearing from the world of realities without leaving a trace behind.

If we stay to consider for a moment that philosophic view of which the dominant characteristic is a vivid consciousness of human meanness and transitoriness, we see plainly that it is hardly entitled to speak at all of aims in life. Its scientific teachings have indeed gone so far as to dissuade men from all carking care concerning such aims and all supersensuous interests in general, and to recommend them to restrict themselves to a regulated satisfaction of natural wants.   But they have seldom gone further, and have hardly ever succeeded in silencing the opposition of a better feeling which always sets itself against such a reduction of life to the condition of a sort of peaceable and aimless vegetation.   On the one hand, they have had to give way to human nature so far as tacitly to allow to the knowledge of truth, the charm of beauty, and the majesty of moral commands, that superiority to all mere natural impulses, however urgent, which the mind is accustomed to attribute to them, allowing it in spite of the fact that the superiority is not intelligible on their principles; on the other hand, they have never been able to put a stop to practical efforts which far transcend the needs of a mere vegetative existence.   Although in theory men would often have denied the existence of this inextinguishable feeling of being bound up with an imperishable world, yet its activity has been shown again and again, sometimes in provident care for the wellbeing of a distant posterity—a care which seems to spring up spontaneously in men's hearts—sometimes in the intense interest taken in the general improvement of mankind ; and, how often, in outbursts of ambition which have disturbed the world !   The individual soul that considers itself to be a mere passing production of Nature is seldom altogether indifferent to future fame, and yet in what would the attraction of such fame consist if it were merely attached to a name which no longer had an owner !   In all these manifestations there is revealed the suppressed belief in a world of spiritual interests, a world to which its individual members are indissolubly united, far as we may yet be from any clear idea of the way in which what seems so transient becomes endowed with eternal existence.

§ 3. But in the mysterious compound of feelings of which we are continually conscious, that particular feeling of the nothingness and forlornness of our earthly existence is not always dominant.   Over against the prose of this resigned mood

stands the wild poetry of *superstition*, as a second great manifestation of human self-consciousness. It has been long ago
remarked how surprisingly near the rankest superstition is to
unbelief, and how it seems to arise out of it. And, in fact,
the thought of the common and natural transitoriness of the
individual and of the perdurableness belonging only to the
dark and unfathomable Eternal are like two notes that ring
out together; a gust of wind may make now one and now the
other swell fuller and overpower its fellow. But all superstition depends upon this, that the activity of that Infinite
Substance which at first was regarded as guiding the course
of individual things only indirectly and from a distance, as
it were with calm indifference, suddenly comes to be considered as immediately present in all the most insignificant
affairs, permeating the whole frame of phænomena, and connecting its parts together with the mysterious force of an
all-pervading fervour, from which the individual creature,
surrounded and caught on every hand, is never able to escape.

This belief that we are encompassed on all sides by a
supersensuous world, among the clouds of which the near and
sharply defined outlines of our lives become lost, indiscriminately
and past recognition, is also a mood of thought which has, on
the one hand, predominated during long periods of human
development, and on the other hand, is in all periods ready to
come to the front again in isolated manifestations. This mood
has influenced life in different ways, according as the temperament and disposition of nations and their greater or less
appreciation of the clear factual relations of experience and
the primary moral demands of the soul have disposed the
imagination either to a calm receptive temper, or to a gloomy
or immoderate enthusiasm. Oriental extravagance endowed
its picture of the world with a wide background and luxuriant
wealth of colouring; it introduced notions of the beginning of
the world, of the good and evil principles of all things, of the
fall of man from his first estate through Satan, of a history of
the world, in the sense of a coherent development of all visible
and invisible reality; for it all these supreme thoughts which

the human mind elsewhere only approaches with timidity, appeared above the mental horizon of everyday life, wearing the familiar aspect of well-known stories; they were retained there by innumerable ceremonies—sometimes by monstrous expiations, by which men imagined that they won back sanctification and a power over Nature (of which recovery, however, unprejudiced observation would not have been able to point out the slightest trace)—sometimes by detailed precepts which, petty, vexatious, and useless as they were, hampered the most spontaneous movements of common life by reminders of their pretended dependence on mysterious bonds of the great universe itself. Grecian mythology took a different course; not without loss of instructive content, but with an increase of gracious and artistic development, it restored to freedom the greater part of human life, delivering it from the rank oppressive growth of a mysticism which darkened the world from pole to pole. Different times and different modes of life have favoured different developments of this temper of mind; but wherever our earthly existence has been penetrated by the conviction of a close and thoroughgoing connection between this existence itself and an universal cosmic life, and the conviction has been systematized by attempts to establish a mystic and theocratic regulation of common social relations, the natural course of development has been hindered by the imposition of artificial and to some extent unintelligible tasks, which have thrown into the shade the true physical and moral interests of unperverted human nature.

There arose from this source not only distorted theories, which unconcernedly contradicted the most ordinary experience, but also a series of gloomy ascetic struggles, which are among the most noteworthy phænomena in the world's history, and which in the interests of an ideal end inaugurated an express combat against just those natural foundations upon which the existence of the combatant depends. But on the other hand, where a more propitious course of events has given greater development to men's taste for daily labour and for the pursuit of commerce and manufacture, interest in the system by which a clear division of daily labour is marked out for different individuals throws into the shade anxiety concerning the connection of our life with an invisible and mysterious order of things; and this anxiety only reappears in isolated manifestations of superstition, which persistently contradict experience, without, however, producing much effect on the whole. In this way of looking at life there is a general preponderance of melancholy; and superstition, believing itself to be everywhere encompassed by the immediate presence of the most profound cosmic relations, feels this encompassing to be for the most part as a continual suspicion, temptation, and menace, with which men are hemmed in by some dark and destiny-laden power. But there is bound up with this gloomy view a higher estimate, unconscious and involuntary, of finite personality. The mysterious connection of things seems to be everywhere concerned with this personality, and to hold it fast; and for that very reason it seems that this cannot be a commonplace, transitory, and insignificant element which the course of Nature makes and then again unmakes, but must be an indestructible and real being that of its own choice and free will ponders the perplexing questions of the universe, and is in a position to incur ineffaceable guilt by its own election. Thus superstition is full of the idea of responsibility, an idea which cannot be recognised by the view which regards every finite being as a mere insignificant production of the Universal Substance.

§ 4. I now hasten briefly to a conclusion which is only intended to form the starting-point of our final considerations. From fluctuation between the two views of life which I have been describing, there arises a state of equilibrium which, though not unattainable for man, is perhaps only fully reached in rare and favoured moments. We would distinguish this third mood of thought as *Religiousness*. In this stage consciousness of our own weakness is bound up with the belief that we are called nevertheless to an imperishable work in the world; and the conviction of an intimate connection between our earthly life and the mysterious whole of this universal frame no longer interferes with our care for the small tasks of daily life. It is not the power of larger knowledge which accomplishes the union of these conflicting thoughts, but the power of a larger and more living faith, which attributes to the voice of spiritual experience and of conscience as great importance as to the testimony of the senses, and at the same time does not twist this testimony in order to make it accord with a pretended higher knowledge, being content to believe that God has reserved to Himself alone cognisance of the day and hour in which all our longings and presentiments are to be fulfilled. The function of earthly life in the coherent infinity of existence seems to be of the nature of a preparation, of an educative probation, not aimless and empty of significance as a vanishing present unconnected with any future, but on the other hand, not to be an end in itself, or of such binding force, that every error of the school-life must have the influence of an irrevocable fate. From this mode of thought arise the conscientiousness, the earnest endeavour, and the patient love which the mind ought to bring to bear upon the tasks of earthly life, together with that still greater earnestness of mood and calm peace which come to us from feeling that the imperfection of earthly effort has the sting taken out of it; for it is not the outward result achieved (which may be insignificant), but loyal honest labour, which is both the end of such effort and the vocation to which we are called.

But it is after all only for brief moments that we really feel this sense of peace. I am not here referring to the conflicts and disturbances, and the ever-recurring unrest which arise from the differences, smoothed over, but not reconciled, between the conclusions of faith and the importunate objections of science ; for it is a keen sense of these differences that is at the foundation of our attempt to get a clear idea of the position and destiny of man. The less, therefore, do we need to point out again in this place what violent disturbances our peace of mind is subject to from this quarter. But there is another human imperfection which we have

often referred to, and must refer to once again at the end
of this survey of the moods which characterize our inner life;
I mean that unsteadiness of our thoughts and feelings which
so seldom allows us to hold fast that which belongs to
our peace, and to make it sound on in deep unbroken harmony.   Sometimes we think of the ends alone and forget the
means, sometimes we are absorbed in the treatment of the
means themselves, and lose all remembrance of the end;
what is exalted dazzles us, and makes us lose sight of small
duties, and no less does the consideration of small things
blind us to that which is great; tension and relaxation alternate here as in bodily conditions, and our thoughts are not
the same on Sundays and on week-days.   How much of that
which in hours of thought we acknowledge as our earnest
conviction seems for long periods together to slip out of our
recollection, being like a hoarded treasure which it is
enough merely to possess—and how rare are the moments in
which that supersensuous world in which we believe is
present to our consciousness as a living truth that really
touches our life itself!   What we so often see in great
matters, delay in carrying out good resolutions, is of almost
universal occurrence in small matters; with an honest belief
in the unity of our work, and of the connection there is
between all human efforts for the fulfilment of one and the
same destiny, we yet put off the consideration of many
questions, and our activities seem to work independently
and in isolation in the most various directions.   Thus the
whole circle of the sciences, and each science in particular, lose
all conscious reference to their common centre, as though each
constituted an independent and self-sufficing sphere of
interests, and it is the same with art and the industries
which minister to the wants of external life; so that while
on high days and holidays we recognise the supreme and
absolute end, we work on week by week for mediate ends,
separated by several removes from the final end.   In saying
this we wish not so much to express a serious reproach, as
to indicate an imperfection from which human nature cannot

quite free itself by the mere force of good intentions.   And
the confession of this very imperfection is just the concluding
duty of this sketch, the business of which has been, not to
describe ideals which we have to pursue, but to set forth the
opinions which as a matter of fact mankind are accustomed
to entertain regarding their ideals, and the efforts which they
actually make to approximate to these ideals.

CONCLUSION.

THE point at which we have now arrived is not a final resting-place, but an inclined plane, along which we have to proceed further, and from which we now make a hasty survey of the whence and whither of the path we have been travelling.

The first important section of our considerations only brought us to an unsatisfying conclusion; it seemed that Man was merely one among countless examples of what can be accomplished by the universal order of Nature's mechanism. We saw, indeed, that laws alone never in any case produce a real being; they produce such only by means of a pre-existent Real, actual, manifold, and primary, which subordinates itself and its working to these laws, its capacity of action being merely directed and regulated by them. But the whole wealth of reality which we have thus to presuppose seems at first to be a mere scattered manifold of fortuitous facts, not joined by any bond of living unity so as to form a second great department of the universe, in the same way as the individual laws of the mechanical order of Nature harmonize together so as to make a first fundamental department. Since experience shows traces not only of a sub-ordination of all individual elements under similar universal laws, but also of their co-ordination into a systematic whole, the parts of which are complementary to one another, this harmony came to be perversely regarded as a blind outcome of the original nature and collocation of cosmic elements; these, it was held, must have a nature and position of some kind, and having just that which they have and no other, must necessarily result in this order, and not in permanent chaos. The pertinacity of this unsatisfactory view was overcome at last; and it was obliged to confess itself as being in fact only the disguised and unwilling expression of the acknowledgment that the final, the most comprehensive and the fundamental fact of reality, is the unity and inner coherence of creative Nature, which did not throw into that realm of necessary laws an unconnected multitude of examples to be experimented on, but set before them the hidden germ of an ordered world, that they might develop it. And if reflection thinks beforehand of the subsequent combination of its individual conclusions, it will add the thought that, speaking generally, this system of law—to which reality seems to submit—is not in truth a pre-existing necessity to which reality, being of later birth, thus accommodates itself; that on the contrary the creative Nature which seems to adapt itself to mechanical requirements, is the first and only Real, this mechanism being merely the form in which its activity flows forth; and in consequence of the thoroughgoing unity and consistency of this activity, the *form* of it can be abstracted from particular examples, can be isolated as though it were a universal necessity, everywhere the same, and finally can be conceived as a foreign and independent limit of that of which it is the very nature.

It is this living reality that has been the subject of our consideration; we have sought to find in it *Man*, and the position occupied by his special nature as contrasted with the equally special natures of other beings. The result, however, which we have arrived at as the conclusion of our considerations is almost wholly negative. Extensive as we found the influence of universal and uniformly acting conditions upon the development of human existence to be, we found also that it never suffices to explain this development without predispositions to civilisation of the most special kind which it encounters in the human creature, but does not first produce in him. But when, on the other hand, we attempted to determine positively the connection of this human nature with the whole of reality and its significant position in that whole, our reflections resulted in doubts and obscurity. We know not what there is hidden from us in the countless stars which touch our lives only when a ray from them reaches our eyes by night; how then should we know our place in the whole great universe, with only a small fraction of which we are acquainted? We, living on the surface of this planet, find ourselves at the head of an animal series the perfected type of which is reached in our organization, but of what import is this dignity in the animal kingdom, a matter of which we hardly ever think during life, and which is of no advantage to the progress of our development? Finally, we feel ourselves divided mentally from this animal world by a great chasm; but pursuing ideals which concern us alone, on the one hand we find that we almost everywhere fall short of that in which alone we believe that there is worth, and on the other hand we remark how there vegetates around us simultaneously that other kind of animate life which knows not these ideals. Our own ends are not clear to us; innumerable things exist outside of us, the meaning and destiny of which we know still less; he who would know himself must divine the plan of the whole great cosmic frame which includes such various constituents.

We shall attempt in the last part of these considerations to develop as much of this plan as has been made plain to us by our survey of history, and by the connection of Ideas which the intellectual labour of the human race has gradually attained—thus uniting scattered threads of reflection, and reconciling many an apparent contradiction.

BOOK VII.

HISTORY.

CHAPTER I.

THE CREATION OF MAN.

Obscurity of the Beginnings and of the Future of Man's Life—Nature and Creation—Steadiness of Development in Nature, and Arbitrary Divine Interference—The Sphere of Nature and the Sphere of History—The Genesis in Nature of Living Beings and of Man—Impossibility of setting this out in Detail.

§ 1. FROM all of us the beginnings of our life are hidden, and beyond the few recollections of early childhood which we venture to trust, there settles down a wide and unknown background of profound obscurity. Yet an eye which could penetrate the gloom would certainly not find it empty; the most plastic period of our life has doubtless been influenced by innumerable conditions which have left behind them results that still continue to operate in us. It may be that these blind and involuntary beginnings of development become comparatively unimportant beside the deliberate self-education of later life; but, for good and for ill, we owe to the impressions stored up in this prehistoric period many a vague propensity of which we are conscious, and which we reluctantly acknowledge, and many a lofty aspiration which we obey as the voice of something higher than we ourselves. And the future as well as the past is hidden from us; we know not whither our course will impel us. A glance at the proximate objects which we have set before ourselves, marks out some part of the path which stretches into our future, but as we travel further along it innumerable unexpected impressions throng upon us, distracting, enticing, suggesting new aims, awaking fresh endeavours, and at the end of our way we find ourselves at a spot quite other than that to which our earliest desires pointed, and unable even to understand much of that which once filled and stirred our whole soul. So

strange is the constitution of that *Ego* which the finite spirit
accosts as its Self, and speaks of as its Self.   In the full consciousness of inalienable self-identity such a spirit believes
that it moulds itself and its nature from the very foundation
by its own activity, and does not see that even at the times
when it is most conscious of development it does little more
than labour at modifying the surface of a germ which, unwit-
ting both of its origin and of its future, it finds implanted
within itself.

The same spectacle is presented on a larger scale by the
history of mankind.   Neither the progress of exact science
nor the wider view afforded to human reflection by the ever
higher standpoints to which it gradually attains, lightens the
obscurity that shrouds both the origin of our race and the
final outcome of its development.   We have only learnt that
there has taken place an inevitable and irreparable dislocation
of those graphic representations of the beginning and end of
all things between which, as between two fixed limits, the
boding imagination of men was wont to believe that the
swelling tide of human destiny could be hemmed in.   And
perhaps the failure to hem it in thus is due to a feeling
which is the heritage of humanity and which humanity itself
secretly wishes to retain—the feeling that there are in the
world immeasurable regions which are veiled in twilight,
and a sense (felt by men who are midway between the two
profound abysses—safe because hidden—of past and future) of
rejoicing in the limited illumination which opens up, over
some few centuries of human existence, an outlook that is
much interrupted and fills men with forebodings.

To us at least it almost seems as though men's imagination
delighted to dwell on the great enigma of our origin and
destiny only because it is assured beforehand of failure, and it
would perhaps recoil with dread if a bold leap were really to
lead to a solution of the questions with which it timorously
and yet rashly meddles.   As long as these outermost regions
are wrapped in total darkness we may interpret the outlines
of that which is hidden, in accordance with the longings of our

own hearts; if light were to break in and convince us that it is not as we had thought, it might easily be that the prospect thus opened before us would seem too boundless, the distances too immeasurable to afford us any longer the unreflecting security which had previously made us feel quite at home in the great universal frame.

But we need not speak of this as of something that might happen in case very unlikely conditions were to be fulfilled; the fact rather is that the discord to which we refer has actually been produced by the initial steps ventured by science in the endeavour to throw light upon the origin of mankind. Therefore we must so far yield to the longing which continually draws us to these mysteries as to try and separate between the possible answer to a general question and the impossible satisfaction of a curiosity that extends to details.

§ 2. It was at any rate only among the most unintellectual nations that opinions concerning the origin of the world were due merely to the unrest of ordinary curiosity which (without any sense of the different degrees of importance attaching to different questions) seeks to satisfy itself about all objects of experience, small or great, by a circumstantial account of their origin. In all cases where cultured intelligence has set forth in poetic legends the beginning and end of things, it has been moved by the deeper longing to show that the enigmatical fraction of cosmic order which constitutes earthly history comes forth directly from a higher world, and that after fulfilling its appointed tasks, it will return again whence it came. We have been brought up to believe the most exalted of all these accounts. According to our faith the earth and its denizens were the direct creation of the divine hand, the earth being the only abode of life in the immeasurable extents of space; and the last day will give back into God's own hand the results of earthly history, which is itself the sum of all history, and which has at no moment of its course escaped the vigilant eye of Providence. Creation and judgment bound the changing panorama of history and satisfy our hearts with a sense of the unity of that unchanging Being in whom are comprehended all the mutations of circumstances.

Is it true that this wide scope of thought has become impossible for the spirit of modern science ?   Or has it (as often happens with great thoughts) only taken on an unaccustomed form of expression, under which guise it continues to exist in its integrity ?   Modern science starts no longer from the " without form and void " over which the Spirit of God broods, but perchance, from a sphere of heated vapour which with countless others is whirling round in space; it no longer marks off periods of the world's formation as the work of different days of the divine creation, but measures them according to the decrease of radiated heat, the formation of liquids, the solidification of the earth's surface and its mani- fold fissures ; it no longer deduces the origin of living creatures from an immediate interposition of God, but ascribes them to the gradual evolution of those productions which were brought forth by the inherent powers of primitive matter, being at first simple and becoming increasingly complex.   Does all this really decide the great question, Do we owe our existence to Nature or to Creation ? and does it decide it in a way unfavourable to the aspirations of faith ?

I think not; on the contrary, the longing to emphasize ever more and more the unmediated creative activity of God, to the exclusion of all natural means, must admit that it does itself only bind this activity the closer to limiting condi- tions, after the inappropriate pattern of human action.   It is not enough that the evolution of Nature takes place according to the will of God ; governed by a secret conviction that there may be something which resists this will, if only through inertia, this temper of mind desires to see the very application of God's hand by which He either makes nothing into some- thing, or introduces order among the formless elements of things.   But such actual application is necessary only for feeble creatures whose will can of itself move nothing, and who must therefore endeavour to accomplish a mediated result by setting in action limbs of a body with which they did not endow themselves according to laws which they did not set up. Such extremely undisguised anthropomorphism, and limitation of divine action, will indeed, no doubt, be readily given up, or even eagerly rejected; but the more refined representations which take its place are still influenced by the working of the same mistaken idea. If God did not form the world by the might of His hand, must He not at least have breathed into it the breath of life—must He not have spoken some *Let there be*—must He not have given an external impetus of some kind, without which His will could not have been communicated to things ? How obstinately does our imagination cling to such requirements ! And yet all the time we are perfectly conscious that it is not in the momentum of His breath, not in the commotion produced in the world by the sound-waves of His voice, that creative efficacy is to be sought; this efficacy resides only in the will of God itself, and things do not need to be made aware of this, as of something external to them, by physical hearing and feeling, in order to obey Him who fills their being.

Now, if that which formed the world were neither the visible hand of God, nor the breath of His mouth that might be felt, nor His word that might be heard, but only His will, silent and invisible, what kind of spectacle would have been presented to a mind that had been so fortunate as to witness the process of creation ? Nothing but the spectacle of things that seemed to arise spontaneously from nothing, or that spontaneously condensed out of invisible diffusion into visible form, since no audible command called them forth from a pre-existing storehouse—nothing but the spectacle of movements which seemed to spring spontaneously from the elements themselves and their invisible action and reaction, since they were not communicated by any perceptible breath from God's mouth—nothing, finally, but the spectacle of bodies which, as no visible hand put together their constituent parts, would seem to be produced by the reciprocal attraction of the elements. Therefore the process of the formation of the world would appear in no way different to him who conceived

of it as pervaded by the creative activity of God, and to him
who could see in it nothing but a successive evolution according
to natural law.   If, therefore, we, setting out from experience,
feel ourselves compelled by scientific consistency to trace back
the chain of such developments to the very beginning of the
world, we need not fear that we shall on this account be
necessarily driven to adopt a conception which excludes the
dependence of the world upon God.   On the contrary, we
arrive in the end at just the same conception that should be
presented to us from the beginning by faith in a divine
creation, if such faith understand its own aim.   For the purer
and grander our conception of this creative activity is, the less
shall we expect at any moment a special manifestation of the
finger of God in the phænomenal world; but we shall, on the
contrary, believe that His almighty power is present in the
constancy of Nature's regular working, invisible, but not therefore less efficient.

§ 3. But—it will be objected—does this set our doubts
at rest?   Is the bitter thought taken away, that what is
great and what is small, what is exalted and what is mean,
all proceeds indifferently from the inherent powers of the
material elements?   Was there no more express divine
volition exercised in the production of living creatures which
are destined to the passionate struggles of an historical
development than in the formation of the inanimate surface
of the earth upon which their life is to be lived?   Did no
specially solemn circumstances distinguish the beginning of our
own existence, did no interposition of powers superior to the
uniform course of Nature mark a division at the point at
which creatures endowed with mental life appear upon the
destined theatre of their activity?

In mentioning this last requirement I am not jesting;
we are all subject to fancy that great events are not quite
complete unless their entrance upon the stage of life is
glorified by a striking transformation both of the stage itself
and of the actors; and even in the present case we are
subject to this fancy, although we must admit that here the

splendour of the new scene would be wasted, no one being
in existence upon whom it could make an impression. This
being so, we can with all the more force meet the first
objection to which we referred, by asking what is meant by
that inherent power of the elements to which men are so
reluctant to attribute the origin of the animate world ? The
fact is, that those who with pitying consideration would
convince us how utterly impossible it is that the beauty and
significance of living creatures could have arisen from the
mere action and reaction of the elements, combat us from
positions which we believe that we ourselves hold more
strongly even than they. For it is they whose view betrays
the erroneous presupposition that there could be action and
reaction of elements, whilst these elements are regarded as
isolated, and not comprehended in the One, and that such
action and reaction might lead to definite results. And
having inconsiderately abandoned to this mode of being and
of action (which they regard as possible) the one part of
Nature, they seek, arbitrarily and too late, to withdraw from
the same influence the other part of Nature, being alarmed
by an exaggerated estimate of difficulties which it seems to
them that nothing but a direct interposition of divine power
can remove. Too late ; for if elements, through their own
nature and without any concourse of God, are capable of
exercising certain activities, how are they to be subsequently
made dependent on divine government ? If the divine will
makes any call upon them for action which does not follow
from their very nature, will they not oppose to such calls,
not only mere passive inertia, but also all the resistance of
which an independent and active being is capable ? And
how could this resistance be overcome unless both God and
Nature were embraced by a higher law valid for both, which
should guarantee to the divine will a definite measure of
obedience on the part of Nature ? If one seeks to heighten
the idea of divine governance by representing it as acting
from without upon a spontaneously active world which is
opposed to it, and by ascribing to it forms of activity other

than those according to which this world itself acts, one is inevitably led to the conception of divine action above indicated, a conception applicable not to the infinite God, but to a restricted and finite being.

But it would be wrong to regard the mode of thought discussed above as the only one which is opposed to our own view.   On the contrary, those who agree with us in recognising God's working under the forms of Nature's activity, may yet doubt whether this working restricts itself to such forms and spends itself in them.   The rejection of the figurative representation of the application of God's hand will not be considered a sufficient refutation.   For your imaginary observer—it might be said—there may indeed have been no divine hand specially visible among the phænomena of the genesis and formation of the world, but all may have seemed to him to result from invisible powers of spontaneous growth.   This, however, would by no means prove that every single moment of such development contained within itself all the necessary conditions for the production of that which should follow, and that there was no need of divine aid in order to complete the conditions necessary to a result apparently, but only apparently, caused by the complement of phænomena.   We should be making an arbitrary assumption if we supposed that after the creation of things and the regulation of their evolutionary relations, God would withdraw Himself for ever from the world; but, on the other hand, it would be possible and probable that at every subsequent moment He should require from things actions which were not contained as self-evident consequences in their previous performances; and, finally, we could not doubt that these commands of God would be unhesitatingly obeyed, just because the nature of things and their capacity of action are a nonentity without Him.

But, we would reply, that completion by divine aid must either be something which is according to rule, and the addition of which at a definite point in the order of the world had been determined by God from the beginning, in accordance with the eternal consistency of His being; or it must be something which is not according to rule, something which He adds without finding, in Himself or in the phænomena to which He supplies it, a reason for choosing this particular kind of completion and no other. In the first case this divine help is included from our point of view in the enlarged idea of natural order, since we hold that Nature never works without the concourse of God; in the second case (which, indeed, is that which common opinion prefers), we have to ask, What is the worth of the advantage which is to be secured by such a view, and which is advocated with jealous preference? Shall we regard God as greater, if we believe that He governs the world by a series of disconnected commands? or Nature as more exalted, if we believe that, as a whole, it is at all times—or even only occasionally—inadequate to produce the phænomena of the next moment? Whence comes it that the other form of divine activity (that of the steady development from within of a pre-existing germ) always has to fight for acceptance in our minds with a preference for uncertain repeated interpositions of divine activity coming from without?

As a matter of fact, it is the ascription of this very consistency to the divine activity which is repugnant to a secret craving of our souls. To make all subsequent resolves only the necessary results of one primal resolve, and all subsequent activity only the inevitable result of an original creative volition, involves a denial of freedom of action which seems to us incompatible with the idea of a living personal God. Our view threatens irresistibly to issue in a superstition which regards the world as being merely the unintentional necessary development of a spontaneously expanding primal being, to which, at the same time, all history seems meaningless, since that which had once been included in this being at the beginning, as something which must necessarily follow, could have nothing essential to gain in the course of events in which it should undergo a special process of production. The capacity of doing what without

such doing would never have happened, of preventing what
without such prevention would inevitably have occurred, the
possibility of gaining in insight and in range of will, and
of ceasing to desire that which had previously been desired,
and finally the consciousness of a capacity of independent
determination, not only as regards the future form of the
external world with which our action is concerned, but even
as regards the consistency of our own nature—all this it is
that we seek in a living personality, that we think we find
in ourselves, and that we miss in a representation of divine
action which exhibits it as always bound by its own special
law.   To secure these treasures of freedom and vital action
for God as well as for ourselves, we have recourse to modes
of representation that labour under obscurities and contradictions of which we are not ignorant.   This is why we
prefer the thought of an uncertain and disconnected divine
activity; for truly to us finite beings it seems as though
our freedom were most clearly certified by the inconsequence
with which we can alter and break off the course of our
development.   This is why we do not even shun the danger
of degrading divine activity to the external elaboration of
a material world existing from eternity; for we even fancy
that we have a fresh proof of our freedom and capacity of
arbitrary choice in the opposition which the inherent activities
of the external world offer to our exertions.   This is why
we so often renew the attempt to reduce as far as possible
(since we cannot altogether deny) the sphere of development
according to natural laws, and to draw a sharp boundary
line between *Nature* as the *realm of necessity*, and *History* as
the *realm of freedom*.

In both there lies before us a succession of changing
events.   But as far as Nature is concerned we should be
quite satisfied if it were only a collection of occurrences
which without being connected in systematic and progressive
development were merely confirmatory and concrete examples of the steady validity of certain universal laws.   It
is only in the mental development of the human race that

we feel a primary need of comprehending the series of events as a history of which the end is more worthy than the beginning, and the whole of which would be worthless if it were merely a repetition, in time and destitute of freedom, of that which already existed—not subject to temporal limitations and prefigured in full completeness—in its causes. All the lavish passion of longing and remorse, love and hatred, with which history is filled, we are unwilling to regard as wasted; and it would be wasted—yes, and the very existence of mental life would seem to us an incomprehensible anomaly— in a cosmos in which there was nothing to change, and which, undisturbed by all this struggle of souls, was entirely taken up by the leisurely development of already existing conditions. And now having reserved for the history of this spiritual life that freedom which it seems to need, we once more extend our demands beyond our requirements; we will not cede to the sway of that detested natural necessity even our physical existence or our origin. We would much rather owe them to the fiat, *Let us make man in our image.* Even in such a representation the creative activity of God seems to us more near and intelligible, more full of life and warmth, and our own existence seems to have a nobler and happier origin than if we believe that we, like the rest of Nature, have been produced by an unresting coherent development.

Now this distinction between Nature and History certainly points to real mental needs, the satisfaction of which we shall consider later. But we can agree to the separation of these two departments without acknowledging the false boundary line, which, needlessly and contrary to experience, marks off the origin of mankind as not belonging to the sphere of natural development. According to the present course of man's life, experience shows us that wherever it is connected with the external order of Nature, it is wholly subordinated to the rules of this order. Races of men arise and pass away according to the same laws, and after the same fashion, as races of animals; the external powers of Nature are not more forbearing towards the pre-eminent creature endowed with a rational mind, than they are towards the irrational animal; their destructive influences affect the life that is historically significant with the same impartial indifference with which they dissolve combinations of lifeless matter; finally, nowhere does Nature quit, for the gratification of rational minds, the paths of her accustomed activity, rejoicing our hearts with the wonders of a Golden Age in which everything happens for our satisfaction, instead of merely that event happening which is the inevitable result of previous causes; there is no way of bringing about transformations of the external world corresponding to our inner life, except by our activity availing itself of natural means in obedience to the laws of Nature. Thus we, being in our life, our sufferings, our achievements, altogether holden by the power of natural necessity, should gain but little by rescuing the origin of our species from the grasp of this necessity. The freedom of such a distant past could be no compensation for present constraint.

And just as little do we feel that our claims to freedom are necessarily demolished if we give up this attempt. For we originally desired this freedom only for our inner life, and indeed only for a small part of that. This spiritual life, receiving stimulation from Nature, and limited in its reaction to natural means, is not itself directly included in the order of Nature. Between this stimulation and these reactions is interposed, as a department *sui generis*, the internal elaboration of the received impressions. There may take place here innumerable occurrences which are more than the steady continuation of effects initiated in us by the external world; there may take place innumerable connections of received stimulations, in accordance with points of view which altogether transcend Nature, resulting in the production of impulses to reaction to which mere natural order would never have led without this complementary interposition of mental life. However highly one may rate this free action of mental power in human nature, it will always receive due estimation as long as it is limited to the world of thoughts; but only in subordination to certain laws will the cosmic order admit of its efficient access to external Nature. And however specially we may imagine the history of mankind to be guided from the loftier standpoint of divine wisdom, from a higher plane than natural evolution, we may be quite satisfied if this guidance takes place through action and reaction between God and the spiritual nature of man, in such a way that the thoughts, feelings, and efforts thus aroused and developed, also alter the external position of mankind, to the same limited extent to which our action is able to change the physical conditions of our existence. Thus within the realm of Nature with its uninterrupted coherence, there is certainly a possibility of history, and we are neither justified in maintaining nor bound to deny, without proof, that to this history freedom appertains; but the external destinies of our race only belong to history in as far as they depend upon our own actions.

§ 4. After these remarks we may return to the two questions which we mentioned above. We can now answer the question which refers to the general process to which we trace back the origin of living creatures in general, including the human race. This occurrence also we unhesitatingly conceive as a necessary result, which at a definite period of the earth's formation arose from the then existing collocation and reciprocal action of matter, with the same inherent necessity which now connects the continued existence and the reproduction of living creatures with the present distribution of material masses and their relations to one another. The course of Nature, indeed, from which we believe that living creatures have sprung, is in our view something richer and fuller than that small fraction of it which is known to science ; so far, such a course of Nature is not confined to working upon lifeless matter, but presupposes inherent activity in its elements, and it will perhaps be the glory of the future to define the special characteristics of this activity, and to determine the laws of its influence upon the external operations of things. Moreover, we do not maintain that all which the elements can accomplish is to be measured by the narrow possibilities still left open by the rigidity which the most essential natural relations have now attained. In earlier stages of cosmic development, when (everything being yet in process of formation) there was both greater celerity of change and also a prevalence of modes of connection which did not afterwards recur, it may perhaps have been the case that the elements produced effects different in nature and magnitude from those to which the present course of Nature gives rise, limited as this is to the maintenance of uniform conditions. However, we do not by any means mention these fluctuating and never definitely circumscribed representations in order to embellish our own view in the eyes of our opponents, but rather for the sake of pointing out that none of them can mitigate the rigour which causes so much alarm. For if there is one thing that we shall always hold fast by, it is that even these creative habits of the primal course of Nature were events governed by law, and proceeded from an activity that in its own course laid fresh foundations, by means of the productions of its early periods, for the more intense and complex activity of later periods. Nature works from the beginning according to laws which either (1) are unalterable, or (2) themselves alter regularly, as the conditions alter which have arisen under their sway, and are therefore to be regarded as regular and ordered functions of their own results.

On the other hand, it is altogether impossible to answer the particular questions prompted by curiosity concerning the circumstantial course of events from which there gradually arose the structure of organic beings and of man himself. A view which does not attribute this occurrence to supernatural and therefore in itself indescribable influence, but makes it dependent on the concatenation of innumerable details, will inevitably lay itself open to the reproach of rash and arbitrary invention if it attempt to enumerate all these details, for the real determination of which our own range of experience is very far from furnishing adequate analogies. This fate has overtaken all attempts to exhibit the gradual evolution of the higher forms of living creatures from the lower, and the origin of these from the immediate action and reaction of the elements. But there are two considerations which we desire not to withhold from the notice of those who would found an objection against the general conclusions of natural science upon its incapacity to exhibit the details of these conclusions.

In the first place we may, without much difficulty, convince ourselves that this difficulty in describing first beginnings is a misfortune by no means peculiar to our theory, but is one which it has in common with all others. It certainly sounds passing strange when a daring investigator of Nature describes the protoplasmic cell, which, having been formed in the ocean and slowly borne to land, is there developed into a quadruped or a man; but the poverty of this attempt lies rather in the total ineffectiveness with which it addresses itself to the insoluble problem, than in the fact that different assumptions might lead to a better conclusion. Hence it seems a matter of indifference whether we attribute the origin of animate life to the natural action and reaction of the elements or to a peculiar vital force; any representations which we can frame of the gradual concrete progress of its formation will be just as strange and untrustworthy in the one case as in the other. If, according to the first view, the elements combine spontaneously to form a protoplasmic cell, or a germ, which then goes on to further stages of development, according to the second view the vital force is just as shy of revealing its mode of operation. For naturally we shall not believe that this vital force forms the finished creature with all its parts in an instant from the elements, and if we seek to show how it works by a progression from the simpler to the more complex, the cell or the germ (from which in this case too we have to set out) seems no better endowed and no more probable than the cell and germ which in the previous case we derided. The Mosaic account of the creation employs two different representations of the way in which things arose. First God says, *Let the earth bring forth all manner of herbs.* Would the results of the command to produce plants, thus communicated to the forces of the soil, have differed in appearance from the conception of natural science, according to which the separate elements of the soil first developed into germs, and these again into plants ? The attempt to work out this idea in detail is as hopeless as all others of a similar kind.   Man, on the contrary, is formed by God's own hand ; but we do not need to repeat how unsatisfactory is a comparison taken thus directly from labour of the most ordinary kind.   It therefore appears that all these modes of thought are involved in equal difficulties when they attempt to give sensible representations, that shall be credible and probable, of processes which are separated by a gaping chasm from the sphere of our own experience.

The other point that I wished to notice is, that we are accustomed to estimate one and the same idea very differently when it comes before us as a conjecture, and when it is offered as the expression of a fact.   What a succession of minute and interdependent events is presented by the intricate processes of formation, fructification, and development in the seed of a plant !   How complex, and in many of its features unintelligible to us, is the development of animals by division and coalescence, segregation, and aggregation, and various changes of an apparently supplementary character in the relative position of parts—some of which seem to waste away after having rendered their mysterious service during a definite period of development !   Now if any one, unsupported by the testimony of the microscope, should have conjecturally described the multiplicity of arrangements which that instrument actually reveals, how those who consider animate life to be only comprehensible as resulting from the misty and magic sway of a single impulse, would have found fault with him for advocating a mode of thought at once rash, tedious, and intellectually poverty-stricken !   The fact of alternate generation among the lower animals having been established by observation, scientific speculation finds it by no means difficult to discover retrospectively ingenious theoretic grounds of interpretation, whereas beforehand any conjecture that such variation might occur, would have been rejected as an impossibility, contradictory of the idea of sex, and of the whole economy of natural history. Whether the original production of animals and plants by the conjunction of inorganic elements will ever be proved as a fact which still takes place, we do not know; but if a day should ever come when it is proved, then people will suddenly remember that it was a thing always possible in the very nature of it, and that it never involved the absurdity that people see in it as long as it is only a scientific conjecture that is inconvenient to various prejudices. Let us therefore trust our question to the future; let us leave science to make further investigations; if it should ever succeed in drawing a more definite picture of the origin of animate life, people will accept with equanimity realities coinciding wholly in essentials with processes which, now that they can only present themselves as possibilities, are peevishly rejected as wretched inventions of a low and unworthy mode of thought.

Such being our views, we regard as useless any further lingering in these outer courts of history, in which science can discover merely shadowy outlines and no clearly defined forms. We will not follow the astronomical investigations which seek to discover how the world was formed, and to decide whether the distribution and movements of the heavenly bodies make it probable that there is a common centre of this universal frame, or whether it is more likely that many stellar systems, each independent in itself, circle round a merely ideal centre of gravity by the force of reciprocal attraction. As much as is certain in these considerations only confirms what we knew otherwise, namely, that it is upon a small eccentric spot, lost as it were in the immensity of the whole, that this human life is developed, with all its passion and lofty aims—a brief and serious monition which points out to us an abyss of unknown possibilities, and warns us that we should not take it for granted that earthly history is equivalent to that of the universe.

Neither will we enter into geological investigations, and immerse ourselves in a consideration of the different periods of the earth's formation, and in discussions as to how the gradually altered condition of the atmosphere and of the solid surface of the earth, furnished at different stages the conditions of the production and maintenance of various successive organic creations. The magic spell which descriptions of this vast and obscure past always exercise upon our mind, would give to my colourless picture a charm which I find it hard to renounce. But these investigations proceed upon many uncertain assumptions and are laden with sources of error; and they are therefore specially unsuited for the confirmation of definite results at the present moment, when many noteworthy discoveries have wakened attention without having caused any decided clearing up of difficulties. Yet it seems that man is one of the most modern denizens of the earth; indubitable remains of our species have not been found deeper than the later alluvial strata, which are still being slowly and steadily increased in low-lying levels by progressive deposition of the matter of abraded rocks which is carried down by the current of swift streams. Therefore it seems that man was not produced before a time in which existing climatic distinctions prevailed, and the vegetable and animal kingdoms had developed in all essentials the forms which we now see around us. We must leave it for the future to prove whether this limitation can be removed and a much longer vista be opened before us, in which there may perchance be hidden many beginnings of races of men differing widely from one another. Without at present declaring for this view as the more probable, we may yet feel that we ought to be prepared to accept both it and the altered position which the small section of historical development at present known to us would occupy in such an enlarged life of humanity—a life which to our imagination would be almost boundless.

And, finally, we should not lay too much weight on presentiments as to the future to which we may be tempted by that insight into the connection between the different forces of Nature which has now been attained. Whether reciprocal transformations of energy or a consistent consolidation of all particular results of the course of Nature, will gradually produce a permanent preponderance of such conditions and modes of motion in matter as are incompatible with the continued duration of animate life, or to what other fate this earthly sphere is destined—these are points concerning which we can no more look for certain information than we can regarding the very first beginnings. Let us therefore bid adieu to these insoluble riddles, and turn from the external history of the human race to that inner history of humanity which, with its manifold changes, is included in the slower progress of external Nature.

CHAPTER II.

THE MEANING OF HISTORY.

What is History?—History as the Education of Humanity—History as the
Development of the Idea of Humanity—Conditions necessary to make such
a Development valuable—Concerning Reverence for Forms instead of for
Content—History as a Divine Poem—Denial of any Worth in Historical
Development—Condition of the Unity of Humanity and of the Worth of
its History.

§ 1. NOW what is the significance of this inner mental
history of the human race ? What are the laws
of its course, or the plan which connects into intelligible
unity the varied wealth of its phænomena ? Our age boasts
as its prerogative that it knows an answer to this question;
but however dangerous it may be to rebel against modes of
thought to which vigorous and brilliant intellectual essays have
accustomed us, we must still confess that in regard to history
there is no lack of the most contradictory opinions, each of
which disputes even the elementary assumptions of the others.
I will not linger over the cool assertion that everything has
happened already and that there is nothing new under the
sun ; but remark that in opposition to the willingly accepted
doctrine that the progress of humanity is ever onwards and
upwards, more cautious reflection has been forced to make the
discovery that the course of history is in spirals ; some prefer
to say epicycloids ; in short, there have never been wanting
thoughtful but veiled acknowledgments, that the impression
produced by history on the whole, so far from being one of
unmixed exultation, is preponderantly melancholy. Unprejudiced consideration will always lament and wonder to
see how many advantages of civilisation and special charms
of life are lost, never to reappear in their integrity, when
any form of culture is broken up. Subsequent ages may com-

pensate the loss by other and indeed by higher advantages; but this does not alter the fact that the earlier ones have passed away never to return ; that which past times have toiled for and won can never be inwoven with the work of subsequent ages with the completeness necessary for continuous and steady progress, but nearly everywhere the new life arises out of the ruins of the old at the cost of painful sacrifices. This melancholy impression received from history as a whole is not much mitigated by well-meant reference to the fact that in individual life too the bloom of youth must be sacrificed to the strength of manhood, and this again to the wisdom of old age, and that it is only the most favoured lands that are permitted to see fruit and blossom and bud simultaneously on the same plant. Do not all these comparisons only increase the grounds of our complaint ? If, however, they comfort any one, is not the comfort they bring derived from the thought that human history is itself only a natural process to which we must accommodate ourselves, and about the right and end of which it is of no use to ask ? But for him who clings to the belief in a guidance which is ordering this confusion of human destinies to some higher good—how is *he* to interpret the spectacle which history presents ?

§ 2. That history is the *education of humanity,* is the first phrase with which we provisionally pacify ourselves. And indeed unfathomable designs of educative wisdom must ever be a fruitful source from which to derive all the astonishing turnings and twistings of the course of history. But if we are not wholly satisfied with this general consolation which would allay our doubt with the bare assurance that a solution exists, if we seek to trace at any rate in the great outlines of history that educative plan, how many hindrances do we meet ! We know sometimes what has happened, and can see how it led necessarily to the subsequent condition of things ; we may often be certain of the greater perfection of what is later in time, and even a dull mind may often perceive some arrangement by which the new condition of things will draw advantage from the old ; but who can calculate with certainty what would have happened if particular circumstances had been different, or can say what possible greater good may have been missed by the actual course of events leading to something that was less good ?

I wish, however, to speak, not of the difficulties of carrying out this view fully—such difficulties being in fact very great for every view—but of the doubts which are raised by the application of this idea of education to mankind.   Education is only intelligible to us when a single individual is concerned ; when it is one and the same person who becomes better, who bears the penalty of his mistakes and enjoys the fruit of his repentance ; and who, if in the progress of development he has to sacrifice some good which he possessed, may yet keep the memory of it as something which he has himself enjoyed. It is not so clear how we are to imagine one course of education as applying to successive generations of men, allowing the later of these to partake of the fruits produced by the unrewarded efforts and often by the misery of those who went before.   To hold that the claims of particular times and individual men may be despised and all their misfortunes disregarded if only mankind improve upon the whole, is, though suggested by noble feelings, merely enthusiastic thought-lessness.   The humanity which is capable of progress can never be anything other than the sum of living individual men, and for them nothing is progress which does not mean an increase of happiness and perfection for those very souls which had suffered in a previous imperfect state.   But the humanity which is opposed to individual men is nothing but the general concept of humanity ; this concept, however, which can neither suffer nor experience anything, nor undergo any evolution, is not the subject of history.   Only individual specimens of humanity, humanity of different periods, can, when com-pared together, show a steady progress towards perfection ; but the earlier know nothing of those which succeed them, and the later know little of the earlier.   What then is it that justifies us in regarding these disconnected members as one humanity, and what is the meaning of an education which does not do just that which is the very business of education—which does not attempt to replace what is more imperfect by what is more perfect in the same pupil, but throws aside the half-educated scholar in order to bring forth better results of culture in another ?

And the same difficulty at once recurs if we look not at the succession of ages, but at each particular age itself. There has never been a period of history in which the culture peculiar to it has leavened the whole of humanity, or even the whole of that one nation which was specially distinguished by it. All degrees and shades of moral barbarism, of mental obtuseness, and of physical wretchedness have ever been found in juxtaposition with cultured refinement of life, clear consciousness of the ends of human existence, and free participation in the benefits of civil order. Humanity, at the different moments of its historical progress, is never like a clear and even current, of which all the molecules move with equal swiftness ; it is rather like a mass of which the greater part moving on thick and slow is very soon checked by any little hindrance in its course and settles into inactivity ; there is never more than a slender stream which, glancing in the sunlight, struggles on through the midst of the sluggish mass with unquenchable life and energy. It is true that sometimes this stream widens out, and then occur those favoured periods in which, at least for us who stand afar off, a general enthusiasm of culture seems to seize a whole nation. That it does not indeed really extend to all, even we who live later can see ; that it does not exclude very dark shadows of sluggishness, of debasement, and of misery we should observe more clearly if we stood nearer.

Now nothing is simpler than to give an explanation of this if we regard history as merely a course of events arising from the concurrent action of external circumstances and the laws of mental life. A culture which does not merely mean natural goodness of disposition, but includes also knowledge of things, estimation of the tasks and circumstances of human

life, and consciousness of the connection between the individual
and society and between society and the universe, is not
conceivable apart from the most varied influences of education
and of continued intercourse with one's fellows ; but the
hindrances which have their origin in the external circumstances of existence, and which always stand in the way of a
general prevalence of such favourable conditions, are unfortunately too obvious to need further mention.   Thus the existence
of a vast spiritual proletariat, which there seems no possibility
of removing, is an objection which the idea of history as the
education of mankind must find it hard to overcome.   Human
action must be content to attain its end only in part ; but it
is not enough that the divine guidance of history should
accomplish its aims only on the whole or in the majority of
cases.   Conditions of mankind which, independent of individual freedom, follow with inexorable necessity from external
conditions, should be susceptible of interpretation as instances
not of the failure of this guidance, but of ends intentionally
aimed at by it.   And in fact such an interpretation has not
been wanting.   As different trees, it is said, have different
bark, and each, whatever its rind, grows green and blossoms
in content, so mental endowment and external good fortune,
and with them the degree of culture attainable for men,
are variously distributed ; there is progress enough if, notwithstanding all these irremovable differences, mankind as
a whole wins higher standpoints ; enough even, if while the
mass of mankind remain ever in an uncivilised condition
the civilisation of a small minority is ever struggling upwards
to greater and greater heights.   In answer to such a view
what can we say except that it sets forth a condition of
things which, alas ! we cannot question, but that it neither
offers any explanation which makes this condition more
intelligible or more endurable, nor shows us how, upon such
assumptions, we can be entitled to speak of an education of
mankind.

Let us, however, for the present reckon as among the many
puzzles which we cannot solve, this inequality in the endow-

ment and good fortune of men, and content ourselves with
the progress of the few. But however great this progress
may be, we would, finally, ask of this view which we are
calling in question, why precisely it was necessary that there
should be an education of mankind resulting in progress,
and why an end should have been set before us which could
only be reached along the tedious path of historical development? And it will not satisfy us to point out that the slow
course of gradual improvement was the only possible way left
open by the nature of mankind and the constitution of the
external conditions of life. The divine power, which is
supposed to direct this education, created the world, and man,
and all the conditions of his life; it was open to it to
order them all according as it would. If, then, it chose to
educate mankind by way of history, it did not so choose
because hindered by the disfavour of circumstances from
endowing us with perfection in the beginning, but because it
willed that history should be, and willed to bestow upon us,
in gradual development, a greater good than that would have
been which it withheld.

This inquiry has indeed been so often and so unanimously
answered that we shall perhaps give offence by approaching
with such circumlocution and delay a philosophical question
the reply to which seems thus certain. Man, we are told,
must become in knowledge that which he is in fact; it
is not enough that he should be and remain in unreflecting
simplicity that to which by his mental constitution he is
destined, but he must realize it gradually and consciously as
his own work. The dignity of man lies in this, that he
does not (like the lower animals) with unconscious impulse
work out ends towards which uncomprehended motives
and favouring external circumstances mysteriously concur,
but that doubting, erring, and improving, he learns to know
his destiny, his duties, and his powers.

A survey of our own individual life will certainly easily
convince us that such development from unreflective existence to explicit self-consciousness is a mental gain of a

unique kind; but can we in truth transfer to the whcle of humanity the value which we see that it has for the individual, and is there not in such a transference an inexactness similar to that which made the notion of education inapplicable to a succession of different individuals taken *en masse*? For can this inner work of development (in the comprehensive and self-conscious remembrance of which the moral enjoyment of life consists) be carried out vicariously by one individual for another, or by one generation for another? Or does history perchance exhibit such a steadiness of connection that the minds of later times pass at least in outline through the same evolutional struggles by which their ancestors were stirred?

It seems to us that nothing of all this happens. In the first place, each individual enters into life without any conscious connection with the past, but with those natural capacities, wants, and passions of his species which are little changed in the course of history; and which, in as far as they are changed, are yet for him who is born with them just as much an unmerited and unconsciously received endowment of Nature as the dispositions of our forefathers were for them. Thus furnished each goes through the experience of his life, each passes through his own evolutionary struggles, and all these also are essentially similar. The influence of history first begins when the individual encounters the results of the labours of his immediate predecessors in the conditions into which he finds himself born, to which he has to grow accustomed, and which he has to use and to combat. Without doubt the form of development which the individual passes through is modified in the course of history; but it is not by any means modified in such a way that every one who comes later has a view of the course of human development which is fuller and more conscious in proportion as the time is longer during which past ages have been endeavouring to struggle upwards through individual stages of evolution. For by this spiritual labour, which wins positions from which it can itself make a fresh start, conscious knowledge is propagated either not at all or most imperfectly ; what happens is that its finished results enter as a great aggregate of prepossessions, of which the foundation is forgotten, into the culture of him who comes after. They may in this way often make it possible for him to mount higher than those who preceded him ; but nearly as often they are, as inherited limitations of his intellectual horizon, hindrances in the way of a development which would have been possible for him if this historical dependence had not existed. But in both cases the way in which the culture of past times is for the most part handed down, leads directly back to the very opposite of that at which historical development should aim ; it leads, that is, to the formation of an *instinct* of culture, which continually takes up more and more of the elements of civilisation, thus making them a lifeless possession, and withdrawing them from the sphere of that conscious activity by the efforts of which they were at first obtained. No fortune, it is said, is transmitted undiminished to the third generation ; and this is very natural ; for the first inheritor is born and brought up in the presence of the activity by which the fortune was accumulated, and if the desire to increase it leaves him, the desire to preserve it generally remains ; the second inheritor born in full possession of the wealth knows nothing of the worth of the labour which created it ; thus the third has to begin the same cycle afresh. The same thing happens with the store of culture which history accumulates. It is true indeed that the results of the latter cannot be so easily dissipated, as on the other hand they cannot be so completely transmitted ; but the elevating freshness and joyousness, full of prophetic insight, that distinguish an age of invention and discovery, are not transmitted to the ages which are its heirs. Scientific truths, hardly-won principles of social morality, revelations of religious enthusiasm and artistic intuition, are all subject to this devitalization ; the greater the amount of this wealth which is transmitted to later generations the less is it a living possession, even when outwardly recognised and retained.

which it not always is.   That which once, when it first arose upon the intellectual horizon of the past, was in truth a living enlargement of the soul, and a perception, full of meaning, of some new aspect of human destiny, is, in the hands of later generations, like a worn coin which one takes at its nominal value, but without knowing what are its image and superscription.

In no department is the progress of mankind more unquestionable than in that of science, although even here it has not been continuous, interruptions caused by long periods of barbarism having often made necessary the rediscovery of forgotten truths.   But first of all we may note that this progress has brought about the strange result that the whole field of knowledge has become too vast to be within the grasp even of those who are expressly occupied with its cultivation How odd and yet how accordant with fact it is to speak of "the lofty position of science now-a-days."   What is science ? Not truth itself, for this existed always, and did not need to be produced by human effort.   So that science means simply knowledge of the truth ; but this knowledge has become so vast that it can no longer be comprehended in the knowledge possessed by any individual.   Such is the strange life of science now-a-days ; it exists, but for any individual it means only the possibility of investigating and learning to know each of its parts ; in no mind does it exist in completeness, approximately in but a few, and hardly at all in the mass of mankind.   We see that now, as in all former ages which were in possession of extensive and varied scientific knowledge, individual men take up particular branches, and on those small battlefields fight out the most passionate combats, combats which sometimes seem to jeopardize all that has been gained by human culture.   The progress of science is not therefore, directly, human progress ; it would be this if in proportion to the increase of accumulated truths there were also an increase of men's interest in them, of their knowledge of them, and of the clearness of their insight concerning them.   Without denying that some periods of history have to a certain extent fulfilled these requirements, we can hardly say that, looking at history as a whole, it exhibits a steady improvement in this respect.

But it will be objected that the progress of mankind towards perfection is to be sought not only in the advance of conscious knowledge, but also in the beneficent effects upon men's condition which science leaves behind even when it has itself passed out of consciousness. These effects have been eloquently described, and we willingly admit that even in the more tangible deposit of material improvements which everyday life owes to advancing knowledge, there is, besides the mere convenience and the increase of comfort, also a certain mental gain and a certain civilising power ; the mere presence of refined surroundings may have a modifying and elevating influence upon those vague general moods which make as it were the background of all our endeavours. But while we do not deny the value of this progress, neither would we overestimate it. Custom soon diminishes it. A new discovery excites lively interest for a time, but it soon falls back into the rank of those natural objects and events by which we are always surrounded, the mysteriousness of which no longer has any exciting effect upon us, owing to the lack of novelty. At the most now and then in a moment of passing absorption in a thing, we think, After all how striking this or that discovery is—or, How it has helped on human intelligence. But most commonly it happens that men thoughtlessly enjoy the fruits of inventions with a certain coarse unthankfulness, without a gleam of interest or curiosity with regard to the mental labour which produced them, and as though it were a matter of course that their poor life should be adorned by such uncomprehended blessings. Hence we are justified in affirming in conclusion, that however great human progress may be, yet at all times men are but very imperfectly conscious of this onward movement, of the point in the path of advance at which they may happen to be at any moment, and of the direction whence they came and whither they are going. If it is

their destiny to become conscious of that for which they
are designed, it may indeed be that they attain such a consciousness, but they attain it without themselves noticing or
feeling its gradual awakening; it cannot be said that men
grow to what they are with a consciousness of this growth,
and with an accompanying remembrance of their previous
condition.   Therefore the notion of education, when transferred from the individual (with reference to whom it is intelligible) to mankind as a whole, solves none of those doubts
which the consideration of history awakens in us.

§ 3. Will they be any better solved by another theory, the
favourite of the immediate past, which has long been im-
patiently awaiting our consideration ?   According to this
theory the education of mankind is an antiquated and unsuitable phrase, although what it is intended to express is the
truth.   This phrase gives the idea that God arbitrarily sets
before men ends which He might have refrained from setting
before them, and leads them in paths for which others might
have been substituted.   Hence the education theory involves
us in the misery of attempting to show the significance and
importance of a series of events which yet as products of
arbitrary will must remain inscrutable to reason, which can
comprehend only necessary consequence.   Whereas, in fact,
the history of mankind (like all genuine evolution) is but the
realization of its own concept.   All true existence, it is said,
manifests itself by emerging, as life, from that condition
of natural determination in which it originally is, unfolding
itself in a wealth of change and varied manifestation; and
finally returning as it were to itself deepened and enriched,
and enlightened concerning its own nature by the work of
development which it has passed through, and the fruits of
which it retains.   It is by this law that mankind are stirred
and impelled to historical development.   As the self-development of the human mind, and as the very destiny and inner
necessity thereof, history can neither be a course to which we
are impelled by the arbitrary choice of an overruling purpose,
nor one to which we are impelled by the unintelligent activity

of external facts. But it becomes intelligible by reference to the idea of humanity; not only does this contain the ground of temporal succession in general, but we may deduce from it, for each and all of the stages of historical development, the strict and complete formula which constitutes the explanatory principle of all the peculiar features of these stages; finally, this law teaches us to understand not only that progress which is the rule, but also the strange retrogressions and eddies by which the continuity of this progress seems to be interrupted.

But in our opinion this last-mentioned service is *not* rendered by the view now under discussion; the fact rather is, that the way in which it admits incalculable chance and arbitrary will in history alongside of the strict development of the idea of humanity, is that which first gives us occasion to test the validity of its confident assertions.

With regard to all phænomena we feel that we have a twofold task—we have to explain step by step the possibility and mode of their occurrence, and we have to unravel the rational signification which is the justification of their existence and of all the assumptions which they presuppose. The philosophical view which gives rise to the above-mentioned conception of history, does not conceal its conviction that the Meaning or the Idea, to the realization of which every chain of events and every creature is destined, constitutes its real being, and that to search out this innermost fount of life is the supreme task of all (even of historical) investigation. But it cannot at the same time conceal—however willing it may be to do so—that it lacks a definite notion of the relation of the Idea to the practical means of its own realization. It must allow that all which happens in history is only brought to pass by the thoughts, feelings, passions, and efforts of individuals, and that the ends towards which all these powers with their living activities are striving, do not by any means necessarily coincide with those towards which the development of the universal Idea tends. And the only addition which in the last resort it can make to this confession is that the Idea does yet prevail—nay, does on the whole ex-

clusively prevail — notwithstanding, and in, and with, and
among all these confused, conflicting, and discordant struggles,
whose powerlessness easily leads to contempt for that which
thus cannot be turned to account.   Hence this view has in fact
often enough declared that individual living minds really
count for nothing in history, that they are but as sound and
smoke, that their efforts, in as far as they do not fall in with
the evolution of the Idea, have no worth and significance in
themselves, and that their happiness and peace are not among
the ends of historical development.   The course of history is
as the great and awful and tragic altar on which all individual
life and joy is sacrificed to the development of the universal
Idea of humanity.   And it is just here that we find the expression of the essential difference which distinguishes this view
from the preceding one, with which in other respects it has so
much in common.   He who speaks of education naturally means
the education not of a concept, but of some living thing which
is only marked out and named by the concept, and which alone
could be capable of rejoicing in its own development.   This
interest in an attainable good which history is to realize, and in
a realm of living creatures who can enjoy the happiness of this
realization, we must, if we have not got rid of it already, learn to
sacrifice to our veneration for the Ideal-development theory.

How much we have it at heart to oppose this theory will
be readily understood.   Above all, we must note that only
he who would reverence history as an enigma without seeking for its solution, can be satisfied with the mysterious con-
cord between what is required by the evolution of the
Idea, and the results of individual efforts which are independent of it.   On the other hand, he who looks for a
solution may take either of two courses; whichever of
these he may choose, he is bound to begin by stating
clearly who or what the mind of humanity is of which history
is the development, and where this mind is to be found.

The first course begins with the statement that it exists only
in the countless multiplicity of living men, contemporaneous and
successive, of whose nature it is the common feature, and that

it has no independent existence outside of, among, or beside them. From an analysis of this general character of humanity (for this is what the present view comes to), and at the same time of the external conditions presented by the earth as the stage of life, we should deduce the consequence that the kind and degree of civilisation which would furnish the greatest possible amount of development and satisfaction of all human capacities would not be attainable in the course of a single life, but only in a series of generations of which each would start in its course from the stage of development reached by that which had preceded it. Then we should bethink us that this development would be worthless if it took place with the unfailing regularity of a natural process, and that living minds were not formed to realize a steady progress determined in complete independence of any free choice on the part of the agents, even supposing that such a progress were in itself desirable. We should expressly point out the unconstrained freedom of all the living elements, the action and reaction of which does, notwithstanding, form the foundation of a steady course of history. Now natural science sometimes shows that the irregular minute and conflicting molecular movements of a mass not only do not affect the uniform molar movement of the whole, but are, for intelligible reasons, incapable of altering it. In the same way we should have to show that the irregular will of the individual is always restricted in its action by universal conditions not subject to arbitrary will— conditions which are to be found in the laws of spiritual life in general, in the established order of Nature to which this life is bound by its immutable wants, and finally, in the inevitable action and reaction between the members of a soul-endowed community. This problem is not new, nor have there been wanting attempts at its solution. Indeed, this is the sense in which the calm and practised observer of men and things is accustomed to understand history. By the nature of men's minds, which is always essentially the same, by the sameness of their needs, and by the constant similarity which exists between the circumstances of different lives, an obstacle is, sooner or later, opposed to the flood-tides of caprice, and only those less violent movements can continue which corre-spond to these conditions with their gradual changes.   In this view, then, history is regarded as a development of the concept of humanity, not only in the self-evident sense that nothing can happen in the course of history which did not pre-exist as a possibility in the general character of the human constitution, but also in the sense that in general and on the whole only those phases of development are durable and succeed one another which correspond to the destiny which is appointed for the spirits of men.

The view which we are combating scorned this course.   It was unwilling to regard history as merely the result of a multiplicity of forces working together; it preferred to con-sider it as proceeding from the unity of a single impelling power, pervading the whole course of historical development. In that case the mind of humanity, of which history is to constitute the self-development, must certainly be differently defined.   It will not help us here to give it the name of Infinite, or Absolute, or the Universal World-Spirit, in as far as this, being engaged in the more comprehensive work of its own development, takes on the form of human existence in order to pass through the series of phænomena which are necessary to it at this stage of its course.   For if this world-spirit is dispersed about in innumerable individual men with-out existing complete in any one of them, how can it guide the reciprocal action of all these (for their power of free choice is not to be denied) in such a comprehensive fashion as to bring about a development conformable to its own concept ? It would clearly contribute to this result in us far as it is present in all individual men as that mental organization which is common to them all ; but it would thus only *confine* their development within the bounds of what is possible for such a constitution, without positively marking out the course and the definite forms of the development. If more than this is intended, the higher unity of history can only be reached if that one spirit which ought, with deliberate forethought, to pervade history and to interpenetrate it with the unity of its own aim, is regarded as being in truth an actual living spirit, having an existence of its own, among, or beside, or beyond, or above individual spirits, and not involved in the necessity of their development as being the substance which undergoes development, but enthroned above them as the power by which they are produced. In other words, this second path leads back to the idea of history as a divine education of mankind, as on the first path we were led to regard it as a natural process in which everything happens which logically results from previous circumstances. The doctrine of the realization of the Idea in history appears in these two distinct modes of thought; but the adherent of the doctrine will doubtless continue to maintain that it presents not a confused blending of the two, but their combination in a higher speculative unity.

But in sober truth this view, with its low estimation of individual life as compared to the development of the Idea, gives us but a stone in the place of bread ; and we must consider this point more in detail, since we foresee that very many will be honestly inclined to profess the opinion which we censure. There are no errors which take such firm hold of men's minds as those in which, as in this, inexactness of thought and lofty feeling combine to produce a condition of enthusiastic exaltation.

For clear knowledge it is necessary that to every concept we should add in thought all those connections without which its meaning would be unintelligible ; but owing to the eager haste of thought and speech these connections are very commonly passed over unnoticed. In our varied and complex civilisation there are many thoughts which seem to have a stamp of intellectuality and a certain striking elegance and simplicity, because they detach from the soil of common experience and transplant as it were into empty space, apart from all explanatory surroundings, ideas familiar to us in everyday life, where we observe, patiently and minutely, all the conditions on which their validity depends. This fate has overtaken the idea of *phænomenon* or *appearance* among others.   It is plain that in order to be intelligible this idea must presuppose not only a being or thing which appears, but also, and quite as indispensably, a second being by whom this appearance is perceived.   This second being may be called the necessary place of the appearance, for nowhere except in it does the appearance take place, being never anything else than the image which the perceiving being, in accordance with its own nature, draws for itself, of that other by which it is affected.   But this reference is almost wholly suppressed in ordinary speech ; and when being and appearance are contrasted, nothing is thought of but that one being which emits the appearance as an emanation from itself—the emanation being supposed to exist and appear on its own account, without needing a second being, as a mental state of which only can it attain reality.

Of course any mode of speech is harmless if men understand what it really indicates, and limit its applications and the deductions from it accordingly ; but both this understanding and these limitations are wanting in the present case. What is called phænomenon or appearance is at bottom only the process which may become, or may cause, a phænomenon as soon as it affects a being capable of perception ; this process is not the phænomenon itself.   Now to the true notion of phænomenon there attaches a value which can by no means be transferred to the process which precedes it ; that a being not only exists, but exists for another, is not merely a fact like other facts, but includes an element of pleasure ; it seems to us that the worth of a being's existence (though not, of course, that existence itself) is heightened and doubled when its image is reflected in another, or when, speaking generally, its content is not only there, but is recognised by some mind and is advanced to be the object of some enjoyment, though it may be only the enjoyment of understanding.   He who asks, Would a being exist if it did not appear ? can hardly mean merely that the real existence of a thing consists in its going out of itself, and in the emanation from it of an activity that is directed outwards. This going out of itself will rather be understood as an emergence from the deafness, and blindness, and night of a state in which it is uncognised and forgotten, into the full clear day of awakened consciousness, of being named and being known. For the poetical apprehension of Nature the rising of the sun does not merely mean that the sun which before was below the horizon now rises above it; it means also that it becomes itself visible and renders other objects visible, and floods the world with an enlightenment which, since it makes all things exist for one another, itself constitutes day and awakening, and in fact the full reality of that which before was, as it were, only potential. In the same way, that *appearance of a being,* which we value and of which we speak as of some great good, signifies always the entrance of something real into a consciousness which takes pleasure in it. *This* kind of appearance cannot be conceived as the mere emanation of some being from which it flows forth as a medium that shines by its own light—a kind of light, in fact, the business of which is to give light to itself and to the darkness, and of which this philosophy knows so much, and optics nothing whatever. For an error it is and will remain to treat that shining of light which exists only in the perception of the percipient, or that semblance which exists only in consciousness, or that pleasure in a phænomenon which can be found only in conscious perception of it—to treat all these as if they were occurrences that could take place in empty space, merely proceeding forth from one being without being received into any other.

Here we have to renew our old conflict with this mode of thought. He who sees in history the development of an Idea is bound to say whom this development benefits, or what benefit is realized by it. I do not, of course, mean that there should be merely pointed out to us in the later stages of development, as the fruit of such development, some blessing which was not previously extant, but that we should be shown that the higher good consists in the previous absence of this blessing, and in its gradual attainment by way of this evolution.   But if we agreed to find enough happiness in the mere spectacle of a developing Idea, and to renounce any further advantage to which it might conduce, yet even the review of these thoughts as they march past would presuppose a world of spectators by whom it would be witnessed. Who, then, are the spectators ?   Either mankind themselves while they are developing are conscious of their development and enjoy the pleasure of this consciousness; or God alone surveys history while mankind undergo it unconsciously ; or finally, there are individual human souls which are conscious of the historical progress of the Idea, while the rest only experience it as their fate and their lot in life.

The first of these answers cannot be given.   Unquestionably mankind have in every age had some notions concerning their own being and their destiny, notions which have come to them from the conditions in life and the experiences which fell to their share.   We would not scorn these notions because they do not constitute a collective consciousness, but merely an energetic mental bent which at the most is only intensified to full reflection on particular occasions, and even then only to one-sided reflection.   But the mass of mankind remain quite ignorant of the historical foundation of this feeling which pervades their life, and of its significant place in the whole of historical development.   Obscure traditions of the " good old times," or unsatisfied longings for a better future, unsupported by any knowledge of facts worth mentioning, are all the philosophy of history with which the majority are acquainted ; the subtle succession of the different phases of development of the historical Idea is displayed quite without effect as far as the consciousness of mankind on the whole is concerned.

The second answer will be more readily given and more willingly received, because it is apt to be understood as being better than it is.   For what view is there that might not join in the modest confession that it is God alone who perfectly understands the meaning of history ?   But more than this is involved.   History being understood as the development of the concept of humanity which is cognizable by God alone, it must also be the case that this development alone is the end and aim of history, while all which finite beings do and suffer, hope and fear, strive for and avoid, attain or fail of, is but as part of the machinery and trappings which the divine mind employs in order to bring before its own view this spectacle of the evolution of the concept. I know that no one will lightly profess this view in its undisguised repulsiveness as his own conviction ; still in reality it is to only too large an extent at the foundation of philosophies of history. It is not indeed conceivable that in surveying the tragic course of events the soul of the observer should remain wholly unsympathetic and not be, at least occasionally, surprised into warmth of feeling ; but how often have we been admonished to rise superior to the softheartedness of this sentimental mode of regarding history, and to learn that it is only the necessary progress of the concept that is of consequence, and not the happiness or misery of men ! And further, what is repulsive in the picture which we have drawn is certainly less striking from the fact that it is seldom God who is spoken of as the spectator of this show, but generally a World-spirit, or an Absolute, or a self-conscious Idea. The unbearableness of an egoism which could use a world of sensitive creatures merely as material for its own refined amusement is, of course, softened when the nature of the egoist is so obscurely conceived, and so removed from all similarity to ourselves, that we are left without any standard for the estimation of moral worth. And for the rest we gain nothing by this change of expression. For an inscrutable impersonal primal being in the place of the living God might indeed govern the world and us as a supreme power, but could not be the source of any obligations or any duties. Therefore the assumption of such a being, even if it really explained the external course of history, would deprive the inner development of history of a most effective spring. For however large a share chance may have had in determining the course of events, something at any rate is due to the honest efforts of mankind who with a sense of sacred duty towards posterity have laboured to preserve and to increase their possessions.   If we were forced to believe that all personal life is but a stage of development through which an impersonal Absolute has to pass, we should either cease our efforts, since we could discover no obligation to co-operate in helping on a process totally indifferent both in itself and for us, or—in case we held fast the treasure of love and duty and self-sacrifice of which we find ourselves possessed—we should have to confess to ourselves that a human heart in all its finitude and transitoriness is incomparably nobler, richer, and more exalted than that Absolute with all its logically necessary development.

We may pass over the third answer very briefly.   No one can seriously believe that history takes place in order that it may be philosophically understood by philosophers; the fact is indeed that there is not even a philosophy of that which has taken place.

But there is another consideration which will be opposed to our rejection of all these answers.   An Idea, it is said, not only exists in the consciousness of him who apprehends it or reflects upon it; it is also really and effectively present in things themselves and their connections.   It is present as an existing condition before the attention of thought, which comes later, has been directed to it; and it is plain that it would continue its previous existence, and that its validity would suffer no detriment, even if the gaze and the reflection of a thinking being should never be directed towards it, making the content of the Idea an object of its own consciousness.   If, therefore, only a few individual minds, or even if no one at all, were conscious of the Idea which is operative in history, it would nevertheless continue to exist in order that, unconscious and unknown, it might guide the destinies of the human race.   Mankind as a whole would then be comparable to an individual man who is unceasingly conscious of pain or pleasure, or some other sensation resulting from his bodily organization, without knowing the Idea or plan in accordance with which the forces of his organism are combined to reciprocal action. We ourselves, however, may be compared to physiologists who investigate the laws of this action, and we should not regard the Idea which orders the system of vital functions as being the less efficient or the less worthy of investigation because the living man generally remains unconscious of it, and because it was unknown to us up to the moment of its discovery.

This analogy, which is a just one, needs only to be pursued in order to refute the objection which it is brought forward to support. For surely we should hardly hold that those relations of organic forces can, while they remain hidden, constitute the aim of life, or that the living body is destined merely to realize ordered activities working altogether in obscurity. In the sensations which we experience in some way not yet understood, in the pleasure and displeasure which are the final result of some secret action of our organs, in the supple activity of our limbs, and the joyous sense of that power over them which is ours we know not how—in all this it is that the life of the body consists. On the other hand, all that unknown activity is to be reckoned as part of the mechanical means which exist, not on their own account, but in order that these higher results may be realized. In this sense the secret development of an Idea may indeed be considered as the guiding clue of universal history, and this clue may remain for ever unknown, provided only that the succession of benefits which are attached to it, and which go on increasing, are enjoyed and known. But a view which accepted this interpretation would not differ essentially from that which regards history as resulting necessarily from the co-operation of the spiritual nature which is in us and the material conditions of life which are without us. It would be distinguished from the latter view by only one peculiarity, and that one of very doubtful value—it would believe, that is, that the manifold impulses which have their source in the human mind, and are operative in history, can be comprehended under the one name of the *concept of humanity*, and that the separate investigation of those gradual changes which these impulses undergo in course of time, may be replaced by the one general formula of a development, assumed to be logically necessary, of that concept.

But just this interpretation, which we allow, is by no means contemplated by the views referred to; they imagine that they have found in that hidden self-development of the Idea not a serviceable means, but the final sense and aim of historic evolution, not a guiding thread on which are gradually strung the substantial goods of life, but the Supreme Good itself. And to this we must unceasingly renew an opposition often offered before.   In the order of the world a never-to-be-explained mystery may possibly shroud the *means* used to attain the ends aimed at, or the *laws* in accordance with which these means work; but it would be the most preposterous form of mysticism to suppose that there could be *ends* in the universe which, although no one knew of their content or fulfilment, should yet continue to be ends, or blessings which were so mysteriously hidden that no one could observe them or rejoice because of them, and which should yet continue to be blessings, and indeed to be the greater and more sacred the less this incomprehensible veil was ever lifted from them.   That which is to be a blessing has its sole and necessary place of existence in the living consciousness of some spiritual being; all that lies outside of spirits, external to them, between them, before them, or after them, all that is mere matter of fact, or thing, or quality, or relation, or event, belongs to that impersonal realm, through which indeed the way to blessings may lie, but in which blessing can never be. As long as we have breath we will strive against this superstition, which though so calm is yet so frightful, spending itself wholly in veneration of forms and facts, knowing nothing whatever of true, warm-hearted life, or overlooking it with incomprehensible indifference, to seek the innermost meaning of the universe in observing a secret etiquette of evolution.   And yet how often do we encounter this superstition !   We have seen it shrink back—like a sensitive plant at a touch—when natural science has cheerfully enlarged upon all the efficient means upon which depend the joyousness of animal life, its abundance of physical satisfactions, its sense of vigour, its joy in the varied changes which it experiences. What this superstition thinks of importance is not that there should be a vigorous, joyous, self-conscious reality, but that there should be a show—that everything which exists should recall symbolically something which itself it is not, should ring in unison with activities which it does not exercise, with destinies which it does not experience, with Ideas of which it remains ignorant. And when in history the rich-hued ardour and passion of human life are unfolded before the adherents of this doctrine—the inexplicable peculiarities of individual minds, the disturbing complications of human destinies which, in many respects alike in their outlines, are yet inconceivably various in their individuality—when this great picture is opened before them, then they rise up and ask if there is no way of reducing this grandeur back to something poor and small—of reducing *back*, in sober truth, for we go backwards and not forwards if we allow the tedious emptiness of a logically necessary development to be imposed upon us as the final meaning and end of the universe. And therefore will we always combat these conceptions which acknowledge only one half, and that the poorer half, of the world; only the unfolding of facts to new facts, of forms to new forms, and not the continual mental elaboration of all these outward events into that which alone in the universe has worth and truth—into the bliss and despair, the admiration and loathing, the love and the hate, the joyous certainty and the despairing longing, and all the nameless fear and favour in which that life passes which alone is worthy to be called life. And yet no doubt our combating will be wholly in vain, for those whom we oppose will ever seek afresh to cover the imperfection of their ideas with the cloak of a generous putting aside of self; they will always be ready to profess anew that there is a meaning in saying that phænomena *happen* even when they are not seen, that symbols *are emblematic* even when no one understands them, that Ideas are expressed by matters of fact even when there is no one upon whom the expression could make an impression. This sounding brass and this tinkling cymbal will ever be struck anew; or rather this brass which does not sound and this cymbal which does not tinkle, for sounding and tinkling have their purest and highest value for this mode of thought, when considered as what they are in themselves when no one hears them.

§ 4. But are we not mollified by another conception, which does justice to the incalculable variety and wealth of history and redeems it from the poverty-stricken condition of being a mere logically necessary development of a concept, and according to which history is a divine *poem*, produced by God's creative fancy, with the spontaneity and life of a genuine work of art ? One might be in doubt as to the class of artistic productions among which this poem should be reckoned; to some it has seemed to have the uniform flow of an epic, to others to be as full of catastrophes as a tragedy ; again, it has not unfrequently been regarded as a comedy by mocking philosophers in sardonic moods; and each of these views has seemed, to those who held it, to have something in it. Meanwhile it is plain that the phrase contains, in the first place, merely a comparison of the impression made upon us by history with the similar impression which we receive from poetry. The peculiar character of the impression is made clearer by the comparison, but not so the causes by which in both cases it is produced. Perhaps we might more justly and more usefully make the converse statement, and say that poetry derives its power from its similarity to history. For art is never a mere playing with forms; it is true and genuine only when we recognise its forms as the same as those upon which the cosmic order is based, and according to which those events happen which, taken as a whole and in the breadth of their simultaneous complications as well as in their temporal succession, are just history itself. Because the epic brings before us with simple clearness this vast and wide and variously agitated stream of human destinies, without offering instructive solutions of particular difficulties, it has

the same effect upon us as history itself, which with equal
reserve hides the secret of its whole significance under a series
of sharply defined events which stand out in strong relief.

So far the comparison of history with poetry is nothing
more than a graceful play of thought, going from one
unknown to the other, and expressing each in terms of
the other without really making either of the two more
plain. But the comparison has something more in view. It
aims not only at comparing the finished poem with the course
of past history, but also at comparing the production of the
work of art by the imagination of the artist, with the origin of
history, due to an equally incalculable spontaneity of the divine
mind. Something would indeed be gained if the essential
peculiarity of that artistic imagination could be defined in a
way that might be understood without again having recourse
to imagination. We do not know that this has been done.
For if we consider the information which we have concerning
this mental activity—concerning the spontaneity with which it
produces what is fair and what is repulsive, inventing examples
of the application of necessary laws with boundless licence
—concerning the perceptible justice with which it proceeds
in the combination of these arbitrarily constructed events,
without our ever being able to take a comprehensive and
intelligent survey of the whole—we find that in these characteristics and others which have often been noted, the mystery of
history is reproduced in all its features, only it remains, unfortunately, just as much a mystery as before. We receive no
enlightenment with regard to the origin of this divine fancy
or its ends, nor with regard to the way in which the conception of it may be combined with our other ideas of God,
or with the rest of our philosophy. Therefore, though we
willingly agree with this view in what it denies, we are in
no wise enriched by what it affirms.

§ 5. And now, after so many vain attempts to interpret the
progress of history, we will consider that opposite opinion
which altogether denies history in the sense of a progressive
development on earth. This view, too, is by no means a mere

peculiarity of mistaken thought, making a casual appearance now and again; in ancient as well as in modern times it has reached the point of the most pronounced aversion to every-thing mundane, an aversion which has been enthusiastically carried into practice. Innumerable heathen penitents and christian hermits have retained in their solitude a deep and pervading conviction that human life on earth does not, as a whole, progress towards any ideal of perfection which is here either attainable or even only aimed at, but that everything is vanity. They regarded only the constant and unmediated return of the individual heart to God, and its exaltation to the supersensuous world as progress, and all other earthly life as but a continual repetition of the old imperfections. This, too, is a philosophy of history. It is probably based upon less profound combinations of thought than the opinions which point to a progress which is supposed to be perceived; but, on the other hand, innumerable sacrifices have proved it to be a most living conviction, and it will continue to receive fresh proof of the same kind; for it is ordinarily our last con-fession when we depart from life and leave behind us all the plans, the carrying out of which once seemed to us a work of such greatness and importance.

Shall we give ourselves up without reserve to this denial of earthly good ? Would there not hence result an inactive contemplative disposition which, by causing too early a renunciation of all mundane gain, would abolish the conditions of struggle after that which is supramundane ? Such retire-ment from the world is conceivable only as retirement from a world which one has known, from a life in which one has partici-pated. It is only a remembrance of the wealth of mental life, of the happiness and misery, the hopes and illusions, which the social interweaving of human efforts includes and produces, that can afford to solitary contemplation an object of reflection in considering which it may develop its ideas concerning the supersensuous life. He who has experienced nothing is made no wiser by solitude, and communion with the phænomena of Nature, and with the thoughts which would be possible for a mind altogether withdrawn from human society, could lead to no better peace than that which the inferior animals possess.

But, as a matter of fact, it was not inevitable that depreciation of what is earthly should be intensified to such contempt for all living activity. Men may recognise that the social relations of human life offer the sole though intractable material by elaboration of which they are enabled to work out the ideals towards which they struggle and aspire; and this recognition may lead them to devote themselves with all their heart and soul to the tasks of earthly existence. We show the perverse pride of human exacting-ness in only taking pleasure in work, and only valuing it, when we are assured that the results of our activity will hold a lasting place in the history of the universe, and will have imperishable value. If we estimate more modestly our performances here, regarding them as mere prentice work, then we can in all seriousness combine with the preparation for a higher end that calm resignation which will patiently endure that our attempts here should be without progress or lasting results. In proportion, then, as we estimated more highly the immediate relation of each individual soul to the supersensible world, the value for mankind of the coherence of history would sink; history, however it may move forward or fluctuate hither and thither, could not by any of its move-ments attain a goal lying out of its own plane, and we may spare ourselves the trouble of seeking to find in mere onward movement upon this plane a progress which history is destined to make not there, but by an upward movement at each individual point of its course forwards.

And, it is asked finally, is it not this unhistorical life that is actually lived by the greatest part of mankind? For the unrest and variety of revolutions and transformations, the meaning and connection of which we are seeking, is yet, when all is said, the history of the male sex alone; women move on through all this toil and struggle hardly even touched by its changing lights, ever presenting afresh in uniform fashion the grand and simple types in which the life of the human soul is manifested.   Is their existence to count for nothing, or have we only for a moment forgotten its significance in scholastic zeal for the Idea of historical development ?

By such considerations the inclination to an unhistoric conception of human destiny is strengthened; still this does not overcome the opposition of a moral sentiment which warns us against giving up everything that we cannot understand, and admonishes us to esteem the temporal advance of history as a real good.   Even that which holds us back from this recognition, when we are considering its course scientifically—that is, the unequal distribution, among successive generations who know not one another, of an ever-increasing quantum of good —is not felt as a misfortune in actual life.   On the contrary, that universal absence of all envious feeling towards future generations which coexists with so much selfishness in detail, is one of the most noteworthy peculiarities of the human mind.   And not only do we not in the slightest degree grudge to this future the greater happiness of which we ourselves can only have a prophetic foretaste, but it is further the case that a vein of self-sacrificing effort for the establish-ment of a better condition of things in which we ourselves shall not participate, runs through all ages, having sometimes a noble, sometimes a commonplace aspect, at one time appear-ing as the conscious devotion of affection and at another as a natural impulse, unconscious of its own significance and of any definite aim.   This wonderful phænomenon may well tend to confirm our belief that there is some unity of history, transcending that of which we are conscious, a unity in which we cannot merely say of the past that it is not—a unity rather in which all that has been inexorably divided by the temporal course of history, has a co-existence independent of time ; in which finally the benefits produced in time are not lost for those who helped to win but did not enjoy them.

This view will certainly not escape the reproach of marring one of the fairest traits of human character by assigning to it a basis of selfishness ; nor will it at the same time escape the suspicion of demanding from human hearts the magnanimity of motiveless self-sacrifice when such self-sacrifice results from love for others or for mankind without any thought of selfish advantage. But these reproaches would show a misunderstanding of the subject under discussion. We, too, would hold such a thought of selfish gain far removed from the motives of our action, but we cannot in the same way exclude it when we are considering the structure of the universe. While we lay great stress upon maintaining the principles of our conduct in all the purity of unselfishness, we feel it equally important that the world itself should appear to us as a significant and worthy whole. We require our own happiness, not for the sake of our happiness, but because the reason of the world would be turned to unreason if we did not reject the thought that the work of vanishing generations should go on for ever only benefiting those who come later, and being irreparably wasted for the workers themselves. All human longing to find a guiding thread in the confused variety of history springs from the unselfish desire to recognise a worthy and sacred order in the system and course of the world. This longing has impelled some who held different views from ours to sacrifice the substantial happiness of all individuals to the constant and uniform development of a universal; but as we regard such attempts as a misdirection of thought, we are impelled by it to the opposite demand for a lasting preservation of that, the continual destruction of which would render fruitless all effort to develop even the universal itself. Each, in order to keep his own thought pure from selfishness, may exclude his own happiness from this demand; but he cannot avoid requiring the preservation of the happiness of others, unless the world itself, and all the flourish about historical development, are to appear as mere vain and unintelligible noise.

This faith, being the interpretation of the results of historic life, is connected with the self-sacrificing and provident love which is the noblest spring of that life. The presentiment that we shall not be lost to the future, that those who were before us though they have passed away from the sphere of earthly reality have not passed away from reality altogether, and that in some mysterious way the progress of history affects them too—this conviction it is that first entitles us to speak as we do of humanity and its history.   For this humanity does not consist in a general type-character which is repeated in all individuals, no matter how many they are, or have been, or shall be ; it does not consist in the countless number of individuals who are only brought together by our thought into a unity which they have not in reality, since as a matter of fact they are dispersed and some would still be if the rest did not exist ; but it consists in that real and living community, which brings together into the reciprocity of one whole the plurality of minds which are separated from one another in time, and in the particular place of each in that whole being marked and reserved beforehand, just as though the whole number had been already reckoned over.   And history cannot be a mere slender ray of reality slipping on between two abysses of absolute nothingness, past and future, ever consigning back to the nothingness in its rear that which its efforts had won from the nothingness in its van ; there must be a pre-established sum, in which the flux of becoming and of vanishing away is consolidated to permanent existence.   Where the human mind fortifies itself in its efforts by an appeal to the spirits of ancestors or to future renown, it does it with this idea ; an appeal to what is non-existent is powerless—no appeal can be of any efficacy which is not strongly penetrated by this thought of the preservation and restoration of all things.

Such a faith is not easy in all ages.   As long as the limited purview of mankind embraced only the near distance of a known past and the familiar surroundings of home and clan, there was a powerful attraction in the thought that this simple life, bounded at the one end by creation and at the other by the last judgment, was a probation at the close of which would begin the happy communion of all those who had been divided from one another by the lapse of time.   Our extended intellectual horizon embraces a multitude of unlike nations.

the indefinite ebb and flow of a far-flowing historical stream, the ever uniform working of Nature, and the immeasurable extent of the universe, and we can neither be satisfied with such a brief and homely solution of complications which have become infinite, nor can we find some different conception capable of meeting our own more exacting requirements and giving a clear representation of the ideal of which we are conscious. Yet, notwithstanding, we hold fast the primitive faith, and do not find that we can replace it by explanations which have seemed more acceptable to the culture of our age; on the contrary, it is only by presupposing the truth of this belief that modern views can free themselves from the internal contradictions in which we found them involved. For no education of mankind is conceivable unless its final results are to be participated in by those whom this earthly course left in various stages of backwardness; the development of an Idea has no meaning unless all are to be plainly shown in the end what that development is of which in past time they had been the ignorant subjects. He who seeks a plan in history, will find himself inevitably compelled to acknowledge this faith; he alone can feel no need of it who sees in history nothing but examples of universal laws of action, each example due to the impulse of anterior forces, and not to the attractive power of ideals as yet unattained.

But in truth our presupposition suffices only for the removal of inner contradictions; neither it nor our empirical knowledge makes it possible for us to exhibit the plan which history follows. Not our empirical knowledge; for we are well aware how small the sum of our knowledge is when compared with all the wealth of life of which our planet has been the scene, and how little the fragments which we know make us capable of discovering the path that may have been taken by the course of earthly history as a whole. And if we did know all this which we do not know, it might still be doubtful how far this earthly life could be understood as a whole in itself and without needing the help of anything else to explain it; and our scientific insight is infinitely far from penetrating all the ramifications of the connections by which it may be bound up with a vaster universe, which perhaps contains material for its completion.   Thus history still seems to us, as it has seemed in all ages, to be a path which leads from an unknown beginning to an unknown end, and the general views as to its direction which we believe we must adopt, cannot serve to indicate the course and cause of its windings in detail.

CHAPTER III.

THE FORCES THAT WORK IN HISTORY.

Theories as to the Origin of Civilisation—Theories of a Divine Origin—Organic Origin of Civilisation—Instance of *Language*—Importance of Individual Persons—Laws of the Historic Order of the World—Statistics—Determinism and Freedom—Uniformities and Contrasts of Development—The Decay of Nations—Influence of Transmission and Tradition.

§ 1. EVEN in antiquity reflection was in many ways directed to the origin of that ordered life, in the enjoyment of which men then found themselves, and there appeared even then the same extreme views by which opinion is now divided. Human civilisation as a whole seemed so wonderful when first apprehended that its origin appeared incomprehensible except as an express divine institution. Pious legends very early sought to find in the benefactions of the gods the source of the commodities of human life, partly of those whose origin is still an enigma to us, and also of many others which would not seem to us to exceed the reach of easily comprehensible developments of human powers. The sense of the evil in society came to strengthen the melancholy notion of a past Golden Age in which there lived innocent men, with simple hearts, at peace with each other and with the world, under the protection of the gods, until growing knowledge of the world brought coveting and strife—or perhaps it was that these latter awaked men's slumbering capacities for knowledge. With this picture of a fair beginning and an ill continuance was soon contrasted that of an origin of brutal savagery, from which mankind, schooled by suffering and experience and making good use of their lessons, gradually advanced to the rich complexity of their contradictory, wonderful, ill-fated. civilisation. Both conceptions have been repeated with innumerable modifications by succeeding ages ; generally with a leaning to assumptions which inter- fered with impartiality of judgment.

Even the old view, which opposed the theory of earthly development to that of divine origin, set out from declared hostility to all religious contemplation ; the rationalistic *Enlightenment (Aufklärung)* which long governed opinion in modern times, was equally prone to express depreciation of all which pointed to something more, in the dim beginnings of history, than lucky chances and the ingenuity of busy brains. This Enlightenment traced back the beginning of political life to a convention entered into by honest men of remote antiquity; language they traced to an agreement to use certain sounds as the most appropriate means of communication ; the maxims of morality were attributed partly to a general recognition of the usefulness (accidentally discovered) of certain kinds of conduct, partly to the precepts of far-seeing teachers ; and finally, the origin of religion was referred to men's natural inclination to superstition and the artful use of this by priestly cunning. In all this, deliberate calculations, such as are known only to a somewhat advanced civilisation, were made the producing causes of civilisation itself, by the Enlightenment—which thus failed in finding the solution of its problem. But it is not this failure, destined perhaps to befall other attempts of the same kind, which has sharpened the aversion of the present generation towards this mode of looking at history ; it is the obvious endeavour to represent all this (which must indeed come to pass *through the instrumentality* of men) as though it were the arbitrary product of human action. We cannot, however, deny that the theory we are considering was due to real need of enlightenment although it sought to satisfy the need in a very inadequate fashion.

When the opposite view was revived, it exceeded all modera- tion and all necessity by connecting the early history of man- kind with supramundane beginnings, in ways which could not afford the expected advantages even if motives for preferring them, which were absent, had existed. In combating these views

I would not refuse them the consideration which is their due. That historical life was preceded by a primitive state of moral holiness and profound wisdom, and that all succeeding ages were taken up with the decay of this glory and a struggle against the decay—such a wholly perverted view of history as this will hardly find advocates in the present day.   But if there were such they need not be alarmed at the objection that it is only development from the less to the more perfect, and not progress in an opposite sense, that has all natural analogies in its favour.   He who has once come to regard history as something more than a mere natural process, who has made up his mind to regard it as part of a great and divine plan of the universe, will also be secretly convinced that to understand its course something a little more profound may be needed than the simple formula of progress in a straight line.   That course may perhaps involve many windings which are only dimly intelligible to us, but which if clearly understood would disclose a striking and living meaning of infinitely higher value than the barren conceit of a continuous advance uninterrupted by catastrophes.   It is not in vain that various ages and nations have worked out, with devotion and longing, ideas of a fall from some better state of existence, of temporal life as a penance, and of a final reconciliation and restoration ; by doing so they have borne witness that if the mind does not (thanks to material analogies) forget its own being and nature, it is capable of believing something differing widely from a progress which (having no loss to regret) is busied in producing with its own hands all the goods that it requires.   But historical investigation, however far it has advanced, has come no nearer the discovery of the existence on earth of an ideal primitive state, and has in fact left it hardly disputable that our civilisation must have grown up from simple and indigenous beginnings along the path of a gradual and much interrupted development.

§ 2. Such an admission, however, does not exclude supernatural beginnings, only that in the place of an ideal

condition of primitive men there would have to be substituted
the thought of a divine education by which men's natural
powers should have been guided up to a point at which the
species had become capable of its own further development.
The addition, expressed or understood, of the opinion that
from that time forth the divine guidance ceased, shows us
that men imagine such guidance to have been exercised in
primitive times in a more express and striking way than in
that later progress of history which it is just as impossible to
withdraw altogether from its influence.   In order to estimate
this opinion we will consider it as manifested in more definite
views.

No one will attribute the beginning of human education to
intercourse with angels who walked in visible form upon the
earth.   We find in primitive times, not infallible wisdom
which could not have been acquired from a merely human
standpoint, but signs of an active curiosity which sometimes
hit and sometimes missed the mark ; not a complete
systematization of society which would seem referrible to
divine arrangement, but simple forms of life easily explicable
as the result of natural relations and natural sociality, and
more complex forms presenting a very human mixture of
pride and fear, cunning and violence ; not a faith the otherwise unattainable truth of which must have come by revelation, but religions in which aspirations after an ideal had
developed conceptions of very various worth ; finally, no
primitive speech of divine construction, but from the beginning a number of different manifestations of the common
faculty of speech.   Faultless perfection in all these cases
might make it necessary to seek an explanation by reference
to constant intercourse with superior beings; what we actually
find, however — mental activity generally, inventiveness of
intellect and vigorous constructive faculty, but not the
exclusion of error — all this does not demand such an
assumption.

But for this inapplicable conception may be substituted an
influence of the Godhead upon the human mind just as

immediate though more hidden. We do not, it may be said, seem to find in the course of psychic life as at present constituted the conditions necessary for the initiation of a civilisation capable of being hereafter transmitted with ease. A state of the mental capacities differing generally from that which we now see must have been the basis of such a beginning, and this may perhaps have been transformed to the existing constitution of mental life by the very reactions naturally accompanying progress. This view takes two different and more definite forms, neither very probable. That the general laws according to which the events of psychic life are combined in men and animals were different in primitive times from what they are now (which is the one form), is a supposition that to us seems incredible, and that can in no case lead to any useful results. For other laws of the train of ideas, if not reinforced by other and copious sources of knowledge or by extraordinary mental activity, would either (1) not lead to new and otherwise inaccessible developments, or (2) would lead to developments merely strange and singular; they could not lead to those from which our historical civilisation has in fact grown up without any substantial interruption. And the same would hold of that other interpretation which sets forth that it is the moods, the inclinations, the receptivity, and the aspirations of the soul—which are subject to the general laws of mental life as being the living objects to which these laws apply—that it is these, and not the laws themselves, which were once constituted and combined in a fashion different from that which obtains in existing human nature. No doubt this significant psychic nature may be very different in different individuals, since its manifestations are not produced by general laws, although they are formally determined by such, and the development of their results similarly regulated; but he who would exaggerate the peculiarity of men's primitive, as compared with their present, mental state, likening it to the instinct of brutes, to demoniac possession, or to the twilight of clairvoyant somnambulism, forgets that what we seek in this primitive

condition is not wildly aberrant and extraordinary phænomena, but the beginnings of our own familiar development.
Therefore, without denying that the mental life of the earliest
antiquity may have been so different from our own that we
cannot fully realize it, we yet hold that the assumption of
unlikeness above referred to is not particularly useful even
when kept within the limits of moderation, and that when
carried to excess it is of no value whatever for the explanation
of that which we want to have explained.

I am compelled to regard with the same scruples a view
which seeks to find the *nidus* of that primitive mental condition specially in the religious life, or in God's presence in
the devout consciousness of man.   Certainly like-mindedness
in religion is one of the most essential bonds upon which the
union of a people can depend, and the greater the contrast
between the faith of any people and that of their neighbours
the more stubbornly often has such a nation kept itself
uncontaminated.   But we should not be justified in asserting
that without the religious bond all other natural inducements
to social life would only suffice at most to constitute a horde,
not a nation.   That language should have been the same for
all mankind in primitive times is not made comprehensible,
with regard either to its origin or its construction, by the
supposition of unanimity of faith ; and we are equally in the
dark as to what must have happened for a division of faith
(due to unknown causes) to have led to a confusion of tongues,
through which new and varying appellations were given to all
those objects of common life which were not in intimate
connection with the sphere of religious thought.   It is easy
to give the general answer, that there is nothing so separate
and isolated in human life as not to be affected by religious
belief and its peculiar character.   But if one is not satisfied
with the vague devotional thrill caused by this indefinite
expression of a true thought, one sees what degrees and proportions there are in this connection of human things with
divine.   Neither in life nor in science is it possible, necessary,
or desirable that true religion should strive to exhibit what is

secular—the course of Nature and human freedom—as the immediate shadow and reflection of what is divine; that it should deny or grudge to these the comparative independence with which, by native strength in the first place, they produce their own special results.

§ 3. We have yet to glance at a view, a favourite of modern times, in which the idea of a mysterious beginning of human civilisation approximates to the thought of natural development. The rationalistic fashion of explaining every coherent department in the whole frame of civilisation as constructed out of a multitude of separately insignificant accidents and inventions, having fallen into disfavour as a caricature of mechanical action, it has become customary to ascribe the forms of society, the growth of morality, the construction of language, and the coherence of religious belief, to organic development. Two points become prominent when we ask what meaning can here be assigned to this term *organic*—for which a long defence will have to be made if at the last day account has to be given for every idle word. In the first place, that which has an organic origin, being withdrawn from the region of conscious invention and free choice which belong to us as men, is supposed to grow necessarily out of the innate constitution of our mental being. And on the other hand, that also which is realized in the intercourse of different individuals as an advantage of civilisation in which they all participate, is held not to result from reciprocal action of which they are conscious or which can be pointed out, but to be the immediate product of a mind that is common to them all.

Now the rule within us of an unconscious necessity needs no demonstration. Each individual sensation in us bears witness to it, for we do not choose what the sensation shall be with which we respond to the external stimulus; every feeling of harmony or discord which we experience is the involuntary expression of something that takes place in us without our comprehension or co-operation; if a melody to which we are listening is broken off unfinished, we are driven to seek for its conclusion, not because we understand at all why the conclusion should be added, but because our soul, with un-comprehended power, struggles to emerge from the state of having begun some movement but not carried it out; and it must be in the same way that in the case of more com-plicated processes, causes of which we remain unconscious, arouse our efforts and guide them with sure and arbitrary power.   Scientific research may perhaps some day succeed in clearing up these obscure processes; but however much may be accomplished in this direction, the difficulties connected with the beginnings of human civilisation would not be lessened thereby.   These difficulties are to be found in the fact, not that a coherent whole of mental life is developed in the individual soul, but that such developments occurring in different souls coincide to form a common intellectual possession.   And it is plain that those who can find the explanation of this in the notion of organic origin, labour under a delusion.

Let us look at language for instance.   Each individual may be forced by an unconscious natural impulse to manifest his mental condition by definite sounds; but this manifesta-tion becomes language only through the comprehension and recognition of the hearer.   Now capacity of excitation, structure of thought, and connection of ideas, may be as like as you will in members of the same tribe, but this harmony would never impel them to choose with mechanical uniformity the same sounds for the same ideas, and the same inflections to express the same relations.   For the spoken word is the immediate reflection not of objects, which are the same for all, but of the impressions produced by these, which are different for different individuals.   Indeed, in the same individual the same stimulus does not produce at all times the same impression, owing to his varying moods; and language as it grew up would greet objects with ever varying names if the name once given did not blend so completely in our remem-brance with the idea of the thing itself that later, even when we learn to know the thing from quite a different point of view, the name recurs to us as one of its most constant and important properties. And certainly also, with whatever solemn obscurity we may imagine the organic speech-impulse to operate, every sound must have been pronounced for the first time by some individual mouth with lips thick or thin. Originally it belonged to him only who had framed it; it could only become common property when others divined its signification and repeated it with the same meaning. How this happens is shown in a general way by the ease with which children of very ordinary abilities master the materials of speech without express learning, and grow familiar with inflectional analogies. But the first origin of language still presents special and unsolved difficulties.

If a great number of individuals with equal claims to consideration had simultaneously taken part in its formation, there would have been a variety of quite independent names for some ideas, and hence a superfluity which would only have been reduced by the subsequent necessity of reciprocal intelligibility. This did perhaps actually take place to a certain extent; the heterogeneous store of roots which we find in languages may be the result of a mutual adoption and surrender of words formed independently by different men. The same simple idea seems to have been originally denoted by several distinct roots of different sound, which later (because the supply was in excess of the need) came severally to express the different shades of meaning attaching to the idea; thus it happens that there are not connected series of words coresponding to connected series of ideas in such a way as that, for instance, the names of colours should be more like one another than like the names of impressions of other kinds, or that the appellations of trees should have a greater etymological resemblance to one another than to the appellations of birds. This systemless incoherence of the material of language would indeed result if objects affected the linguistic imagination of a single individual not similarly, in as far as they were similar, but in a way varying according to accidental and varying conditions; and we see that if language grew from the concurrent contributions of many persons, there must have been still more reason for this variety. It would have increased past all possibility of comprehension if (as we suggested above) the number of equally influential language-builders had been considerable.

But there is no doubt that language did not spring into existence like the statutes of a suddenly formed society, but that it grew up gradually within a family, or clan, or tribe ; and that as one generation succeeded another in the natural course, the store of words already formed would be transmitted with the same authority as other traditional arrangements. The creative impulse soon dies out in any department when it finds patterns provided, by imitating which its wants may be satisfied. Therefore an existing word prevents others from springing up to express the same idea; or if they do spring up, they disappear like the numerous words invented by children, which are lost when their mode of thought grows into harmony with that of adults. So it happened that only so great a variety survived as resulted from a process of mutual accommodation between the contributions of those families (not very numerous) who had been independent constructors of language.

But in this way we reach merely a generally used store of words and not the grammatical construction of language. There are very many different rules for denoting different relations by compounding, blending, and modifying roots, and each of these modes, again, allows of course of an innumerable variety of applications. How among this abundance of possibilities a logical construction of language could have grown up is an enigma. Besides, one cannot believe that such a construction could be produced in short time and by few men ; but if we allow a long time, this does not make it easier to understand how amidst the succession of different generations and among a very numerous people, just one single plan of construction out of the many possible, should have gained universal recognition and mastery. One would conjecture that in such a long course of time very many varying attempts at construction would be made by many different persons, attempts which could hardly have been consolidated to the unity of one logical construction even by the compensatory process of mutual accommodation. But do we find this logical consistency existing throughout in the grammatical construction of language, or are there here too traces of a complex origin? Do not most languages make simultaneous use of different kinds of construction, using root-modifications together with prefixes and suffixes? Are there not various forms of declension and conjugation having all the same meaning and value? In this abundance of forms—forms which in all developed languages are the last to experience the transforming influence of the principle which has come to be predominant—we may perhaps find survivals of constructions which were originally diverse. Is the superabundance of cases, of tenses, and of moods really to be ascribed to an inexpressibly delicate sensibility on the part of those with whom language originated—a sensibility that from the very beginning and as it were at one stroke provided, with systematic completeness, for the expression of the finest shades of thought—or can we not rather trace in these various forms the remains of originally diverse attempts at formation of language, which attempts—since they held their ground— came as a consequence of their superfluity to be used for the denotation of those fine shades of thought? Recent progress in the investigation of language makes me feel more sure than I did formerly that many of the latter questions may be answered in the affirmative, and that many of the examples adduced may be really conclusive; meanwhile what I have said here is said not so much for its own sake as in order to explain what that is which we are seeking, and which a practised eye might perhaps really detect under other forms.

And however it may be in the special case of language, our assertion will yet hold good in general. The origin of every mental possession held by men in common supposes a period in which by reciprocal appropriation, surrender, and accommodation, the contributions brought by individuals and resulting from an organic necessity of their nature, have become blended into one coherent whole.   It is only individual living minds which are centres of action in the course of history; every principle that is to be realized and to become a power must be first intensified in them to individual activity, and then, through a process of reciprocal action between them, become extended and generally recognised. How commonplace this remark is—yet it almost seems as though through the unintelligent use of that comparison of organic origin we had come to think that, when language began, individual words fell ready made like snow-flakes from the atmosphere of a general consciousness upon the heads of individuals, or as if works of art, the results of national imagination, could arise like clouds in the sky and grow larger by the spontaneous addition of formless vapours.

§ 4. But this organic view of history would banish from human life not only the mechanism of reciprocal action, but with it also every element of chance.   Among the most choice accomplishments of the theory is the demonstration (*post facto* indeed) that events must necessarily have happened as they did, and that being logically consistent developments of the spirit of the age they could not have been prevented by any exercise of individual free will.   Now certainly no individual power can make itself felt in history unless it knows how to subserve some prevailing motive of action, or is capable of in some way alleviating human suffering. But on the other hand, those mighty men who through inventive genius or obstinate constancy of will have had a decided influence upon the course of history, are by no means merely the offspring and outcome of their age. In most cases the general spirit of humanity, the organic evolution of which we extol, has produced no more than a feeling of present pressure, a yearning mood, or a devout desire for change.   It has stated the problems, a solution of which was wanted; but the fulfilment of these desires and the special mode of fulfilment are works the doing and desert of which belong to a few individuals.   In other cases there has not even been this precedent sense of helpless want, but the heavy unintelligent opposition of the majority has been laboriously overcome by the successful mental effort of a few, who have thus given to that majority new aims of action. And finally, where individual strength has actually taken up the tasks of the age, there has perhaps seldom been an exact accomplishment of what the moment required, no more and no less; in most cases there has been added much both of good and bad which, extremely effective in itself, yet went beyond the immediate need, or was altogether beside it. In innumerable cases the anticipated development has been interrupted; the skilful calculation of far-seeing minds has often been perverted by some strong tide of feeling from its original purpose, and for long periods been used for artful ends. Modes of thought which under appropriate conditions were adopted by men of genius, have withstood progress for centuries with incredible tenacity. Forms of art worked out by great minds, but not of universal validity, have continued to maintain their predominance when they had become out of harmony with the altered dispositions of mankind; and even in science inherited errors drag on like a slow disease. What we can thus observe now in history we would also claim as explanatory of its beginnings. It is of course true that all men had in early times similar capacities and wants, but all did not take an equal share in satisfying human impulses; the germs of civilisation did not, like the upward growth of a young forest, shoot forth simultaneously over wide extents with organic necessity and regularity, but the wandering, incapable, uninventive impulse of the whole was indebted to individual happy strokes of genius for its first distinct ideals and the first satisfactions which paved the way of its advance.

Meanwhile this influence of persons no doubt varies in magnitude in different domains of human activity, and according to the divergent characters of different periods and the multiplicity of conditions on which may depend the action and reaction between individual force and the mass of mankind. It is dependence upon Nature which most universally

rouses the inventive ingenuity of men, and the thoughts
which here help them to obtain what is most necessary arise
from such simple combinations of ordinary experiences that
the elementary furniture which we find among the most
different peoples — weapons, implements, woven stuff, and
ornaments—is easily intelligible as the production of a general
instinct without any special invention by individuals. But
all those higher and more refined aids which have led to a
more productive command over Nature, are connected with the
names of individual discoverers; between its first beginnings
and the period of universally diffused culture to which we
are perhaps approaching, life has in this respect too had its
age of heroes.   And as in other departments so here also
there is a gradual transition from one stage to the other.
When any sphere of thought (as for instance Natural Science
in the present day) has reached a grade of development which
furnishes not only innumerable factual items of knowledge,
but also general forms of investigation and clear indications of
the regions in which answers to yet unsolved riddles must be
sought, then the current of inquiry once set in motion produces in swift succession a multitude of useful inventions,
which seem to spring from the general mind.   This seems to
be the case, because the multitude of individuals actively
interested, and the vigorous action and reaction between them,
throws into the background the particular contribution of each
several person.   Further, the general laws which science shows
to be at the foundation of the vast commerce of modern
times, are familiar to every one in their application to the
simple relations of ordinary everyday life; the ill results of
acting in opposition to them are so obvious in the case of
individuals, that a great number of slight modifications of a
man's course of action are the immediate result of any unsuccessful attempt on his part to contravene them.   Thus
it seems that the whole system of our arrangements for the
satisfaction of men's wants goes on improving progressively
by its own inherent force, and without needing to be pioneered
by the inventions of individuals.   Nevertheless these laws,

like all simple truths, become hard to trace when with increasing intercourse they have to be applied to a group of relations which are very numerous, and perhaps themselves either unknown or modifying one another after an unknown fashion. To have shown that these laws are valid, and how they are valid, even under such circumstances, is unquestionably a great achievement of science, and it has not been accomplished without help from the creative genius of individual persons. The arrangements of social and political life have also passed through the two stages of development which we are here distinguishing. The universal homogeneity of human nature and its wants no doubt lead in the first place with uninventive necessity to rules of intercourse which develop in the same way and succeed one another in the same order everywhere. But even if the purely indigenous development of a society could be left altogether to the organic interaction of its own individual forces, the political guidance of the society under difficult external conditions, and the choice of the right path at the right moment, would be always dependent upon the wisdom or folly of individual men. Hence it was that antiquity always set at the beginning of its political histories the name of some individual lawgiver, not that they might derive from the individual power of some master-mind, the first foundation of order—since this indeed could of necessity only be developed by means of the reciprocal action of a number—but that they might derive thence the first firm consolidation of that order, and such accommodation as had been arrived at, of difficulties occurring in the application of law to concrete cases. We scarcely need to add in conclusion, that though often ill-defined forms of enthusiasm seem to be of obscure origin, yet this is not the case with religions, which never appear in history without some founder; here too it falls to the concentrated strength of individual minds to satisfy wants which under similar circumstances are always alike among the homogeneous masses of mankind.

The incalculableness with which, for human eyes at least, individual greatness influences history may seem to threaten

the logical consistency of all historical development, and to
reduce it to a continual fluctuation in different directions.
Yet any personal power requires for its efficacy the receptivity
of the masses ; the want of this or the presence of a hostile
disposition prevents the working out both of all the good and of
all the bad effects which a remarkable mind tends to produce,
and prevents likewise the realization of all the good exclusively,
or all the bad exclusively ; this is, of course, especially the case
with respect to anything which is in opposition to the requirements of the hour, or foreign to them.   The more active the
reciprocal contact of men in society is, and the more intricate
their exchange of thought, and the larger the bodies of men
are among whom this contact and this exchange of thought
prevail, the more are those circumstances changed by which
the influence of individuals is conditioned.   The scene of their
possible action is certainly enlarged, but the probable magnitude of their influence is decreased with regard to all that is
not a direct continuation or fulfilment of projects already
begun and wants already felt.   For it is only where this is
the case that a man can reckon upon the collective strength
of a public opinion and sentiment which has already taken
into consideration all possible circumstances of life, and made
up its mind about them somehow, and which is not likely to
let itself be easily detached as it were from the soil to which
it clings by so many roots, and carried away by the arbitrary
will of a single individual into some new order of development.   Thus as the ascendency of leading characters seems,
even on an external view of history, to disappear as their
number multiplies, there arises a general activity of stimulating and stimulated elements, presenting the appearance of
organic growth.

§ 5. Now the more the wholly incalculable disturbances
caused by free individual minds are in the end outbalanced
by the opposing invariableness of that human nature which
always remains the same, and those conditions of earthly life
which are always alike, the more are we entitled to inquire
for universal laws to which the historical course of things is

subordinated. The assumption of their existence is not incompatible with the idea of a plan by which history is guided. For though such a plan presupposes a unity of history, involving the condition that each member of the whole series can occur but once, and that no two are interchangeable, yet it may be that the above-mentioned similarity between all the subjects of human history, and the parallelism between the forces operating upon them, may produce resemblances between the course of one individual stage of development and another, while if we take the whole series we find that these resemblances are gradually repeated on higher and higher levels, and are thus really specially distinguished one from another. However, the attempt to mark out these resemblances according to general historical laws is very much impeded by the difficulty of determining the transforming influence which the peculiarity of each member of the series has on the course which we should expect to be taken by those events with which he is connected, if we were guided by the analogy of other examples. Hence, though history is so much extolled as the teacher of men, but little use is made by men of its teachings. Every age thinks that it must regard the peculiarities of its wants and its position as new conditions which abrogate the applicability of those general points of view that are due to the reflection of previous ages. And, indeed, many historical laws which have been spoken of are of very doubtful validity, and are hardly transferable from one period to another. They are often only applicable when all the conditions of the individual case from which they have been abstracted are restored; and when that is done they cease to be laws, and become mere descriptions of that which has happened under certain circumstances, and which we are by no means justified in expecting to happen again under similar circumstances. This inexactness appears in all cases in which people, without being able to go back to the separate effective elements of a complex event, attempt merely to discover the final outcome of the course of events, by a comparison of experiences in the gross; the inexactness can only

be avoided in these cases in the same way as in other cases.
We want a Social Mechanics which can enlarge psychology
beyond the boundaries of the individual, and teach us to know
the course, the conditions, and the results of those actions and
reactions which must take place between the inner states of
many individuals, bound together by natural and social
relations.   Such a psychology would furnish us, for the first
time, not with graphic pictures of individual stages of historic
development and of the succession of the different stages, but
with rules which would enable us to compute the future from
the conditions of the present ; or to speak more exactly, not
the future from the present, but a later past from an earlier
past.   For even in the construction of ideals it is best not
to be exalted above measure ; we shall never bring any such
mechanics to so great perfection as to be able by it to sway
the future ; it will be enough if it enable us to explain the
concatenation of past occurrences when they have occurred,
and if with reference to the future it establish probabilities,
action in accordance with which is wiser than any other
course.

Now it is natural that we should first seek to establish the
rule of such universal laws within short periods, during
which we may regard the whole sum of conditions upon which
the course of events depends, and which we cannot analyse
exhaustively, as an unknown factor which remains almost
invariable.   And here men think they have discovered that it
is only where our view is bounded by a strictly limited
horizon that the appearance of freedom and indefiniteness is
presented to us ; that if in dealing with events, we take large
numbers and wide surveys, we find that not only does the
physical life of mankind proceed with well-established regularity in life and death, in the relative numbers of both
sexes, and in the increase of population, but that also the
manifestations of mental life are determined by universal
laws, even to the number and nature of crimes committed in
equal spaces of time.   Not indeed by immutable laws ; for
just as there is a slow change in the sum total of unknown

circumstances by which events are conditioned, so also there is an alteration from time to time in the formula which expresses the law of their occurrence. There is nothing, however, to prevent our conceiving of these very alterations of laws as themselves subject to another and more comprehensive formula, since the changes of that sum total of conditions on which these laws depend are due almost entirely to the effects of those states of human society which themselves come and go according to law. If by the method of taking large numbers it has been made out at what age, on an average, great poets produce their greatest work, what is to hinder us from seeking to discover, not only how many remarkable men of every kind (expressed either in whole numbers or in decimals) appear in every century, but also how in the course of thousands of years this proportion alters according to some law ? We may easily imagine how in this way all kinds of formulæ may be arrived at, expressive of the acceleration and breadth and depth and colouring of the current of historical progress—formulæ which if applied to particulars would be found to be utterly inexact, but which can yet claim to express the true law of history as freed from disturbing individual influences.

Very closely connected with this way of regarding the matter is one of the very worst of all the views which banish freedom from historical development. That veneration of forms instead of content—itself one of the most dangerous errors to which our thought is liable—which is vindicated by the view alluded to, could not be exaggerated in any more senseless way than by the final acceptance of a mere realization of statistic relations as the aim and the informing Idea of history. He who, following oriental Pantheism, believes not only that he encounters, as a matter of fact, in the order of the world, an eternal alternation of genesis and dissolution, but thinks that he may also regard this form of occurrence as being itself the most profound meaning and the true secret of reality—he can at least give himself up with misty feelings of enthusiasm to the awful and exalted pleasure which the thought of such a course of events produces in us. He who after any other fashion believes that he finds in history nothing but the rule of an iron necessity, must hold that this is in itself full of meaning; he seeks to find this meaning in some kind or other of justice, according to which the content and nature of any condition of things being what they are, allow and demand the effect which takes place. To such a concatenation in thought, the motives of which at least are reasonable, the mind may conceivably sacrifice the idea of its own freedom if it finds in this scheme no place for it. But on the other hand, it would be an instance of unparalleled perversity to see the guiding ideals of the order of the world in the establishment of regular numerical relations, or in the fact that events happen in accordance with such relations. And yet here I am not altogether beating the air, and my fear that even this attempt—the attempt to make us thus believe in such " shadows in the cloud " and nothing else—will be essayed, is not quite without foundation. For we do actually meet, not infrequently, with what is the beginning of this very error. It is with some pride, and not without something of the thrill of awe which may accompany the discovery of an ultimate mystery, that people caricature careful investigations (the value of which we do not depreciate), declaring that the tale of yearly crime is paid by mankind with greater regularity than that of governmental imposts. It is plain that in saying this they think they have affirmed not a mere fact resulting from unknown conditions and changing as these change, but a fundamental law which with mysterious power can always find the means of its realization, and work itself out whatever may be the opposition of unfavourable circumstances.

This erroneous view will indeed hardly be put forth as a doctrinal assertion concerning the meaning of history; but it secretly disturbs just judgment in the matter by causing a confusion of thought, and this the more easily because it is not equally wrong with regard to all departments of events. For among those phænomena of human life which show such regularity in their recurrence, we may certainly regard some as being subordinate ends of the cosmic order, or merely means to the realization of higher ends, and that will hold of them, to a certain extent, which we denied to be of universal validity. Most of such phænomena, however, may be compared to the impeding friction which, though it is no part of the designed performance of a machine, must yet always bear a certain determinate proportion to the size of the machine as long as the work of this can only be accomplished by mechanical means. But it is worth while to investigate a little further the insignificance of the extent to which this additional determination does away with existing difficulties.

The equality of numbers of the two sexes may certainly be reckoned among those arrangements of Nature in which we see means designed for the attainment of the higher ends of life. But as even the causes are unknown which in any particular case determine the sex of the child, so, much more, are those circumstances unknown which determine these causes (that lead to different effects in the different cases) in such a way as to obtain the unvarying gross result. The logical rule which directs us to anticipate that diverse possibilities, when there is no actual reason why one should occur more frequently than the others, will all be realized with equal frequency in the future, is no doubt for us a necessary subjective maxim—and we have to regulate our belief in the probable future occurrence of these cases by this maxim, for the sake of practical ends; but it contains no shadow of explanation concerning the mechanism of those conditions by which the equal frequency of two events is really brought about in the cases in which it happens. And we get no help from our general presupposition that the very possibility of all reaction is based upon an essential and inherent connection between all existing things. This presupposition does indeed provide us with a general formal reason for expecting that anything which happens in one part of the world will react in accordance with some law on every other part thereof; but—just because it seems so unquestionable that all things in the universe are connected with one another—we only remain all the more at a loss to explain the particular and favoured connections which are closer and more effective between some portions of the world than between others, and upon the presence of which each individual determinate event must depend. It therefore continues quite obscure by what determinate arrangements mankind comes to form a complete whole of such a kind that a preponderance of one sex which has accidentally happened here, calls forth there, simultaneously or subsequently, a counterbalancing increase of the other sex, the external conditions of life being so very dissimilar, and we being entirely destitute of any idea of how the necessary action and reaction could take place. And yet not only does the fact exist, but we are doubtless justified in considering that in it (if in any case whatever) one of Nature's ends is attained—an end for the fulfilment of which preordained means will not be wanting.

The course of the spiritual life of society is still more obscure. We believe that from the number of actions of a particular kind observed in a certain period which has just elapsed, we can conclude to a certain number of similar actions in an immediately succeeding period of equal length, only because the sum total of natural and social conditions, upon which they depended in the former case, alter but slowly, and in short periods imperceptibly. But where such change occurs spasmodically, we do not expect that a forecast made in reliance upon the past will be applicable. Still this caution does not remove all difficulty. Even the modified statement would be fully justified only if we could regard the sum of unknown conditions as a compelling force which would itself command a definite result in a definite time ; which further, finding the total resistance opposed to it to hold always a similar relation to its own magnitude, would be capable of exercising in every unit of time one and the same fraction of its energy ; which could then moreover always make actual use of this capacity by ever seeking and finding, like the pressure of a compressed fluid, the points of non-resistance, wherever those may be ; and which finally, for every portion of the result already produced would lose a corresponding portion of its potential energy. Now in the case before us, how many of these conditions are given ?

Let us take as an example offences against property. The evils of the existing distribution of goods in a society have active force only in as far as their pressure is felt. If then we make not poverty but the feeling of want our point of departure, can we say of this active force that there corresponds to it as its natural effect a certain number of thefts without any regard to the total amount of unlawful gain ? If it further happened that in a certain condition of civilisation, this power always encountered equal resistance, what would be the explanation of the fact that it always finds for its exercise the same number of favourable opportunities, and that these should always be presented to persons incapable of resisting them ? If, on the other hand, we suppose that there always occur a great many more opportunities than are taken advantage of, and that the numbers of those accessible to temptation are equally in excess of those who actually offend, it becomes only the more difficult to understand how the number of offences already committed can so restrict the number of those yet to come as to cause the attainment of a definite sum total. So the connection of events which produces uniformity in the numbers of such actions, is altogether unknown to us.

Just as little are we satisfied by the numerous attempts to make the validity of such laws harmonize with individual freedom of will. If (as has been done) we regard the commission of a certain number of offences as an inevitable necessity imposed upon society, it does not help us at all to add that this necessity only necessitates the actions but does not predetermine the agents. If human freedom cannot get rid of the sum total of offences, the fact that the particular agents are not predetermined does not leave individuals free—the only thing that still remains doubtful is, whose unfreedom will be taken advantage of next? It has been said that if an insect were to creep over any part of the circumference of a circle drawn with chalk, it would see all round it nothing but irregularly distributed molecules of chalk, though for an eye that took these in all at once, from some distance, they would be arranged in the regular definite order of a circle.   If these dots were beings endowed with souls, it might be imagined that taken separately they had scope for free choice of their position in the circle, while taken altogether they were bound to contribute to the formation of a predetermined outline. We reply that if an orderly arrangement of many elements actually exists (for the circle has been drawn), it is indeed easily intelligible that this arrangement can only be fully taken in from particular points of view.   But the unorder of the elements when looked at from other points of view, is not by any means the same thing as the freedom of those elements.   All those dots of chalk are perfectly fixed in such relations as are necessary for the structure of the whole ; they all lie in a narrow ring-shaped zone confined both internally and externally by a bounding line that has no breadth.   How they are grouped within this zone is, as regards the form of the whole, to a certain extent indifferent, and it is just to the extent of this indifference that they are indeterminate.   Now if the dots were living beings, this comparison would only teach the simple truth that they had freedom of action in those directions in which nothing had been fixed by general laws ; thus if it chanced that such a law required in any society a certain number of thefts, the agents would be free not with regard to their thievish resolutions, but with regard to whether for instance their thievish exploits should be accomplished on horseback or on foot.

The dislike with which we hear of laws of psychic life, whilst we do not hesitate to regard bodily life as subordinate to its own laws, arises partly because we require too much from our own freedom of will, partly because we let ourselves be too much imposed upon by those laws.   If we do not find ourselves involved in the declared struggle between freedom and necessity, we are by no means averse to regarding the actions of men as determined by circumstances ; in fact all expectation of good from education and all the work of history are based upon the conviction that the will may be influenced by growth of insight, by ennoblement of feeling, and by improvement of the external conditions of life. On the other side, a consideration of freedom itself would teach us that the very notion is repugnant to common sense if it does not include susceptibility to the worth of motives, and that the freedom of willing can by no means signify absolute capacity of carrying out what is willed—either of the carrying it out in conflict with the obstructions of the external world, or of that other and internal carrying out by which the will suppresses the opposing movements of the passions. Therefore not only the possible objects of men's endeavours, not only an idea of the means to their attainment, are suggested to the mind by a number of stimuli involved in the culture of the individual and of society, but also that effective strength of the free will by which it withdraws itself from being determined by passionate impulses, is dependent upon the collective culture of society. Hence there would certainly be no irreconcilable contradiction between the assumption of freedom of will and the other assumption that the sum of active conditions which operate in any given state of society, hinder to a certain degree the effectiveness of all free action, and produce a pretty uniform amount of mere instinctive action.

It would notwithstanding still be wholly incredible that the struggle of will and moral consciousness against all these obstructive elements should be as exactly predetermined with regard to its result as those statistical laws indicate. For the fact is that these laws do not measure at all that which we should expect to be so predetermined. Such laws originating for example in a comparison of tried and sentenced offences presuppose that the number of crimes which become known bear an unvarying relation to the whole number of those committed, and of this primary assumption no proof that is by any means cogent is possible ; indeed, if they are designed to prove anything with regard to human freedom, they must further show that also the number of crimes committed bears just as constant a relation to the number of those which have been resolved upon or prevented, or have miscarried, and indeed to the whole multitude of more or less serious temptations that have arisen in the recesses of men's minds. Not only do they not do this, but deeds of murder and manslaughter being counted by the hundred, there are grouped together under those class-names cases of the most various degrees of moral turpitude, the mere number being no criterion of the sum of evil committed in a given time by a given society in any direction. Only that such evil being a kind of friction inseparable from the life and progress of society, we may assume this sum to be connected by some definite law with the amount of movement in any society; but this would by no means hold of the mere number of cases in which the incidental ill effect takes tangible form under definite heads of crime. Therefore even if the constancy of this number should be confirmed by a fresh appeal to experience, we should still have to regard it as a fact of which we do not comprehend either the mode of production or the significance; we should never think of regarding it as an historical law in the sense of a predetermination of that which is to be. However, the fresh appeal itself (which has been quite recently made) convinces us of the extreme overhastiness with which the statistical myth has been built up from deductions which cannot be relied upon. We have yet to obtain from exacter investigations the true material for more trustworthy conclusions—material which should take the place of the statistical myth above referred to.

§ 6. The investigations of which we have been speaking referred only to limited periods of time. The succession of longer periods markedly different in historical features has seemed to reveal not less definite laws, which I may here pass over more briefly. They are of interest only in as far as they have reference to the individual tendencies of human life, which we shall have to consider later; the more widely they attempt to formulate the progress of humanity, the less real explanation do they generally contain. Thus one man talks of a law of uniformity in development, another of its sharp contrasts ; others prefer the trinity of thesis, antithesis, and synthesis. It seems clear that all these are not modes of occurrence to which events are bound to conform, as if there were in the mere forms themselves something which it were worth while to realize. Rather in as far as they have any real existence, they are ultimate forms which appear as social action and reaction progresses, from causes for which we have yet to seek. If we attempt this search we shall find that the significance of such laws is partly very unimportant and partly not of demonstrable universality. Thus it seems hardly worth while to decorate with the name of a law of uniformity the very simple observation that the culture of a later period is commonly a further development of the impulses received from preceding periods ; at the most it is only useful as emphasizing briefly the limiting condition to be found in the fact that the actual transmission of what already exists must precede further development. For historical progress is not (as people sometimes fancy) to be compared to a miasma that hovers in the air and seizes humanity unawares, either all mankind simultaneously, or particular sections by turns ; it has always taken place only within that narrow circle where favourable circumstances permit the regular transmission of attained civilisation, and of efforts directed to the relief of permanent wants ; and it has only spread as far and wide as geographical conditions, accessibility of countries, facility of communication, density of population, and multifarious intercourse between men in war or peace have given occasion.

The law of contrast that people sometimes, without any difficulty, allow to have validity at the same time as the law of uniformity, without drawing any boundary line between the conflicting claims of both, is not less simple. Speaking broadly, it only applies where simple forms of life, which in themselves admit of unbroken uniformity of existence, are in any way disturbed, and men's minds have become agitated by the longing for new satisfactions. Then their inventive power produces peculiar forms of civilisation, corresponding to the

momentary wants of the people and the mental temper of
the time, without satisfying in an equal degree all the
wants of human nature. The longer and the more fully
any such characteristic civilisation has stirred up, satisfied,
and exhausted all the receptivity towards it of which men's
minds were capable, and the more widely it has set its stamp
upon all external social relations and customs of life, the
more sensibly do men feel the pressure of its one-sidedness;
and the more vigorously do there come into prominence
those spiritual pretensions (still fresh and unsatisfied, and
seeking to impose a different mode of life) which this one-
sidedness had forced into the background. But the articulation of any civilisation of long standing forms a whole
that is too far-spreading and too widely rooted for newly
arisen tendencies to overcome it in all points, and to set
up easily in opposition to it a new and different and
consistent philosophy. Generally the influence of such new
tendencies is disintegrating and destructive; it is only after
a long interval that a new system is established—a system
that is not now the opposite of that which preceded it,
because the time that has elapsed between the two has
smoothed down the more extreme contradictions. In reference to individual departments of life we see more clearly
the need of change which impels the human mind not
only to continual removal of narrow and one-sided arrangements, but also to an aversion for truths that have grown
old. As one gets tired of a good garment that one has
been wearing for a considerable time, and finds that another
which has been long laid by seems to have a wonderful
charm of restored novelty, so the satiation of one side of
our spiritual nature produces a burning thirst for just as
one-sided satisfaction in another direction; and not only so,
but there comes in addition a general inclination for para-
doxical return to long-forgotten standpoints, and thus moods
and opinions are kept in a continual state of fluctuation.
Steady development belongs almost exclusively to those
sciences which are capable of practical application in

ministering to our wants, and in which unrestricted change of "modes of thought and points of view" would produce painful consequences. On the other hand, men's views of life, the tone of society, artistic ideals, opinions concerning what is supernatural, views of history, taste in the enjoyment of Nature, and forms of religious worship—all these are subject to the influence of constantly changing moods—sentimentality or noisy activity, prophetic enthusiasm or realistic moderation; and it often seems as though the most profound penetration were shown in seeking the truth just where no one suspects it, that is in errors which the previous generation had succeeded in refuting. Thus there arises the alternation of characteristic forms of civilisation in history, and thus we understand how it comes to pass that in the course of progress not all the several charms of life, on the exclusive development of which earlier times may have expended their whole strength, can be preserved and handed down in equal vigour; on the contrary, they often have to be sacrificed altogether to other requirements of human destiny, on which succeeding times rightly lay stress. This surrender of previous gains is explicable to us as a result of human weakness; that it is not merely a partial failure of historical progress, but also an essential feature in the course that this progress must take, according to its very meaning, is an assertion which we can only regard as resulting from that perversion of thought, which undertakes to justify everything that actually exists.

More strange is it that not only the forms of civilisation but also the torch-bearers of civilisation change as history goes on. Not only are mankind as a whole never found moving forward together at the same stage of progress, but also the nations—at least those of antiquity—which have blossomed into civilisation have without exception sunk back from the summit they had reached to varying depths of barbarism and commonplace. Men are certainly in too great haste if they found upon these facts the historical law that each nation, like each individual, has its life, in which strength first increases and then decreases;

and it is still worse if, supported by this comparison, they
venture to pronounce sentence on the future of nations which,
having passed through some phase of their culture, are seen to
be making new and tentative efforts.   It is not clear either
what reasonable signification this growing old of nations can
have for the plan of history, nor what is the inner connection
by which, as a mere matter of fact, it is universally brought
about.   Since in an individual man it is one and the same
organism to which must be referred all impressions from without, and all reactions of its own activity, we can understand
how in such a case there may be certain relations between the
acting and reacting parts which would necessarily make that
summation of experienced states result in the gradual alteration
and disintegration of him who is the subject of them.   But
why the vital strength of a nation cannot always remain
vigorous seems from this consideration only the more
obscure; and certainly it does remain vigorous in those who
for centuries have gone on in the monotony of some simple
civilisation.   The growing old of nations is plainly not
included in the idea of a people as a predetermined necessity
of development; and where it takes place it is the result of
particular conditions of life, due not only to the peculiarity of
the stage of civilisation which has been arrived at, but also in
part to external circumstances.   Nature strives to furnish
afresh each new generation with the old capacities of the race,
and ever to present anew vigorous and unspoiled subjects for
further development.   She does not altogether succeed;
bodily vigour and mental power may diminish through the
fault of a dissolute past; even without such fault, long
habituation to some definite form of national culture may
gradually transform the mental dispositions of a people in
ways which we are unable to trace, and make it difficult to
find a new equilibrium of healthier conditions of life when the
internal corruption of that culture works out, and causes its
disintegration.   But nowhere do we find justification for
assuming that these national diseases are incurable, and that
when one flower has faded a second cannot follow.   If the

nations of antiquity have not fulfilled such a hope, the reason was that not only did their culture become disintegrated through its own inner deficiencies, but also their national integrity was broken up by the destructive conquests of enemies more robust than themselves. That any national civilisation may flourish, it requires both political power and material wealth; but when the general condition of the world does not allow of the reinvigoration of its power, or when the opening of new roads of commerce and the abandonment of the old ones dries up the previous sources of wealth, and innumerable incitements to industry fall away, then the nation which any such fate befalls will seem to pine and dwindle incurably. Yet it will be capable of reviving to fresh life if the wheel of fortune makes a new turn favourable to it.

We must add to these considerations yet another question concerning the forces operative in history. Shall we attribute the similar elements which are found in the customs and legends of different nations, to transmission from one to the other; or shall we regard them all as indigenous productions which having sprung up anew, again and again, have everywhere assumed similar forms on account of the essential sameness of human nature? No one will really doubt that, taken broadly, both views are to be accepted to some extent; the real difficulty is where to draw the line between the claims of the two. Both modes of thought have often tried their strength on one particular example, the legend of a flood, which is spread far and wide among peoples very remote from one another. Riverside valleys subject to frequent and considerable inundations have been the homes of all the earliest civilisations; nothing could seem more natural than that this supreme danger threatened by the elements should be everywhere recorded in national legends. It is not quite so easy to explain, without supposing transmission, the great, although not absolutely uniform, similarity of the particular details with which legend fills in the history of the occurrence; hence people have been inclined to believe in a common origin of the different Asiatic accounts, and to assume that these were subsequently varied. But the American Indians, too, relate the same story; it was surprising to find that in one of their legends, Tespi—the man who was saved from the flood like Noah—when the waters begin to abate, sends forth first a bird of prey; this bird does not return, because it is feasting on the dead bodies of the drowned; then Tespi sends out other birds which also do not return; it is only the humming-bird that comes back with a leafy branch. The correspondence is striking enough to make one suspect communication, perhaps at a very late date; but at the same time the whole character of the legend is so thoroughly Indian that its being of native growth is not in the least improbable, and if chronology would permit we should perhaps be more inclined to think of Indian traditions having influenced the Mosaic account than of the converse having happened. So it still does not seem improbable that even such striking coincidences may have arisen independently in many unconnected mythologies.

And yet I confess that I regard with mistrust the unrestricted generalization of this way of judging. It is true that the natural surroundings of all nations are pretty much the same; but it does not follow so clearly that the impressions produced by them must, on account of the sameness of men's mental nature, everywhere lead to the same estimation of events, to the same trains of thought, and, finally, to the employment of similar artistic and figurative expressions. The points of view from which men, notwithstanding their human likeness, may regard Nature are manifold enough; the possible impressions produced by the same event may vary infinitely with mood and circumstances; the direction which may be taken by the course of thought that they stir up is incalculable; every correspondence that goes beyond the most inevitable deductions from facts seems always to require an individual proof of having arisen without communication or transmission. Appeal has indeed been made to general psychological laws in accordance with which the impression produced by facts, the reflection following the impression, and the final expression by figure and comparison, must be connected together; it has been attempted to interpret the course of all human fancy by a kind of general symbolism supposed to produce similar embodiments of similar thoughts in the most diverse mythologies; but here too the question recurs, Are we not, in the cases which this assumption seems to confirm, mistaking the effect of secret transmission for a proof of independent correspondence ?

The general scope of tradition in history is difficult to estimate. The very existence of complete and flourishing civilisations is forgotten in lands which were their home, and only a fragmentary remembrance of them preserved in the records of neighbouring nations, and for us great spaces of past time are wholly blank. On the other hand, isolated features (neither the most important nor the most common) of earlier civilisations have been saved amid the general wreck, and reappear among the most different nations. Our nursery tales contain echoes from the very earliest antiquity ; the same fables that exercise our own reflection in youth were once told in India and Persia and Greece ; many popular superstitions of to-day have their root in heathendom. With regard to much of this we know how it has been preserved and communicated, with regard to much we do not; and hence we not only learn to appreciate the great amount of transmission which has gone on imperceptibly, but we also remark that (as in all ruins) it is not always that which is the most imposing and the strongest and the most coherent that has been preserved, but that very often individual fragments of what was once the common property of mankind—fragments which look strange in their isolation—may unquestionably be dispersed among the widely differing civilisations of later nations.

CHAPTER IV.

EXTERNAL CONDITIONS OF DEVELOPMENT.

Common Origin of Mankind—Assumption of Plurality of Origin—Variety of Mental Endowment—Guidance of Development by External Conditions— Geographical and Climatic Furtherances and Hindrances—Examples of Peoples in a State of Nature.

§ 1. HISTORY, in the sense of a coherent development, connects but few sections of mankind. The west of Asia and the seaboard of the Mediterranean were the only places in which during thousands of years varying forms of civilisation followed one another, each transmitting to its successor its own gains and impulses to fresh progress. Outside this focus of civilisation innumerable other nations have either gone on living again, century after century, the common life of their kind and nothing more, among favourable or unfavourable surroundings, or they have perhaps struck out particular forms of development, but without connection with the favoured nations, and without contributing in any essential way to the further progress of these when they came into contact. Hence if we take a survey of history, it is presented to us not under the image of a single stream embracing all mankind and carrying them forward with steady action and reaction, though with different velocities, in the same direction ; it rather seems to us as though various currents flowed from various sources, remaining long without any reciprocal influence—until now in our own age all nations begin for the first time to be brought within view of one another, and the way begins to be prepared for a universal reciprocity of action between the different sections of mankind.

Even classic antiquity had this impression of the condition

and fate of the human race when political conflicts and the
curiosity of travellers brought to light upon the narrow stage
of the then known world many peoples differing widely in
appearance, language, and customs.   In this impression there
was nothing that seemed strange to the mind of antiquity, to
which human existence appeared to be merely a production of
the great mother, Nature, coming forth from her infinity and
returning to it again ; in the view of antiquity the numerous
races of men (destined merely for the passing joy of life, and
not for the accomplishment of tasks of eternal significance)
may have sprung each from the soil of its native place, without any original connection, and as manifold witnesses of the
inexhaustible fertility which Nature displays in her productions.   It is only where some individual race in the course
of its social development has acquired a sense of the lasting
connection of its members, that national tradition seeks to
strengthen this feeling by the supposition of a common origin;
but the thought of a comprehensive unity of mankind was so
far from these times that if two nations were found to have a
connected origin, it was thought to be quite a discovery, just
because it could not be in the least presupposed.   It was
Christian civilisation that first developed with decisive clearness the thought that all nations made part of one whole, and
that evolved from the concept of the human race the concept
of *humanity*, with which we are not accustomed to contrast a
corresponding concept of *animality*.   For the name humanity
expresses just this, that individual human creatures are not
mere examples of a universal, but are preordained parts of a
whole ; that the changing events of history which men experience are not mere instances of the similarity or dissimilarity
of results which spring from similar or dissimilar conditions
according to the same universal laws of Nature and of life, but
sections that have their place in a vast coherent providential
governance of the universe, which between the extreme terms
of creation and of judgment allows no part of what happens
to escape the unity of its purpose.   While Christianity
developed this conviction, it at the same time connected it

with the Hebrew account of man's origin, in which early and cognate views (strange to classic antiquity and far above it) had prevailed, without, however, having got rid of all the narrowness of exclusively national conceptions.   The wish to hold all the numerous races of men together by the bond of likeness in kind and species, became intensified to the desire to trace back their origin to a single ancestral pair.   Even this duality seemed to the Mosaic record too wide a beginning; according to this record the mother of the human race came forth in a wondrous fashion from the one father of us all, who was himself made directly by the hand of God.

The beauty and religious depth of thought from which these representations sprang will never fail of their effect upon our mind; but if the necessary development of human imagination involved a representation of the beginning of our existence under such figures, then it may be doubtful whether the Mosaic picture reveals an historical reality, or whether it can only be justified as affording satisfaction to an inevitable craving.   The doubts which have long assailed this interpretation of our primitive history justify the brief consideration which we now subjoin.

If the human race has really descended from one pair, what moral results would follow from the fact, and at the same time become impossible if it were denied?   In the course of propagation the splitting up into plurality by which the unity is succeeded is as much a fact as the unity itself.   Hence as long as we are in the habit of making historical facts the sources of moral commands, the second fact would bind us to division just as much as the first one does to unity, and indeed even more so, since the plurality increases as time goes on; and it is the future and not the past in which the theatre, or at any rate the objects of our action, are to be found.   On the other hand, if mankind arose from many unconnected beginnings—being however, as it now is, such that the different races, endowed with capacities similar yet not altogether the same, can only find full development and perfect satisfaction through the reciprocal action and reaction of all upon all—even if this were the case, should we not be equally justified in assuming that the moral destiny of men must be fulfilled by the union of all in one humanity ? Undoubtedly we should : men would still be brothers in the same sense in which they would in the contrary case ; for since they certainly are not brothers in a literal sense, the name signifies merely the recognition of that spiritual organization which is given to all of us alike, and of the worth of that personality which we have to reverence even in its most insignificant form. According to these facts, which are actual, and in as far as they are actual, we have to regulate our conduct ; but we never have to regulate it according to uncertain historical circumstances which, perhaps *have been*—the reality of which would not in the least increase the imperativeness of our obligations, while a successful refutation would necessarily plunge into confusion the mind which had based its sense of obligation upon them.

But among the things that *have been* we deliberately reckon that singleness of origin, supposing it to have been a fact. The more earnestly we seek the unity of mankind, the more must we desire that that which we find should be real and living and eternally present ; for him who only seeks it in the first pair, the unity must always be something that merely has been. For the influence of this unity has nowhere continued to operate in history. It has not held mankind together, and has neither insured to them as a whole a steady common development, nor to the different branches knowledge concerning each other; scattered abroad, in parts of the earth's surface most remote from one another, the different nations have passed their life, each unacquainted with the existence of the rest. But, in fact, wherever any of them have early come into contact, we find national hatred existing as the guardian of national peculiarities which no race is willing to sacrifice for the sake of another; the earliest times are filled with incessant conflicts of races, even of those whose actual relationship could be historically proved ; as one wave of the sea makes way for the next, so in this wild tumult one nation after another has been swept away.   So little has that assumed community of origin worked in the outward destinies of the human race; and it has been just as little active in men's feelings.   In the most ancient times a foreigner was regarded as altogether without rights ; it is only very gradually, and in proportion as history gets further and further from the beginning, that there are developed ideas of humanity as a whole, and of the regard which we owe to its representatives.

A glance at these facts leads very naturally to the question, Should we not place the unity of mankind in the future as an end of action to be sought after, rather than seek it in the past, where it can never be more than an ineffective and ornamental beginning of our existence ?   What is it that we should lose if we had to sacrifice a unity of beginning which subsequent progress has everywhere contradicted ?   It certainly would not be difficult for poetic fancy to imagine a chain of events which would exhibit man's original lapse from unity as a significant part of some secret purpose in the divine governance.   But while we fully admit the worth of the religious thoughts that can be embodied in such representation, we should yet, when they are put forward as history, require proof of their truth independent of the proof of their significance.

The assumption of originally distinct races of men, differing mentally as well as in bodily formation, each arising in a region suited to it and attaining the kind and degree of civilisation which its capacities made possible, has not unfrequently been opposed to the theory of mankind's original unity as corresponding more naturally with the view which history presents to us.   This assumption has been set forth in various forms, of which each has its special interest.

It has been found necessary, in the first place, to distinguish two great families of mankind, the active family of white men and the passive family of coloured men.   It is supposed that the latter, dreamily patient and inert, loving home and inaction, possess nothing of the ever active inventive restlessness which is the heritage of the white race; that it is this latter race alone which, impelled by the spirit of progress, has spread over the world in all directions, stirring up, educating, subduing, and supplying the sluggishness of the coloured nations with germs of civilisation which they would have been incapable of producing themselves. Indeed, the latter have been compared to the monocotyledons of the vegetable kingdom, those grasses and reeds which growing in countless multitudes give a green hue to uninteresting landscapes, with their monotonous luxuriance; the white races, on the other hand, which alone produce individuals of historic importance, who are of account taken separately, are compared to the class of dicotyledons which, and which only, produces trees with their picturesque individual forms. How easily this comparison — by reference to the vast pine forests of the North and the isolated palms of the South—might be so elaborated as to give it quite another meaning! We should learn that the external conditions of habitat and climate may degrade even dicotyledons to a homogeneous crowd that is counted only by thousands, and that even monocotyledons may under a favouring sky develop to forms which excite our admiration. Though we may, however, admit provisionally that this bifurcate view, without being applicable in detail, yet expresses on the whole a real historical fact, still we consider that it is illogical if it thinks that it can hold fast the unity of the human race, while it separates a branch *ab initio* useless from the only fertile one, by a chasm greater than that which generally exists in Nature between two species of the same genus, supposing neither of these species to have been influenced by culture and discipline.

According to another and more self-consistent theory which gives up the bond of a common origin, the different families of men sprang up independently of one another at different spots on the earth's surface; besides the Caucasians perhaps only the Mongolian race being indigenous to Central Asia, whilst burning Africa produced the Black man, and America fostered the Red man from the first, the islands and coast lands of the South Sea and the Pacific Ocean having been gradually peopled from some unknown centre in the neighbourhood of the Sunda Islands. This view again has, so far, neither been able to prevail, nor have others prevailed over it, it has not even succeeded in defining exactly the content of its own doctrine. For neither has it been conclusively shown that the different races could not have sprung from one root, nor are the difficulties which stand in the way of a wide diffusion of mankind while yet in a helpless condition, so great as to prove the necessity of the isolated origin of each nation in its own native place. On the contrary, there still come within our experience many facts which establish the possibility of migrations to great distances by land and sea, even under the most unfavourable circumstances. But on the other hand, there are wanting a sufficient number of clear indications concerning the actual process by which mankind was divided into unlike sections, and concerning the paths by which they were actually dispersed over the earth. There seems so far no prospect of the discovery of one single primal language; the similar elements of civilisation which are found among nations separated by great distances from one another may indeed to some extent point to early intercourse and communication of thought, but cannot prove the common origin of those between whom the intercourse took place. Reasons and counter-reasons being so evenly balanced, it must be left for the future to decide whether the assertion of an independent origin of different races deserves to be accepted; on the other hand, however, the content of the assertion itself has remained hitherto somewhat indefinite, on account of the uncertainty which exists as to the number of primitive races which should be assumed, as to the way in which these became mixed, and as to the degeneration which occurs to a limited extent. The choice of the five races above referred to, was perhaps arbitrary; it is possible that others may have just as much right to be brought forward;

and it is just as arbitrary to consider, as this view generally does, that the appearance of the different races should be conceived of as an almost simultaneous creation of all mankind ; it may be rather the case that each race belongs to a particular geological period. If this were so, those which we know may have been preceded in the very earliest ages by many others of which we know nothing, either because they have left no traces behind them, or because the monuments which testify to their existence are buried in the soil of the great continents of Asia and Africa, the palæontological investigation of which has hardly yet begun. The present state of science does not allow of any decisive judgment on these matters ; our views are kept in a state of continual fluctuation by unexpected discoveries which throng one upon the other, and which we shall not be able to interpret with certainty until their increased number has made their connection clearer. Sometimes we seem to get glimpses of an immense vista rousing vague anticipations, extending to prehistoric times of our species of which we have at present no knowledge ; sometimes these avenues through which we had had a glimpse seem to shut again, and the far-reaching views which they opened before us, close in, leaving nothing but representations of trivial events that have taken place within the short historical period which we know. At such times it is useless to insist upon having, at any price, some decisive answer ; the only thing that is of use is to look steadily at the various possibilities, and to forecast the consequences which a future confirmation of any one of them would have for our philosophy as a whole.

This we have attempted, and though we believe we have ascertained that the original unity of the human race is not one of those thoughts the truth of which is necessary for the satisfaction of our soul, on the other hand we by no means share the hostile feeling that we so often see displayed in disputing this unity, which after all may possibly be a fact. Mankind would really lose nothing whatever by the establishment of the one-origin view, in which (though it is by no means indispensable) long ages have believed and rejoiced ; and just as little would they gain anything if by proving their dispersed and plural origin their fate should be externally assimilated more closely·to that of the grass of the field, the blades of which we cannot count so as to form a unity.   We cannot share, we can only understand, that hostility ; it will very naturally arise wherever a mistaken zeal for certain forms of conception (in which religious truth is supposed exclusively to reside) attempts to settle without reference to the verdict of science, certain questions which ought undoubtedly to be submitted to scientific judgment guided by observation.   This zeal, while it injures science, gains no advantage for itself ; for since it cannot avert the coming results of investigation, it will at last find itself in the disagreeable position of having to regulate its faith according to the discoveries of the hour. It would escape this fate if it were more clearly conscious at the outset that the real treasures of faith are independent of any special forms of the historical course of events, and above all cannot be exclusively attached to any one form in particular.

§ 2. With the assumption of a plural origin of mankind is commonly combined the other assumption of original differences of endowment of the different races.   This combination finds special contradiction in a view which, without caring about the original unity of mankind historically considered, believes that it must hold fast the unity of kind and the original equality of all men's capacity for civilisation.   According to this view, to trace back the varieties of development which individual nations have experienced to innate and permanent differences of their bodily and mental organization is, as it were, a shortening of the arm of science, the business of which rather is to explain the divergence of mankind from one another as regards their way of life, by pointing out all the natural and social influences which have worked upon the originally similar natures of men.   It is certainly not necessary to lay special stress upon the truth which is unquestionably involved in this demand upon science; it is perhaps more to the purpose to observe that even this correct principle of investigation may be carried too far.

The principle is fully justified in the investigation of all those phænomena which we may still observe recurring and reproducing one another, in connection with and conditioned by other phænomena, but cannot on the other hand impose on Nature itself any greater simplicity of origin than Nature really has. He who assumes a peculiar vital power for the phænomena of organic life may soon convince himself that the supposed activity of this power is determined on all sides by physical conditions, and here it is that it becomes necessary for him to explain the consequences of this power as results which spring from co-operating causes in accordance with the same universal laws to which those external influences are subject. But he who traces back all plants to one primal plant, and all animals to one primal animal, on the one hand diverges from experience which presents no facts that require such a supposition, and on the other hand affirms a process which, even independent of experience, is by no means necessary on general grounds. For that Nature itself, like human thought, should in working progress from the imperfect to the perfect, from the simple to the composite, from the homogeneity of the universal to the manifold variety of the particular, is only a probable conjecture, in as far as Nature requires to utilize the imperfect, simple, and homogeneous, in producing the more complex perfection of the individual. Where we cannot assume that this real and solid advantage accrues from Nature's following such a path, we have no reason to attribute to it as necessary the same course as is taken by our own thought in observing, comparing, and classifying the perfected reality upon which it comes to work. Nature does not make first things and then their attributes, first matter and then the forces inhering in it; just as little is it necessary and self-evident that it should, in the first place, embody in some single primary form the universal generic concept under which our thought may subsequently group together a plurality of species, effecting later the historical development of the species from this primary form, by the supplementary influence of further conditions. Rather does Nature (not destitute of the necessary and essential means of direct production) undoubtedly begin with all the rich variety of creatures which are equally possible as embodiments of the universal.

Though, however, we vindicate the possibility of this assumption, yet we do not recommend that it should be thoughtlessly employed. The principles according to which we must estimate mental life in general would above all things never permit us to deny altogether to certain races certain mental capacities, attributing to others an exclusive possession of them. The most general laws, according to which the events of mental life happen, are valid alike for men and animals to such an extent, and the connection between the different forms of mental activity is so close and many-sided, that if we take two kinds of mind which, in relation to many departments of this activity, present such a perfect similarity as we find in the different races of men, we shall see that these two cannot well be separated (with regard to any other department of the same activity) by the existence or defect of some innate capability. If there is a difference of original endowment, it is without doubt to be found in that which most strikingly distinguishes from one another even individual members of the same race; that is, in disposition and not in the nature and mode of operation of the mental powers in general, which are common to all. By disposition we mean that particular combination of impulses by which the mental powers have the direction of their activity determined, as well as their ends, and the vigour, variety, and constancy of their exercise; and all this may be different in different races, partly on account of inherited peculiarities of organic formation, partly on account of original idiosyncrasies of mental nature. And it is this which also determines the amount of attainable development, according to the direction in which it predominantly guides the interest of the whole mental life, and according as it makes the mind more receptive towards relations of things the observation and treatment of which must inevitably lead it further, or causes it to find satisfaction in occupations and forms of life which contain no living germ of progress. The attempt to extend higher civilisation to nations which have hitherto remained wholly strangers to it, has been frustrated much more by the difficulty of arousing lasting interest in the benefits of our own culture than by want of the insight necessary for understanding it.

Now whether in point of fact these varieties of disposition are irremovable differences of original endowment, or whether even they are but the accumulated effects of constant external conditions, is a question which historical experience so far can hardly decide. The nations which hitherto have had a long term of life constantly reveal to us, amid all the striking changes of civilisation which they undergo, a tenacious persistence of peculiar characteristics which often merely change the scene of their manifestation. However estimable may be the attempts made to explain the varieties of human development by reference merely to the effect of those circumstances by which life is conditioned, they have not hitherto enabled us to dispense with the assumption that there are special variations of generic human nature which were given as the material upon which those conditions had to operate in the various branches of mankind.

Our judgments, however, on all such questions are never based altogether upon scientific grounds, but depend also on unspoken moral needs and doubts. Even the aversion to allow, in the case of mankind, the possibility of original variety, depends on a reason of this kind. If different creatures differ from one another by an altogether distinct generic stamp, it is not thought surprising if some lack the advantages of others; it seems that each should be contented with that with which his nature has provided him. The different races of men, however, seem to be as near one

another as possible, on account of the predominant similarity
of their most essential characteristics, and what is of still more
importance, they are capable of a common life of reciprocity
in work and enjoyment; here a difference of mental endowment, which would be not merely a difference, but also a
gradation of more and less, would seem like an unjust
abridgment, as regards the less gifted races, of the means
necessary for the fulfilment of that business of human life
which is common to all men—means upon which all therefore
seem to have an equal claim.   This consideration is not
without weight; indeed, we willingly allow that the supposition
is inexplicable that a race of men may be for ever hindered
by some concealed defect of their organization from reaching
a civilisation for the attainment of which they seem to possess
all externally cognisable capacities; yet the enigma referred
to is so often suggested to us by history, and in forms so
obtrusive, that our failure to understand it must not lead
us even in this case to deny its existence.   For still
more inexplicable than those inherent natural hindrances to
progress are the numerous cases where, on the one hand,
individuals of the most favoured races remain far below
the general level of endowment of their race, and on the
other hand, whole nations are for centuries hindered by
external circumstances from attaining a degree of civilisation
by no means beyond the reach of their actual mental capacity.
If we can neither alter nor deny this fact of the tyranny
of external conditions, we have just as little reason for
regarding the limiting power of original natural endowment as
inconceivable.

§ 3. The aversion to allow innate differences in the dispositions of nations is not obscurely connected with that
increasingly popular mode of thought which would dispense
with all predetermination of future development in the human
mind, and would leave it, as selfless and plastic material, to
be altogether formed by external conditions.   As men's taste
varies in art, so it does also in the way of looking at history;
and although we may easily admit that each of the opposed

views may be justified within certain limits, yet in neither case are there wanting unjustifiable transgressions of the limits of validity. Early historical idealism often proceeded as if the spirit of man dwelt upon earth devoid of wants, and in an atmosphere of purest ether, and as if, following nothing but the impulses of its own nature, it produced the melodious succession of its own significant developments, unabridged by any opposition, and only incidentally condescended to the prose of mundane circumstances in order to transfigure them to a reflection of its own splendour. In opposition to this idealism, the realism of our own time asserts, and rightly, the stimulating, restricting, and guiding power which those same mundane circumstances exercise upon the uncertain and want-laden nature of frail humanity. But neither is it necessary that the idealistic view should be held in the exaggerated form which we have just noticed, nor has the opposed view either the duty or the privilege of carrying its necessary and well-founded warnings to the point of that mephistophelic scorn with which men sometimes dispute the efficacy of all nobler springs of development—of all except such as depend on imperative need.

Only plants are destined to live by the favour of external circumstances, without reactions that bear the stamp of living activity, and accommodating themselves to moderate change of such circumstances, but helplessly succumbing to the effects of greater change. Hardly anywhere in the animal world is the satisfaction of natural wants attained without some individual effort on the part of those satisfied, and in some kinds of animals this activity is so developed as to have become an instinct of co-operative labour. But in these very operations—to which, indeed, the animals are stimulated by outward impressions, but the mode of which is determined by themselves in accordance with an unalterable impulse of their nature — the agents seem to us less free and active than in those less striking performances by which they (within narrow limits) modify the operations referred to in accordance with changing circumstances. Mankind, not being directed and restricted to one definite occupation by any similar prompting of Nature, see before them the whole earth as the sphere of their activity, and must find out by manifold experience that which Nature itself has imprinted on the souls of brutes; that is, necessary ends, efficient instruments, and the most useful division of labour.   They do not come to this task unprovided, but they come without having received an impulse from Nature to use these means in some one direction only; with an unbiassed sensuous receptivity and a capacity of bringing received impressions into reciprocal relations of inner connection, they are forced by their wants to seek out unknown sources of satisfaction.   It is certainly true that instinct leads more easily to the satisfaction of wants than the reflection following experience, which errs in a thousand ways; but every error that fails of the end at which it aimed, finds in its path truths which would have remained undiscovered if an infallible natural impulse had led the mind straight to its goal.   Hence even the simplest occurrences of daily life develop in the most uncivilised nations at least much skill in using the properties of things according to general physical laws, even though these laws (as *e.g.* those of equilibrium or of the lever) may never become to them explicit objects of consciousness.   And all knowledge thus gained, just because it did not exist as innate endowment of the mind, but came to be formed through contact with things, and thus was matter of living experience, is felt to be, as it were, the production of our own activity.

At first the individual may, by a kind of superficial and hasty construction, gain shelter and support from his immediate surroundings; but a growing society, with its ever-increasing multitude of wants and the fresh demands which it develops, finds itself obliged to appropriate also, by well-considered division and combination of its powers, the less obvious utilities of natural products.   By bringing large extents of land under permanent cultivation—by connecting distant regions for exchange of commodities—by increasing the value and convenience of their immediate surroundings through manifold elaboration of the material they have appropriated—the members of such a society are ever transforming larger and larger extents of the earth's surface, as it were to another and more home-like Nature, to the scene of a life of social order. In proportion as this happens, the dependence of man upon the elementary material world that surrounds him becomes lessened ; he becomes accustomed to have most of his wants satisfied, not by direct application to this external world, but at third hand by the co-operation of social labour ; he with his ideas, feelings, cares, and plans belongs far more to this new and secondary order of things, to the concatenated whole of human society, than to primitive Nature, which, while it is the basis of his existence, seems ever to withdraw further and further into the background.

It is, then, only when this early progress has transferred the centre of existence from the natural world to the artificial world of society that distinctively human life begins, and the possibility of its further development. For the inferior animals are as capable as ourselves of enjoying without preparation that which created Nature freely offers ; it is the distinctive task of humanity to create for itself the world in which it is to find its highest enjoyments. In order to do this, mankind had to restrict the manifold possibilities of existence and action contained in the course of events and of our own impulses by thoughts of what is right and fair ; they had by multiform elaboration to transform the productions of Nature, together with the soil which brought them forth, into a world of commodities, the attainment, preservation, and use of which combined the dispersed powers of individuals to a connected whole of occupations depending upon one another ; out of the social contact which occurs in the course of Nature, and which is increased by the dawning community of labour, a community of life had to be developed that sacrificed many a liberty which Nature allows us, and imposed on itself many an obligation for which Nature gives no reason. So the human mind reared above the tangible sensible world of that which actually exists the not less complex ramifications of a world of relations which ought to exist because their own eternal worth requires their realization. And this whole artificial order of life which man had to create in addition to created Nature has only in isolated moments of despair, due to conscious failure, seemed to the minds of men to be an arbitrary and revocable structure of their own invention; on the whole, social order has appeared to the human mind to be an altogether irrevocable natural necessity.

Now plainly we could not expect the construction of this spiritual cosmos to result from spontaneous evolution of the human mind without the stimulating and guiding influence of various external causes. It is not as though there were in us a natural impulse to progress which, like a pressed spring, strains to the rebound; but like bodies that cannot of themselves quit their state of rest, or that, when once set in motion, exert force upon the obstacles which they encounter, so the impulse to progress in the human mind, and the direction which it will take, are due to the velocity of the evolutionary movement in which the mind is already involved. It is certainly true that we may regard the ideals of the Beautiful, the True, the Good, and the Right as an innate possession of our soul, but only in the sense in which it is generally allowable to use this expression *innate*. They are not presented to our consciousness from the beginning as distinct representations, but only after our moral nature has been stirred on many occasions to approve or reject various modes of action do we think of them and recognise in them the principles according to which our judgment had previously proceeded. And if they had actually been innate in our consciousness as living representations that were in it from the beginning, of what value would they be for our development? The comparing activity of thought may, indeed, separate the feeling of reverence with which everything that is beautiful, right, or good inspires us from these particular occasions of its exercise, and attach it to the general notions of beauty, right, and goodness; but as none of these ideals has reality except when embodied in definite examples, to which it gives significance, so neither would any of them have for us a definite content if we were not able to recollect individual instances of its realization. And if we think of the worth of all ideals as united in the unfathomable wealth of the divine nature to a blessedness which was before the world began, we are used to expect, even of this nature, that it should manifest itself in the creation of a world of varied forms; it is this which seems first, by means of perceptible relations between its different elements, to give to the hitherto formless universality of the ideal content an abundance of definite characteristic manifestations, and thereby a fulness of reality which it had seemed to lack while it remained self-contained. The human mind cannot accomplish such a mysterious creative act; it would have been vain to set it the task of excogitating with inventive fancy from the formless tendencies in which its innate ideals must have consisted, a multitude of cases of their possible realization. For in fact our whole existence, historical and unhistorical, is occupied in receiving the influences exercised upon us by the circumstances of the material world in which we are placed, these circumstances constituting the stimulations which first call forth our activity, the guiding conditions which fix the possible aims and content of our being, and finally the material on which we are continually impressing, in individual and limited forms, the image of the ideal. Much that is beautiful, much that is good, much that is just, admits of realization; but only such beauty, such good, and such justice as may be contained and comprehended in this world of sense and the relations subsisting between its perishing inhabitants. He who desires to see realized the beautiful in itself, or the good and the just as they would be in themselves without the realization being at the same time occasioned and restricted by some actual relation for which it is valid, desires something as contradictory as he who wishes that the speed of any movable object, the movement of which is only made possible by that contact with the ground which at the same time retards its motion, should be accelerated by the total removal of this resistance.

Human development, then, requires occasioning causes, and historical idealism is wrong if it takes offence at this dependence of the progress that has been made upon conditions which the human mind has not devised for itself, but has found upon its path.   But the conditions of the begizning and of the progress of culture are not quite the same.   When mankind have actually reached any stage of civilisation they are generally urged on by the impulsive force of this civilisation itself to a further stage, in which, with an already awakened consciousness of ends to be attained, they seek the satisfaction of yet unsatisfied wants; on the other hand, the first steps towards development can only be made possible by the favour of natural circumstances, by which also their direction is in the first instance determined.   No rules of justice are conceivable at the beginning of civilisation without direct reference to objects of need or enjoyment, the use of which must be determined by a consideration of various claims; but it is Nature that by her niggardness or bounty must fix the worth of the productions which become the first objects of the dawning sense of justice with its regulative activity.   There can be no individual development to a fixed order of life without connected labour; and it is external Nature which, by the special character of its products, and by the necessities which it imposes, determines how great a share of life is to be devoted to the task of mere self-preservation, and how much is to be left for enjoyment; determining also by the kind of work which it allows or requires, whether the human mind shall be pent up in a narrow circle of ideas and activities, or shall be spurred on to a life of many-sided and inventive action.   The development of artistic and religious views depends only to a smaller extent, and not in its most essential features, upon the immediate impression which natural surroundings make upon human imagination; yet mediately the influence of these surroundings is great; for upon the geniality, ease, and elasticity of customary life, and the forms of intercourse which they allow of, depend the variety and vigour of that mental reciprocity within a society which is indispensable for the formation of any coherent philosophical views. As, finally, the individual's sphere of thought becomes impoverished when it lacks the stimulating interruption of intercourse with others, so also for the progressive civilisation of nations, it is necessary that their different modes of philosophic thought should be brought into contact, and perhaps also, according to an oft-conjectured natural law, that there should be a physical blending of races not too alien from one another.   Where the nature of the country affords means of communication that facilitate this reciprocal action between nations, we see the civilisation of mankind fall earliest into a course of coherent progress ; on the contrary, it has remained for thousands of years in the same uniform condition in regions whose boundaries, inhospitable and difficult to pass, have restricted the inhabitants to a constant employment of the same means to their ends and the same conditions of life.

All these thoughts, even in that more detailed presentation of them which we must here renounce, have long lost the charm of novelty—they have lost it since the time when the modern realism of historical investigation began to make the dependence of progress upon the geographical conditions of the earth's surface a favourite subject of its inquiries.   Meanwhile, however thankworthy these may be, they do not quite suffice to explain the capricious course that history has actually taken.   Mankind cannot accomplish that which is impossible ; hence we see how it is that a country of which the poverty and ruggedness make life difficult can produce no indigenous culture, but can only adopt one which has germinated and grown strong elsewhere.   The presence of favourable conditions in other places, however, by no means explains how it is that they are made use of.   The human mind is far from being so desirous of development from the very beginning as to be hurried away, by the favour of natural circumstances, to make all the progress which these render possible.   Men may for long periods of time use with careless indifference natural products which seem directly to suggest some definite application of their powers, without discovering this application ; not even necessity is the mother of invention in the sense of leading men generally to seek satisfaction of their wants by reflection which may be the herald of subsequent progress ; on the contrary, so great is men's natural sluggishness that, satisfied with warding off the most extreme misery, they will long endure the continual recurrence of sufferings which it would be by no means difficult to avoid by a moderately intelligent use of means which are actually at their command.   We deceive ourselves therefore if we think we see in favourable geographical conditions — the advantageousness of which is immediately obvious to our practised observation—an impelling power which without reckoning upon happy receptivity of disposition in men could force them to develop, as if by natural necessity, in some definite direction and at some definite rate. And least of all can the special colouring which growing culture has taken among different nations be altogether deduced from a corresponding speciality of external conditions. We must admit that similar conditions have produced different results, the germ of which must be sought for first in the historical lot of nations, and last in the incalculable aggregate of those inner springs of action which stirred their spiritual life and in turn helped to determine national destiny.

§ 4. If, without any pretensions to completeness in the enumeration of infinitely varied facts, we now take a glance at those nations whose life—either unhistorical, or if historical interrupted—will afford us no opportunity of considering them more in detail at a later stage, we shall find that their fate is partly, but only partly, explicable by reference to the circumstances of their external condition.   Without a certain density of population which brings men with their wants and claims, and their varieties of temperament and experience, not only into frequent contact but into lasting intercourse, both hostile and harmonious, the growth of higher civilisation among men is impossible.   It was but few climates that afforded to infant societies the favouring conditions necessary

for this—making life easy by spontaneous fertility of soil, by
a mixture of good and bad in the climate arousing wants
without making the satisfaction of them very difficult, and
finally (by the variety of the products and impressions which
it afforded) establishing a sufficient variety of mutually complementary occupations and dispositions.

The frigid zone cannot be like a home to its inhabitants, in
whom want does indeed rouse ingenuity in satisfying the most
pressing needs, but at the same time frustrates every effort
after beauty and fulness of life. Forced dependence upon
those few productions of a niggard Nature which it is possible
to reckon upon, makes the labour of preserving life difficult
and too much alike for all. One can hardly imagine what a
Greenlander's life would be without the seal. Shapelessly
huddled up in furs of seal and reindeer, and tied into the skin
covering of his kayak, a narrow pointed hunting boat capable
of holding only one man, he navigates the Arctic Ocean in
pursuit of the seal with inimitable skill; then he creeps back
into his winter hut, constructed of stones, driftwood, turf, and
skins, and feeds upon his greasy spoil, by the light of lamps
that are always burning, the moss wicks of which are fed by
seal fat; and the subject of conversation is a description of
the hunt, graphically given and attentively listened to—" Thus
he sat—thus he stretched himself out and threw the harpoon."
And in the happier future world which he supposes will be
in the depths of the sea, he expects a superabundance of
birds, fishes, seals, and reindeer; and it is only in his hope
that the short summer and sunshine of his present home will
there be continuous, that he betrays his sense of the climatic
burden under which he bends. This gloomy picture of a
miserable existence is pretty much the same for all the
northern coasts of the old world, amid local differences of
position and instruments ; these wildernesses have nowhere
been able to produce a higher condition of human life, and to
those races which by some unknown fate were driven into
them they have only left the remnants of civilisation attained
previously in more favoured abodes. The small amount of

subsistence furnished by a wide extent of country has every-where prevented the density of population necessary for the beginning of political organization ; all having to work at very similar occupations, and being separated from one another and from all foreign culture by the difficulty of intercourse, the scattered families have neither been able to advance to an educative division of labour, nor had they any motive for the development of social forms and ideas of right for the application of which no cases occurred.   Natural goodness of disposition and various mental gifts have not been sufficient to prevent men in this existence of constant bodily hardship from coming to regard the coarsest enjoyments of the senses as the only really good things in life.

The people of the South Sea islands, though in a graceful instead of a repulsive fashion, are really quite as backward. When they were first seen, happily disporting themselves in the sea with easy agility—behaving with hospitable and gracious sociality on land—passing away the time in dancing, round games, songs, and cheerful talk—not given to assiduous labour indeed, yet managing their small plantations with skill—hardly needing clothing or shelter, but showing taste in what they had—healthy, strong, and active, even their most aged men contented and good-tempered—when they were seen thus, they seemed to have retained a paradisiac condition.   A nearer acquaintance showed the dark side of this fair picture. The confined extent of the islands had indeed caused greater pressure of population and hence active commerce; but the fineness of climate had made work too little imperative, and the uniformity of weather and natural products had caused the lives of all to be too much alike.   The islands were too small to be the scene of great enterprises, and there was no large continent accessible, capable, by the foreignness of its natural features and its inhabitants, of giving to the minds of the islanders a stimulating enlargement of their intellectual horizon ; their isolation in the midst of the ocean could hardly develop anything beyond a peaceful and unprogressive existence.   But such a simple idyllic life is a defensible mode of existence only when considered as a temporary withdrawal from some familiar civilisation : where it is everything it is not an existence worthy of mankind. Where each individual brings into circulation as his contribution only the natural capacities of his kind, without having worked them out to an individuality which is all his own by some special labour of development, each will be esteemed as nothing more than a mere example of his kind, that may be used and worn out ; and the life of the whole, like that of a herd of animals, only with the higher mental characteristics of the human race, will in the end have no higher sources of enjoyment than those with which it is furnished by Nature. Hence neither science nor art nor morality has been developed from the not inconsiderable mental capacities of these islanders, and it is but few who have lived through that idyllic life in innocence— with all the prevailing good nature and friendliness, it was possible for societies to exist, formed for the indulgence of immoderate sensuality and pledged to child-murder, and there was wide-spread cannibalism. So, like other fair products of animate Nature, they sported together with all the gracefulness of their kind, only to devour each other at last.

There was added another source of misery unknown to the polar nations. It was said that at an early period there had come from the north-west, from the mythical island of Bolotuh, where the gods feed upon ethereal swine, a light-complexioned race which spread over the islands and supplanted and enslaved its original inhabitants, who were of darker colour. By innumerable intermarriages, the external differences of the races were obliterated ; but a strict system of caste was kept up, not founded upon differences of culture and hardly upon differences of occupation, but upon degrees of purity of descent. This system gave to the nobility, the Eries, rights without duties, and to those of lower rank duties without rights ; to the former immortality and deification after death, while to the latter it did not even allow a human soul during life. Jealously guarding their rank among themselves, the nobles on the whole treated the people without

cruelty, although occasionally murdering these soulless beings
without hesitation ; and still more inexhaustible than the
arrogance of the Eries was the patience of the subject caste,
any of whose possessions a noble, by the *taboo* which contact
with him could impart, might appropriate to himself and
make it unlawful for the former owner to touch.   Secular
power was overridden to some extent by priestly influence,
as is the case with all uncivilised nations among whom
pretended mental pre-eminence is, on account of its greater
rarity, more highly esteemed than bodily vigour, which is
common enough; but here as in the north, this priestly
influence represented not moral truth but that superstition
which arises from a dread of the unknown powers of Nature,
and which, driven by an ill-regulated imagination into erratic
courses, has led nearly everywhere to a multiplication of
horrors but nowhere to any wise regulation of life.   So that
here we find subtle complications of social order, attractive
simplicity of life, and complete absence of all the higher aims
of existence combined into a whole that abounds in contradictions.

The vivifying contact with foreign nations, customs and
views, which the Polynesians lacked, was enjoyed in vain by
the Negro races and the Indians of North America.   The
shores of the Mediterranean beheld one after another the
brilliant civilisations of the Egyptians, Phœnicians, Greeks,
Romans, and Saracens; they were certainly separated by a
wide wilderness from the country inhabited by the Negroes,
yet for thousands of years active intercourse was carried on
notwithstanding this obstacle.   All this influence of cultured
nations, which certainly extended far into the interior of
Negroland, produced no civilisation among the black tribes,
neither the formation of great states, nor a dawn of native art
and science—at the most nothing more than some scanty
industries for the adornment of life.   The same passions which
move men everywhere, in Africa too caused wars and the
successive predominance of the various tribes, from very early
times; but whilst in the history of white men the dominion

of each great nation has been perpetuated in lasting and characteristic monuments, and has marked a memorable stage of social conditions, all these changes have been without result for the black races, and the tide of their national existence, after the waters had been disturbed for a moment by some unusual undertaking, went on rising and falling again just as they had done before.

In the explanation of this great historical fact, opinions are still found violently opposed to one another. The assumption that black men have less capacity of development is scarcely worthy of refutation, if it is understood in such an exaggerated sense as would justify the abomination of slavery. There has been sufficient experience, even under this unfavourable condition in America, to forbid us to regard a fixed limitation of intellectual endowment as a permanent hindrance to the development of the black races. It would be only in peculiarity of disposition which everywhere determines the force and direction of the application of mental capacity, that we could seek for conditions that have made an independent beginning of civilisation impossible for the Negro, and the appropriation of an alien civilisation difficult for him. To say the least, good nature, by which he is distinguished, is in the early stages of history never inventive; it is far more the evil desires of ambition and of unscrupulous egoism that nerve all the forces of the mind to attack, and induce men to search out every means of defence. White men have conquered the world, not by their superior morality, but by the obstinate perseverance with which they attacked all those who could only oppose passionate ebullitions and unconnected sacrifices to their merciless penetration and the consistency of their well-laid plans. The Negro's temperament gives no promise of any such results. Sanguine and changeable of mood, he is excited and diverted by every fresh impression, and is just as little disposed to steady labour as he is to pursuing chains of thought along those important intermediate links which do not charm by their own interest, and yet are indispensable for connecting that which is in itself more valuable. His warmth of heart makes him accessible to religious awakening; but from the unruliness of his imagination, even these feelings are more likely to be the source of acts of isolated self-sacrifice than of a course of life ordered in detail. In a temperament of this description there are, without doubt, many features unfavourable to an independent commencement of higher civilisation, but also some which are sufficiently favourable to subordination under powerful and originative minds, to justify us in expecting either an imitation of foreign culture, or a gradual indigenous development under the consolidating pressure of an intelligent despotism. But hitherto neither of these events has taken place. The incapacity of the negro state of Hayti to attain a condition of permanent order has certainly too many obvious causes in its hasty formation amid a population vitiated by slavery, to prove conclusively that all similar attempts of coloured men must be equally resultless, supposing they were made under more favourable circumstances, such as have hitherto been lacking. On the contrary, in Africa itself the existence partly of despotic and partly of more democratic polities, shows that an external formal regulation of society is not wholly incompatible with the genius of the race, only there lacks that content of life which alone is worthy of man, and is capable of high development by means of these forms. That the Negro did not borrow this content from European civilisation is explicable partly by reference to the hostile fashion in which this came to him, and partly by the too great violence of the contrast subsisting between the complex variety of this civilisation and his own simple way of life. We see the hard-living masses of the white nations retreating with a similar lack of receptivity before the culture of the higher classes, as though it were a manner of life belonging to a different species of animals, and living on according to their own fashion, which they can understand. Finally, that in their native country Negroes have never by any progress of their own developed germs of higher civilisation, may be to an important extent, though hardly altogether, explained by the geographical conditions of that country. We find these conditions partly in the enervating effect of the hot climate, which does not admit of the vigorous work either of body or mind that is possible in a more temperate region, partly in the natural fertility of the soil, which too easily affords satisfaction of the few wants which men feel in tropical countries, partly in the early age at which bodily maturity is attained, the period of education being thus abridged, and independent life allowed to begin too soon. Finally, we certainly find one of these conditions in the difficulty of carrying on communication over the unbroken stretches of the African continent, a difficulty to which it is due that the different tribes with their various fashions and customs (which, however, do not differ to any great extent) cannot come much into contact either with one another or with the views of men of different race. Whether that temperament which has made the Negro nations so little fitted to advance has resulted from these circumstances, so that under better climatic conditions generations which have had time to get rid of their inherited native temperament would be capable of much progress; or whether there is in their organization some impassable barrier to high development, which will compel them always to remain at a low level—these are questions which can only be decided by the future of the race itself. It would certainly be unfair to conclude from the past absence to the necessary future absence of historical development, and to seek the ground of this absence only in natural incapacity without having regard to obstructive influences; but when men (carried away by the certainly not inevitable assumption that all mankind are similarly organized, and by horror of the abominations of slavery) forthwith conclude that the Negro race will in the future reach that higher development which has not been attained all through the many centuries of past history—then, on the other hand, it seems to us that the conclusion reached is not convincing. With regard to morality, by which the laws of our future conduct are determined, this last assumption may be preferable, since it is one which cannot do harm. As far as the consideration of past history is concerned, the point in dispute is not so interesting; for an originally existent capacity, which has for thousands of years been so obstructed by unfavourable conditions that it could never attain development, is in an historical point of view no less a puzzle than an originally poorer endowment of the race would be.

§ 5. For the most part the Red men of North America have resisted European civilisation even more expressly than the Negroes, and have not themselves developed any that is of much importance, although their condition may have been better before their social relations had been altogether disturbed by the ascendency of the whites and their perfidy. The superior appliances of European civilisation, matured under more favouring conditions, have made North America a rich country; the densely-wooded nature of those regions, with lack of water in the west and cold in the east, put difficulties in the way of an indigenous civilisation. Our cereals were not produced, and the scanty crops of maize in the north did not lead to permanent cultivation of the soil; potatoes and the domestic animals of the old world were unknown, and the allied native kinds of animals not very easy to tame. But there was a superabundance of game, and the hunter's life, everywhere for the sake of self-preservation the primitive form of existence, continued here to be the sole form. This was unfavourable to civilisation in every way. Without other sources of supply, even the best hunting grounds could support only a few persons to the square mile; populations never reached the degree of density necessary for the development of society, and were kept from the stationary form of life and its educative influences. The tortures of hunger, which are depicted terribly enough in their legends, made the care of a family a burden; the noble liberality which the less skilful hunter expected from the more fortunate, and which the latter cheerfully exercised, deprived the unskilful of motives to greater exertion, and the skilful of that useful egoism which attracts to further enterprise by the pleasure which increase of gain awakens. The necessity that the men should always be ready to fight caused all the ordinary work of life to fall upon the women, while the poorness and scantiness of the goods which they had to take care of, afforded them no opportunity of making their womanly guardianship of much account. The wide dispersion of the population and ignorance of the use of metals prevented any great development of manufacturing industry. Restricted to the most childish modes of sticking and joining things together, even fastening the laboriously cut stone heads of their axes into the cleft of the wooden handles with strips of leather or fibres of plants, they busied themselves only in the weaving and plaiting of ornamental stuffs, which was an affair of patience and of simple taste. The only things on which they expended labour were arms, ornaments, and the most indispensable implements, being generally more inclined to suffer than to take much trouble for their own relief. The custom of shedding blood in the chase, and the unavoidable disputes concerning the boundaries of hunting grounds—a serious matter for people to whom hunting was a bitter necessity of existence—gave them a fierceness of disposition which led to mutual destruction. Thus their life went on without historical progress, like the movement of a man who is swimming against the stream—movement which suffices indeed to keep him up but does not carry him forward.

They are not universally ignorant of the sources of their ill-fortune. "Do you not see," said one of their chiefs, "that the white men live upon corn, and we upon meat—that meat requires more than thirty moons to come to maturity, and often fails—that each of the wondrous grains which they plant in the earth gives them back more than a hundred-fold—that the animals upon whose flesh we live have four legs to escape with, while we have only two to follow them with—that wherever the grains of corn fall, there they remain and grow—that for us winter is a time of toilsome hunting, for the white men the time of rest? That is why they have so many children and live longer than we. Truly before the cedars of our village die of old age, and before the maples of the valley have ceased to yield sugar, the race of the corn-sowers will have supplanted the race of the meat-eaters, unless the hunters make up their minds to sow too."

It was but few who did make up their minds.   The free life of the wilderness has often had a permanent attraction even for Europeans ; the real benefits of our mode of life pass away almost untasted even by many among ourselves, being buried under a multitude of small restraints ; to the Indian especially the latter must have been more obvious than the former.   And strange enough in other respects is the temper with which he meets foreign influence, whether this temper is an original endowment of the race or results from the long-continued action of the circumstances of his life. The silence, the reflective humour, the immovable pride of the red warrior may have been produced by the hunter's life, with its requirements of patience, attention, and foresight, of presence of mind under surprises, of fortitude under suffering ; but both the customs and legends of the Indians show an inclination to fanaticism which does not seem to result altogether from these habits, nor to be due to the mere brooding of an unoccupied mind.   " Ah, my brother," said a chieftain to his white guest, " thou wilt never know the happiness of both thinking of nothing and doing nothing ; this, next to sleep, is the most enchanting of all things.   Thus we were before our birth, and thus we shall be after death.   Who gave to thy people the constant desire to be better clothed and better fed, and to leave behind them treasures for their children ?   Are they afraid that when they themselves have passed away sun and moon will shine no more, and that the rivers and the dews of heaven will be dried up ?   Like a fountain flowing from the rock, they never rest ; when they have finished reaping one field, they begin to plough another, and as if the day were not enough, I have seen them working by moonlight.   What is their life to ours—their life that is as nought to them ? Blind that they are, they lose it all !   But we live in the present.   The past, we say, is nothing, like smoke which the wind disperses; and the future—where is it? Let us then enjoy to-day; by to-morrow it will be far away."

This is not the language of stupidity. On the contrary, if it were presented to us in Greek verse, we should admire in Latin commentaries the fineness with which it derides the perversity of the white men of whom so many in their haste to get forward lose all remembrance of their goal. But it is certainly true that this mode of thought could not be favourable to the development of social life, as long as it held its ground and was supported by the combined influence of all surrounding circumstances. Meanwhile the attraction southwards which animated the migrations of the European nations, moved these tribes also in ancient times, and whilst North America saw no indigenous political development, we are dazzled by the splendid spectacle of the kingdom of Mexico in the central region of the great continent, and numerous ruins bear witness that there once flourished other centres of civilisation, of which the history is lost to us.

The mild climate of Mexico, where the land is narrowed between two great oceans, the four-hundredfold return which maize not unfrequently yields, and the banana, which in a given space of ground produces twenty times the nutritive matter of wheat, here admitted of a settled population increasing till it became very numerous. Life was divided between work and leisure, and division of labour became possible; wants grew with the production and offering for sale of goods; there came into existence nearly all the arrangements which conduce to social intercourse and luxurious enjoyment of life. A disposition to cultivate flowers began to appear in addition to careful husbandry and orcharding and culture of medicinal herbs; the weaver's art produced magnificent garments of gorgeous colouring composed of cotton interwoven with feather-down; gold ornaments and precious stones faultlessly cut might be put on before obsidian mirrors. At feasts the tables were decked with costly utensils, and these feasts were conducted according to a complicated ceremonial, and with all the adjuncts of civilised entertainment; the general tone of society was courteous, and the morality of domestic life (which was held in great honour) was marked by propriety and moderation and was a subject of instruction. The exchange of products was accomplished by means of markets held at fixed times. At these times in the large and populous towns, of which more than one seemed to the Spaniards to emulate Granada in its palmy days, many thousands of persons moved about among the various stalls which belonged to different trades and were arranged in orderly fashion, and in this busy mart there was wanting neither police supervision, nor a special Court of Justice that sat continuously for the settlement of any disputes that might arise.

According to the Toltekian legend, the founder of this civilisation was the hero Quetzalkohuatl, with fair face and long beard, who came to the country from some unknown and distant region, accompanied by many followers clothed in long garments. Whatever may be the historic kernel of this tradition, the limitations of Mexican civilisation seem to bear witness to its native origin. Quetzalkohuatl was said to have come over the sea; but the Mexican merchants, in other respects so enterprising, did not navigate the ocean; there had not come into the country from over the water any of the domestic animals of the old world, nor even the thought of taming native species; the bales of goods were conveyed by human carriers along the broad highways; our cereals remained unknown, maize being the only grain until the Spanish conquest. The Mexicans did not know how to obtain iron, they worked the land with implements of copper and bronze without the help of draught animals, setting not sowing the seed, providing for irrigation by dikes and trenches; finally, they did not adopt any of the modes of writing employed by earlier civilised nations, but developed for themselves a system of written signs. Thus none of the elements which are generally most easily communicated by foreign civilisation came to them from without, and we may regard their civilisation as the native development reached by the genius of the Indian race under favourable climatic conditions.

On the other side of the equator similarly favouring natural conditions enabled the seaboard country of Peru to attain a remarkably flourishing civilisation ; but the pastoral nomads who in the old world seem to have been the first to undertake the task of bringing several centres of civilisation into communication with each other, did not exist in America, and no intercourse took place between Mexico and Peru. On the other hand, in the great eastern half of South America the spirit of man was cowed by the overpowering might of natural phænomena. Monstrous rivers with resistless inundations, vast and trackless forests, the irrepressible vegetative vitality which causes every cultivated piece of land to be quickly overgrown with a rank luxuriance of weeds, the number of large beasts of prey, and the countless multitude of insects, winged or creeping, which speedily devour a whole harvest—all these hindrances still stand in the way of development in Brazil, notwithstanding the European industry which has long flourished there, and much more must they have frustrated the early attempts of isolated tribes.

If it were necessary to make this hasty sketch complete, Europe and Asia might increase the aggregate of unhistoric life by the addition of many nations who still live on in their old abodes with the same manners and customs which they had at the beginning of history. They would thus confirm afresh the impossibility of speaking of a past History of Mankind, since it is only among a small fraction of the human race that that connected series of events has occurred, which with an unwarrantable generalization we sometimes call the History of Mankind, and sometimes under the name of Universal History regard as signifying the development of all reality and the unfolding of the World-spirit. From the future, however, we may expect, as the best which it can bring, the diffusion of European civilisation over the whole earth. For the only native dawn of development of the coloured race in America was completely destroyed by the bloody hand of Europeans, before the time to come could decide what were its capabilities of further development; and no one will imagine that the Negro race, being everywhere exposed to the influences of European culture, is now likely to develop a special national civilisation. But the Negro has at least some reason to hope that his race will be perpetuated, while according to a very general opinion Indians and Polynesians are doomed by the very genius of history to die out before the higher race of the Caucasians. The truth is that those coloured races were reduced to such an extreme degree of weakness simply by the frightful cruelty of their white conquerors and the numerous diseases which they introduced, or which—from some unexplained causes— are usually developed when races of men that are widely different first come into contact. In the Middle Ages a similar fate befel European nations more than once; but they had time to recover, for there was not in their rear any race still more Caucasian than themselves, seeking with the same consistent cruelty—partly natural and partly doctrinaire—to execute upon them a supposed sentence of history. Where such a chance of recovery has been given to the coloured races, they also have begun to slowly increase again; where they are really melting away like snow, there are to be found, first and foremost, frightful secrets of European colonial government—but the fulfilment of an historic doom will be found only by him who counts every accomplished matter of fact among the necessary phases of development of an Idea that rules the world.

CHAPTER V.

Stationary Civilisation and Nomadic Life in the East—Semitic and Indo-Germanic Races—Ancient Greece and Rome—The Hebrews and Christianity — Character and Early History of the Germanic Nations — The Germanic Nations in the Middle Ages—The Characteristics, the Problems, and the Difficulties of Modern Times.

§ 1. IN the old world, too, we see how the beginnings of human civilisation depend upon the favour of natural circumstances. It is between the Yangtsekiang and the Hoangho, in the lowlands of the Indus and Ganges, in the plain that lies between the Euphrates and the Tigris, and in the valley of the Nile, that we find the nurseries of the earliest civilisations. Fertilized by regular inundations, in restraining and utilizing which men's powers were for the first time combined for the co-operative production of careful hydraulic constructions, these river-lands brought forth in luxuriant abundance the vegetable products that were sufficient for human support in those climates which by their mildness reduced all physical wants to a minimum of complexity. In China and India the yield of rice was far above a hundredfold; the quantity of fruit borne by the date-palms in Mesopotamia and Egypt was enormous; Herodotus extols the splendid crops of corn and barley in the Babylonian plains; he is silent, he says, regarding the wonderful growth of millet there, because he does not wish to be disbelieved. Such an abundance of edible natural products, besides which each country possessed also some special advantages, favourable to civilisation in other respects, allowed these countries to attain a density of stationary population which early led to a complex development of social relations.

The accounts given by ancient writers, and a consideration

of the monuments which have been discovered, equally convince us how early the civilisation which grew up in these
countries attained that perfection in the adornment and
regulation of the surroundings of man's life which we sometimes consider to be an exclusive privilege of the enlightened
present.   Of the dim shadow that in our thought is wont to
lie upon the gray and distant past, not much could have been
observable in that past itself; it was bright and noisy, and in
many places the externals of civilisation were developed with
a perfection which could only be attained in an age sensible
of having awakened to full consciousness, in contrast to the
unawakened life of the past.   Partly collected in large and
populous towns, clothed in garments of cotton, or silk, or
linen—sometimes simple, sometimes a marvel of taste and
splendour—these nations walked the earth with a most lively
susceptibility to all the grace and beauty of existence; the
habitations of the rich were devoid neither of the variety
of household furniture, which self-indulgent ease requires,
nor of the mere embellishments of luxury, and the thousand
charming trifles which imagination asks for the beautifying
of life; their social meetings lacked hardly any of those
means of amusement with which modern times are familiar,
nor was their intercourse devoid of that ceremony which
distinguishes human converse from the gregariousness of
beasts.   But all this brightness was not without its shadows;
on the contrary, even in those times, the splendid remains of
which we admire, men suffered under the pressure of the
same social evils from which in the later periods of history
they have never been able wholly to get free.

The fewer the indispensable necessities of life are, the more
easily they are satisfied by the spontaneous productiveness of
the soil, the more mildness of climate tends to make these
natural productions sufficient, and the less—in fine—general
civilisation (as yet undeveloped) requires provident care for
the future and for descendants: so much the more rapid will
be the multiplication of an impoverished population, who will
be forced by every temporary deficiency of their ordinary

sustenance, and by every unusual disaster, to offer their services to those who have property, each underbidding the other.   Even if it had ever been the case that a society, of which all the members had perfectly equal rights and claims, had shared equally in the means of production in a new country, the natural course of things, by the different increase of different families and a thousand other accidents, would soon have introduced inequality of fortune.   But it hardly seems that this ever has been the case, the first permanent settlements having apparently grown up under other conditions more adverse to equality.

Those favoured river-valleys of which the luxuriant fertility invited to steady cultivation, are in Asia separated from the inhospitable north by an extensive zone of steppes and pasture lands which, solely by their innumerable flocks of tameable and useful animals, afford support to a numerous population.   Men have dwelt here from time immemorial; pastoral tribes who still in many particulars remind one of the customs with which their most remote ancestors first appear in history.   Made hardy by the discomforts of their roving life, and brought up to warlike vigour, and many of them being tribes of horsemen, in ancient times, as now, they moved about as nomadic hordes among the settlements of fixed civilisation.   The chief towns of the latter were secured by impassable mountain boundaries from the continued repetition of petty attacks, to which perhaps they would have succumbed: but any considerable natural calamity which lessened the number of the flocks upon which the nomads depended, or any increase of population making richer sources of supply necessary, induced large bodies of the war-like shepherd tribes to make incursions into the countries of developed civilisation.

The history of Asia is full of the conflict between these two forms of life.   Often in ancient times have the rich lands of Western Asia been trodden under foot by hordes of mounted Scythians; the growing prosperity of China was threatened by Mongol attacks; the already highly developed civilisation of Egypt was subject for centuries to the assaults of the Hyksos; it was with the warlike nomads of Central Asia that there began that migratory movement which, after the fall of the West Roman Empire, initiated a new period of European history; and not much more than five hundred years have passed since there broke upon the eastern confines of Germany the last billows of that tremendous storm which the mighty spirit of Genghis Khan, supported by the united strength of all his tribes of wild horsemen, had brought upon the world. Thus the impulse which the unceasing restlessness of these nomadic races has communicated to the external destinies of mankind seems to be extraordinarily great; but on the other hand, in the history of civilisation no reminiscence of progress is attached to their name. In this region they have only made destructive incursions, and then have either sunk back again into their unhistoric existence or have fallen in with the civilisation of the nations with whom they mixed, without giving it any new direction. It was only the Arab nomads who were of another complexion—burning religious zeal transformed them with amazing rapidity into conquerors of a great part of the civilised world. Without possessing advanced native civilisation, they appropriated many elements of western culture with a receptivity due perhaps to their southern origin, and gave to that which they had appropriated the characteristic stamp of their own mind.

These occurrences of later times must have had their analogues also in the earliest historic ages. Most civilised nations, according to their traditions, consider themselves as settlers and not aborigines in the countries which they have made famous. In many cases they came with an already developed civilisation to these countries, and found them inhabited by aborigines who, notwithstanding favourable natural conditions, retained the savagery of their primitive condition. So the Aryan Indians, when they spread south of the Himalayas, drove out a native race of blacks, who retreated to the most inaccessible mountains of the Deccan; and so in Egypt some Negro race may have

enjoyed the first-fruits of the rich soil, though the development of its historical life may have begun with the immigration into the country of men of Caucasian race who
later regarded themselves as autochthonous; and traditions
concerning the settlement of the Mediterranean coasts are
full of the struggle between alien civilisation and aboriginal
barbarism. But the converse process has also occurred; it
has repeatedly happened that tribes from pastoral districts or
mountain regions, men of natural vigour and capable of
development though as yet undeveloped, have fallen upon the
more enervated inhabitants of the plains, and have carried on
in their own name the civilisation which the latter had first
established. It is not the more frequent but the rarer case
when those nations which have first expended their labour
on the soil of any country, have subsequently maintained
themselves in possession of it, and kept in the van of the
civilisation which its gradually unfolded resources have made
possible. These circumstances have been influential in the
formation of social order.

Tribes of hunters and of nomads are apt to develop at a
somewhat early stage of civilisation an aristocràcy of rich and
leading families; and just as naturally are they inclined to
regard mental endowment which boasts connection with an
unseen power, with greater awe than bodily strength and
warlike courage, which for them are quite in the ordinary
course. Nomad life offers but few inducements for developing
these differences of social consideration into really valuable
privileges; but in the transition to stationary life, the heads
of tribes and the priests have always drawn tighter the loose
reins which they held, and have succeeded by various means in
bringing the fertile land entirely into their own power, and in
compelling the great majority into their service as unpropertied
labourers or dependent tenants. Among nomads the interests
of all are too similar, and their simple way of life too readily
scrutinized by all for it to be easy for budding despotism to
make the individual members of the tribe permanently serviceable to its own ends; but a settled population involved

in a multiplicity of complicated relations, soon becomes unable
to take a comprehensive view of its own capacities and wants,
and the difficulty for each individual of reckoning with
certainty upon the intentions of others causes them collec-
tively to fall an easy prey to the narrow class-interest of the
few who understand one another.   So it came to pass that in
the most fertile regions, stationary life fell under the power of
the priesthood, and of an hereditary nobility belonging to
the order of chieftains; where the nature of the country was
favourable, the next step in advance concentrated the secular
power, which is always jealous of partners, in one person,
and produced the knitting-up of the spiritual power (which
is everywhere conscious that it can only be effective as a
combined unity) into an orderly system of strong corporations.
The inequality of splendid and wretched lots, which thus arose
in society, was finally only intensified when a conquering
nation oppressed the conquered with the right of the stronger
and the pride of nobler blood.

Hereditary callings are natural to dawning civilisation.
Partly with the object operated upon, as in the case of tillers
of the soil, partly with the instruction which coincides with
family education, where the transmission of knowledge by
schools separate from the home is as yet non-existent, the calling
of the parents is transmitted to the children; free choice of
some other employment is prevented both by the narrowness
of men's intellectual horizon, which embraces only that which
is familiar, and forces them to attach themselves thereto, and
by the natural jealousy with which not only the different
classes of society, but also the various trades, strive to keep
themselves exclusive.   These customs have, moreover, swayed
in many ways the civilisation of later times.   They occurred
in the dawning culture of Egypt and India; but it was only
in India that the contrast between the conquered race and the
native population (which here was even greater than in the
valley of the Nile), and, moreover, the influence of priestly
views, developed such customs into those irremovable distinctions of caste, which, while they made certain callings

obligatory, oppressed all lower castes with the graduated contempt of those which were above them. China alone never laid these fetters of caste and status on its industrial population, knowing no hereditary differences of rank and calling, and all being under general state guardianship; this was perhaps a happy incidental effect of the absence of religious fanaticism and warlike thirst for glory. It was only here that access to learning (though to learning of not much value) was thrown open to all, and instruction early diffused and favoured. In India there stirred an infinitely deeper intellectual life, that with its strange mixture of extravagant imagination and penetrating subtlety, embracing the secrets of heaven and the vanity of earthly life, affected only the favoured upper classes of society; in Egypt and Babylon science and writing, the laboriously developed means of communication, were in the hands of the priests. Common life lacked the stimulus which might have been given to it by the wisdom which was kept secret, and this in its turn certainly lacked quite as much the impulse to progress which it might have received from intercourse with the thought of the people. Industry was not backed by any knowledge worth mentioning of the efficient powers of Nature; it was facilitated by but few technical artifices, and animated by no spontaneous artistic impulse. Astronomy alone became early a subject of instruction, but it teaches only what happens and cannot be altered; a knowledge of mechanical forces which man may use for his advantage was still wanting; lucky discoveries might be treasured and transmitted, but no knowledge of the principles of mechanical action invited men to progressive improvement in practice. The want of instruments similar to our machinery obliged actual manual labour to be employed everywhere, with a disproportionate expenditure of time and strength, and however great might be the luxury of the wealthy, the growing increase of remuneration could not repay the arduous labour expended on the productions required. Artistic activity was soon drawn into the service of religion; and hence, and from love of splendour on the part of despots, it was stimulated to great works, though limited to certain established forms.   Some few of these, as waterworks and roads, were of general utility ; most, like the Pyramids of Egypt, and in the new world the Teocallis of Mexico, and enormous temples and palaces, only bear witness to the harsh oppression which, when there was no advanced knowledge of practical mechanics, extorted such prodigious results by lavish expenditure of human strength.

It is with varying feelings that we transport ourselves into these times.   As long as it is only their productions that are before us, we admire ; these seem to our imagination to bear witness to a mighty creative impulse in which all men with one accord must then have revelled.   If we consider the means by which it was all produced, then it seems to us that any state of society must have been unspeakably miserable which allowed the sorely oppressed majority to be tyrannically used for the satisfaction of the aimless fantastic vanity of a few, which abolished the natural equality of men by cruel distinctions, and restricted their activity by innumerable checks and hindrances.   But it may be doubted whether history would have made any progress if its beginning had been a quiet and peaceful sort of life in which every individual produced and consumed undisturbed whatever was necessary for the satisfaction of his frugal wants ; mankind needed to be made aware that their vocation is not the mere supply of physical needs.   The systematizing division into castes certainly restricted men, but then it also first brought into the world the idea of a vocation, and it taught men not to think that in merely being men they had attained the end of their existence.   The iron oppression of despotism used men as mere instruments, but it also was first to combine them together as members of one whole ; the extravagant pride of rulers dragged men away on expeditions that aimed at conquering the world, but this thought of the sovereignty of the world was perhaps the only way in which hostile tribes, still in conflict with one another, could be brought partly to the enjoyment of comparative prosperity by the attainment of external order and security, and partly to a feeling of the connection of all mankind—a connection which, as with a binding law, overrides the caprice and hatred of individual races. And finally, the petty restrictions with which priestly ordinances beset life in all directions have in the East given rise to and maintained in the most effective manner the feeling of a constant connection between earthly existence and a universal history extending beyond mundane limits. The school of this first stage of education was hard and bloody; but on the one hand the progress of mankind for a long time dragged on the same social evils under other forms, and on the other hand without such a school the beginning of civilisation is even harder to conceive than its progress. It was through it that there arose for mankind the first really valuable content of life; extolled by one, cursed by another, having for the great majority the imposing aspect of natural necessities of inscrutable origin, established social organizations captivated the imagination by the splendour of their monumental constructions, and the will by the force of their attraction.

This it is that we are accustomed to point out as the characteristic feature of the East and of its philosophy. The ordering of life which men established seemed to them—that is, the thing created seemed to the creators—to be a self-evident and unconditional necessity, and the freedom of the individual seemed to be swallowed up by the superior power of that universal the outlines of which each individual must help to fill in. Social arrangements were regarded not as historical and alterable human constructions; all seemed to bear the stamp of supramundane sanctity; whether the whole order of existence appear as in China to be an impress of the being and rule of an impersonal Supreme, the copying of which restricts all caprice of personal activity to a faithful following of ancient customs and transmitted wisdom; or whether, as in India, acquiescence in the melancholy condition of oppression was due to the mystic tradition according to which different sorts of men proceed from more and less noble parts of the deity; or whether, as in the pompous inscriptions with which the kings of Egypt and Persia used to cover the rocks, the ruler, as the direct representative of the Most High God, considered his commands to be binding on all the world.   And as each private individual was reckoned as of little account in himself, so it was even with these rulers; it was not as persons but as office-bearers that they stood at the head of humanity.   In the East, when the insignia of supreme power have passed from one person to another, obedience and submission have always been transferred along with them, apart from any fidelity to individuals. This sense of being embraced in a vast and predestined order was on the whole undisturbed by any spirit of disintegrating criticism; the vast extent of the countries, the difficulty of communication, and the want of means of intellectual intercourse prevented this feeling from being opposed by any flexible and progressive public opinion.   Customs and systems of thought were maintained unaltered by tradition; morality and secular law were not separated from religion and worship.   Great as was the division of industrial labour into distinct callings, in practical politics the most diverse governmental functions overlapped; general abstract points of view for the treatment of similar problems were not developed, and even in its most craftily contrived arrangements the oriental art of government (like the lives of individuals) shows a matter-of-fact simplicity which aims solely at its particular end, without any attempt at shortening the way by the help of general maxims.

Though this is the general character of the impression produced on us by a consideration of eastern nations, yet that impression must, of course, include many strong contrasts and counter-currents, since the men who lived there and then were in all respects the same manner of men as ourselves. The oriental character was not so wholly immovable and torpid as it seems to us at this distance of time.   The ancient civilised states of Asia were not without mental revolutions, which for us, indeed, do not materially alter their general aspect, but for the men who experienced them were just as much periods of active progress as European development is for us. Our attention is diverted from these circumstances by the consideration that but very few of them have been of service to the subsequent progress of mankind. Almost all those civilisations, shut up in themselves, passed through their various phases of development in isolation; China, on the eastern edge of the continent, was from the beginning out of communication with the rest of the world; India did indeed come in various ways into contact with other countries, but without any important effects; it was only Egypt and Asia Minor that gave to the West most of the elements of their civilisation.

§ 2. Only two great families of people—long in conflict with one another—the Semitic and the Indo-Germanic, have been instrumental in the further progress of history; and even of them many branches have diverged from the main line of development, some continuing the practice of old accustomed forms of life, some in course of time disappearing altogether. In ancient times the south of Western Asia, from the mountains of Armenia, belonged to Semitic races. And even if we leave undecided whether the primitive culture and the language of Egypt were attributable to them, yet the high development of Mesopotamia, the mighty Babylon, remains an early monument of their strength. From the narrow coastland of Phœnicia Semitic merchants went forth on bold and adventurous voyages to all the islands and shores of the Mediterranean Sea, and beyond the Straits of Gibraltar, and the traces of industrial settlements which they left behind in then obscure Europe may have guided in many ways the later civilisation of the Grecian world. When the rich cities of the little Phœnician mother-country had fallen from the giddy height of luxury and self-indulgence which they had reached, and had succumbed to an invading and hardier race, the colony of Carthage, the mistress of the Western Mediterranean and its coasts, long withstood in tremendous conflicts the growing might of Rome; when this struggle, too, was decided, and the secular power of the Indo-Germanic races was firmly established in Europe, the whole of the western world gradually submitted to the spiritual supremacy of Christianity, which took its rise and found its first advocates and ambassadors among a Semitic people. And even once again, in the Middle Ages it seemed doubtful whether the van of historical progress would henceforth be led by the still oriental genius of the Semitic race through the incursion of the Arabs, or by Indo-Germanic vigour, which had first attained full development in the West.

Whether the nations which now possess Europe were preceded by an aboriginal population of different race, we know not. Comparative philology teaches that the European nations are, with few exceptions, branches of one stock, which more than four thousand years ago fed their flocks in the favoured regions on the western slope of the Himalayas. One branch of this stock of Aryans, the " excellent," as they called themselves, gained possession of the land watered by the Indus fifteen hundred years before the commencement of our chronology, and about the same time another branch developed into a well-ordered and flourishing nation in the more westerly Iranian highland. India soon dropped out of the course of history in the isolation of its own fantastic development; on the other hand, the Iranian tribes succumbed to the attacks of their Semitic neighbours on the west, before the permanent supremacy of their race was established in the great Persian Empire. If we lack historical information concerning even this first division of the two tribes which were nearest to one another locally, and which likewise continued to be in language and thought most closely allied both to one another and to the parent stock, still more obscure are the times at which and the paths by which the migrations of others to the far west took place. The Celtic tribes which pushed on as far as the Atlantic Ocean, and hence were probably the earliest among those who immigrated westwards through the continent of Europe, have won no special place among the great civilised nations. Their development (in which at one time in their Gallic abiding-places they were certainly in advance of their

Germanic neighbours) was interrupted by the impulsive force of Roman civilisation; the remains of their dialects and customs are dying out. Later the Germanic immigration, and later still the Slavonic, reached Central Europe; earlier than this, the as yet combined Greek and Roman branches of the Aryan stock had spread over the Ægean Archipelago, the Hellespont, and the shores of the Black Sea, and split up into those two nations to which belongs the first brilliant instalment of European development.

§ 3. The great Asiatic civilisations have, it is true, developed many a treasure of knowledge, of order, and of beauty; but it was among the Greeks that mankind first opened their eyes full upon earth and heaven with that fresh, lucid, priceless awakedness of the whole living mind that we ourselves can feel and sympathize with. The various tribes of the Greek race lived through a long period of somewhat slow development, the beginnings of which are obscure, until the exertion of united strength, called forth by the pressure of foreign power, accelerated the onward impulse of that marvellous civilisation which, though its vital strength was soon exhausted, long continued to scatter its blossoms far and wide over the world.

As blazing suns may have been produced by the condensation of fiery vapour, so in Greece we see the immensity of oriental dimensions reduced to moderate and proportioned forms instinct with the most intense life. The theatre of development was a small district that could never boast anything like the number of inhabitants that an oriental monarch would have been content to rule over. Greece did not possess the fantastic wealth of alluring and terrifying wonders, which in the East had an enervating influence on organized energy, and amid which imagination ran riot; the nature of the soil— which yielded a good return to labour without being luxuriantly fertile—accustomed men to industry; a mild climate and bright atmosphere were favourable to fine physical development and to the training of the senses to accurate observation. The conformation of the country, which was

broken into deep valleys and numerous mountains, caused
small communities to be shut up together in the closest
proximity; the unequalled extent of coast-line and the
rich profusion of islands were favourable to intercourse
between the inhabitants, whilst here — as everywhere —
nearness to the sea was decidedly inimical to lasting union
under one government.   Thus did this land, a rare jewel of
terrestrial conformation, nourish many independent communities, within the narrow bounds of which the awakened
nationality of the Greek-speaking race early developed
extremely active public life — the Greek mind esteeming
comprehension by means of language and knowledge of
causes to be the crowning excellences of man, and social
communion and intercourse with one's fellows to be the very
flower of life's happiness.   The age which regarded the
heroic times as having immediately preceded it, and which
celebrated in song the deeds of the heroes, was not without
graceful forms of intercourse and demeanour; the continual
friction and reciprocal action produced by interchange of opinions caused the nation to withdraw itself ever more and more
from the yoke of transmitted custom as it gained new points
of view, and it began to reconstruct with conscious art all its
social and political relations ; soon, having become accustomed
to doubt and to critical analysis, it called in question all the
foundations of ordered human existence, and was ruined by
a sophistical excess of free thought, which here rose supreme
over all constancy of existing relations and duties, just as in
the East the traditional objective order of things had fettered
all freedom of subjective conviction.

To indicate to some extent in a single phrase the historical
position of a phænomenon so complex and full of life, is what
we can hope to do only if we attempt not to exhaust its
many-sided content, but merely to emphasize the difference
between it and preceding times.   Considered in this restricted
sense, those no doubt are right who find in Greek life the
first youthful self-comprehension of the human mind and the
first dawning of that light of self-consciousness by which man

examines both his own destiny and the claim which existing
natural relations have upon him.   In the most various departments we see both this critical impulse and its youthful
freshness.

However much of knowledge and of skill and of wise
maxims earlier nations may have possessed and employed
both in the regulation of social relations and in systematic
art, the thought of seeking out the very grounds and bases of
our judgment of things, and of combining them demonstratively and deductively in a system of truths—the foundation
of science in fact—will for ever remain the glory of the Greeks.
The immortal services which they rendered in this direction
belonged certainly, then as now, to individuals, not to the
crowd.   However, to have produced the individuals—and
of them not few—who aimed at and accomplished such
great things, belongs, whether as good fortune or as merit,
to the historic idea of the Greek nation.   Among the
special national characteristics of the Greeks were always
that active insight and dispassionate spirit of investigation
which examines every fact on all sides, tests every dictum,
analyses every prepossession, and by an ineradicable inclination to try and understand every particular by reference to
general causes and in its connection with the whole, led to
the conscious formation of general notions, to proof, to classification, and, in short, to all those methodical forms of thought
by which the theory and science of the West will be for ever
distinguished from even the most imaginative sagacity and
the most intellectual enthusiasm of the East.

They brought this spirit of investigation into all departments.   Not only did they lay the foundations of logic and
mathematics with remarkable exactness, but at the same time
they interested themselves in the exhaustive treatment of
domestic economy, the organization of the body politic, and
the problems of moral education, as subjects of systematic
science.   A quick and unbiassed eye for matters of immediate
experience helped them to free themselves from slothful
acquiescence in inherited prejudices and the unreasoning

passion of superstition, which, mixing things human and divine with confused ardour, furnishes neither peaceful faith as regards the one, nor intelligent equanimity as regards the other. They shook off ever more and more the influence of oriental mysticism, which, with rank growth, everywhere sees and shuns incomprehensible and oppressive secrets in the smallest trifles, and created what was to this as prose is to dithyrambic verse. I do not mean prose composition, which also they did at last laboriously develop, but the judicious way of looking at the world which receives that which is inspiring with enthusiasm, that which is sober with sobriety, that which is earthly as earthly, that which is mechanical as mechanical—which does not treat everything with the same excitement and grandiloquence, but calmly estimates different things according to their degree of importance. Thus they early separated the secular life from the religious, as far as the two can be separated, and freed themselves from oriental theocracy; thus their impulse towards political freedom suppressed by degrees all those differences in the rights of individuals which they had received by tradition; thus in art much which was great and splendid, which the East cherished passionately but expressed chaotically, they preferred to leave altogether, in order to devote themselves to more manageable tasks in which they could make the special orderly rhythm of beauty as supreme as they tried to make the laws of truth over the facts of science.

But this spirit of investigation is in its very nature of double significance. It must assume an unconditioned and objective truth in things themselves, for without this its critical labour would be aimless; at the same time it must be the individual subject who by his recognition and confirmation first establishes this truth. The Greeks were not able to escape the influence of the double impulse here involved—the impulse on the one hand to reverence for that which is in itself true, and on the other hand to the ever-busy search for a truth that is yet more true; and herein, as in their bright artistic freshness of life, they exhibit the youthful age of the human race. For youth in struggling upwards, often—when it has thrown aside the dreamy prejudices of childhood—becomes presumptuously doctrinaire, over-estimating, in the consciousness of growing insight, the instruments of knowledge, thinking little of the immediate and indemonstrable evidence of obvious truths and feelings; and while seeking ideals, unable to recognise as ideal anything that it cannot by proof and deduction transform into a product of its own reason. This over-estimation of pure thought and its instruments, logical forms, itself in many ways impoverished the science of the Greeks; they too often thought that they knew the thing itself when they had merely analysed the movement of thought by which we seek to approach the thing. In practice, however, reverence for individual dexterity of thought, and for dialectic skill in dealing with things, far exceeded respect for the nature of things themselves. The active Greek mind had discovered in rapid succession a multitude of standpoints from which to estimate all human affairs, and sometimes the establishment and development of art, sometimes any novel paradox was held to be of more consequence than the approval of an incorruptible conscience, the simple sense of duty, or immediate conviction. They thought that they could everywhere begin afresh from the very beginning, and that they both could prove everything and needed to do so; they connected moral teaching with theoretic speculation and its uncertainties; they had little feeling for historical relations which cannot be charmed away by the magic of a theoretic dictum; every fresh fancy to which any logical support whatever could be given seemed to them entitled to be tested as a new principle. We often hear them enjoining upon one another respect and reverence towards ancestral traditions and the historical continuity of social conditions; but a glance at the multitudinous variety of political, social, and ethical experiments made by them as time went on shows how little these admonitions were attended to; and when by some chance they were attended to, this was due to their having the attraction of presenting some other momentarily new point of view.

It had not always been so. Before the Persian wars the undeveloped state of society, and the prevalence of a busy, hard-working way of life had counterbalanced this excessive mental activity; but at that time the Greeks had not yet reached the turning-point of the historic race they had to run. The score or so of years that elapsed between their conflict for freedom and the Peloponnesian war comprise the time of short but brilliant bloom when the Greek spirit of liberty in its onward evolutionary struggle had not yet developed pernicious fruits. But lasting prosperity was impossible; the distinguishing excellences of the people were ruined by their unbridled sophistry. None of their virtues touches us more or was more of a novelty in history than their patriotism, and their readiness to sacrifice themselves for the good of a commonwealth that was founded on freedom of intercourse between the citizens and on comprehension of the benefit resulting from participation in common joy and labour and recreation and danger. But however highly they esteemed their fatherland and national freedom, yet each one understood national prosperity after his own fashion, and sought to realize this ideal after his own fashion; there were incessant revolutions, and these caused the rights of individuals to be in a state of constant fluctuation, and often produced such terrible crimes that the bloody history of real events forms a melancholy contrast to the splendid insight which we admire in the works left by Greek genius to posterity. Without the individualist spirit which impelled single towns to emulation for the palm in civilisation and artistic distinction, Greece would not have reached the eminence which she did; but when there came changed conditions, not admitting of such a dissipation of strength, the Greeks did not learn to suppress that selfish envy which had everywhere associated itself with the less ignoble form of the affection. Their imagination, indeed, continued susceptible to the great national thought of the freedom of all Greece; but they knew too many points of view from which it was possible to justify anything and everything, and they had lost the simple sense of duty which robs all sophistry of its strength. They had early allowed to the Persian king an influence in their internal affairs which Rome never granted to the Punic enemy of her kingdom; and Greece—abounding in examples of treachery on the part of her distinguished men, depopulated by constant dissensions and by immorality that was sometimes sophistically justified and sometimes practised shamelessly, and lacking steady discipline—fell an inglorious prey to the attacks of Italy.

§ 4. We are accustomed to regard the Roman nation as the potent temporal power which, after it had destroyed the independence of the Greeks, afforded protection to Greek genius, enabling it, as it were, to concentrate itself, and laying at its feet a conquered world. And, in fact, what the Romans contributed from their own resources to the treasures of civilisation may soon be reckoned up; but the worth of what they did contribute is not lessened by its lack of variety. They accomplished a work which was impossible for the Greeks; they combined the nations of the earth in the community of a vast political life, and in the most diverse countries left seeds of civilisation which were not slightly sown, but took such deep root that their living branches have ramified through the whole history of later times. When Alexander of Macedon, leading the combined forces of Greece, and dreaming of a union between East and West, sought in his rapid triumphal progress through conquered Asia to spread Greek civilisation to the confines of India, the dazzling splendour of his individuality, so strange and full of genius, blinds us to the hopelessness of an undertaking of which very soon the only traces were to be found in legends in which the wondering nomads of Asia praised the hero who had come from afar. The Romans never indulged projects to be carried out at such a distance from their natural sources of supply; after they had, in hard-fought struggles for their own independence, subdued Italy and warded off Carthaginian supremacy in Europe, they progressed but slowly—impelled by circumstances and lingering by the way—to that universal dominion which, when once established, was maintained for centuries. Such great historical results indicate the historical significance of the nation itself, and indeed, compared to Rome, Greece lived from hand to mouth, passionately pursuing immediate ends, while the political activity of the Romans was guided by a wider view, taking in the future, in which they were conscious that their destiny lay.   The Greeks lived, as on some terrestrial Olympus, only for the sake of beauty and of working out their own development ; to the Romans the known world, all the countries bordering the Mediterranean Sea, seemed to be an actual field of labour, setting before them definite tasks of acquisition, guardianship, and government. From ancient Italian civilisation they had received the idea of a mysterious lapse of time through ages marked by distinct characteristics ; they felt themselves to be the bearers of this historical development and co-operators in its production, and their poets are hardly so loud in praising Rome's existing greatness as in emphasizing perpetually its undying future. And the result has proved that they were right.   Greece, having perished as a terrestrial power, still lives on in the mind of the civilised world, though without any striking influence upon the conditions of our lives ; but a countless number of our social and political arrangements, and a great part of our mental life, may be traced back along a line of unbroken tradition to Rome ; and to places where there are no flourishing towns that owe their origin to her, modern civilised nations have carried with their language the lasting influence which they themselves received from her ; Latin words and forms of speech are heard on the banks of the Ganges, and mingle on American plains with the labials of Indian dialects.

Human action is either guided directly by the idea of some desired result, and then easily comes to consider the means as sanctified by the end, or it follows general principles of universal validity, and will refrain from carrying out an intention as long as this can only be done by transgressing them. The artistic bent of their minds inclined the Greeks to the first way; the Romans are distinguished by the conviction that a valid result can only be attained by respecting the fixed relations prescribed by the nature of those elements which co-operate in its production. We shall have occasion later to see how far this thought penetrated their whole life and action; in the present rapid survey we only wish to recall to mind the pearl of Roman civilisation—the development of law. For knowledge of the truth that is and operates in things and events, the Romans, as compared with the Greeks, did nothing; but the thought that the world of relations brought into existence by our actions, is just as much governed by a complex and inviolable order independent of our will as the forces of external Nature are in their general statical and dynamical relations—this thought owes its existence to the Romans. They did not, like the Orientals, regard existing relations as irrevocable decrees of fate; neither did they, after the fashion of the Greeks, consider actual rules, established institutions, and acquired rights as having the pliability of wax, and a capability of being moulded differently according to men's caprice, if they hindered the realization of an ideal; the Romans regarded both—both the variation which the needs of human nature demand, and the fixed condition which refuses change—as two valid forces between which men had to steer by means of law. They did not begin at the apex of the pyramid—at the ideal or desirable form of the state as a whole, logically deducing from this the just rights of the citizens, but they first of all established on general principles those relations between individuals which arise in the living intercourse of daily life. It was real needs, the requirements of circumstances, which subsequently impelled them to limitations of those private rights, in order to attain the prosperity of the whole which is itself the sum of all the individuals; and the final form of commonwealth aimed at was in every age that which combined in satisfactory practice respect for transmitted rights, provision for new wants, and the conditions required for the growth and continuance of the whole. Thus there arose that unparalleled social struggle between the patricians and plebeians, in which violent passions—on the one hand a haughty insistence on privileges, and on the other a consciousness that participation in these privileges must be got by fighting for them—were held in check by regard to the necessary stability of political life, by recognition of the sacredness of law, though it were only formal law, by unswerving obedience towards governmental authority which had once been recognised, and finally, by a stern patriotism from which all thought of treachery was far removed.

The results of this evolutionary struggle were not as fortunate as the character which found expression in it was noble. The inadequacy of the republican political construction which had been suited to earlier and more limited relations was only compensated, as the state grew and enlarged, whilst the great men—of whom the patrician race produced many—used the space for independent action which was left to them to show brilliant examples of self-sacrifice and inherited political wisdom. This famous aristocracy fell into the background, as circumstances came to require rather the concentration of power in one hand than a general distribution of rights. In contrast to a new nobility of wealth without ennobling traditions that began to arise, the numbers of the unpropertied increased. The almost uninterrupted state of war which marked the early days of Rome had never favoured peaceful labour and industry; when at a later date Greek civilisation and acquaintance with the customs of so many different nations had undermined the old simplicity and strictness, when the treasures of the East and the products of the pre-eminently industrial countries poured in, and swarms of slaves in the palaces of the rich practised every kind of craft, a class of free labourers could find neither respect for their position nor a market for their products; even the ancient agriculture of Italy and the independence of the country population suffered from the accumulation of enormous wealth in the hands of individuals, and the expenditure of this wealth on useless luxury. Between the inordinate self-indulgence and ambition of the aristocrats, and the beggar-liness of a populace that could be won over to aid in any destructive project, the order of free citizens, which was the strength of the state, disappeared ; and after long and bloody conflicts, the republic, shaken by the unprincipled struggles of individuals after power, fell under the dominion of the emperors, with undeveloped or impoverished forms of government.

For centuries as these rulers succeeded one another, mad-ness alternated with discretion, cruelty with clemency, and Roman civilisation had an opportunity of showing what power of endurance and of resistance there was in it and in its creations, even after the animating impulse had died out. Whilst general enervation went on increasing, the discipline of the Roman armies still continued for a long time victorious over external foes ; under the pressure of arbitrary political rule, the legal consciousness still went on developing to scientific clearness and completeness ; amid the decay of morals there shine forth many examples of noble manliness that bear witness to the enduring power of a great past; and by similarity of regulations, by great roads of communication, by the general diffusion of one language and of one culture taught in numerous schools, all the countries bordering the Mediterranean Sea were connected together into one great whole of common life, which in the isolated happy intervals of peace and benignant government might with justice rejoice in the consciousness of such a degree of human happiness as had never before been attained. If, however, this state of society still contained the seeds of permanence, yet as far as human eyes can see, there were in it no elements of fresh progress ; it was from outside the circle of nations which had thus far developed civilisation, that there came, through Christianity, the shock with which ancient history concludes and a new period begins.

§ 5. Among the theocratically governed nations of the East, the Hebrews seem to us as sober men among drunkards ; but to antiquity they seemed like dreamers among waking folk.

With thoughtful imaginativeness these latter had considered
the causes of the world and the sources of their own life and
death ; and feeling themselves to be parts of the great divine
universal frame, they accompanied with wild rituals of sensuality or self-torture all the convulsions of its mysterious life—
the yearly change of decay and revival in Nature, the struggle
of the bright and beneficent with the dark and hostile powers ;
and over and above this wisdom which was current in daily
life, the exclusive learning of the priests seemed to hide
innumerable further secrets.   All this was regarded by the
Hebrews with the most extreme indifference ; the mighty and
jealous God who desires uprightness of heart, who pursues
sin, and is avenged on iniquity—He indeed it is who has
created the world and has caused all kinds of herbs and
animals to spring up, and has formed the stars of heaven,
because He willed that everything should be very good.   But
the imagination of the people was not absorbed in the contemplation of this creation, in which His glory was expressed
only as it were by the way; to them God was a God of
history, to whom Nature is as the mere footstool of His power,
but the life of men, the life of His chosen people, the one
object of His providential care.   The whole superfluity of
mystic natural philosophy, which so uselessly burdened the
other religions of antiquity, was cast aside by the Hebrews,
that they might devote themselves to the great problem of
the spiritual world—the problem of sin and of righteousness
before God ; they felt themselves involved, not in the whirl
of everlasting natural cycles, but in the advance of historical
progress ; they did not trouble themselves about secrets which
concerned only past events, but all the more deeply were they
interested in the problems of the future; and these problems
were not to remain hidden, but the prophets were impelled
by divine inspiration to announce to all, for their comfort the
final attainment of a heavenly kingdom, for their repentance
the commands of God.   After the times of the first patriarchs
with whom God had entered into covenant, the national mode
of life had undergone many changes.   The patriarchal

shepherds of the early times had, after the Egyptian oppression, become a warlike nomadic race; they had then formed
permanent settlements and cultivated the land; finally, they
were inspired with the commercial spirit of their Semitic
neighbours, and, like the Phœnicians, became scattered over
all parts of the then known world; the fundamental thought
of their national life—their covenant with God, the consciousness of an historical destiny, and the hope that this would be
realized—they had not forgotten, but on the contrary had
become more and more confirmed in after many waverings at
the outset. The civilised nations of antiquity, whose ingenious
mythology and philosophical notions of divinity lacked nothing but simple faith in their reality, began to have their
attention drawn to a nation that possessed in so high a degree
the living conviction of which they themselves were destitute,
and to which the ideas of God and His kingdom were not the
mere ornamental poetical framework of a wholly secular view
of life, but the most deep and serious reality. In the
gradually sinking Roman Empire the Jewish faith gained
consideration and adherents, although its national character
was a drawback to it. But now the predictions of a Messiah
had been suddenly fulfilled; the new covenant with God was
proclaimed by enthusiastic disciples as an historical reality,
and not merely a new doctrine added to the many other
doctrines of the past; and the tenor of their announcement
did not contradict the hope of finding the true satisfaction of
longing desire in the final union—of which the secret had
been long lost—of mundane and supramundane existence.
The excellences and the weaknesses of existing Roman civilisation combined with some special historical circumstances to
favour the spread of Christianity; but of more efficacy than
all these was its own inherent power, due to its startling
contrast with the hitherto received view of the world, and its
consoling agreement with the secret thoughts that had been
wont to rise in rebellion against that view.

Everything which a religion has to give it offers to the
understanding in doctrines, to the heart in its characteristic

tone, its consolations, and its promises, and to the will in commands. The original doctrines of Christianity were not very multifarious. All those questions concerning the origin, coherence, and significance of Nature which Judaism had already passed over were also left undecided by the Gospels. Speaking only of the kingdom of heaven, it exalted the community of spiritual life as the true reality, in the glorious light of a history embracing all the world, and let Nature and its evolutions quietly glide back into the position of a place of preparation, the inner regulation of which will be revealed in due time. Neither did it speak of divine things as if it would measure out the Infinite demonstratively in concepts of human reason; all questions concerning the relation of God to mankind, which had already exercised in various ways the ingenuity of ancient culture, it passed lightly over with figurative phrases borrowed from human relations. Thus it seemed to reveal even less than that culture had already discovered. But in speaking of the sacred love which wills the existence of the world for the sake of that world's blessedness, and has its justice restrained by pardoning grace, it emphasized so much the more certainly that one thought the unconditioned and ever self-asserting worth of which can do without the confirmation of proof (which is very foreign to the nature of religion); and the content of that thought as the only thing that is really certain, at the same time guides the activity of sagacious investigation in a definite direction.

So Christianity offered infinite stimulus to the understanding without binding it down to a narrow circle of thought; and to the heart it offered full as much. For, according to Christianity, the sole truth and the source of reality with all its laws was something of which the eternal worth must be felt in order to be known; from the reality thus known through feeling, man's understanding can reach back to that which is divine, and can very often conclude from it to the divine, as from the ground of demonstration to that which is demonstrable. In this it met the eternal longing of the human heart, and satisfied it in a fashion wholly new.   The consciousness of finiteness has always oppressed mankind; but however much moral contrition we may find in the enthusiasm of the Indians, however much dread of self-exaltation in Greek circumspection, however much fidelity to duty in Roman manhood, yet everywhere this finiteness was felt to be merely a natural doom by which the less is given into the power of the greater, and its existence irrevocably confined within limits, whilst within these limits the finite is destined to attain by its own strength its highest possible ideal.   The Indian sought to extort eternal life by frightful penances; the Greek was afraid of rousing the envy of the gods by pride, but he aimed at perfecting himself as man, and it seemed to him that virtue might be taught as any craft may be ; the Roman, knowing nothing of a blissful life of the gods beyond his own, went self-renouncingly to death for duty's sake, an honest man whom yet no god had helped to be what he was.   The characteristic of humility and submission, that is lacking even in the most mournful expressions of this sense of finiteness in antiquity, was brought for the first time by Christianity into the heart of men, and with it hope came too.   It was a redemption for men to be able to tell themselves that human strength is not sufficient for the accomplishment of its own ideals ; hence from this time mankind no longer seemed to be an isolated species of finite being, turned out complete by the hand of Nature, and destined to reach unaided, by innate powers, definite goals of evolution. Freed from this isolation, giving himself up to the current of grace, which as continuous history combines infinite and finite, man is enabled to feel himself in community with the eternal world, which he must stand outside of as long as he desired to be independent or believed that he must be so. And since the mere belonging to a particular race was now no longer a source of justification or condemnation—salvation needing to be taken hold of by the individual heart, which must be willing to lose its life in order that it might find it again—there now began to be developed for the first time that personal consciousness which thenceforward with all its problems—freedom of the will and predestination, guilt and responsibility, resurrection and immortality — has given a totally different colouring to the whole background of man's mental life.   This momentous content has indeed never reached the clearness of calm comprehension in the minds of all mankind to whom it was proclaimed; but even those who tried to resist it have never been able to get rid of its influence; it has remained the centre about which the civilisation of later times has always revolved, in hope or doubt, in assurance or fear, in zeal or scorn.

To him who so regarded the eternal connection between earth and the kingdom of heaven, all earthly history must seem but as a preparation for the true life, not valueless, since it aims at this goal, nor yet burdened by the tremendous seriousness of absolute irrevocability.   Therefore Christianity proposed to the will only such commands as require permanent goodness of disposition; from the ordering of human affairs by ceremonies, law, and government, it stood indefinitely far.   It could do without that which the heathen theocracies were compelled to demand; since what it asked for God was God's, it could give to Cæsar that which was Cæsar's.   As for it God was not primarily revealed in Nature in the manifold forms of His creation from which the grounds of reverence might be deduced, so life was not primarily an established order of moral relations within which man might walk with a sense of security along paths definitely marked out; but to man's inner life was entrusted the work of gradually raising the forms of society to relations which were in harmony with his spirit.   Therefore the attitude of Christianity towards the external conditions of mankind was not that of a disturbing and subversive force, but it deprived evil of all justification for its permanent continuance.   It did not forthwith abolish the slavery which it found existing, but in summoning all men to partake in the kingdom of God, it condemned it nevertheless; at first it let polygamy continue where it existed; but this must necessarily disappear spontaneously when the spirit of Christian faith made itself felt in all relations of life. And this conflict is still carried on in many directions, for the perversity of human nature, which is ever much the same, opposes to the better way all the resistance of which it is capable; but there is one permanent advantage by which the new age is distinguished from antiquity. That which is better and juster did indeed make a way for itself in ancient life, but almost exclusively in those cases in which the oppressed struggled manfully with the oppressor; the provident humanity which, without seeking its own happiness, takes the part of the suffering section of mankind, and requires and exercises deeds of justice and of mercy, was something very foreign to the ancient world, and in the new world it has no more powerful source than Christianity.

In conflict with mundane circumstances and human passions, and yet linked to both as the instruments of its realization, no ideal can, in the course of its historical development, remain faithful to its full perfection. Christianity, forced to justify itself to the civilisation of the ancient world, became entangled in the attempt to establish dogmatically the articles of its belief, in the hopeless effort to force upon its professors, instead of the inexhaustible fulness of living thoughts which the gospel can arouse in each, a complete system, many of the regulations of which were as barren in regard to practice as the productions of ancient sophistry. The simple division of labour which had arisen in the primitive Churches from the duties of the society in regard to worship of God and ordinary life, was transformed into a gradation of fixed offices as the diffusion of Christianity increased; in opposition to the universal human priesthood of the gospel, there was a fresh separation from the laity of an order of priests, and in the edifice of the hierarchical church the empire of the Holy Ghost stiffened into a slavish, earthly mechanism. But these deformities of Christian life, which a later age might undertake to rectify, were but as the tough rind which alone enabled that life, amid the ruins of the falling Roman empire, to save itself for its future.

§ 6. The Germanic nations — the often victorious, often conquered, but never subdued enemies of Rome — at last completed the work to which they seemed destined, by disintegrating that empire of ancient civilisation which had lasted for a thousand years. But they were not in a condition to substitute from their own resources a new civilisation for that which was passing away. The long death struggle of the Roman empire—which the Germans themselves, as the most valiant of the auxiliary troops, prolonged for a considerable time—had indeed brought them into many-sided contact with the elements of ancient civilisation and the teachings of Christianity, but the mass of that great people which spread victoriously over the Roman provinces had yet remained true to the simple life which they had lived on without historic record from time immemorial. No one knows what events filled up the long succession of centuries which lay between their first detachment from their original abode in Asia and their appearance in the history of European civilisation. It is probable that being long without a settled home, harassed by tribes who were pressing on them in their wake, they maintained their valour and warlike vigour in the struggle for existence; but that in the northern settlements, where they finally established themselves, they made little progress towards polite manners. At the time that the Roman empire began, and was revelling in all the treasures of the known world, the Germanic tribes still lived by the chase, by the produce of their herds of cattle, and by a somewhat rudimentary agriculture; they no longer roamed about homeless, but had fixed dwellings; as, however, their settlements were much dispersed, and they had no towns, they had none of that industry which is developed as a consequence of density of population and division of labour. Accustomed to hard simplicity in food and raiment—having even to economize the iron which they used in their weapons, since they did not know how to procure it for themselves—they braved the inclemencies of the weather in rude huts, being sometimes driven by hard winters into subterranean caves. Inclined to sociability, they yet found little to occupy them except fighting games, carouses, and listening to heroic songs which repeated the great deeds of that same simple life. But with this meagre culture they yet combined qualities of character which were destined under more favourable future conditions to bring special benefits to mankind in the course of history. They possessed in high measure the love of freedom which contents itself with guarding from foreign influence its own liberty of choice in the conduct of life, but they had not the envious impulse towards equality which cannot endure that others should have the advantage in anything. It seems as though for the sake of that independence they had purposely refrained, in the simple arrangements of their society, from numerous steps in advance which, while bringing greater fulness and development of life, would have prejudiced the independence of many; but they submitted to the superior power of gifted leaders of their own free will, and with the most perfect fidelity; and, without recognising hereditary sovereignty or nobility, they yet had high respect for the heroic blood of famous families. This trait of willing service and absolute personal devotion is widely noticeable throughout their history, and as this is only possible in personal relations it has in later times always made the German nations more disposed to associate in somewhat small circles than to combine into one great whole. In the same way it always remained difficult for them to become enthusiastic about general principles which were not presented to them embodied in some personal form; but when such an enthusiasm did take possession of them, it was all the more lasting, for it was a long time before they came to know how to take up any cause half-heartedly. They were bound to give their whole soul to anything which they took in hand. It may be admitted that with such a disposition they were well prepared for the reception and inner elaboration of Christianity, without denying that in the early ages of the Church it was the more southern nations of the Roman provinces that produced those men of lofty enthusiasm and deep earnestness who, as Fathers of the Church, were the forerunners of Christian life in the north.

The tremendous movement of national migration now caused the Germanic peoples to spread, in successive great waves, repeatedly breaking one upon another, over all the provinces of the Roman empire. They were not able to hold any of these southern conquests, being everywhere in a minority compared with the native population; but for a long time the union of the civilised world was broken by them, and over the rich countries bordering the Mediterranean Sea, which Rome had brought together in the noontide light of organized intercommunication, there fell a long twilight, in which some countries disappeared from the view of the others, and many elements of a previous common civilisation were lost.

The great and varied admixture of peoples and modes of life which the increase of the Roman empire and the growing development of intercourse had produced, had already in the closing period of ancient civilisation begun to disturb the simple, pliable, and self-confident spirit of antiquity, the one condition of which had been an isolated national development in accordance with a natural bent. Before the period of this disturbance a system of consistent philosophic views had been instrumental in causing the production of finished works of art exquisitely proportioned; clear and definitely determined tasks had given harmony and character to life itself; notwithstanding the inexhaustible variety of detail, reality as a whole—with its store of attainable good things and those desirable forms of human life to which it gave scope—was spread before men's eyes with the perfection and completeness of a well-arranged picture. Yet this whole mode of thought had certainly rather suppressed than satisfied the wants of the human heart. The self-distrust which had earlier overtaken Greek life found its way into the Roman world too in the time of the emperors. Unquestioning faith in the supremacy of Rome gave way to cosmopolitan considerations; the narrow but robust system of thought which constituted national morality was invaded by philosophic reflection; artistic imagination, which suffers most of all from mental indecision, changed its calm mirroring of reality for dissatisfied and passionate flights beyond the world of fact, and commingled accepted forms of representation in attempts of a new kind. Religious belief had long since lost its certainty; with the most baseless superstition was combined a restless longing to win back from any known or unknown worship prevailing among men the certainty which had been lost. Then Christianity came, and the new spiritual growth had to force its way up through the rents of the ancient system of thought, the external integrity of which was finally destroyed by the invading torrent of the German barbarians. If this blending of all imaginable forms of life could not fail to change fundamentally the genius of what remained of the ancient nations, it could likewise not fail to be difficult for the conquerors to know what attitude to take towards such boundless variety. These conquerors came down upon the Roman empire without any definite aims, partly yielding to necessity, partly urged by the struggles towards expansion of a strong nature that sought to appease its impulse to action by violent and powerful but yet objectless exercise. Now there lay before them the down-trodden classic world, with all its rich treasures of Nature, of art, and of life, and with the countless elements of civilisation which it still contained; in exercising themselves upon this battle-field they for a long time gave to history that stamp of adventurous romance which—with its wealth of free and original powers, its inharmonious struggle after great and passionately pursued yet mutually inconsistent ends, its variety of strange forms of life, and its incoherence—distinguishes the Middle Ages from the period of ancient history.

§ 7. When after three centuries the stream of national migration had come to a stand, there had become united under Frankish rulers districts in which indeed Germanic blood preponderated, but the inhabitants of which could hardly feel themselves bound together by any common tie except when they were obliged to take the field together against an external foe. Especially in those German countries which had only come into contact on their confines with Roman dominion, the absence of towns caused the continuance of that old life of meagre social intercourse natural to a sparse and scattered population. The differences of disposition of different races, the lack of common administrative interests, and the difficulties in the way of exchange of thought, prevented the development of any active public spirit. Charlemagne was able by his individual power to hold all these provinces together by help of arms, and in peaceful activity to enrich them with the germs of a subsequent flourishing civilisation; but to breathe the vital strength of a self-maintaining political whole into a society of which the constituents had so little need of one another and so little dependence upon one another, was a task beyond his strength. Hence, when the re-establishment of the Roman imperial dignity in him once more gave a supreme ruler to the world, the new unity of the human race was just as much the imaginative ideal summit of a not yet existing society, as previously the first institution of the same dignity had been the natural conclusion of a long social history, from which it grew without any appearance of novelty. And this character the empire of the Middle Ages maintained throughout. It only temporarily possessed the power corresponding to its ideal position; but though this was a merely imaginary picture, it yet really lived in the imaginations of men; the thought of the majesty of a single temporal government was by no means an empty dream, even although it could not be carried into effect, but—like conscience, against which the passions are always in rebellion without being able quite to silence its enunciations—this ideal picture, while lacking actual power, hovered before men's minds in the Middle Ages, and reverence for it always kept much self-will within bounds and called forth many an act of self-sacrifice.

As a matter of fact, the real articulation of life did not start from this point, and hence work itself out to unity, but it worked up from below, developing into innumerable small circles, with different degrees of slowness and difficulty in different countries. Italy, with its long cultivated soil, with many ancient towns still existent though depopulated, with its commerce which partly had been preserved and partly was growing up afresh, and with the civil organization of its communities which had never been quite destroyed, was the first to collect together these rich remains of former culture, and developed a vigorous intellectual life in numerous small states, the emulation of which was favourable to culture whilst it hindered political unity. The great inland countries of the Continent, on the other hand, suffered from the ungenial nature of their more northerly climate, from the difficulty of internal communication, from the want of great social centres, from the inconvenient character of their medium of exchange, in short, from all that torpor of existence in which consists the darkness that generally seems to us to brood over the Middle Ages. Thus the inland countries too, like Italy, but from different causes, were at first able to form only small states.

The original communities had consisted of free owners of the soil ; in conquered territories the victors were rewarded and their wants supplied by enfeoffment of tenements and lands ; the undeveloped state of society made it necessary for the guidance of affairs that there should be personal representatives of the supreme power, and these held office at first temporarily and afterwards permanently ; they too were provided for partly by property in land, partly by rights over certain districts ; finally, in developed feudalism, the once homogeneous community was transformed into a complicated and graduated system of persons endowed on the one hand with privileges, and on the other hand burdened with obligations, both privileges and obligations binding those to whom they attached to some definite parcel of land. The country was covered with countless strongholds of the feudal lords ; in the solitude of these the sense of family unity, of honour, of purity of blood, and of reverence for tradition grew; the position of wives and mothers increased in importance; a feeling of solidarity among persons of the same rank—carrying with it in the knightly order a consciousness of having some duties with regard to human culture—bound individuals together into a certain community of life; traditions of romantic reverence and of uncompromising manly fidelity gave some moral content to life; and there even revived a taste for poetry. But neither general culture nor the development of public life made much advance under this form of society. National life had ceased to exist; the chasm between the feudal lord and his vassals was bridged over by no recognised law and seldom by kindly care; between the individual communities of serfs there existed no bond of common consciousness or of legal connection. Even the order of feudal lords, united by social intercourse and similarity in mode of life, felt only that they were an order, not that they were part of a political whole for the benefit of which it was their duty to make sacrifices. Few territories were large enough for the development of a civilised life of their own; the co-operation of several was hindered by the independence of the lords—the obscurity of their mutual obligations—the lack of a general and unquestioned system of law which as these obligations gradually grew up should have developed along with them—the impossibility of carrying out sentences, when they had been pronounced, in any other way than by the exercise of force—and the ease with which a number of individuals about equal to one another in power could combine to resist legal force, which could be brought to bear only with extreme difficulty. It was only within very small communities that definite and intelligible relations existed, the state as a whole possessing only the most unwieldy machinery; care for the general welfare was crippled by the want of an established and regulated system of taxation, and external policy by the lack of a standing army and by the intricate arrangements of the feudal host; for the administration of justice there were wanting established tribunals representative of the general sense of justice, and in almost all cases legal jurisdiction was disputable, or actually disputed, or owed its recognition to force.

In this state of things, notwithstanding all its disorder, a certain characteristic sense of justice was not lacking. The Germanic nations, with no inherited treasure of ancient civilisation, no gift of abstraction due to such an inheritance, no eye for principles, had been placed historically in circumstances which forced them to rapid development. They could not discover the universal principles of justice offhand, but every relation which had become historical forthwith seemed to them, whatever its irrationality, to be *de facto* just; it would not have arisen if it had not at the time corresponded to existing needs. Added to this was the fact that Christianity appeared to them less as a body of doctrine than as a history of past events—as among those transactions by which Providence, and not the nature of things themselves, gives laws to the course of human affairs. All that we are now accustomed to judge by universal laws of morality and justice was regarded in the Middle Ages as dependent upon divine institution, upon human appointment, upon investitures and treaties, upon the significance of particular occurrences. On account of the continual change of circumstances such a foundation for the arrangements of human life could not fail to be a most fruitful source of incessant opposition to justice which had become unjust; it produced the countless outbreaks of unbridled caprice which mark the Middle Ages. But where the opposition took a more peaceable course, this too did not proceed from abstract principles, but sought to meet the requirements of the hour by transforming particular existing laws through fresh enactments, which were themselves of equally restricted application. This kind of procedure pervaded in the most various forms every department of life. When towns began to flourish, and redeemed their territories from complicated obligations towards the feudal lords, and love of work and the moral deepening of character gained in busy spheres of labour became the fairest adornment of the closing period of the Middle Ages, then we see this full life crystallize into a multitude of sharply defined corporations, each having its own internal system and legal relations to others, both regulated by contract, and all surrounding themselves with innumerable trade customs and symbols, and as a whole developing into organisms of which the real significance was sometimes clouded by numerous irrational additions having a merely historical justification, yet—taken altogether—becoming individualized into a most intense life.   And in the imagination of the Middle Ages it was not men only but things also which had special rights —rights which were not merely measurable by natural qualities, but were in a sense historic ; to times and places were attached privileges, obligations, and liberties of all descriptions.

Within this world of external life mental culture was for a long time attended to only by the Church.   The Roman empire, after the recognition of Christianity, had begun to give important political posts to the clergy, who were gradually forming themselves into a separate body ; their activity, stirred up by lively enthusiasm for what gave so much worth to life or by aspiring ambition, in many ways took the place of the slack civil authority ; rich endowments gave them independence and the means of doing good works. Although it was a long time before the hierarchical edifice was complete, the authority of the Roman chair soon took firm root in the West, and the numerous missions which went out from every newly-established settlement felt themselves to be members of one whole.   Without having been thus organized into a church, Christianity would hardly have weathered the storms of those times, and could have exercised but little of its beneficent influence upon temporal life.   By the help of transmitted culture, and through the resources (whether its own or not) which its authority enabled it to command, the Church was able partly to keep invading barbarism at bay, partly to press forward itself and fill the still darkened northern countries with those churches, monasteries, episcopal residences, and agricultural settlements from which there were diffused not only the art of husbandry, but also that of gardening, not only the elements of knowledge, but also those of technical crafts, and under the walls of which gradually reviving trade held its markets, whilst within their gates the sick and weary found tendance or healing. Thus in the early period of the Middle Ages the Church was in many respects at the head of progress and of civilisation; from it proceeded the majority of such establishments as were of general utility; from it the ignorant sought teaching, for it alone possessed the treasures of transmitted learning; to it alone could the longing go for consolation and for the resolution of their doubts, for it alone had studied all the relations of human life, and with active enthusiasm, combined the results of its reflection into one comprehensive philosophy; finally, it was to the Church that the oppressed appealed for help, for it was the Church alone that, amidst the general licence and the thirst for adventure, recognised and taught a truth that was valid for all men and a divine order of things independent of all human caprice, obeying these in a life of strict discipline, and not unfrequently asserting them with courageous self-sacrifice in defending the weakness of the oppressed against the violence of the strong.

Passing lightly over the eventful history of the Church during the Middle Ages, we find that at the end of this period its relation to secular life had very much changed ; whilst the latter was making remarkable advance, the Church had fallen into the rear, and had become a hindrance to progress. It no longer led the van of science ; the religious philosophy which formerly, in contrast to the scattered and wholly secular culture of antiquity, had so beneficially striven to grasp all reality and to embrace and classify all knowledge, was, after the slow decay of that culture, incapable of giving any satisfactory insight into the connection of the external world ; and at the same time the secular learning of antiquity which continued to be propagated, being merely transmitted and not cultivated with that zealous interest which has re-creative efficacy, lost in breadth and precision : whilst in secular life new relations were being formed and new facts discovered, the ecclesiastical sources of instruction were becoming impoverished.

Even the cure of souls had lost its energy. With penetrating zeal the fathers of the Church had once defended the faith against all the doubts of ancient culture ; and it was certainly advantageous to Germanic barbarism that there should be presented to it some definite profession of belief, but the hard and fast formulation of dogmas which thus became the cut and dried content of tradition, diminished even among the clergy the intensity of spiritual life ; and the people were deprived of the little that still remained of such activity, by the use of the Latin language and the care with which the Church reserved to itself the secrets of religion and the administration of the means of grace, no longer preaching to the laity of the inner life of faith and of a new birth of the soul resulting from its own struggles, but denying them. Grievous faults had also appeared in the lives of the clergy, and they were no longer either the recognised pattern of conduct or the hope of the oppressed. They had not indeed become an hereditary ecclesiastical caste, but recruited their ranks from among the people, although no longer by means of congregational election ; but the inferior clergy who lived among the people were wanting in influence and insight ; those who were invested with superior dignity, and as feudatories occupied many political posts, often favoured insubordination to secular rule, but not the freedom of the laity in ecclesiastical relations.

There had never been any lack of vigorous struggles between these two great powers. The conflict between the empire and the Roman Church had led to no decisive victory of the one or the other. The empire, with its claim of sovereignty over nations between which there was no bond of union except Christianity, could not on such grounds be triumphant over the Church which demanded the same supremacy in the very name of Christianity ; the Church had on its side the naturally unifying power of religion, and used the national differences to which in their secular development freedom should be allowed, as an instrument against the defectively established supremacy of secular power. But when the empire had been obliged to let its claims drop, secular life had attained an importance of its own in a number of national developments, as the natural representatives of which the princes of the different countries could more efficiently resist the encroachments of the Church. The opposition of these temporal powers to the attempts at renewing a theocracy succeeded in proportion as they identified themselves with the national life of their respective countries; they disabled themselves where they joined with the spiritual power of the Church in the obstruction of progress. This progress itself was due partly to a further development of previous conditions which had gone on unnoticed, and had also been favoured by a striking succession of historical events and discoveries. Unceasing wars, which no longer had the character of national migrations, had kept the nations in reciprocal contact; the internal action and reaction of society was increased by the revival of trade and the growth of flourishing towns; the Crusades had for a long time united Christian nations in common enterprise; not only were Italy and Byzantium, with their inherited culture, again brought into contact with the more northern nations through these causes, but the East also, with its different customs and its treasures and marvels, roused in the nations of Christian Europe a spirit of emulation and a doubt as to the exclusive validity of the state of things which had been established among them by custom and tradition; the geographical horizon was still further enlarged by the discoveries of the Portuguese navigators; and finally the discovery of America presented to human imagination, to the spirit of adventurous enterprise and to industrial activity, openings undreamt of before, and which were to help men to become both externally and mentally wholly detached from the traditions of antiquity. At first, indeed, they tried to bring their new life into connection with antiquity, whose treasures of thought had never quite vanished from human memory; but now, on the one hand, they entered with greater zeal into the growing activity of mental life, and on the other hand the increasing danger from the Mohammedans with which Byzantium was threatened, and its subsequent fall caused what remained of Greek learning to be transferred to Italy. Then began that revival of learning which first restored to thought (which had grown stiff and clumsy) formal flexibility and adroitness, and inundated life at once with great Ideas, with comprehensive views, with critical contempt for all existing goodness and beauty, and with an audacious imitation of the errors of antiquity. The creative force which might have given worthy content to the new forms was very backward in most directions; in Italy alone the confusion of social conditions was to some extent compensated by a magnificent flight of creative art; yet there were laid those foundations of higher mathematics and of natural science which were destined to produce the most important instruments of the new civilisation. Finally, the torpor which had long hung about the exchange of thought was removed by the discovery of printing; from that time public opinion could exercise its influence upon all the relations of life, and the awakening spirit of criticism which was to distinguish the period just beginning was armed with its most powerful weapon.

§ 8. The various germs which the end of the Middle Ages had produced, gradually bore fruit in a succession of great revolutions. They did not develop simultaneously or altogether in harmony with each other; the human mind in its onward struggle is capable of the inconsistency of maintaining in one department the same new views which in others, yielding to old-established custom, it eagerly persecutes. But amidst all such contradictory and retrogressive currents there developed, with ever-increasing power, as the distinguishing characteristic of the new age, that Enlightenment, destroying in order to reconstruct, which sought *to break the dominion of all prejudice* and *to undermine every ill-founded belief.* The spirit of modern times, to which it is essential to be constantly reflecting upon itself, has often enough used these phrases as watchwords indicative of its own characteristics, and the indication is perhaps accurate for good as well as for ill. For both the strength and the weakness of our position, both our hopes and our fears as regards the future, depend equally upon that unchained spirit of criticism which, investigating all the relations of life with self-conscious purpose, more easily accomplishes the inevitable demolition of error, than the reconstruction of truth, and, in the zeal of its analytic incursions, runs the risk of injuring unperceived the most necessary foundations of ordered human existence. We have, perhaps, reason to give more scope to hope for the future than to fear; but above all it seems clear to us that we have not yet seen the conclusion of the developmental struggles into which the impulses of the immediate past have plunged us.

It was religious needs that first kindled the flame. The Reformation sought to lead men back from the secularization of the Church and the externalizing of ecclesiastical life to the purity of primitive Christianity. Though the positive teaching of the Reformation, far from professing to be a production of individual reason, was in fact mere submission to the authority of revelation, yet being in declared opposition to the existing order of things, it could not avoid formally recognising individual examination and decision as the starting-point even of religious life. It freed conscience from the obligation of submission to commands (proceeding not from the gospel but from tradition and from ecclesiastical speculation) which it was attempted to force upon men; and laid upon them instead the obligation, which was at the same time a privilege, of appropriating to themselves the content of faith by their own struggles towards development and their own inner experience. In doing this it ventured to hope that the result of this struggle would be agreement with that which it esteemed to be eternal truth, and to which it held fast; but it was bound to acknowledge that, though it might lament, yet it could not condemn the opposite result. The principle of free investigation of the gospel could not escape expansion into perfect freedom of conscience, in the acceptance or rejection of all Christian and finally of all religious truth whatever. For a long time the Reformation, conscious of the value of its faith, struggled against this conclusion; to it too the disposition to

persecute for faith's sake was not unknown, and when the
battle for the freedom of personal conviction had been fought
out, there remained doubts as to the legitimate sphere of this
freedom.   And these occurred first in the renewed Church
itself.   The very investigation of Scripture as the sole foundation of faith required the co-operation of subjective interpretation; a Church which adopted this principle could neither
exclude all variation of dogmatic conviction, nor could it
easily mark out definitely the limits within which such
variation should be allowed for the future.   In such doubts
we ourselves are still involved; the only men who are sure of
themselves are those who hold the most extreme views, either
demanding a stricter unity of the Church at the expense of
individual freedom, or an atomistic dispersion into innumerable small communities in favour of individual freedom at the
expense of the universal Church.   And yet between these two
extremes Protestantism has gone on living and developing; for
in holding fast to the principle of free investigation notwithstanding all the perplexities and difficulties of its ecclesiastical
polity, it has secured the adherence of all the rich culture
which has arisen from the stimulus given by itself and from
the schools which were for the most part established by it.

The relation of religious profession to the state was affected
by the changes which the state itself experienced, or through
which it was first developed.   In the Middle Ages influential
connection between the different departments of life and
the consciousness of solidarity occurred almost exclusively
in individual minor communities, the praiseworthy and
active public spirit of which could not make up for the
absence of important and varied relations, and the external
connections between which remained uncertain and unorganized.   From this incoherent condition there sprung up
the formally systematized State, with its comprehensive
administration of differently endowed and mutually complementary districts, and its regulated employment of means.
It arose first in the form of that absolutism which regarded
the country and the people as the private property of the

ruler, and either used them despotically for the glorification of the throne, or filled the part of guardian towards them with well-meaning carelessness. Certainly in the suppression of innumerable petty sovereignties by a few great ones there was a gain in general order and security; but, on the other hand, the pressure exercised downwards by these great powers was continued, and the independence of the several communities disappeared before the centralization of national power. The age of the Revolution in shattering despotism shattered also those limits of free movement which it should have allowed to remain; in demanding equal justice and equal rights for all, an unlimited field for all activity and an open course for talent of every description, it took a hostile attitude towards all specialities of historical development in which it saw only hindrances to that freedom at which it aimed, and it carried on the work of centralization to the point of planing down as far as possible all characteristic differences. After men had seen how in the wide workshop of America success had followed the attempt to build up a construction of social order without restraint from historic tradition, and guided purely by the needs of the moment, without any greater limitation of personal freedom than those needs made necessary, and after France had gone back to the universal rights of man for the foundation of society, and had broken with history even in the externals of life, it seemed as though for the future the State would be only a great society for gathering in the treasures of Nature and carrying out the exchange of varying productions, established and governed by the will of all, and really without any moral duty of self-preservation, being indeed entitled to dissolve itself at any moment; yet with all this the fact was that the real freedom of individuals was tyrannized over by the common will of the majority. But the glory of the tremendous results which France achieved in its defensive struggle, soon brought back, in the national pride which it stirred up, a new and deeper consciousness of political coherence; other countries had not to atone so severely for the mistake of setting equality above personal freedom, but they attained more slowly to the development of this freedom and to the rejection of many limitations which had grown up historically, and, without any absolute right, obstructed social movement.

The history of these struggles, which is full of vicissitudes, does not come within our present hasty survey; that they are not even yet ended is a wide-spread and oppressive conviction of the present age. The spirit of criticism which called them forth has triumphantly maintained many general principles, but has not been very happy in the discovery of living forms in which these principles might receive a satisfying realization in fact. It has been established that the outline of the State is not irrevocably sketched out beforehand by history, to be merely filled in by the living activity of the people, but that the State is rather the comprehensive final form which social order has to take on in order to satisfy those aims of national life which are historically possible—that State guidance and administration must always have regard to the changing needs of the hour, as well as to that connection with the historic past by which the nation is constituted a nation—that there is necessary a division of power which on the one hand allows to existing men (who have a right to live) a modifying and innovating influence, and on the other hand allows to the representatives of the permanent element in historical development a restraining and guiding influence—that as much scope must be given to voluntary combination and the self-government of communities as is necessary for the production of all the commodities and the satisfaction of all the wants which they are naturally able to produce and to satisfy, and that just as much must this freedom submit to the limitations which the safety of the whole requires. But in the representative constitutions of our own time political art has either not yet attained to adequate forms, capable of ensuring the fulfilment of these ideal ends, or the forms appeared too soon, before the spirit that knew how to make a perfectly right use of them was developed. And as an effect of the oppression that has gone before, mistrust and not trust still continues to be the soul of constitutional life; the jealous guarding of formal political rights still outweighs understanding of and sympathy with the real ends for the attaining of which the existence of these rights is necessary; qualification for taking part in public business has not increased in the same proportion as the extension of the right to do so. Neither life nor education accustom the people sufficiently to the consciousness of important national ends. Skilfulness of co-operation in the prosecution of particular undertakings has no doubt increased; but the nature of trade, which connects the subsistence of the individual with a wide-spreading ramification of remote and foreign conditions, uproots the sense of citizenship which existed in earlier times, and which, arising from having all the interests of life in common, bound together the members of local communities; the diffusion of information has certainly made progress, but the inner progress of knowledge has been all the greater, because notwithstanding this diffusion the greatest part of the culture of which nations are proud remains wholly unknown to the majority of the people. How very indeterminate the line still is between what government should reckon among its duties and what should be left to the voluntary activity of the subject is shown by the unsettled disputes about free education, the political rights of different religious professions, and the necessity or dispensableness of a coincidence between political boundaries and the geographical limits of unmixed nationalities.

Not only life, but science also, has felt the influence of awakening criticism. During the Middle Ages minds had been ruled by traditions handed down from antiquity, and for a long time but little fresh result of investigation was added to them. From this time forward there comes out in ever growing strength that critical impulse of the Enlightenment, which indeed could never be so wholly absent in science as in other departments of life; the ingenuous setting forth of truth of which men believed themselves to be in possession gave place more and more to questions concerning the general cognisability of truth and the final principles of all judgment.

Science now first began to assume the character of an investigation which tests with careful exactness the worth and trustworthiness of its sources, considers the possible paths of progress, and is anxious to confirm its results by proofs and counter-proofs of every description, estimating even the amount of error which it is in danger of making in these proofs themselves, and allowing for such error in its deductions.   By this procedure science has introduced into even the most familiar departments of human thought the idea of universal laws to which reality is obedient in all particulars, and a lively conviction that results can only be obtained by using things according to these laws.   In doing this it has been able not indeed to *destroy* superstition, but to set bounds to its public and formerly bloody activity; by its astronomical discoveries it has given to imagination a new and enlarged background for cosmic theories; and by the development of mechanics and chemistry it has produced a boundless supply of instruments for the production of new commodities and the enlargement of commerce, and hence for the enlargement of men's intellectual horizon altogether, and for the increase of general wellbeing.   And whilst finally it came to make not only external Nature, but also the course of events in history more and more the object of reflection, and sought to trace back to universal laws the action and reaction of human activities, and the production and exchange of commodities, it gave rise to that progressive spirit of conscious calculation that is not content to continue passively in any merely instinctive condition of being or doing, but must actively mould the future by independent use of all available means.   Even within the range of this cheering human progress, sceptical and materialistic ideas and the dreams of socialism and communism show that neither firm foundations of knowledge nor practicable plans for the removal of undeniable social evils have as yet been in all cases discovered.

§ 9. The hasty survey of the external course of human development upon which we have ventured has convinced us how far hitherto human conditions have been from attaining that satisfactory state of equilibrium which may be regarded as the completion of historical development needing only to be kept up and worked out, not to be wholly transformed. Will this development progress steadily, or will it share the fate of those great civilisations which have preceded us in history, and which, destroyed partly by internal dissolution and partly by external force, have had a fertilizing influence upon the renewed attempts of later times only when they had fallen into ruin, and even then very gradually ? No one will profess to foreknow the future, but as far as men may judge, it seems that in our days there are greater safeguards than there were in antiquity against unjustifiable excesses and against the external forces which might endanger the continued existence of civilisation.

The civilisations of antiquity existed in national isolation ; the general difficulty of intellectual intercourse diminished, in the Middle Ages, the benefits which might then have been derived from the unifying power of faith ; now at last the different divisions of the world which have so long lived on in separation are striving to be something to one another; and the all-pervading current of interested traffic and of zeal for discovery is beginning to establish that external coherence of the human race by which the hitherto disconnected development of different sections may in the future become combined into a history of mankind. Already the wide diffusion of a culture which is on the whole homogeneous, and in which so many nations with all the varieties of national temperament participate, will prevent disturbances of development which may befall any of them in particular from becoming hindrances to human progress in general. And thus the power of barbarism over culture is broken. In consequence of the defective development of their knowledge of Nature, the civilisations of antiquity had not the weapons which would have enabled them in all cases to defend their intellectual wealth successfully against the savagery of the uncivilised world ; modern culture has through the progress of the technical arts become so well armed and so warlike that the inundation of civilised countries by tribes in a state of Nature has long ceased to be a probable danger; on the contrary, the assuredness of the influence exercised by civilisation as a whole upon the destinies of all parts of the world grows from day to day, though the regions thus affected may be too extensive to be as yet thoroughly pervaded individually by such influences.

And if by this extension in space, human culture has become established on too broad a basis to be easily washed away altogether even by a tremendous wave of barbarism, it has also attained internally, as the result of all its evolutionary struggles, a balance which throws its centre of gravity deeper than in the past, below the surface depth which is commonly disturbed by sudden currents. From the best features of many scientific researches which have failed in detail—from the increasing clearness of our retrospective survey of history and of human error—from the experiences of life itself which teaches us, in the exchange of necessaries, to have a due appreciation of what is foreign—from the wonderful advance in interchange of opinion which disturbs the one-sidedness of narrow intellectual views, bringing many currents of thought into beneficial mutual action, and unceasingly urging men to the exercise of comparison—from all these roots there has grown up, in the spirit of the present age, that peculiar temperament or dominant mood which we may distinguish by the name of Modern Humanism.

The difference between human development and the mental constitution of the lower animals consists chiefly in this, that the soul of animals is roused directly by a limited circle of perceptions to sudden and disconnected action; whilst the human spirit, far less endowed by Nature with instincts consciously directed towards their ends, has first to collect a copious store of experiences in the daily school of life, and by calm elaboration of them to work out gradually the motives of coherent action. An intensification of this self-control which distinguishes human activity as a whole from animal impulse is in a certain sense, and to a certain extent, a distinguishing characteristic of modern civilisation. Not indeed by any means because greater thoughtfulness is among the special merits of modern men and women, but because without any merit of theirs all the circumstances of life, education, and tradition under the influence of which they find themselves, are full of motives adverse to precipitate action, exercising externally as much influence in hindering the unrestrained outbreak of individual desires as they exercise internally in diminishing the effect upon the mind of innumerable exciting impressions. After all imaginable interests in life have been discussed and criticised from the most different points of view, and all these discussions and criticisms have, however much weakened and obscured, become part of the common consciousness, the world is less easily interested and less credulous than it was before; always indeed fertile in the production of strange views and heady schemes, but more moderate in its admiration for and its devotion to the improbable. In its bad form—that *used-up* condition in which all higher aims and all motives to action generally have lost their stimulative force—we may find this peculiarity of our own age repulsive, and all the more so in proportion as we know it only in the present and from living experience ; but as a matter of fact it is the case that this *aweary-ness* of a great part of mankind has not been lacking in any age which has produced a multiform civilisation abounding in sharp contrasts. And it has never either now or earlier taken possession of the whole race; but now more than previously there has developed alongside of this sterile passionlessness an allied but more earnest temper—tolerant, circumspect, and self-controlled—which among so many un-finished social constructions yet makes possible for us a life abounding in worthy pleasures, and keeps up our hopes of continuous progress.

This refined conscience of modern society makes itself felt in the most various departments of life. Not that it is able to get its commands obeyed without any trouble, or that the men of to-day are incomparably superior to those of the past in the excellence of their private morality; on the contrary, human nature is ever the same, and continues to resist the restraints imposed upon it with all its inherited passion and perversity, and evil and folly. But now it feels the reins drawn tighter; while every new generation is born with the old impulses and the old imperfections of its kind, each is forced to recognise the truth of the progressive moral insight with which growing civilisation gradually interpenetrates all the relations of life, as with a conscience that is ever becoming more fully awakened, and the utterances of which force themselves even upon the unwilling. Perhaps modern humanity falls further short of the increased demands of this conscience than the humanity of previous times did of the simpler and less complex demands of the conscience of its day, and a desponding view may attempt to depreciate modern civilisation even in comparison of the natural open savagery of past times, regarding such civilisation as mere surface polish and hypocrisy; but to us it seems that the very fact that hypocrisy is needed is a mark of progress, and that much that is base is now obliged at least to cloak itself, whereas formerly it would have ventured to show in its true colours. Upon the steady progressive development of this conscience, upon the pressure which it exercises on willing and unwilling alike, our hopes for the future rest; to a certain extent human action will be obliged to conform to it. Ambition with its lust of oppression will always remain; but the days are numbered in which men will attempt to justify slavery as such in the eyes of public opinion. The political destiny of nations may yet have many melancholy revolutions in store, for in order that practical injustice may be effectually prevented, comprehension of the existing position of affairs in any particular case and the improvement of favourable opportunities must be in accordance with the general conviction; still it is to be hoped that sentence of condemnation has already been passed on all invasions of the freedom and honour of individual life. Many attempts to interfere with liberty of conscience, to re-establish exploded religious dogmas, and to revive strange forms of worship may yet be made ; but they will never permanently succeed beyond the lines which some will find drawn by the spirit of independence, others by scientific taste, and the rest by the general sense of moral fitness which belongs to modern Humanism.

Such are our hopes for the future ; but what is the end of all ? Is there any such end in the sense of a goal which is to be reached, of a state of perfection which will be the conclusion and as it were the final accomplishment of all preceding historical struggles—and if such a perfect condition of things should be reached, will it last on to all eternity ? Or is there no such goal, and will the progress of mankind cease for no other reason than that of having exhausted all external means of advance, and will the imperfect condition then reached (which the inherent defects of human nature will not permit it to transcend) present that action of mankind (at last become uniform) which it is destined to carry on *ad infinitum ?* Or, finally, may not things go on for ever as they have done in the course of history hitherto ? Will not every civilisation that seemed to have been destined for eternal duration always be brought to ruin by some unexpected fate, and with every advance in one direction will there not be bound up a loss in some other direction, so that the sum of human perfection and of human happiness may always be a tolerably constant quantity, if we take, one with another— success and the exertion it necessitates, gain and loss, the growing wealth of civilisation and the increasing difficulty of full participation in it ?

The boastful days are over in which speculation flattered itself that it possessed the answers to these questions. Our intellectual horizon has gradually become wider again. We have bethought us that the history to which we can look back as sufficiently well known to form a judgment upon is of very limited extent; it embraces the classical nations, the European Middle Ages, and the immediate past. In this small and coherent fragment of development in which the parts are con- nected by tradition, it may well be that we can trace a

progressive advance.   We do indeed all lament that the
beauty of antique life has passed away—a beauty which men
have never been able to recover in modern times, in which
more northern countries have become the scene of the most
active development; but seeing that the ruin of antique life
lies before us as an accomplished fact, we might easily point
out the defects of civilisation from which that ruin proceeded.
These were only partially avoided in the Middle Ages—a period
which, notwithstanding its want of political and social stability
and unity, notwithstanding its strange mixture of profound
mental life and indescribable barbarism, yet shows us a splendour
of Christianity and a variety of individual development with
which we can sympathize; and shows them as being, though not
perhaps themselves actually higher stages of development, yet
hopeful steps toward such.   Far other was the aspect which
this period wore in its own estimation; more than once it
seemed to the minds of men, horror-struck by the boundless
misery which existed, that the end of the world must be close
at hand.   The gradual development of the modern European
political systems and of modern society was without doubt
another swiftly advancing wave of evolution, when looked at
in comparison with the immediately preceding period; borne
upon its summit the speculation of the age might momentarily have taken a view of history according to which it
would seem that no further development was to be attained in
the future, but that the evolution of the human race had, in
kind at least, reached its conclusion, and that the only growth
remaining for it was an extension on all sides.   But since
then we have become more cautious with regard both to the
past and to the future.

Growing acquaintance with pre-classical civilisations is
already beginning to arouse in us misgivings of having under-
valued them in many respects.   It is certain that they exhibited
such a full and complex and active life that it is impossible
to regard them as a mere unimportant prelude to European
history.   Our acquaintance with them is still but too meagre,
since their literatures, which are the only thoroughly trustworthy

witnesses of the depth and character of mental life, are partly lost to us and partly are difficult of access ; hence we are now unquestionably as much in danger of over-estimation as we were formerly in danger of inconsiderate neglect. But a philosophy of human history can give no satisfactory results concerning the course and the amount of its actual progress before these long ages of past time have become known, and their performances been compared with what we have hitherto regarded as the advances of later periods. On the other hand, the progress of mechanical art which has provided new means and resources, and of the economic sciences which have produced a better adjustment to human needs of the means for their satisfaction, has caused our attention to be directed more than ever towards the future ; the aggregate of all that it has to do, alter, procure, and arrange has never been present in such distinctness and importance to the consciousness of any previous age ; no time has lived so fully as the present in definite plans for the future ; we feel ourselves more stirred up to try and promote progress for the future than to investigate the steps it has made in past history.

So now again the future stretches out before us, more full of significance than ever, and we can fill it with dreams of boundless progress. But the course of history has already been so long that in looking back upon it we shall soon find ourselves obliged to confine our hopes within a narrower compass ; for plainly the regions within which there is any great probability of unlimited progress are very definitely circumscribed, and for all others the probability is but very slight. The splendid initiation of the rule over matter and its forces which rejoices us in the natural sciences, having been made once for all, we may reckon upon a rapid succession of new discoveries. From these may be anticipated a varied increase in the conveniences of life, greater facility in the satisfying of our wants, and purposive alteration of many of our customs ; the enriching of some favourably situated countries by increased use of natural resources and the addition of others to the abodes of civilisation ; increase of the population of the earth, and a manifold heightening of the activity of commerce. All sciences which combine facts of experience according to clear and simple laws of thought have the prospect of making continuous advances towards perfection; they will not only extend their knowledge of particulars, but will also learn by the discovery of new laws to understand better the coherence of all reality. These general results may be expected to exercise a favourable and gradually increasing influence even upon those sciences which, transcending experience and real existence and searching after God and divine things, early accumulated a store of valuable thoughts, but during the thousands of years that have passed since then have not been able to make any important addition to their early stock; and the progress may also be shared by that practical wisdom which has to deal with the necessary aims of our action, the binding commands of conscience, and beneficent social constructions.

But whilst this world of truth and of Ideas increases, human nature will not change, and life will always remain a long way behind the ideals that are set before successive generations. There will never be one fold and one shepherd, never one uniform culture for all mankind, never universal nobleness; but strife and inequalities of condition and the vital strength of evil will always continue. And we do not think this prospect desperate; for it does not seem to us that all history is so bounded by the limits of earthly life that we needs must see the dawn upon earth of its brilliant closing scene, that golden future which we dream of. On the contrary, as long as men are bound by their bodily organization to the material wants of life, their perfection and happiness must also be bound up with imperfection and ill, just as inevitably as any of our modes of progression both presuppose and at the same time overcome external friction. Both our virtues and our happiness can only flourish in the midst of an active conflict with wrong, in the midst of the self-denials which society imposes on us, and amid the doubts into which we are plunged by the uncertainty of the future and of the results of our efforts. If there were ever to come a future in which every stumbling-block were smoothed away, then, indeed, mankind would be as one flock ; but then, no longer like men but like a flock of innocent brutes they would feed on the good things provided by Nature, with the same unconscious simplicity as they did at the beginning of that long course of civilisation, the results of which, up to the present time, we shall now briefly consider, as a sequel to the review we have already taken of the external destinies of the human race.

BOOK VIII.

PROGRESS.

CHAPTER I.

TRUTH AND SCIENCE.

Stages of Philosophic Thought: Mythologic Fancy; Cultured Reflection; Development of Greek Thought; Science—Over-estimation of Logical Forms and Confusion of them with Matter-of-Fact—Philosophic Problems of Christian Thought — Limitation of Thought to the Elaboration of Experiences—The Exact Sciences—The Principal Standpoints of Philosophy, and its Efforts in trying to reach a Knowledge of the Nature of Things—Idealism and Realism.

§ 1. THOSE various embryon impulses from the development of which all human civilisation has grown up, have always sprung to life simultaneously as products of one common root, the unchanging nature of mind. Different periods of history may be pointed out in which one after the other, religion, art, science, law, and social problems, have become for the first time so distinctly present to the consciousness of mankind, that they seem to have been then first discovered or invented, to the advantage of future ages; but even in the very beginning of civilisation there could not have been altogether absent any one of those activities of the human soul which later became more clearly differentiated one from another, taking separate paths to various ends. And all are in continual mutual action as far as their requirements and results are concerned; and this most actively in just those times of dawning civilisation in which as yet none of them have found either cause or possibility of independent further development, in the possession of the wealth of some special department and in the peculiar mode of procedure made necessary by the nature of that department.

So if we try to survey this complicated whole of human civilisation as far as lies within the scope of our general intention, we cannot follow any one of the stems from which it has sprung without meeting ramifications by which each

communicates with the rest.   Yet still in the history of the
development of the whole mind, the development of scientific
knowledge takes a certain specially favoured position.   Whatever may be the several roots from which spring the creative
impulses of art, or the moral convictions of religious belief,
they are all, as regards the fulness and trustworthiness of their
development, dependent partly upon the extent to which this
knowledge subordinates reality to its sovereign influence, and
partly upon the clearness with which each has come to comprehend itself, its tasks, and its instruments.   To scientific
knowledge therefore as the general form under which all
activities of the mind reciprocally test each other, reflect upon
themselves, and bring their results together for transmission,
the beginning of the present considerations may be devoted.
In view of the immensity of the subject, we shall only briefly
refer to that gradual extension of cognitive knowledge which
with every fresh conquest both furnishes human activity with
new aims and also gives a different colouring to our whole
philosophy.   But even the progressive self-comprehension of
scientific knowledge, and the development of a definite conception of truth for which we are seeking, and enlightenment
concerning the intellectual means to these ends which are at
our command—even these points we shall only be able to consider with a one-sidedness of which we are fully conscious,
selecting a few points of view specially suited to our purpose.

Of three essentially different ways of looking at reality
which the awakening consciousness of mankind has gradually
come to adopt, we find the earliest in that mythologic philosophy to which at the very beginning of this work our attention
was directed by more restricted considerations.   Intensifying
the impressions of perception so as to influence the whole
mental mood, imagination—here going beyond perception—
makes to the reality which it finds those additions which seem
to be demanded by the vague feeling of a contradiction
between that reality and the tacit presuppositions of our
minds.   For every myth which gives a new and poetic
form to some phænomenon, bears witness to the activity of

human cognition, that can seldom be satisfied with direct perception because the content of this but seldom harmonizes
with those unanalysed demands which our mind brings with it
to the comprehension of reality, whether as innate endowment
or as the rapidly acquired fruit of previous experience. But
mythologic fancy has not a clear consciousness either of the
full content of the truth which it thinks it must recognise in
phænomena, or of the definite contradictions of truth which cause
mere facts to seem to demand some mythical and explanatory
transformation. The soul rejoices in the enjoyment of its own
activity, and is without suspicion of the numerous conditioning
causes which go to produce its happiness—a happiness which,
though it seems to arise without trouble and as a matter of
course, is yet a result laboriously produced;—it is accustomed
to see changes of the external world arise from its own
activity, and hence as yet it knows no truth other than life, and
no problem of cognition other than that of recognising in all the
forms and events of Nature an energy analogous to its own. It
seems to it that nothing has a claim to exist except that which,
if not itself mental energy, may yet be understood as the
action of some mind or as the traces left by some such action :
only those qualities and events seem to it natural which
have sprung from the activity of a living soul, or which
have arisen in some course of events incidentally set going
by spiritual activity intentionally or unintentionally. It is
true, indeed, that the unfamiliar character of particular natural
phænomena may cause the attention of the imagination to be
specially directed to them, but that which incites men to give
them mythic expression is to be found not so much in the
particular characteristics which constitute their unfamiliarity
as in the fact that they seem to appear without any explanatory history, which by connecting them with spiritual life
should afford a justification for their existence. The notion
of an unconditioned factual self-dependent existence remains
unaccepted; unrecognised the thought of a *nature of things*
which, independent of all spiritual life and preceding it as of
much more primary necessity, should produce the succession of

phænomena as its own inherent logical consequence. Not that the assumption of such a necessary connection of things has not constantly afforded secret aid to mythology in the combination of its personifying ideas. For in fact the briefest account cannot explain any striking natural phænomenon by a history of how it arose without assuming that the connection, transition, and succession between any two events which it brings together are to be comprehended by reference to an order of events which is of universal validity. But fancy (whilst in all its flights it tacitly relies upon that necessary connection of all things upon which also ordinary practical life must depend at every step) altogether overlooks this part of its own procedure, and is not conscious of the indispensable help which this nature of things affords in giving reality to imaginative constructions; for such a philosophy anything which seems full of meaning and significance has within itself all necessary guarantees of its truth and reality; and it is that which is living, or produced by what is living, that is pre-eminently full of meaning.

If this way of looking at the world were something that merely *had been*, it would be hardly worth this renewed mention; but the same impulses which led to it at the beginning of civilisation still continue to influence every human mind, even after the discovery of other points of view. In all ages the popular imagination explains the phænomena of Nature as resulting from something that had previously occurred. Since this or that happened, the bird sings such a song, the blossoms of such a plant are white instead of red; since something else, the bean has been slit in two and the salamander has had a spotted skin. But this tendency of thought, which in such examples pleases us as poetic licence for which we make allowance, has a much stronger hold upon us in other ways. There comes to all of us a time in our life in which a general dissatisfaction begins to overshadow the reality which we had previously accepted and enjoyed in all simplicity, while yet a hitherto hidden light seems to shine through the gloom. Innumerable particular perceptions

which we have not specially noted have filled us with a
feeling of the surpassing reality of beauty and goodness and
holiness ; innumerable others, just as unanalysed, have
produced in us disconnected impressions of the confusion and
uncertainty and evanescence under the burden of which all
reality suffers. And now this world of perception is to us no
longer the world of truth, but a mere world of perplexed
phænomena ; but we are able to look through it to another
and better world of real and ideal existence, to which the
enthusiasm of our soul would fain take flight. We of the
present day, however, are in our education and the conditions
of our life in the midst of the results of a labour of thought
that has lasted for centuries—results which surround us like
an atmosphere of the presence of which we are not conscious ;
and we are thus not likely to be carried by any flight
of such enthusiasm to a mythology which would be dispersed
and dissipated if brought into contact with daily experience,
with which it is in contradiction. But still we find ourselves
travelling the same path which fancy took when it created
such a mythology, in our youthful attempts to transform the
supposed real world of our dreams and forebodings into a shape
in which it may become an object of distinct intuition.

Youth strives to get from particulars to the whole, and not
to the universal ; it seeks more earnestly for the one meaning
of any phænomenon than for the numerous conditions of its
realization ; and it would always much sooner discover the
unity of the thought which binds together the disconnected
fragments of the cosmic course as living members of a beauti-
and harmonious whole, than inquire after the unattractive
conditions, upon the universal validity of which depends the
possibility of all beauty and of all connection of parts into a
whole. Memory will tell each of us that our youthful dreams
took this turn. We should hardly have been able to say why
exactly it was that we were not satisfied with what reality
offered ; still it was the case that reality could not justify
itself to our unanalysed dissatisfaction, and still less was it
comparable with the indescribably fair content of the dream

which hovered before us in indistinct splendour.   And then led away by the splendour of this dream we set to work to, as it were, develop afresh from it the whole fulness of reality ; for what else could the unrest be which filled us and urged our imagination to artistic production, than that very creative principle itself which is embodied in this world of phænomena ?   And what we attempted seemed to succeed ; as note can be joined to note to frame a melody, so one form gave rise to another, and one thought to another, and seemed to interpret to us the secret meaning and the inner connection of phænomena.   With the most unsuspecting confidingness we put our trust in the poetic justice which was the law of our imaginative constructions, and accepted it in lieu of that proof of their truth which we lacked ;   deaf to every reminder of universal laws (which without being themselves the highest, seemed to limit that which was highest), we passed by with utter disregard those actual facts which were in contradiction to our dreams.   Thus we shared the conviction of mythology that that alone which is worthy truly exists ; only that while mythology sought the worth of all existence in the joy of some animate life which it conceived of as similar to our own, the present more advanced development of thought led us to other ways—less obvious, though perhaps not more true—of embodying the ideal, which we reverenced as exercising un-conditioned power over all reality and as the secret source of its evolutionary energy.   And just as mythology forced the analogies of human spirit-life upon natural objects the furthest removed from any likeness to us, so we have imposed upon the nature of things the meaning and connection which our mind in moods of dream and misgiving demands for the satisfaction of its unanalysed needs.   And in this lies the strength as well as the weakness of these attempts, which are not peculiar to youth but are frequently repeated by science, though in the more modest forms which an increased experience of life forces them to assume.   Their strength, I say ; for having sprung from a powerful agitation of the soul, which intensifies all the deepest longings of the mind so that

they become a living mood, these efforts are real experiences
in quite another sense than the thoughts which calm reflection
attaches to phænomena at a later stage, with greater reserve,
and as it were more on the surface; this living intuition
may divine many a truth, many a relation between things
which more deliberate thought would discover either laboriously
or not at all. For in truth it must be even as we were
taught by the feeling which animated our dreams—it must be
that that which is worthy is that which truly is, and there
will come a time when the soul which has learnt to know
itself will be able to return to this re-acknowledgment of its
primal faith. But it will have to overcome the weakness
which led its early efforts astray. Instead of being as it were
mastered by the feeling, it must seek to become master of it;
it must not let the seeds of truth spring up from the soil of a
passionate mood, in a series of poetical developments, along
with seeds of the most casual errors and of "idols of the
cave," but must learn to follow the course of things along the
path which it really takes.

§ 2. Helped by the thought of long ages of past time, a rich
inherited stock upon which we can draw, it is easy for us to give
up an inadequate standpoint—a standpoint, however, after reaching which the historical development of human consciousness
had to traverse a long distance before attaining a more tenable
position. The mythological beginning, both in history and in
the life of the individual, is followed by a period of active and
inquisitive reflection; meditation, no longer supplementing the
world by poetic inventions, gives itself to a consideration of
the course of events, and gradually works out to greater
clearness the idea of a *nature of things*, with regard to which
the proper attitude of the human mind is one of docile
recognition. In mythological philosophy it was only the
notion of Destiny which had any reference to a necessity
regulating the connection of things; but this view of necessity
was not such as to be favourable to the development of knowledge. For Destiny, wholly devoid of cause or reason, did not
bind the course of events to general laws, which as universally

valid truth would rule in unnumbered similar cases, but it connected together particular events by a link which, because destitute of law, must be also incomprehensible. Not knowledge but prophetic inspiration, not thought which from a basis of reason calculates what must happen but intuition that becomes aware by signs of some approaching event, was the faculty to which this necessity revealed itself. Gradually at first, by steps which cannot be historically traced but may be conjectured, a fitful awe of incomprehensible fate passes into the clearer thought of a necessity which as being the *nature of the thing* is no longer regarded as joining things together fortuitously, but as joining, according to general points of view, things which have a connection with each other. This transformation of view, with which for the first time self-existent truth as an object of scientific knowledge is brought face to face with intelligent cognition as the instrument of its comprehension, was no doubt due to an impulse originating in the fact that life itself urges men on the one hand to an industrious cultivation of Nature, and on the other hand to the establishment of social relations. Both were impossible without the practical application of general rules of judgment, of which, later on, dawning reflection had to become conscious as forming the principles of its procedure. And these rules denied equally both the unregulated supremacy of a blind fate, and the self-sufficiency, the power of self-realization, which had been attributed to everything that had intrinsic worth.

In contrast to the temper of youth, this new conception of the world commonly appears in the development of the individual as the culture which results from life and the experience of life, and there is between the two phases an undeclared hostility. The idealism of youth, with its confidence of being able to bring all reality into subjection to its fairest dreams, is broken in upon by the realism of riper age which gives calm recognition even to what is unimportant when it occurs as a fact, as one of the unalterable fashions of the world's course. For there comes a time in our lives when the heart grows weary of fiction, and hungers and thirsts for

reality; there is an indescribable joy in the consciousness of
having gained insight into a part of that which not only stirs
our longing, but surrounds and upholds us with the incomprehensible charm of reality, and the mind of the observer is
conscious that such a feeling raises it infinitely above the
pleasing but unstable moods which once filled its being. To
the reproach of having become unreceptive to the ideals of
youth, it rejoins that it has now learnt instead the virtue of
renunciation, and does not forcibly transfer to the world the
results of subjective intuition, but is content to learn with awe
and humility, from a comparison of experiences, as much of
the nature of things as they themselves reveal. And now
indeed the individual can hardly expect that in his limited
sphere of experience the secrets of the universe should be
fully unveiled to him. Fixing his attention at different points
of experience, he will have to content himself with discovering
the proximate causes of some special groups of phænomena
without reaching the ultimate principles upon which their
whole variety depends. This fragmentary method characterizes the teaching of life throughout. Many trains of
thought starting from particular natural processes, energetically
follow out the connected course of these processes for a time,
but come to an end when they have found the *axiomata media*,
beyond which abstraction from perception cannot proceed.
Various maxims arise from the consideration of conduct, often
bringing together and answering cognate questions with great
acuteness of discernment, but unconcerned both about first
principles and about their own contradictions of one another.
But even in the very want of connection and unity which
marks this living development there is a charm which fills it
with a sense of wellbeing—the charm of half revelation. If
to our view the topmost summits of reality are veiled in mist,
they appear as a whole only so much the vaster and more
infinite ; even the contradictions to which we are led by a
consideration of its different parts strengthen the sense of
submissive security with which we consider, and merge ourselves in, a world so vast as to be able to present to us such

different aspects on the different sides which it turns towards us. Reverence for the inherent truth of anything is greater in this mood than it was in the enthusiasm of youth, and he who has experienced it will find that the suggestive poetry of this prose is more profound and more full of content than the sparkling foam of youthful dithyrambs.

§ 3. In the history of mankind we can trace this evolution of consciousness nowhere but in the gradual development of Greek science. It seems that Greek philosophy, following this path of a living development having many starting-points, was occupied (until it reached its culminating point in Plato and Aristotle) in trying to arouse everywhere a consciousness of the existence of a truth, and of a nature of things, which constituted possible objects of human cognition. Making guesses and using the analogies of perception with more or less penetration, it made repeated attempts to obtain a provisional formula for the content of truth before it turned its attention to consciousness itself, and inquiring into the nature and instruments of human cognition, passed from the fragmentary activity of living development to the coherent method of scientific investigation. For the rapid survey which it is our present object to make, its particular doctrines are indifferent; what is important is the general condition of human culture and insight which its procedure reveals.

Poetry had early succeeded in expressing the results of life's experiences in striking pictures and in general reflections. And when the first Greek sages appeared enunciating gnomes—such as that which blames all excess, or that which connects every suretyship with some fatality, or that which exhorts men to self-examination—what they said seemed to contain much less than was already familiar to the poetic consciousness of the nation, and thus they appeared to be behind the civilisation of their own time. But if this had really been the case, they would not have received the admiration which has connected their names with the dawn of philosophy. The first awakening of the scientific spirit always causes surprise, not by its unusual wealth of new matter, but by its special mode of regarding that which is already known. Compared with the wealth of thought which the national mind possesses in its poesy and employs in life, the infancy of science always appears inexplicably meagre ; it is only a high degree of perfection which enables it (by means of discoveries which it alone can then make) to be supreme among the mental activities of life. The rich variety of Homeric characters and the soul-painting of Sophoclean art had caused the Greeks to see clearly and sympathize warmly with all the depths of spiritual life, long before its dawning speculation could answer (even with the most inadequate and superficial conjectures) the question what the soul is in itself. But such special instances are unnecessary. Language itself shows in its structure and use the great chasm that exists between the wealth of spontaneous living thought and the poverty of reflection which strives to analyse its own procedure. Without the trouble of seeking, and with the certainty of a somnambulist, the most uncultured mind finds and uses forms of expression which language has invented for him, indicating the finest shades of difference in the relations of things, of events, and of thoughts ; but even with the help of the most complete apparatus of words of " second intention," he would be wholly incapable of rendering to himself or others any precise account of the content of the thoughts which he expresses (as easily as he breathes) in forms of language the use of which has become a living habit to him. From this mere thinking life to self-conscious thought a decisive step was taken by those first sages. When they expressed their familiar and to some extent unimportant truths as simple sayings detached from poetical surroundings, constantly repeating them with the same emphatic simplicity, they gave to their content a new form and with this a new value. They roused the attention of the mind to the fact that the general maxims with which so often before it had as it were toyed unsuspectingly are not mere breathing-places for the soul when roused to excitement by a consideration of events ; but that they are in all seriousness real laws of the cosmic order, fragments of that self-existent truth, that nature of things, to the recognition of which the fully awakened consciousness had to apply itself. Hence these sayings, although founded on particular cases of experience and referring to them in their phrasing, plainly had a general symbolic meaning; they showed that in other departments as well—everywhere in fact—similar conditions governed the connection of events.

The study of Nature passed through the same stages as the study of human life. When we see all phænomena derived sometimes from water, sometimes from air, now from fire, and then from the confusion of chaos, or by determination of the indeterminate, we are surprised at the poverty of this conception of Nature when compared with mythology, which knew all this and much more, and which reproduced the characteristics of phænomena with much more penetrating subtlety. We may think it strange that when Anaxagoras declared *νοῦς* to be the principle of the universe, without being able to apply this thought to any particulars of perception, he should have seemed to his contemporaries to be announcing something great and new; for not only had mythology always had the same notion, but it had also been able to show, after its own fashion, how *νοῦς* works in Nature in individual cases. But only after its own fashion; it is easily seen that notwithstanding all the poverty of its content the dawning philosophy was new because of the different mode in which it apprehended things. Whilst fancy hitherto had merely gone on to dream after dream of phænomenal beauty, reflection now became ever more and more conscious of that universal necessity which, as the nature of things, gives order, tension, and stability to the whole world of phænomena; and the unskilful essays which followed one another in rapid succession helped to form ever clearer and clearer notions of primal matter, primal force, and universal modes of motion from which individual creatures and events proceeded, as results brought about after various fashions. But there still went on working the youthfulness of thought, which hankers after intuitive perception and is led away by circumstantial histories of the origin of things from investigating the final conditions of their reality. In order to indicate that content of the really existent which it strove to grasp, the mind turned at first to remarkable phænomena of internal and external experience ; and brought into prominence as the essential principles of the universe those comparatively permanent and universal phænomena by which, as Matter or Cause, the rest were in various ways conditioned. From such notions as that the really existent is water or air, more practised reflection has in course of time risen to more abstract determination ; the Infinite, the One, Measure, Order, gradually took the place of the more sensuous early notions. But all these changing dicta belonged, as far as form went, to the " contingent aspects " of growing development. Of course each of these principles was chosen because it seemed to possess the qualities which the prejudices of natural thought require in that which is to be accepted as the supreme principle. But these principles were not analysed, nor comprehended in all their fulness, and one or another guided individual thought according as it seemed from some accidental cause to be more clear to consciousness ; the particular thought which corresponded to his own obvious requirement was one-sidedly regarded by each as the whole content of the supreme principle, and he thus came to regard the whole principle as being embodied in the phænomenon which rendered that thought most strikingly perceptible to the senses.

In this process of reflection there were traces of recent emergence from mythical philosophy ; from which also another heritage had descended to it—that reverence for symmetrical and rhythmical forms of occurrence in the order of events which would very naturally arise when the mind, though it no longer sought in the world a direct copy of its own spiritual life and its own joy in existence, yet strove to find (as it were in compensation for this) in the independent nature of things which it began to recognise, a perfection peculiar to that nature.

In the real existence which men sought for, the ideas of all goodness and beauty and holiness were so blended with the idea of reality, that the æsthetic relations of form indicated by the former expressions, seemed to belong also to the essential nature of reality itself. This notion of the necessary symmetry of the really existent (which no doubt contains a kernel of truth the more exact determination of which is worth a searching investigation) is an assumption which has influenced the philosophic conceptions of all periods, and it has not lost its power in modern times ; the early ages of antiquity were wholly swayed by it. Long after people had begun to speak of the laws of things, these laws were not understood as general rules of the behaviour of phænomena which did not in themselves require any definite form of phænomenal occurrence, this being determined by the peculiarities of the special cases to which they were applied ; they were, on the contrary, regarded as definite, symmetrically ordered, harmonious rhythms in the occurrence of phænomena— rhythms intuitively perceptible which, since they embrace the universe, determine the direction of every individual's movement. For a long time the tendency of reflection was to class the fates of individuals under great existential habits of the universe ; the attempt to explain the final form of cosmic order as resulting from the reciprocal action of individual circumstances, was made later. At the stage to which we are now referring, the thought of the whole which with predetermined form and development precedes the parts, quite outweighed the thought of general laws, which first enable the parts to form a whole or the whole to be built up of parts.

§ 4. Tradition connects with the name of Socrates the record of the step by which living reflection was first led into the methodical path of scientific cognition. Earlier specula- tion had imagined that it could only discover that nature of things which is the source of all concrete objects by a process of guessing, which had to penetrate through all kinds of phænomenal obscuration, in order that, far behind them it

might find the being itself. The objects of Nature that surround us, and the events that take place among them, were
then connected with the real being, not indeed by something
that had happened, but by something that was continually
happening—whether this was conceived under the form of an
element that, passing through many intermediate stages, had
transformed itself into the multiplicity of individual things, or
as an order and a rhythm that can only be fully perceived
in the great whole, and that seems to vanish in parts and
individuals of that whole, through inexplicable contradictory
fluctuations. Now for the first time it became clear to
men's minds that the nature of things is present everywhere,
that its connection with the existing world is not a connection
dependent on history, but is of the essence of the world—that
this nature consists not in any one element, not in any one
definite form of existence, but in a Truth which, ever the
same in small things and in great, joins all parts of the
world, and is the very nerve of connection between them.
It was from this third standpoint that cognising knowledge
first became possible ; for now for the first time there was a
cause and basis for ever-present necessities of thought, instead
of the merely historical cosmic facts which formerly men had
been able to guess, to describe, or to observe without comprehending them. Yet it was very long before the fruits of
this new standpoint became to any extent matured, and the
injurious effects, which the deficiencies of the first historical
harvest in this field left to posterity, have not even yet all
disappeared.

That the objects of observation and the various images
which fill our thoughts may be co-ordinated under general
class concepts, and that the content of these concepts is
eternally the same, and is what it is, freed from all the
mutation and change to which its particular manifestations
in actual fact are subject—these two apparently insignificant
discoveries mark the beginning of the new period. These
two insignificant discoveries, I say ; for they only revealed
what the living course of human thought had always pos-

sessed; and yet they both had very important results. For long as it was since language had begun to indicate in words the general concepts of things (as indeed it was inevitable that it should), consciousness had still continued unaware of what it was about; and even for the contemporaries of Socrates it was hard to see that the convenience of using a common name for different things arose from their dependence upon something which was common to them all, and in all self-identical. And inevitably as both reflection and practice had tacitly assumed the unchanging self-identity of every notion and every determination of things, yet the theoretic consideration of reality had led to confused ideas of an eternal flux of all things, in which the mutability of reality had come to be regarded as involving also the mutability of truth, and every fixed standard of fluctuating particulars was lost sight of. As opposed to this incapacity and that confusion, the conscious emphasizing of universally valid truth (narrow as the view of its content might as yet be) appeared as the first basis on which a firm position might be taken up, and from which further advances might proceed. After men had long been striving to grasp the highest real existence at one bound, as it were, there began the logical period of thought, in which it became possible for men to attempt first of all to make clear what must necessarily be required in that which was the goal of their desires; and then, and only then, to ask whether that which satisfies these requirements is to be found, and if so, where?

This newly-gained insight set two tasks for further development to accomplish: first, that of becoming conscious of the forms and principles of procedure which are indispensable for observation and for the connection of our thoughts in order to reach that which the thinking mind should accept as truth; the foundations of this logical science were laid in a masterly manner by antiquity, but the science was left far from complete. Just as unavoidable was the second problem— the inquiry as to the worth which all these laws of thought (inevitable for our intelligence) possess as regards the comprehension of truth and acquaintance with things themselves ; and neither in ancient times, nor in the long course of development which science has since then passed through, has this investigation reached a solution of manifold doubts, which in their most general form we must now consider, as far as they can be made intelligible without systematic scientific investigation.

If in common life we seek by a comparison of apparent signs, by the use of numerous analogies, and by inferring back from results to their causes, to ascertain some secret, hidden, or forgotten fact, we do not doubt that all the indirect courses thus taken by our thought are means which are only necessary for *us* who seek ; necessary because of the position we are placed in as regards the object of our search ; we do not suppose that the nature of the thing itself which we are desirous of explaining has gone through a similar series of steps in the course of its development. The course which our thought has taken is therefore regarded by us as merely our subjective mode of procedure, and as the result of this we hope indeed to arrive finally at a knowledge of the nature of our object ; but we do not imagine that our labour, while it is in progress, is an exact reflection step for step of that inner process of development by which the object was formed, or of the inner coherence by which its actual existence is maintained. This notion of the relation of thought to its object, which appears unsought in such cases, contains in combination two assertions which are sometimes separated into two opposed views. Every useful instrument must fulfil two requirements ; in the first place it must be suited to the hand that is to use it, and in the second place it must be suited to the nature of the object to which it is to be applied. Just in the same way the processes of thought must be determined both by the nature of the thinking subject, and also by the nature of its objects. But the peculiarities imposed upon it by these several conditions cannot be quite the same.

The intelligence of finite beings is not placed at the centre of the universe, and cannot grasp at once the whole of reality in the true and natural relations of dependence which subsist between all its parts; placed amid phænomena it finds itself face to face rather with the derivative properties of things than with their nature, and much oftener with results than causes; it is forced to become acquainted successively with the parts of a coexistent whole.   Thus there arise in our thought an immense number of necessary activities of discrimination, combination, and relation, which are all merely preparatory formal means of knowledge, and to which we can by no means ascribe real validity in the sense that their succession presents a reflection, exact or resembling, of the internal processes and reciprocal action upon which the reality and development of objects themselves depend.   But on the other hand, it is just as plain that when these laws of thought are capable of leading to a knowledge of things, they cannot be mere subjective forms in the equally one-sided sense that they arise from the organization of our mind as innate modes of its activity without having any original relation to the nature of the objects with which they are destined to deal. On the contrary, thought and existence certainly seem to be so connected as that they both follow the same supreme laws, which laws are, as regards existence, laws of the being and becoming of all things and events, and as regards thought, laws of a truth which must be taken account of in every connection of ideas.   All reality is connected according to these laws in such a thoroughgoing fashion, and with such unbroken logical consistency, that our thought may at its own choice use any mesh in the network as its point of departure, and proceed therefrom in any direction it will; and as long as on its part it makes those laws the rules of its progress, it will always be sure to arrive at any other point of reality which it seeks, however much the direction and the windings of its own motion between the two points differ from the real connections by which reality itself connects one of its divisions with another or causes one to proceed from another. The calculation of the peculiar properties of a plane figure by means of a diagram may serve as an illustration of this.   To help our demonstration we draw lines in the figure, and the larger the number of equally useful constructions among which we can choose, the less can we regard any of them as essential parts of the figure. We attain the correct conclusion by a concatenation of propositions which does not in the least follow any real process of construction in the object; but so inexhaustible are the possibilities of connection in any geometrical object, that our thought, setting out from any selected point, may take the most various ways of covering the object with a network of relations, and can always rest assured that at every halting-place in its circuitous course it will find some essential relation, and that at the end of the whole methodical procedure it will infallibly reach the truth for which it was seeking.

But this relation of thought to its objects is clear to us only so long as we can keep in view the complicated whole of such considerations, in the order in which we used them in our demonstration; for there arises the appearance of quite a different state of things if we go back to the separate elements of thought from the combination of which that whole has grown up, that is, to the forms of the idea, the concept, the judgment, the syllogism. It will seem to us as though a complex train of reasoning can only take an arbitrary course on the whole (arbitrary, that is, within wide limits), because it directly expresses and realizes in these its component parts those laws which thought has in common with existence; and it will seem that the circuitous course of our thought can coincide finally with the nature of the thing only because those component parts harmonize with it; hence allowing only of such modes of combination as belong to the logical consistency of this nature of the thing, however much freedom there may be in other respects. Therefore when our thought combines individual ideas into one whole, when it integrates many similar ideas to one general concept, joins concepts to make judgments, and judgments to make syllogisms, it will easily believe that in these processes it is copying the very inner relations of its object; and each of these logical

forms will, on account of the mutual relations into which it
brings the component parts of the train of reasoning, be
regarded by thought as a reflection of some element of the
relationship which exists between the constituent parts of the
object.

I reserve for the present the proof of the illusiveness of
this semblance; if for the moment we assume its decep-
tiveness, the injurious consequences in which it involves
us are clear.   For whenever we consider the reciprocal
relations of those parts of ideas which we have combined into
one whole, or the process by which, dropping or adding
characteristics, we transform one idea into another, we shall
be inclined to believe that we are thereby enabled to understand not only the structure of our idea, but also the
inner articulation of the object ideated, not only the
procedure of our own thought, but also the course of the
facts which actually occur, as the object comes into existence
and develops.   This confusion between clearing up our concepts and analysing the corresponding objects is an error of
reflection which is very natural, and recurs in the most varied
forms; and it may be allowed to occupy a certain phase
in which, when men's attention has been first called to the
presence in our mind of a reign of law to which all truth
must conform, they are very easily led to over-estimate a discovery so important.   If we say that the knowledge of things
belongs to Metaphysics, and that the doctrine of the forms
of thought to be used in knowledge belongs to Logic, then we
may say that antiquity has very generally erred in thinking
that it could answer metaphysical questions by logical
analyses of ideas.   And in this lies the cause of the unfruitfulness which strikes us when we look to antiquity for any
furtherance of knowledge as regards facts—an unfruitfulness
which we find side by side with a splendid exhibition of
intellectual strength.   Being quite unable in this hasty survey
to give any account of the latter, we must content ourselves
with indicating some of the by-ways into which later times
have been misled through the influence of antiquity.

Plato's doctrine of Ideas was the first attempt to grasp the nature of the thing in general concepts—a grand attempt which, though unsuccessful, yet exercised an influence for long ages to come. There were strong inducements of two kinds to make such an attempt. In the first place, observation of living creatures has in all times given rise to the thought that nothing but the living generic concept can be the combining force which in every individual unites properties and vicissitudes into one whole of orderly development, causing in each the realization of the same form of life, notwithstanding the transforming influences of varying and casual external conditions. But again Plato's doctrine of Ideas, as opposed to the sophistry which was analysing away all sense of duty, rendered splendid service by attempting (in obedience to the second and equally strong motive) to point out that the worth of human actions is not temporarily determined by arbitrary institutions of local prevalence or changing taste, but that it depends on universal immutable moral Ideas of an absolutely good and just and beautiful, and only exists in proportion as these Ideas which are always self-identical are reflected in the various and changing forms of action. In these two cases the question is of phænomena and events which we can easily imagine to be the work or aim of reality; we find no difficulty in understanding the generic concept of living creatures as a type which the cosmic order seeks to realize in innumerable copies; still more are we inclined to do homage to the other conviction (to which enthusiastic expression has so often been given)—the conviction that universal original types of the Good and the Just and the Beautiful, are to be conceived as the exalted patterns which our actions have to imitate. So that here general concepts seemed to contain the essence of the thing, because this very essence consisted in the universality of an ideal which was intended to be realized in innumerable particular cases.

But it is not all the objects of reflection of which we can frame universal concepts that favour this way of looking at

them ; so that if we disregard the fact that in the instances
cited the nature of the content ennobles the form of the concept, and look at this form as indicating universally the
essential nature of things, then in consistency we must go
further than we would.   That every particular thing which is
beautiful and good and just, is beautiful, good, and just only
through participation in eternal Ideas of Beauty in itself, Good
in itself, and Justice in itself, was a notion which could inspire
Plato with enthusiasm ; but that a table is only a table and
that dirt is only dirt through participation in the eternal Idea
of the Table or of Dirt was a difficulty which Plato himself
encountered but did not remove ; these concepts of commonplace realities which from a logical point of view are just as
legitimate concepts as any others, could not well be reckoned
as imperishable original types in that world of Ideas of which
the phænomenal world is but a dim copy.   These, however,
were just the cases which early directed attention to the fa;t
that the realm of thoughts and concepts with the whole
ordered system of its internal connections is not a reflex of
the realm of existence, but bears to it that different relation
which we referred to above.   Our own voluntary actions
adapt the materials of Nature to our ends in many ways, and
thus among other things produce the table, of which there was
no original type among the integrating constituents of universal
order ; but everything in the world is so connected according
to law and rule, that of these products of art with which we
enrich reality there may be just such concepts as of the
original constituents of reality, and general logical laws are no
less applicable to these concepts than to the Ideas.   And
further, the course of our thought arbitrarily compares together
things which are quite unaffected by the comparison, or brings
them into relations which are quite unessential to them, and
thus produces the concept of dirt, which certainly does not
express the nature of anything ; and yet such a notion is a
help to thought which we are justified in using ; for as long
as we use it with reference to those considerations to the
arbitrary prominence of which it owes its existence, all the

laws that thought prescribes to concepts hold of it, and their application leads to correct conclusions.

Between *truths* which are *valid* and *things* which *exist* Greek philosophy always made very inadequately the distinction which our language marks plainly enough by these two expressions; valid truth always seemed to it to be a particular department of existence. And it is with this very distinction that we are here concerned. It is upon the fact that the same supreme truths hold of the ultimate bases of both thought and existence that the general possibility of their mutual relation depends; but the relation does not consist in this, that a fixed number of concepts *as existing* are to us *things*, and *as thought* are the *ideas of things;* on the contrary, our concepts may be increased indefinitely without any addition to the sum of existence. And further, setting out from innumerable arbitrarily chosen standpoints, we may build up the same whole by constructions of particular ideas, varying according to the variety of these standpoints; and thus there may be many definitions which define the same object with equal accuracy and exhaustiveness. None of these definitions *is* the nature of the object, though each *is* *valid* as to it, because there is no object of which the nature can be conceived by means of an Idea that is isolated, and unconnected with all others, and characterized *only* by eternal self-identity; but each object has its nature and its truth only in as far as there are general laws of reciprocal behaviour which are valid as to it and all others, and according to which it not only is distinguished from others as a coherent whole, excluding all others from itself, but also reveals itself and enters into connection with others. Thanks to these laws, thought can form innumerable new concepts, since under their guidance it makes arbitrary lines of communication between things, and is conscious of each movement which it thus accomplishes as the idea of a certain connection between the things. Of these new concepts, Plato's great successor Aristotle would perhaps have said that they were indeed potential in the nature of the thing, but in point of fact were first made actual by the subjective procedure of thought.   A consideration of this relation would have led in the first place to a clearer distinction between that aristocracy of Ideas on the one hand which (as the generic concepts of living creatures and of determinations of moral value) are among the eternal types that are original constituents of the cosmic order, and on the other hand that proletariat of concepts that increases indefinitely the more curiously thought plays with the infinite possibilities of comparison and connection among things.   But this distinction (as to the first part of which we reserve some important doubts) would have been crowded out by a second, which admonishes us to consider not only the form of the concept, but also the form of thought of the judgment, and to search for the truths—expressible only in this form—without which no intercourse between existing things and no cosmic order is conceivable, one of the things which we owe to this form of judgment being the possibility of valid concepts.

§ 5. This world of concepts not only could not be brought into adequate connection with reality, but further, it did not attain the internal articulation necessary for a typal world of Ideas.   It remained a collection of motionless Ideas between which nothing takes place in the present, and nothing is foreshadowed as about to take place in the future, and which only cohere among themselves by means of logical connections of subordination, and compatibility or incompatibility.   All the transitions from one to another which thought finds or establishes between the objects of perception are but misused by having their meaning likewise petrified into eternal and everlasting Ideas, which take their places calmly beside the rest without thinking that their business was not to be links, members of the series, but only copulas between other members.   Thus the eternal self-identical Idea of identity stands beside the equally eternal and self-identical Idea of unlikeness, and along with them the eternally motionless Idea of movement ; none of them makes an effort to exist after a fashion suited to its content, as a relation of predication between two other points, or as the movement of something in some direction. Aristotle was sensible of these deficiencies; a taste for the observation of Nature, and systematic occupation with the forms of thought, drew his attention to the numerous relations which connect the individual elements of reality into one living whole, and to the ways in which these relations are expressed by our thought. He knew that Ideas are not existent but valid, that a truth is expressed not by a concept but by a proposition ; he searched language for all those expressions by which we indicate the manifold relations between things which we find or assume ; he frequently distinguishes between the dependence upon one another of the different parts of a complex thought and the order in which the elements of the corresponding reality condition one another. But his practical philosophizing no more avoided confusion between the logical analysis of thought and the investigation of things with reference to the form of judgment, than Plato did with reference to the concept.

In the judgment we combine two ideas by means of a third ; we attribute to an object a property or a condition or the manifestation of an activity. As long as these predicates have once for all been received as unchanging and belonging to the subject in its integrity, the judgment expresses no event, but only analyses our idea of an unvarying content ; and as long as this is the case, it may escape our notice that there is need to ask specially what exactly there is in the object itself correspondent to that which (with obviously figurative expressions) we call its possession of some property, its sufferance of some condition, or its manifestation of some activity. If on the contrary we attribute to a subject assumption or loss or alteration of predicates—that is, when we describe an event—we have a more unmistakeable interest in knowing what it is that actually happens to this subject —the very object itself—to justify our imitative thought in now conceiving of it under a second idea which has arisen from a previous idea of the same object by the addition of new or the dropping out of old marks. It would be difficult to show that the Aristotelian philosophy generally satisfies this requirement.   Much occupied with the concepts of change and of becoming, it yet in analysing them makes no inquiry as to what it is that justifies us in their application.   We are told indeed that in change, properties always pass into their opposites ; for a brief moment we indulge the hope that this remark indicates at least the path and direction which are taken when there is alteration, thus revealing a truth which, since it could not have been the product of mere thought, must have been directly gathered from the nature of the thing ; but it speedily appears that nothing more was meant than that naturally nothing can become what it already is, but only something that it previously was not.   Thus this somewhat inadequate information merely expresses the result of an analysis of our idea of becoming, announcing that in it two different individual ideas succeed one another in such a way that when the one comes the other goes.   But what is it that in existence and reality so corresponds to this course of our ideas that we are able to believe that the ideas are a copy of the reality ?   We do not know; the transformations which our idea of an object undergoes when the object changes, are, in the last resort, regarded as if the alterations of the object itself on which they depend were quite similar to them, and as if a knowledge of them could take the place of a knowledge of the objective alterations.   When a white object becomes black then in our representation, in the mosaic of marks which constituted its mental counterpart, we, as it were, erase the mark of white colour, and replace it by one of black ; if we then ask what has happened to the object itself, in virtue of which we have been able by this alteration to make our idea correspond to it again, it seems that the process was essentially just the same; the white departed from it, and the black came instead.   That *properties* inhere in and are connected with the *thing* quite otherwise than the *marks* (or parts of presented ideas) are related to the *concept,* is a fact of which now and then a theoretic suspicion has been expressed, but this has had no important effect upon practical philosophic investigation.

The celebrated concepts of δύναμις and ἐνέργεια, which as Potentiality and Actuality are still favourites of philosophical dilettanteism, bring these barren considerations systematically to the investigation of all objects. If a thing passes from one state into another, the conditioning causes of the later state are contained never wholly but always partially in the earlier state; if they had been contained wholly in it, then the earlier state could never have been, but the later would have existed from all eternity without any need of coming into existence; since they were only contained in it partially, there was something in the earlier state which contributed to the later without actually bringing it into existence. If we compare the two, the ingenuity of thought cannot fail to set down the possibility that the second state may at some future time arise, as an actual mark of the first state. The nature of such an abstract concept as that of possibility, which makes it very difficult to handle, here conceals the barrenness of this procedure which in other and similar instances is very obvious. In any case of a b that was greater than c and less than a, these properties of relation were regarded by the ancients as characteristics originally existing in b, and they greatly wondered how it was that b could be at the same time a greater and a smaller. In the view of modern thought, these same properties of relation belong to b only when it is compared with a and c, being then new expressions for its really unchanging magnitude. It is after an equally shallow fashion that the possibility or δύναμις of the later state is contained in the first. The real task which cognition has to accomplish in comparing the two is to indicate definitely what the earlier state was; and to prove that being what it was it formed a part of that circle of conditions, which (subsequently completed by the accession of other conditions) helped to form the whole cause of the second state, and hence could subsequently produce the realization of that state, which earlier in the absence of the complementary conditions it could not do. On the other hand, it is wholly useless, and merely produces delusion as to the real problems of knowledge, to assume generally for every reality merely a previous corresponding possibility without inquiring what are the actually existing facts upon which the possibility of the subsequent change depends.

It may be objected that δύναμις and ἐνέργεια or ἐντελέχεια are not merely the bare concepts of possibility and actuality but intuitions of something more profound. It is true that as Plato's Ideas sometimes denoted all concepts merely as such, and sometimes denoted a selection of what should be typical concepts, even so that more general signification of the technical terms above referred to which follows from Aristotle's own illustrations, is limited to certain actually favoured cases. For instance, there is nothing to hinder our regarding the state of rest of a system of elements as its ἐντελέχεια, and the motions leading to this as its δύναμις in which the rest is already present but not realized. But this is not Aristotle's meaning. In his view, the mind that can penetrate to essential assumptions concerning the worth of things, regards activity as the sole and only reality which ought to exist, inactivity merely as movement which is as yet undeveloped Thus with the concept of δύναμις as a possibility which in itself may be a capacity, not only of action, but also of inaction, there is blended the concept of force, which is no longer a mere possibility, but an impulse to realization, a living faculty. But this transformation of the concept makes it more seductive indeed, yet not more fruitful; it only beguiles us the more into being satisfied with explanations which are no explanations. The soul is in this sense the ἐντελέχεια of the organic body. If we interpret this to mean that everything which is found in the body as an actual relation of the elements out of which it is constructed, is used, assimilated or enjoyed by the soul according to its worth, significance, and possible results, partly in conscious perception, partly in feelings of pleasure and the reverse, partly in free activity— then we have a proposition which sets forth the problem of psychology, but does not furnish that explanation of it which we desire. For that the facts are thus we all know without the help of philosophy; the work of investigation begins just where this formula ends; what we want to know is, by what concatenation of definite and assignable actions and reactions that fact of the translation of organic outwardness into spiritual inwardness comes to pass. In a similar fashion the logical analysis and comparison of our concepts are but too often proffered as real explanations of their content.

The ancients did not to any extent worth mentioning develop theories which, by the subordination of varying circumstances, present a circle of numerous phænomena as the results of general laws or as deviations from a type. Hence the confusion between logic and metaphysics, which we have already noticed in treating of concept and judgment, meets us again later in full force when we come to syllogism and the systematic connection of objects. For undoubtedly errors are committed in presenting the formulæ resulting from the investigation and disentanglement of a series of events as if they were the very nerves of inner connection between the events themselves—in frequently accepting that orderly classification which facilitates the survey of given reality, as though it contained the essential meaning of the things themselves—in often regarding the insertion of some definition in its proper place in any system as being in itself an addition to real knowledge, even when it adds nothing whatever to the previously known qualities of the object defined. Moreover, the meshwork of the draught-net of method is often taken, without more ado, to be the very articulation of the objects which it encloses; and not a few philosophical works take the grouping of problems for their solution.

This kind of over-estimation of logical forms is perhaps not the least injurious, but it is the most excusable. He who takes the connections between ideas in concept and judgment for real relations between the things presented in idea, regards as a process in things that which by its very nature can never take place in them after such a fashion, and is wholly mistaken. But he who regards the connection of an order which is systematic or regulated by law, and which he can transfer

to given facts as the really conditioning principle of the objective connection of things, only over-estimates the significance
of a proposition which is valid both as to form and content.
For as to form, no one doubts that the form of law and
systematic order is just as binding and valid for the inner
coherence of reality as for the connection of our ideas; the
only question, therefore, is whether the content of the laws and
order assumed by us have such claims to objective value.

Now, supposing that a is the principle—inaccessible to us
—by which the phænomena m, n, o are really conditioned, but
that b is a circumstance accessible to our observation, which
as necessary consequence or in some other way is inseparably
connected with a, we may succeed in representing m, n, o as
dependent upon b, and in doing so continue to be in harmony
with existing facts.   The law expressing this dependence
would be perfectly valid, although in a higher sense it would
not be true, for it would derive the phænomena, not from their
really supreme principle, but as it were from a vassal thereof.
It is, however, such validity as the above, and not such truth,
that we ascribe in a general way to the laws and orderly
classifications of science; in practice they merely lead from
some point of departure in facts to some conclusion in which
there is a return to facts.   It is of little consequence whether
any one thinks that the course of reality itself between those
points of departure and conclusion is also determined by the
law, or that the real inner connection of the manifold is
expressed in systems.   Since one soon sees that many laws
may be expressed differently from different points of view,
and that the same group of phænomena may be arranged
with equal significance in various classifications, this pretension is easily given up.   None of these forms and laws are
held to be expressive of the true order of things to the exclusion of all the other forms and laws, but reality is understood
as a whole that from very different points of view may be
represented in connections ever different but ever orderly
The traveller who goes round about a mountain, if he goes
repeatedly backwards and forwards and up and down, sees a

number of different profiles of the mountain recur in an order which might have been foretold. None of them is the true form of the mountain, but all are real projections of it. But the true figure itself, as well as all these apparent ones, would consist in some relation of all its parts to one another. This true figure, the actual inner relation of things, may perhaps also be discovered, and then, of course, this true objective law of reality would be preferred to all derivative and merely partial though valid expressions of it; meanwhile we comfort ourselves with the thought that the nature of truth is such as to make possible innumerable apparent manifestations of itself, and a valid movement of knowledge from one to the other.

§ 6. It was mythology that first in the exercise of unrestrained fancy added a world of real existence to the world of phænomena which had become enigmatical; with greater moderation the reflection of subsequent wider civilisation opined that there was a nature of things to the heart of which we cannot penetrate by poetic insight, but only touch here and there at the surface by means of a thoughtful comparison of facts; finally dawning science tried to substitute for the uncertain groping of these attempts, methodical investigation, which was guided by a clear consciousness of the conditions under which our thought can contain truth. From this position, which had been won once for all, and could never be given up again, human knowledge was hindered from making further advances by deficient insight into its own relation to that nature of things for which it sought, and it attributed to the movements of thought a significance with regard to facts which they did not possess. It was only at a comparatively late date that this error was clearly perceived and avoided—at least in some departments of human knowledge; the old mistakes have never been universally remedied, and there have never been wanting acute minds which, deceived by the venerable rust of antiquity which has accumulated upon them, have beheld in those very errors the golden grains of a truth to be religiously transmitted and further developed.

Even the ancients made the question whether we are capable
of a knowledge of the truth the subject of wide-reaching and
oft-repeated reflection.   But they ended in scepticism and not
in advance towards a positive conclusion; and even in the
arguments with which they contest or doubt that capacity of
knowing the truth, they frequently betray afresh the habit of
regarding the logical connections between our concepts of things
as real states of the things themselves, thus creating anew
difficulties which would be avoided if the assumptions made
were better grounded.   A renewed and very powerful impulse
towards the prosecution of these investigations arose in the
world of Christian thought, when Christianity had to effect a
reconciliation between the content of its own practical faith
and secular scientific thought — doing this partly in the
struggle with heathen civilisation, and partly as a natural
result of men's inextinguishable impulse towards knowledge.

The contrast of the world of appearance to that of real
existence had among the ancients arisen chiefly from theoretic
considerations; and it was in fact only the really existent
about which human knowledge (which looked for nothing in
real existence but its own concepts) ascribed to itself clear and
exact cognition.   The world of phænomena was consigned to
fluctuating and uncertain opinion.   Christianity developed
this contrast almost entirely from moral points of view; not
as unknown, not as empty form, not as an object of search,
but known through revelation and experienced by faith, the
world of real existence appeared in consciousness, opposed in
its holiness and majesty to the created universe.   Yet known
and revealed only in this its glory, not in the secrets of its
construction; being capable of having its value experienced
in feeling, but hard to be grasped by the thought which
strives to ascertain the conditions upon which this value depends.   And yet the call to do this was more pressing than
ever; the true world was no longer a mere holiday thought
for leisure time, which people might entertain or not as they
liked; and the more tasks it set for men in this life, the more
indispensable was it to investigate its connection with the

everyday world of appearance, which could not henceforth be neglected as simply an object of varying opinion, but had to be examined into as the soul's sphere of work on earth. This new seriousness distinguishes the investigations of the Christian era; notwithstanding the increasing clumsiness of thought, they seem, as compared with the many - sided dexterity of antiquity, like some weighty business of life beside some sport of chivalry by which men's leisure was adorned. Almost wholly occupied with the most difficult problem of thought—the question concerning the connection between the world of worth and the world of fact—this long-continued and mighty effort of the human mind was yet unable to attain its object ; and it was prevented by this predominant direction of its endeavours from providing the convictions which it developed concerning the relation of thought to existence with any positive results.

Conscience and revelation held up to consciousness ideals of action and of existence, the truth and eternal validity of which seemed the one and only fixed point in all the fluctuations of human reason; but the attempt to bring the content of these unchangeable requirements into harmony with the forms of thought according to which we are forced to apprehend reality and its coherence, revealed the impossibility of getting near to that immutable goal by the help of such resources. A number of dogmas arose in which the deep conviction of the worth and truth of an intuition which is rather sought after than experienced, struggles with the incapacity of thought to express without contradiction that which men had in their minds and were seeking after. But the burden of this confusion was laid not upon existence, but upon cognition ; assertions of the absolute unknowableness of God, and exaggerated utterances which seek for the marks of truth in that which is repugnant to common sense, concur in bearing witness to men's conviction that the worth and the essential truth of the higher world are indeed revealed in faith, but that the laws of connection obtaining both within it and between it and material existence remain unattainable by science. In the more restricted question concerning the validity of general concepts which was debated between the different Nominalist and Realist sects, these investigations are brought into closer connection with the questions which we have been hitherto considering. Do the general concepts of kinds and genera exist previously to the individual things, as eternal types according to which the things were formed by God, or did they arise in our minds after the things themselves were in existence? and are they empty names which signify nothing, or do they, without containing the essence of things as their types, yet exist in things after such a fashion that they can spring up in us as valid modes of apprehending things? This last opinion met with acceptance as well as the others; but the germ of truth which it contained remained undeveloped. On the one hand, traditional custom directed attention almost exclusively to the concept, the most unproductive of the forms of thought; diverting it from the consideration of the judgment and the syllogism, which by their mode of connecting their content would have made more clear the distinction between the validity of a truth and its identity with the object; and on the other hand, the investigation of the world of outer experience had not advanced far enough to assist the more abstract course of thought with the illustrative force of analogy. It was not until the end of the Middle Ages that there arose this new kind of science, which, worthy as it was and destined to give a new form to all investigation, remained for a long time restricted to the domain of Nature. Respect for experience, the idea of universal law, and the renunciation involved in accepting the exact investigation of the connections between phænomena by way of compensation for that knowledge of the nature of things which men despaired of attaining, are the characteristics that distinguish the spirit of the new movement.

Experience, indeed, could never have been a matter of indifference to men who have to live their lives and find their way in the world of facts, and the little-regarded wisdom of common life had even in ancient times gained much from experience; but the more exalted wisdom that was transmitted in the schools, in its attempts to build up a copy of the world was not careful to test by observation and experiment the validity possessed by its assumptions in reference to existing reality; it was enough if these could justify themselves to thought, and the conclusion from the conceivableness of a proposition to its validity in the system of the universe was generally drawn without any hesitation. Thus men did indeed recognise that things had a nature of their own, and that it was this which ought to constitute the object of knowledge; but the content of this nature was determined in a one-sided fashion by reference merely to subjective thought and men's sense of probability. There was unquestionably a deeper reverence for truth in the newly attained consciousness that, for the demonstration of any thought its conceivability needs to be supplemented by proof of its efficacy and validity in the world of fact. Men began to feel the charm of reality. The ancients had been puffed up with the strange notion that they had rendered some service by developing a world of pure thought that needed no connection with experience; for this idea there came to be substituted the conviction that knowledge had only been reached to the extent to which those connections of things apprehended in thought could be confirmed by fruitful agreement with the results of observation.

In this the new investigation of Nature was entirely of one mind with religious reflection; it took its stand upon external sensible experience, just as religious reflection did upon the inner experience of the life of faith; that which the eye saw or the heart felt could not be taken away or diminished by any subtlety of thought; on the contrary, the results of all scientific labour must be in agreement with these already established and immoveable positions. But the investigation of Nature had an advantage over the examination of the inner life; there were presented to the senses an immeasurable variety of sharply defined phænomena susceptible of exact measurement; equally perceptible by all, when some easily recognised sources of error had been cut off; recurring in regular series corresponding to their inner coherence, and capable of being freed by arbitrarily chosen experiments from the ambiguities to which direct observation is subject in consequence of the crossing of different series of events. The experiences of the inner life, neither recurring regularly nor separable from the incalculable peculiarity of the individual mind, offered much greater difficulties to investigation; and the believing heart had to be contented to hold them in opposition to the requirements of thought, or without their being in adequate connection with these requirements, whilst the investigation of Nature succeeded in developing positive methods for the reduction of its problems.

The connection of natural phænomena into one coherent whole was a favourite task among the ancients also; but they blended two questions with injurious effect. They sought first of all to grasp some primal activity or primal event which should be not a mere indifferent fact, but should also produce an æsthetic impression of its value; from this beginning the particulars of reality were to proceed in a succession, to the order of which was attributed the double office of showing on the one hand how the significance of every phænomenon depends upon preceding ones, and how on the other hand in its realization it is an effect of these. This mixing up of an ideal interpretation of events with a causal explanation could not afford to antiquity the fruits which in our own time it has always refused. It was only Atomism that even among the ancients took another course; favoured by fortune which is not always gracious to the most deserving, minds of a lower order in this school—minds infinitely inferior to the incomparable genius of Plato and Aristotle — yet hit upon the fertile thought which was to be a lasting gain for all future time. I am not speaking of their direct teachings concerning the nature of things, of the atoms and the void, and of the subsequent rude and unskilful working out of these ideas and of their conse-quences; on the contrary, the only important thing is the fundamental notions of their procedure as regards method. They first of all laid down as their established belief, the maxim that the origin, preservation, mutation, and destruction of natural objects could not be primarily explained by means of Ideas as though mere significance were sufficient to transform a postulate into reality, but that on the contrary everything that happens, whatever its significance and value may be, whether it is great or small, noble or common, right or wrong, depends for its realization on the universal rules of a mechanism working uniformly everywhere. And further, they accustomed men to see in the inexhaustible multiplicity of mathematical distinctions which may be applied to the properties, states, and movements of elements, a middle term (or a collection of infinitely variable middle terms) by which minor premisses may be supplied to major premisses expressing universal laws; these minors affording to the majors not only definite guidance towards the establishment of various results, but also enabling the whole special and definite result to be deduced in each particular case.

Later times learnt the value of these fundamental notions in the development of the idea of universal natural law. For although the general concept of law could never have been unknown to a civilised people, yet its application in the investigation of the existing world required that it should have assumed a particular character which did not belong to it till a late period. If there exist between two real elements connections which vary in such a way that their various values may be measured by a common standard; if further those elements can experience or assume states or properties which in the same way form varying series of members susceptible of comparison, these members having any measurable differences; and if moreover a change in the states or properties of the thing is involved in any change in the connections — then there will either be a constant formula according to which the magnitude of the change of states depends upon the magnitude of the change of connections, or there will be another constant formula according to which the ratio of this dependence itself varies regularly with the change of any condition that admits of degrees. This general expression, to which every natural law is reducible, clearly reveals the limitations which science imposes upon itself, its tasks, and its performances.

And first comes its dependence upon experience. For science cannot guess what elements and what connections between them must be contained in the order of the world; it waits to learn this from observation, and for itself desires to be nothing more than a development of the results which become necessary when circumstances actually occur, the non-occurrence of which would involve no contradiction in thought. And it is not sufficient that experience should show to science determinate elements in determinate connections; for even what will happen under such conditions science has no means of guessing; it is, again, experience which must teach science what kind of change in the states of things is produced by the presence of this determinate connection, and it is the comparison of many observations which first leads to a knowledge of the general law according to which the worth of these results depends upon the worth of their conditions.

But the possession of a general law would be worthless if it only served to sum up the particular cases from which it had itself been abstracted. What is much more important is to comprehend the whole varied content of every complex phænomenon as in the course of events it now arises and now passes away again, owing to the crossing of many and various conditions. Science cannot seek the solution of this problem by reference to that which the inner nature of things requires, or that which is included in the necessities of its development, or in the reasoned plan of the universe. Science does not know what it is that is valid in all these connections. But it knows that the unknown inner being of things (as far as it is revealed in their properties and connections, which are quantitatively comparable) must inevitably have the consequences which accrue, to everything that has magnitude, from the summation of similars, the cancelling of opposite symbols, and the combination of differences so as to produce a mean result. It is only at this one accessible point that science can lay hold of reality, and hence it imposes upon itself the other limitation of being only a mathematical not a speculative development of given data. To an individual connection there attaches as a matter of fact a definite result, arising we know not how, and the magnitude of which is dependent upon the magnitude of the connection; if there is a complication of many such connections, science deduces a new connection as the effect of this complication, and from this proceeds a new result capable of predetermination as regards form and magnitude, and likewise arising we know not how. Thus the whole theory is an investigation of how far the order of the changing course of the world, which springs from the varying action and reaction of its parts, may be apprehended by means of empirically recognised constant connections of unknown elements, without searching into the inner nature of things, and the end to which this nature is destined. As far as variation of phænomena goes, every occurrence is for science a result the producing conditions of which it searches out; as soon as facts and connections which are unchangeable and always valid, are either encountered by science in observation, or found to be assumptions on which existing facts may be adequately explained, these facts and connections are regarded by science as ultimate principles at which its investigations may stop. It does not seek further to deduce this final reality itself, for the domain of that causal connection by which alone it is led, ends where change ends; the coherence which beyond this domain may subsist between the unchangeable elements of reality, could only be such as should have its order and mode of connection justified by the worth of the significance which they possess. Science has not the least reason to deny such a coherence, but its investigations do not refer to this, but to the operative economy by which phænomena must be connected in every case, whether an intelligent Idea prescribes the work of the world, or whether all that takes place is merely the result of causes that lie behind, and does not work towards any goal.

Whilst these thoughts had been gradually developed much had changed. The world of phænomena, once the object of obscure and varying opinion, had become the field of the most exact investigation. Plato and Aristotle—in opposition to the Heraklitean doctrine of the eternal flux of things, which as it seemed to them unjustifiably did away with the validity of all immutable truth—agreed that there can only be a science of that which is eternally self-identical; more modern times emphasized the opposite doctrine, saying that reality is of interest to science only in as far as it changes; of that which is eternally self-identical we can merely have cognisance; eternal truths are of worth not under the form of a motionless order, in which the particular occupies a fixed position of subordination to the universal, but only as principles of change in accordance with which things alter their states. In this contrast, the meaning of which cannot be here guarded against all misunderstandings, is to be found the real advance of science in its new stage; in the admission that only phænomena can be developed from phænomena, and that we remain wholly ignorant of the nature of things, we find the limitation under which this advance is to be recognised. To describe the results which have been obtained in this way would be as unnecessary as it is impossible; it is not only knowledge of Nature, but also mental and social life, which have experienced the influence of the new mode of thought; and even where its more concrete instruments of search have not yet penetrated, it has already introduced its methods and spirit of investigation. The manifold procedure of induction, the subtle devices of experiment, the fertile ingenuity of calculations in probability, constitute the stock of an inventive and active art of knowledge which the energetic and Promethean spirit of modern times has added to the not less admirable structure of ancient logic. By these means science advances, whilst unfortunately the traditional philosophy of the schools knows little of them, and satisfies itself with continually renewed reflection upon the wisdom of past times, pushing aside the problems which this cannot advance. And finally, these investigations which primarily concern phænomena only, have not been unfruitful even with regard to those reflections which we desire should be carried on concerning the world as a whole and the significance of its order, and concerning real existence. On the other hand, it is to empirical investigation and its mathematical interpretation that we owe our only trustworthy view of the magnitude and construction of the universe, the connection of the effects that take place in it, and the complete circle of mutually compensatory processes which actually occur—facts that have not indeed received an interpretation, yet for all that facts—facts the knowledge of which has provided philosophy with a basis for its explanations of cosmic order quite other than that which in ancient times could be furnished by its own assumptions concerning the necessary nature of things, and real existence. To know facts is not everything, but it is a good deal; to despise this knowledge because one desires something more befits only those fools mentioned by Hesiod who can never understand that the half may be more than the whole.

§ 7. Philosophy is a mother wounded by the ingratitude of her children. Once she was all in all; Mathematics and Astronomy, Physics and Physiology, not less than Ethics and Politics, received their existence from her. But soon the daughters set up fine establishments of their own, each doing this earlier in proportion as it had made swifter progress under the maternal influence; conscious of what they had now accomplished by their own labour, they withdrew from the supervision of philosophy, which was not able to go into the minutiæ of their new life, and became wearisome by the monotonous repetition of insufficient counsels. And so when every offshoot of investigation which was capable of life and growth had separated itself from the common stem and taken independent root, it fell to philosophy to retain as her questionable share the undisputed possession of as much of all problems as remained still inexplicable. Reduced to this dowager's portion, she continued to live on, ever pondering afresh over the old hard riddles, and ever resorted to afresh in calm moments by those who held fast to a hope of the unity of human science.

The experiential sciences had investigated the connection of phænomena ; they showed how many and what kind of links constitute the chain of events which connects any cause with its final effect ; but what it is that holds together any two contiguous links escaped them ; they told neither what things are in themselves, nor in what consists that action between them by which alone the condition of one can become the cause of a change in the condition of another.   Religious and moral life had developed the belief in unconditional worth— an unconditional ought, which if there is any meaning in reality must be the most real of all things ; but the world of creatures and of facts in which alone it could be realized was opposed to it as quite alien, neither derivable from it nor, as it seemed, even compatible with it.   This condition of things contained incentives to a constant repetition of two questions —first the question as to the intrinsic nature of existing things whose manifestations to us are the subject of our observation, and secondly the question as to the connection in which this world of existing reality stands to the world of worth, of what ought to be.   And all attempts to answer these two questions always stirred up forthwith a third question, that as to our capacity of knowing truth, and the connection of this capacity partly with existing reality and partly with that which reality ought to be and produce.

Our thoughts receive the stamp of certainty by being reduced to either the already proved certainty of others, or to that of immediate truths which neither need nor are susceptible of proof.   The trust which we repose on the one hand in the laws of thought by means of which this reduction is accomplished, and on the other hand in the simple and immediate cognitions to which this leads us, may be guarded by repeated and careful proof from the influence of prejudices of which the persuasive force is accidental and evanescent ; but on the other hand no proof can guard against a doubt which suspects of possible error that which men have always found to be a necessity of thought. A scepticism that does not demonstrate from individual contradictions which may be cited the erroneousness of specified prejudices, and hence the possibility of correcting them, but goes on causelessly repeating the simple question whether in the end everything is not really quite different from that which we necessarily think it to be, would, in banishing certainty wholly from the world, also destroy all the worth of reality. That, however, this cannot be—that the world cannot be a mere meaningless absurdity—is a moral conviction, which is the ultimate ground of our belief in our capacity of cognising the truth, and in the general possibility of scientific knowledge. But this conviction does not define the extent of such knowledge.

It is only our own existence of which we are immediately conscious; all our information as to an external world depends upon ideas which are only changing conditions of ourselves. What, then, is our guarantee that this image of an external world is not an innate dream? He who is cautious asks whether this is so; he who is incautious asserts that it is; he forgets that our experience must be the same in both cases, whether there be things without us or not; even a real external world could only be reflected by us in images resulting from affections of our own being. Hence the nature of all our ideation being subjective, it can furnish no decision concerning the existence or non-existence of the world which it believes that it reflects. But the attempt to regard the image of the world as a native production of the mind alone has always been speedily given up again by scientific instinct; for in order to attain this end it has always been necessary to assume the existence in ourselves of just as many impulses foreign to our mind, and not derivable from it as in the common view we are believed to receive from the external world. Reserving for future consideration the important points in this view, we now go on to speak of the conviction (to which philosophy has always speedily returned) that our ideas arise from action and reaction with a world independent of ourselves.

But if this is so, can our ideation be more than an effect of things, can it be a copy which resembles them, and can the truth which we are capable of knowing consist in an agreement between thought and thing? We speak of the image of an object when any construction of other material makes the same impression upon our perception which the object itself would have made; thus as far as we are concerned one thing becomes the image of another through having a similar effect. But can the effects produced in us by both be ever so exactly like the things, that the eye of an independent observer would regard our cognition as an image of the object? Wherever action and reaction take place (and cognition is only the particular case of such action between things and the ideating mind), the nature of the one element is never transferred, identical and unchanged, to the other; but that first element is but as an occasion which causes the second to realize one single definite state out of the many possible for it—that state, namely, which according to the general laws of the nature of that second element is the fitting response to the kind and magnitude of stimulus which it has received. Hence definite images in us, and *produced by us*, correspond to the causes which act upon us; and to the change of those causes there corresponds a change of these inner states of ours. But no single idea is a copy of the cause which produces it, and even the connections which we think we cognise between these still unknown elements are not primarily the very relations that really obtain between the elements, but only the form in which we apprehend them—and we do not regard this state of things as human weakness, for it is of the very nature of all cognition, which depends upon action and reaction with its object. All creatures that are subject to these conditions are subject also to this consequence; they all see things not as they are in themselves when nobody sees them, but only as they appear when they are seen.

Though limited in this way to phænomena, yet knowledge

is not devoid of all connection with what really exists. For we are not justified in complaining, as if it were so illusive that a mere appearance only is shown to us, the nature which appears (which is altogether unsusceptible of comparison with the appearance and of which even the very existence is doubtful) lying wholly beyond our intellectual horizon. We cannot regard our fundamental intuitions as merely human modes of apprehension by which things which are in themselves of wholly different form are taken up, and under which they appear to us alone, without admitting that (in order that they may be able to be taken up by these forms) things must have such a relation to them as any object must have to the meshes of the net in which it is to be caught. Or to speak plainly, every appearance presupposes as the necessary condition of its appearing a real being in the inner relations of which lie the grounds that determine the form of its appearance. From the analysis of the forms of intuition under which our perception immediately apprehends its objects, we may easily attain the conviction that these forms do not, in the shape in which they are familiar to us, admit of application to things themselves; but we shall always need to seek, in the nature of things and in their true mutual connections, the conditions which admit of our apprehending them under those forms. Thus it may be doubtful whether space and time do not exist as space and time solely in that ideating activity which can grasp a manifold in one act of apprehension; but we cannot doubt that, if this is so, that which exists must itself be subject to an order neither spatial nor temporal, which acting upon us is by us translated into the form of spatial and temporal order. It is certain that the sensation which any object or event causes in us is not exactly like its cause; but it is equally certain that we shall regard two objects or events as exactly like, similar, or different, if the impressions they make upon us are exactly alike, similar, or different, and we shall estimate their degree of relationship by the amount of difference between their impressions. Thus we inevitably regard the apparent existence and events which we perceive as being proportional throughout to real existence and real events which, belonging to or occurring between things themselves, by no means exclude concepts of truth and order. The attempt to renounce this supposition would produce not any increase of precision but fruitless and self-contradictory agony of thought.

But if appearance indicates existence, it yet indicates only formal relations of existence and their changes; the nature of the things which exist and act under these relations remains inscrutable. And just because the nature of things remains unknown, we are also unable to comprehend the occurrence of action and reaction between them as a result of their nature; it is only appearance, which is the matter of experience, that can lead us to divine this true action and reaction. Thus philosophy takes the same course that we have already seen taken by the natural sciences; it begins with the individual enigmatical and contradictory phænomena which experience offers, and guided by the general laws of thought seeks to ascertain the form of real existence and occurrence which, in order to explain what is strange and contradictory in facts, must be supposed to underlie these as their efficient cause. It must be admitted that some admirable results may be attained by this Realism, which contents itself with tracing back actual facts of appearance to facts of existence which must necessarily be assumed, even when its action is wholly subject to this limitation; not only may it succeed in throwing light upon the efficient connections in particular coherent groups of phænomena, but a consideration of the knowledge attained may also lead it to a view of that which as true reality lies at the foundation of the whole phænomenal world. But even this final result will retain the character of mere fact, and thus Realism will always arouse the opposition of that idealistic bias of the human soul, which recognises real existence not in facts which only are because they are, or because they must be assumed in consequence of the existence of something else, but only in such a fact as certifies by the worth of the thought which it represents, its vocation, its right, and its capacity to appear as the apex and crown of reality, as the final datum and the highest constructive principle.

§ 8. Idealism opposes to the realistic acknowledgment of the unknowable nature of things the bold assertion that Thought and Being are identical. In saying this, it does not necessarily mean (what, however, it is occasionally audacious enough to assert) that human cognition will some time succeed in penetrating by thought the existence of all things, and recreating it in idea; for the narrow limits of our finite nature which hinder this extension of real insight are but too obvious. It means that for a cognition free from these limitations things would no longer be insoluble realities, they would no longer be as unapproachable and incomprehensible for thought, as for instance light is for the ear or sound for the eye; rather thought would recognise them as realized ideas, thus recognising itself in them. So this proposition, understood as not properly an assertion concerning the relation of knowledge to its object, but much rather as a conviction concerning the nature of existence in itself, palpably gives to the existence or nature of things a different meaning from that given to it by common opinion. For a man of ordinary intelligence thinks he immediately knows that matter or content by which a thing as such or such is distinguished as different from some second thing—knows it partly in the impression upon the senses, and partly in ideas which are directly connected with the impression and hold together its constituent parts. And it seems to him all the more difficult to see how it can happen that this content should have the power of meeting him as something existing, independent, tangible, as a Thing in short; he who should discover the secret spring by which the thinkable τὸ τί of existing objects is endowed with the extension, body, resistance and elasticity of Thinghood, would seem to unsophisticated thought to have found the real and very nature of things, not that which distinguishes one thing from another, but that in which they are all alike, the essence of their existence, reality itself. Now can Idealism

maintain that it can solve this problem ?   Certainly not to
any greater extent than Realism, in its own view, is capable
of; in what exactly consists the existence of things, what is
meant by their being connected with one another, finally
how it comes about that anything results from these connections—all this is as impenetrable to Idealism as to its
opponent. Perhaps—to admit the utmost that we may—it may
also succeed in proving that there exists—though it does not
know how—a connection in accordance with which if there
exist (in some incomprehensible way) a being of such and
such a description, there must in an equally incomprehensible
way exist such and such change and activity, and no other; but
even if we admit this, Idealism would only have penetrated
the *meaning* and the intelligible connection of the individual
determinations which under the name of being we grasped
together into one whole ; how this inner connection of
reality could *be* would still remain wholly uncomprehended.—
Yet to do all this was just what was promised by the bold and
striking expression given to the proposition which made
being identical with thinking; it led one to expect that just
that by which being as being seemed at first to be irrecon-
cilably differentiated from thinking or from being thought,
would finally be presented as a vanishing distinction, and that
this being would be altogether resolved into thoughts.   And
now it seems that of the two ideas which we regard as
blending to produce existence, the ideas of the τὸ τί and
of its existence, Idealism leaves that of existence just as
unexplained as it was before.

But just as no end was gained by the reference to Being
in the proposition to which we have alluded, even so is it
beside the point to speak of thought as that with which it
should be identical; as long, at least, as this name distinctly
signifies one activity of the mental life as distinguished from
others.   And yet this seems to be what is meant, for even
the Idealist does not allow that sensuous intuition and
perception can grasp the truth of things ; he abandons both
these, and reserves to thought, as a special and higher activity,

the privilege of searching out real existence, behind the illusions with which sensuous intuition and perception surround us. But his expectation rests upon a widespread error. Men are universally much disposed to regard as a product of thought anything for which language has furnished a name, although what thought has contributed to the building up of the content which it indicates may be very little, and sometimes nothing whatever. As long as we are considering sensuous impressions, we are indeed soon convinced that no skill of logical operations can supply the place of sound or colour to him who is blind or deaf; that thus for instance blue or sweet are not concepts which we think, but impressions which we experience, and their names merely linguistic signs which remind us of a content for which all that thought does is, at the outside, to indicate its dependent nature by the adjectival form which it gives to it. But in the more general concepts which are everywhere interwoven with our perceptions and give them form and stability—in the ideas of Being, of Becoming, of Action, and of every Connection which subsists between any two things—we feel more assured of finding the genuine products of thought, and of thought alone. And yet the meaning of Being cannot by any interpretative activity of thought be made intelligible to him who does not know immediately what it means; all that thought can do is by proceeding analytically and removing all accessory ideas which are not signified to teach us to distinguish that meaning of the word which can only be grasped by immediate intuition. No one will ever invent a definition of Becoming which does not contain (under some other name) as its most essential constituent the idea of passing from one to another, or of something happening; thought can contribute to the building up of this concept only by illustration of the two points between which the nameable but unanalysable enigma of transition takes place. And the concept of Action is equally incapable of being approached by any logical operations. It is easy to fancy that one has traced it back to the more abstract concept of that which conditions—although here it would be questionable whether the converse reduction might not be more correct; but supposing we have done so much, can we then analyse further in thought the real meaning of the idea of conditioning? Apparently perhaps we may, but as a matter of fact we certainly cannot; for in the last resort all that thinking does is to denote by this or that name the idea of a necessary inner connection between different occurrences, which connection it cannot by its own activity produce.

And here it will be objected that I lay useless stress upon that which is self-evident; since it is of course necessary for thought, as the activity which connects and combines, to pre-suppose as given from elsewhere the elements which are to be connected and combined. My real object has only been to make this conviction very vivid for a moment, and to deduce the consequences which it involves. For with a little attention one will soon be convinced that these elements, which thought has thus to take up as coming from elsewhere, comprise nothing less than the whole sum of that knowledge of real existence and occurrence which was formerly ascribed to thought as its own possession. Thought is everywhere but a mediating activity moving hither and thither, bringing into connection the original intuitions of external and internal perception, which are predetermined by fundamental ideas and laws the origin of which cannot be shown; it develops special and properly logical forms, peculiar to itself, only in the effort to apply the idea of truth (which it finds in us) to the scattered multiplicity of perceptions, and of the consequences developed from them. Hence nothing seems less justifiable than the assertion that this Thinking is identical with Being, and that Being can be resolved into it without leaving any residuum; on the contrary, everywhere in the flux of thought there remain quite insoluble those individual nuclei which represent the several aspects of that important content which we designate by the name of Being. It would be more simple and more true to say that Being contemplates itself; we—since we exist—feel, perceive, experience, or know well enough what it is to exist; we—since we act—know well enough what we mean (although it is unspeakable) when we talk not only of a temporal succession of phænomena, but also of the one being conditioned by the other. And in this sense all the world has known from the beginning what is the import of Being or Reality, for all the world has lived the meaning of these words; but if it has always been difficult or impossible to express by determinations of thought that which men have so plainly experienced in their lives, philosophy has not succeeded in removing the need for such expression; all she has done has been to find names for that which men experience; and since it is in a world of names that she lives and moves and has her being, she has sometimes had less vivid experience than others of that which is the object of her efforts.

It will be demanded on the part of Idealism that, as far as all such scruples are concerned, this question should at last be allowed to rest; it is admitted that we do not know how things can be and act, but their nature is said to consist, not in their reality, but in *what* they are and *what* they do. Now is this content of things really more accessible to thought? Whatever else thought may be, it is an activity of the mind; or if not this, it is at any rate a changing succession of states which mind experiences. Now, how can a succession of states copy and reproduce anything except states? Can they represent the nature that experiences the states which are reproduced? They can only do this if we go still further in our assumptions, and regard, not only what things are, but what they experience, as their innermost nature, and as that real existence which philosophy seeks. And thus, by a path the several stages of which we must here refrain from describing, Idealism would reach the admission that in truth it neither knows how things are, nor what they are, but that it does know what they *signify*, and that this, their real existence, is immediately cognisable. What everything is in itself, what its nature is by which it exists and is capable of making its efficiency felt and of being different from other things, this may remain for ever inaccessible to thought. But with regard to the forms of that to which they are destined, the forms of their changes, development, activity, and of their several contributions to the sum of reality—in all these relations things are comprehensible to thought, and are comparable among themselves; the essential significance of each, as far as it consists in these, is in itself susceptible of exhaustive expression in thought, whether or not we men are capable of discovering the thought which does express it. Thus Idealism, like Realism, comes to acknowledge that it is limited to a cognition of what happens in and between things that remain unknown; but it believes that in knowing the import of what thus happens it possesses all essential truth; that it is only for the realization of this truth that things exist.

Religious belief in understanding the world as a divine creation has always cherished and expressed the same conviction in another way. It denies just as vigorously as philosophic Idealism that there is in things a nature (or any part of their nature) which they have of themselves. All that they are, they are by the will and intention of God; the most essential part of their nature consists in what God meant or willed that they should be, in their significance in the unity of the cosmic plan. Religious belief did not maintain that it could penetrate the plan of this unity, but in its representation of God were contained, as it were, centres of light which illuminated each other, and also cast enlightening rays upon the created world. The strict order of its phænomena was regarded as in fitting correspondence with the immutability and justice of the Creator, its beauty with the infinite fulness of His blessed nature, the order of events in the moral world with His holiness. To trace back all the particulars of reality to these creative forces in God was neither attempted nor regarded as possible; it was sufficient to believe in their truth on the whole, unmoved by the apparent contradiction of many perceptions, and, as regards particulars, to be ever drawing afresh from a selection of favoured phænomena the living feeling of their universal and governing efficacy.

Philosophic Idealism tried to outbid this faith in two

directions. It first took offence at the unconcerned way in which religion spoke of a personal God, and regarded Him as creating things out of nothing, and then entering into a relation of reciprocal action with these realities that had been manufactured out of nothing: the metaphysics of all these processes needed to be found and explained. But none of the attempts to find and explain them (which we shall have to consider more particularly at a later stage) attained its end; since they were destitute of all ideas concerning the relation of God to the world (ideas which religious belief had framed anthropomorphically), they have left as their only result the assertion (couched for the most part in artificially obscure forms of expression) that there is a single supreme Idea that penetrates all the phænomena of reality and gives them form and order; but they do not say how it does this. And just because it was at most the meaning of the universe and not the origin of its reality which was accessible to Idealism, everything that might remind men of this problem seemed to fall out of its consideration. God was no longer spoken of, for this name signifies nothing without the predicates of real and living power and efficiency; it was only the Idea that could be spoken of, the content of which was supposed, in some incomprehensible way or other, to really constitute the nature and significance of the world. But the idealists hoped to be able to express the whole content of this Idea completely and systematically in thought, and by this second performance to far surpass religious belief, which only knew in a general way that divine purpose which in particulars was inscrutable.

Even this promise could only be fulfilled by breaking off from the nature of the thing that which remained incomprehensible to thought. For in fact the living forces which had been beheld by faith in God showed themselves as inaccessible to thought as the sensuous impressions which occur in perception; for them, too, we invent names; and their content, too, is known to us through living experience, and not through thought. What is good and evil remains just as incapable of being reached by mere thought as what is blue or sweet; it

is only when we have learnt by immediate feeling the presence of worth and of unworth in the world and the gravity
of the difference between them that our thought is able, from
the content thus experienced, to develop signs which subsequently enable us to bring any particular case under the one
or the other of those two universal intuitions. Can one find in
concepts the real living nerve of righteousness ? Much may
be said of compensations, of the correspondence between conditions originated and endured, of the return of good and ill
to him who caused them ; but what movement of thought
explains the interest which we feel in these forms of occurrence when, and only when, they indicate what we call a retribution ?   Are love and hatred thinkable ?   Can their nature
be exhausted in concepts ?   In whatever combination of
duality to unity, or whatever division of that which might be
one, their significance may be found, the expression of that
combination or unity will never do anything but state an
enigma. For an enigma is the specification of signs which do
not of themselves set forth the whole living content to which
they relate, this having to be guessed because it is not plainly
contained in them.   Now not only did philosophy hope that
it could reproduce in thought all the living content which
was possessed by faith in a personal God, but it imagined
that it was applying a process of ennobling clarification
to Him who is more than anything that can be called an
Idea, when from the dimness of that which is experienced by
the whole heart and the whole soul, it raised Him to the
dignity of a concept capable of being an object of pure
thought.

Both the natural and the moral world received this treatment, which traced back the real content of all things and
events to what was formal in their mode of appearance, and
regarded the things and events themselves as merely destined
to realize these forms. The creatures of Nature existed merely
in order to take their place in a classification, and to provide
the logical degrees of universal, particular, and individual
with an abundance of phœnomena ; their living activities and

reciprocal action took place in order to celebrate the mysteries of difference, of contradiction, of polar opposition, and of unity; the whole course of Nature was destined to represent a rhythm, in the movements of which affirmation, negation, and mutual limitation alternated with one another. Consideration of the spiritual world sometimes in a kind of realistic fit regarded thought and all spiritual life as merely the highest form assumed by those unfathomable powers of affirmation and negation, opposition and its removal; sometimes in a more idealistic mood it regarded thought as the real nature and goal of all things, and those forms of mere blind being and occurrence as imperfect preludes. But it never succeeded in establishing *thought* as what is most essential in *mind*, and *thinking about thought*, the pure self-reflection of logical activity, as what is highest in *thought*. The existence and the worth of the moral world were indeed not forgotten; but even that which *ought to be* had to submit to this reduction to form; it seemed as though it only *ought to be* to the extent to which it reproduced in the forms of its realization those much-esteemed relations which were held to be the real nature of being.

I break off in the midst of an enumeration of these errors. This short sketch has been partial, leaving much unmentioned which within the philosophic school itself is regarded as weighty and important, and laying stress only upon what could serve as an introduction to the end aimed at by our present inquiry. Philosophy is not at present exclusively ruled by the false Idealism with which we have just been confronted, nor is it impossible to avoid the errors which deform it; but this is not the place for developing the conviction which we wish to maintain. Here we can only give it provisional expression, and affirm that the nature of things does not consist in thoughts, and that thinking is not able to grasp it; yet perhaps the whole mind experiences in other forms of its action and passion the essential meaning of all being and action, thought subsequently serving it as an instrument by which that which is thus experienced is brought into the connection which its nature requires, and is experienced in more intensity in proportion as the mind is master of this connection. The errors which stand opposed to this view are very old. It was a long time before living fancy recognised in thought the bridle which guides its course steadily, surely, and truly; perhaps it will be as long again before men see that the bridle cannot originate the motion which it should guide. The shadow of antiquity, its mischievous over-estimation of reason, still lies upon us, and prevents our seeing, either in the real or in the ideal, what it is that makes both something more than reason.

CHAPTER II.

Pleasure and the Means to Pleasure—The Patriarchate—The Adventures of the
Heroes—The Liberal Culture of Antiquity—Slavery—The Growth and
Preponderance of the Industrial Classes—Economic Character of the
Present Time, and its Causes and Effects—The Modern Forms of Labour
and their Social Consequences.

§ 1. NATURE with its unchanging order, and Society
with the variability of its internal relations,
have from the beginning been spread out before men as
the great fields of all activity. It was need — partly the
urgent need of self-preservation, partly the more calm but
not less powerful need of mental satisfaction—which in the
one field as in the other gave birth to the first action along
with the first reflection, and did not permit the deferring of
reaction until the completion of all science. Men were
obliged to begin to work upon things and to use or construct
the relations of human society, while their store of cognitions
was as yet incomplete; but the tentative effort enriched
scientific knowledge by its results, and the increase of knowledge enlarged the sphere of men's powers and the spirit of
enterprise. Thus science and life were developed in constant
action and reaction. It was only while thus occupied with
the whole wealth of experience, that knowledge developed by
degrees all the multiplicity of its modes of investigating,
analysing, and combining; it was only through the wide
extension of its contact with the most varied kinds of objects
that it discovered its own instruments, and learned to comprehend its tasks (which were presented to it at first in
isolation) in that connection which as perfected science it
ultimately seeks to reflect in the form of a systematic combination of all truth. However attractive the history of this

development may be, we must renounce any more detailed consideration of it than has been given in the brief survey which we have just concluded.   Since the general purpose of our reflections has regard to the totality of human development, we have no further space for the representation of the inner regularity and beauty with which the edifice of science—a self-sufficing whole—grows up from its own principles and becomes articulated ; our attention is due in greater measure to the other division of this reciprocal action between knowledge and life—that is, to the fertilizing stimulus which life itself, the customs of commerce, the spirit of social institutions, and the enjoyment of existence, receive from the gradual development of the world of thought.

Human life being dependent upon Nature for its continuance, men had first of all to attend to the business of self-preservation by satisfying external needs, in order that they might then be at liberty to devote themselves to their real vocation in enjoyment of beauty, delight in holiness, and practice of what is right.   Now a consideration of the efforts which have been directed to the production and perfecting, the administration and diffusion of material goods might easily allure us into a wide and brilliant region of scientific development which touches life at innumerable points—might allure us, that is, to the history of the Natural Sciences. Yet we forbear a systematic exploration of this region.   For why attempt to repeat in a narrow and insufficient compass what has already been given in detail in innumerable delineations ?   The triumphs of human sagacity in the investigation of the celestial regions and the remote parts of the earth, in the explanation of the chemical transformations of bodies and of the processes of life, in determining the conditions of action of all forces, and analysing composite forces into their elements—all these are in our times favourite subjects of triumphant exposition and eager attention ; lauded in a thousand ways, it is not they themselves but the blessing that they have conferred on human life which stands in need of mention.   And in saying that this needs mention, I do not mean that it would be worth while to repeat the enumeration of those countless individual benefits, concerning which (after the numerous accounts that have been given) we now know to what principles of natural science and to what inventive application of those principles they are due. Let us suppose the place which I here leave vacant to be filled by one of those easily obtainable descriptions which show us how the progress of knowledge of Nature, lingering at first, has in modern times, advancing with greatly accelerated speed, given new developments to life—how we have learnt to overcome innumerable obstacles which Nature opposes to human activity—and how increased insight into the connection between different effects in Nature has put us in a position to produce with ease, from despised material which in former times was thrown away as refuse, instruments of enjoyment which in those times were either not known, or could only be procured with difficulty from some few sources which Nature voluntarily set at man's disposal. Having supposed, then, that this picture of an increasing dominion of Mind over Nature stands clearly before our eyes, in what is it that the blessing of this dominion consists ? And in asking this question we refer not only to the fact of dominion, but also to the advantage which increased power over Nature affords for the attainment of that which is the special destiny of man.

Unless I am mistaken, the answers to this question will not be harmonious. In moments of deliberation, in which we survey with a comprehensive glance these achievements of human intelligence, the undeniable advance which they show may rejoice us with the feeling of satisfaction which naturally springs from every increase in efficient strength. But if looking at life as a whole we seek there the useful results of this progress, it may seem doubtful whether this greater dominion over Nature of which we boast, does not result for us in a greater dependence upon that power over which we are continually victorious. For every fresh commodity that we produce immediately becomes a necessity, and entangles us in new efforts—on the part of the community to produce and exhibit it, and on the part of the individual to obtain it. Every new discovery of science that has splendidly abridged laborious modes of attaining some definite end, has forthwith exhibited as necessary a multitude of new ends which the new resources tempted men to aim at.   Hence though much labour has certainly been materially simplified, as science taught men better combinations of the means by which all effects are produced, it is plain that, taking life altogether, labour instead of becoming gradually less has become greater. The old complaint that so large a part of men's time and strength must be sacrificed to the mere maintenance and securing of existence, is not allayed but sharpened; ever more and more room is taken up, in our short span of life, by the preparations and equipment required for life itself; the sunny strip of leisure seems ever to grow narrower and further away on our horizon—the leisure in which, in quiet communion with self or cheerful intercourse with others, we hope to enjoy the final net result of so much effort—a result worthy of our human nature.   Thus it seems as though the enlarged possibility of satisfying a multitude of wants, taken in conjunction with the amount of work necessary for the realization of this possibility, did not make us happier on the whole than men were in the times when those wants, the means of their satisfaction and the labour required for this, were all alike unknown.

But equally old with this complaint is the rejoinder that it is erroneous to try and divide labour and enjoyment by a sharp boundary line, as if they were as opposite as commodities and the prices which are paid for them; not only the possession of the enjoyment, but also the receptivity for it, is given to leisure as the result of what has been experienced and gone through in labour; labour is itself a source of enjoyment, and not merely the road thereto.   We do not need to draw out in detail the universal truth of this remark; we have already had frequent occasion to consider how little the spiritual content of life in an unlaborious state of Nature, and the enjoyments of leisure in such a state can be compared to those with which culture rewards the exertions of a life's hard work. The human soul is not like a plant which requires only that the universal conditions of its existence should be favourable in order to exhibit in succession the several beauties of its cycle of development—bud, blossom, and fruit; it is only the ever-changing struggle for external necessities that stimulates us to acquire knowledge, that furnishes our leisure with subjects for reflection, and at the same time deepens the value which we set upon those social relations of which natural order lays the foundations—deepening it until it becomes that refined moral feeling which finds the most stirring interest of life and the most elevated enjoyment in the discussion of varied views of life, and in emerging victorious from its moral conflicts. We desire even for the individuals who are the inheritors of some long-established civilisation, the education which only life can give; the traditional ideals of all that is good and beautiful, although even in tradition itself they have long been bound up with representations of those definite relations of life in which they are to be realized, yet seem to stir the soul vaguely, hovering before it formlessly and without being seriously apprehended, until incessant contact with the hindrances of real life and with the claims of others reveals the full significance of their content—the content that is of the traditional ideals—and makes the contemplation and realization of them a life-work which is self-sufficing and self-rewarding. Without this complication and intensifying of stimulations and hindrances which culture brings with it, the isolated experiences and activities of men would hardly have produced even an indefinite sense of something really worthy. Thoroughgoing, however, as the superiority of culture to a state of Nature is in a general way, yet it is not equally indubitable that its internal progress involves in itself a continuous heightening of the enjoyments of life, and that there is not a point beyond which the increase of labour of all kinds leads men in living and in maintaining life to lose sight of the ends of life. At all events, in all periods of many-sided civilisation there seems

to remain a longing after the simpler conditions of past times—
a proof that it is not easy for men to bring the results of
their own progress into harmony with the wishes which they
call upon life to fulfil.

In the patriarchal state which the Old Testament writings
describe, there is presented to Christians, as it were a com-
pendium of simple and noble life, which, glorified by the
idealizing power of distance and of poetic representation, may
well seem to this retrospective longing to be an exemplar of life.
Certainly traditions of earlier civilisations and the possibility
of contact with the developed culture of neighbouring countries
was, even so early as this, at the foundation of that which
interests us in the patriarchal life ; this life being not so wholly
self-dependent as it seems in the Scriptural picture, where it
is presented in strong relief detached from its surroundings.
But external relations were still so slack that friendly
obscurity veiled the surrounding regions, and all the problems
and all the enjoyment of life remained concentrated within a
narrow circle that could be taken in at a glance. Men's
wants were provided for by a labour that was light, or in
which there was as yet little complication and little division
of employments—labour that consisted chiefly in the un-
irksome tendance of living creatures ; if want occurred it was
regarded rather as the disfavour of Nature than as the result
of social evils. As the division of labour had not yet
taken place, life had not yet the aspect of an uncertain and
ingenious struggle for existence; careers were marked out
upon which each entered with a regularity as great as that
with which Nature develops corporeal life; the differences
of social consideration which inevitably appear at an early
stage were not yet combined with such intellectual and
philosophic differences as might make one man's interests
in life unintelligible for another; connected chiefly with
family relations, they were yet important enough to introduce
into life, instead of an enervating equality of claims, a variety
of reciprocal moral obligations which were profoundly felt.
There were united in the head of the tribe all those functions

of work and action which give worth to human life ; father and master, law-giver and judge, prince and priest, all in one, he experienced in himself the full and undiminished enjoyment of that mental power which lifts man above all Nature, and set before his people this unity of life in visible embodiment. If to all this we add that to the religious belief of this time and of these tribes their connection with God was an experience that was ever being renewed, we may well admit that we find in the patriarchal period a concentration and intensifying of consciousness and of life which prevented the attention of individuals from passing over unobserved any attainable happiness or any recognised duty.

Doubtless this form of life could not be maintained for ever in its completeness ; the greater concentration of population and the transition to stationary life developed new needs and required new kinds of labour, which led to different social arrangements ; also we would not conceal from ourselves that in reality the spiritual content of the patriarchal life must have been poorer than it appears in the poetic representation which emphasizes its bright parts and says nothing of the duller intervals that come between. Certainly the moral significance of all individual relations of life was sounded to its depths and reflected upon with remarkable refinement of feeling, but the relations themselves were too simple to produce that complex and varied wealth of thought, in the possession of which advanced civilisation always feels in the end that it is superior to those simple states of society which in other respects are envied. But the patriarchal form of life, the self-centred completeness and isolation of the family and the home which, being self-dependent to an extreme degree, provides for all its own necessary wants, and is able in its own little circle to find a solution of all essential problems—this form of life must always be regarded by us as the type to which we must seek to revert, in opposition to that unattached condition that in a more complicated state of society makes the individual feel like a lost atom, tossed hither and thither by the wholly incomprehensible forces of a great all-embracing

external world.   Let us now see whether increasing civilisation has brought with it the conditions of an inner enrichment
of this form of life or only causes of its disintegration.

§ 2. To reap without having sown is naturally man's
original mode of existence.   When the simplest appropriation of natural products no longer sufficed, the labour that
tends, transforms, and produces, with all the patience, self-
denial, and steadiness which it requires, long continued to be
held in contempt as compared with the destructive activity
which in the chase, in robbery, and in war took possession of
finished products capable of ministering to human enjoyment.
The period of Life according to Nature was succeeded by the
Heroic Age—an age in which men's mode of life was an
imitation of that of beasts of prey, from the weakness of
admiring which the human mind will never be wholly free.
For, indeed, the struggle in which one's own existence is
staked for speedy gain, and one's whole nature is roused to
all the activity of which it is capable, not only swells the consciousness of the combatant with proud and passionate excitement, but offers to imitative poesy much more picturesque and
intelligible images than the quiet industry which transforms
a peaceful society merely by conquering the inertia of intractable objects.   The ambition of emulating the lion or the eagle
developed indeed all the natural beauty of the human race,
and all those traits of capricious magnanimity and uncertain
generosity which, combined with just as inexplicable fits of
savagery, makes the "king of beasts" such an attractive object of
contemplation to us ; but human capacities were not moulded
by this kind of life for their own special and appropriate
work.   At all times this mode of thought—this emulation of
the beasts—has been powerful enough ; in the most remote
antiquity it shows itself openly in robbery by land and sea ;
sword and lance were to the Greek Klephthen as plough, sickle,
and wine-press with which to sow and reap and press the wine
from the cluster ; the Romans, in their legends, claimed
robbers as their ancestors ; and to the Germanic nations it
seemed unworthy to seek by labour for that which might be

gained by the sword; the highway robber of the Middle Ages, and the runaway vassal, acted from the same feeling. All of them were right in so far as this, that labour is apt to enslave the mind when it requires exclusive occupation with objects to the peculiarities of which the labourer must accommodate himself, by narrowing his circle of thought to but few trains of ideas; on the one hand it destroys receptivity for the various enjoyments of life, and on the other hand may paralyze the elasticity of his powers, which are naturally inclined to exercise themselves upon reality in various ways. But they forgot that, notwithstanding all this, it is only labour which can develop a coherent human character, and that the unrestrained exercise of strength which they thought so splendid is only superior to the savageness of wild beasts when it lays aside that character of adventure which employs strength only for the sake of subjective enjoyment, and takes on the character of protective service, which applies the same powers for the defence of interests that are worthy in themselves, doing this under a sense of obligation.

The ends of human life, and the means of attaining them, were thought over by the Greeks more eagerly than by other nations. In the world of the Homeric poems there appears a dark stratum of labouring bondmen as the foundation upon which rests the serene and gracious happiness of the nobles; but either there is as yet too little difference of needs and of cultivation to embitter this contrast, or else tradition is so obscure that it does not make plain to us the sharpness of the contrast. Of Labour, which had not yet split up into a number of branches dependent upon one another, it was therefore still easy to take a comprehensive view, and it was regarded with honour, especially in as far as it stimulated the early-developed artistic sense of the people, not supplying a foreign demand, but serving to satisfy the needs of a great and self-sufficing domestic economy. When the brilliant development of mental life in Greece began, these relations gradually changed. In proportion as there was an increase in the significance and excellence of the enjoyment which advancing culture promised to him who had time for it, men sought to shorten the labour necessary for supplying the needs of life; human life properly so called had its beginning in leisure, and to learn how to occupy and enjoy leisure in a way worthy of humanity was the business of Greek education, which, in order to attain this end, not only did not shun the labour of severe and long-continued discipline, but even undertook it with eagerness.

I will not here inquire whether the symmetrical development and exercise of all the bodily and mental powers with which Nature has endowed us—in other words, whether being educated to the perfection of human kind—is in reality the whole destiny of man.   But it is certainly correct to hold that the essential difference between the maxims of this antique art of education and of that of modern days, consists in this, that in the education of the ancients the cultivation and perfection of skill were esteemed more highly than the labour to which the skill was applied, and the products of that labour. Every individual was to be formed into a perfect specimen of his race, the race itself having nothing to do but to exist and rejoice in its capacities of enjoyment.   Education fulfilled its task in producing the attitude of perfect humanity— that reposeful and plastic stamp of character which henceforward in all the occurrences of life with which it meets or by which it allows itself to be reached, maintains an unchanged mien, and employs its skill to raise itself to independence of material things.   To this many-sided and self-contained development the spirit of modern education is certainly less disposed; it favours more than is right an extensive acquaintance with facts as compared with general cognitive ability, productive and monotonous labour as compared with the free exercise of all man's powers, the narrowness of efforts restricted to a definite occupation as compared with interest in all human relations.   Yet there is at the bottom of all these errors one characteristic which is not to be despised— the conviction that man's destiny is, not to present a perfect embodiment of all the beauty of his kind, but to develop into an unique individual—a development which cannot be attained by aimless exercise, however splendid, of the capabilities common to all, but only by devoting these in earnest labour to the accomplishment of some individual life-work. Only in such voluntary devotion of the powers bestowed by Nature and developed by education, to the laborious pursuit of some definite end can the individual win as his personal property the endowments of the race, developing them, in a course of evolution which extends through life, to an individuality in virtue of which he becomes something more than a perfect exemplification of a general concept.

We by no means lose sight of the fact that the active political feeling and the love of art of the Greek nation and its receptivity for science provided very worthy occupation for leisure time, and that in the eager and steady pursuit of great enterprises, or the constant but more calm interest taken in public business, life found a sufficing content and vocation. But the contempt which was felt for common, rough, hard work, and the low estimation, extending even to artists, in which all handicrafts were held, did not fail to exercise an injurious influence. Much as men laboured, there was not formed in any degree worth mentioning that love of work which is jealous of the honour of its handicraft, which is able to find sufficient sources of mental satisfaction within the narrow limits of a monotonous occupation, which delights in colouring the whole of life with the ways of thought peculiar to its calling, and loves to glorify its mental gain in song. This was chiefly the reason why there was lacking in public life that fidelity to duty and conscientiousness bordering on rigidity, which is more surely produced by the steady exercise of a modest calling than by the pride of a culture which can take any point of view, and has no moral obligation to take one rather than another. Only where morality requires fidelity in small things can great things be secure. The new culture estranged even family life from the beautiful and simple patterns of the Homeric age. For the more exclusively that culture was directed to political interests and scientific occupations, the further were women from keeping up with it and participating in it.   The society of ancient Greece was exclusively masculine.   It was only in assemblies of men that there was the pulsation of that which we call ancient life; the women lived in domestic seclusion, relieved from burdensome supervision in Sparta only, and even there they did not gain much from that life of the time in which they were allowed to share.   The absence of community of labour entailed also absence of the feeling of equal human rights; and of the gain which woman's mind can contribute to life, little accrued to the Greeks.   I do not mean that the natural good disposition of the people did not afford room for the exercise of all the love and tenderness of family feeling which we admire even in the beasts; but still in common opinion the female sex was regarded as the less perfect creation.   Plastic art knew how to honour its beauty, and poetry its charms; but we need only remember the evil sophisms by which, in the Eumenides, Æschylus (by no means an isolated example) proves of how much less consequence the mother is than the father, in order to recognise the insulting contempt with which Greek civilisation on the whole looked down upon women.   It has nowhere produced a conception which in seriousness and human worth is comparable to the noble ideal of the Roman matron.

The worldly wisdom of the Indian gives to the man the toil and the exciting enjoyment of combat and to the woman hard and stupefying labour.   The Greeks did not, indeed, make such a division ; but not less superficially and mechanically did they solve the problem of determining the relation between labour and a liberal enjoyment of life, since they solved it by the institution of slavery, and this without reference to any natural relation which (as, *e.g.*, difference of sex or of race) seems, to the untutored mind at least, to furnish some justification of such an arrangement.   When Hector and Andromache with foreboding sadness lament the misery of slavery which awaits the widow and orphan, not only are we somewhat reconciled by the melancholy beauty of the verse, but, moreover, in this heroic age such misery appears as an event which naturally occurs in the order of life and for which the as yet unfurnished social science of men knew no remedy. In the noontide of Greek civilisation, a time of political insight and reflection upon the order of society, we are revolted by the calm way in which even the noblest minds regard slavery as being, as a matter of course, a constituent part of their political structure. " When the shuttles set to work of themselves," says Aristotle, " then we shall no longer need slaves." It is not the first clause of this sentence (which has so often been regarded as an inspired anticipation of future machine-labour) which seems to me remarkable ; for Aristotle is here giving expression not to any anticipation but to a recollection of the dædalian works of art which mythology had extolled. But what is remarkable is how wider development (governed by the idea of the advantage to be expected) seeks, under a condition of things in which slavery exists, to realize the contradictory notion of an instrument that acts intelligently and yet remains a mere instrument. With much adornment of logical periphrasis it veils but slightly the aristocratic egoism, which from the self-regard of the favoured individual, from the requirements of the refined and liberal culture of one man, infers the servitude of others as a matter of course. The capabilities of men are various ; Aristotle distinguishes kingly souls which are capable of living nobly and worthily in their own strength, from others which can neither set before themselves any intelligent aims in life, nor if they had such could find the means of working them out. But the moral duty of careful teaching of the weak and compassionate love towards them is not assigned to the strong as a consequence of their superiority ; the title of " kingly souls " once bestowed introduces unperceived into the discussion the claim of sovereignty, and the weak become the chattels of the strong.

Such a foundation would be even worse than the reality. Debt and capture in war were everywhere the most frequent

causes of slavery.   In the second case, the harshness of the
victor may be understood as a result of the hatred which
survives the contest, this hatred at any rate being a passionate
emotion; and in the first case, a series of deductions which
are not without some show of justice, easily leads to the
conclusion that the debtor who is unable to pay off his debt
should with his capacities of labour be made attachable.
Then in order to secure the use of these capacities his
freedom should be restricted, so that finally, in order that
they may be exchangeable for money, his person should be, not
indeed immediately vendible but liable to be bound to render
an equivalent in labour to any third person in return for the
payment by that person of the sum owed by him.   In both
cases there is wanting the indispensable recognition that the
dignity of human personality does not allow either of such
a satisfaction of the victor's passion nor of such a mode
of carrying out legal claims; but the cold-bloodedness of
Aristotle's sophistical deduction is without even the feeble
excuse which may be made for these two historical causes
of slavery.

The harshness of theory was only partially mitigated in
practice.   What was the sign by which those kingly souls
were distinguished from the souls that were born to serve ?
In the first place of course Hellenic pride regarded those who
were not Greeks as destined by Nature to slavery; not because they were incapable of being civilised, for even the
barbarian slaves who had been purchased were educated in
order to make them more useful, but simply on account of
their descent.   In the endless internal wars, however, inhabitants of conquered towns were sold as slaves, Greek was
enslaved to Greek in spite of the condemnatory public opinion
of those not concerned in the traffic and of occasional laws
forbidding slavery or requiring that redemption should be
allowed.   For the rest the condition of slaves was various
enough.   Cruelty and delight in torture were not prominent
national faults of the ancient Greeks, but just as little were
they a tender-hearted race; what was most important was

that their moral principles depended upon the existing condition of their speculative convictions without any active and
immediate sense of duty. Athens treated her slaves mildly,
and it may be that their condition was happier than that of
the free proletariat of more modern times; Sparta had a
doctrinaire tendency to inhumanity due to her principles
of statecraft; the Lacedæmonian youths roaming steathily
through the forests and plains in order to slay secretly
the discontented helots, present in the midst of fair Greece a
dark picture which is genuinely Indian in character.

Upon this foundation of deep dark shadow there rested
the brilliant development of liberal culture which has made
Athens and some other of the Greek states an imperishable example to posterity. The αὐτάρκεια, the self-sufficingness which
Greek philosophy so often extolled as the crown of human
perfection, was by no means to be found in this constitution
of society, for here the enjoyment of some depended on the
labour of others. Therefore, however great the mental development might be which was so won (and it can hardly be
proved that it could have been won in no other way), yet in
the clear recognition by the common consciousness of the un-
suitableness of such a foundation for the highest human perfection there is certainly involved a great and perceptible
advance in human progress—an advance, however, that only
came slowly and that is not yet complete.

In the period succeeding that of which we have just
spoken, the Roman Empire only developed further the pernicious germs referred to. The Italian tribes being actively
disposed, and not much inclined to the cultivation of a
variety of industries, were all the more attached to the
unvarying pursuits of agriculture; to this kind of labour even
the Romans continued for a long time to recur with liking
and esteem. But the continuous wars in which the growing
state was involved prevented manufactures from flourishing,
and gradually led to a habit of taking possession of the
necessaries of life by force of arms instead of producing them;
and subsequently led the Romans to treat the greatest part

of the known world as though it had been a mere store-house for themselves, thus dulling their own liking for labour. The way in which the Roman dominion spread, not through plundering expeditions, but with regular administration and exaction, easily explains how the gains of conquest led to the disproportioned wealth of a few, while the majority became poor. The Romans had to spend their own strength in the labours of unceasing military service, and the home-returned veteran lamented that he could no longer find a clod of earth on which to rest his head, and that there was not even room for him to work for wages, since all labour was in the hands of the multitude of slaves taken in war. Society was shaken by repeated attempts to regain the lost basis of economic equilibrium by means of repartitions of land ; the state was forced to bestow in benefactions of food and money the fatal gift of unmerited alms (instead of wages gained by labour) upon a multitude who soon ceased to demand anything but bread and theatrical representations. Public life certainly continued for a long time to have, in the greatness of political activity, an interesting and important content; the strict family morality of former times long continued to exercise its educative influences ; but rigid legality had in Rome's early days imposed even upon Romans harsh restrictions of liberty and bondage to creditors, and made the power of the father and master unlimited, at least in theory. The same disposition, not softened by any varied and humane culture of native growth, and having once for all missed the true principles of morality, led to the extreme of doctrinaire and systematically regulated cruelty in the judicial and legal ordering of the condition of slaves.

§ 3. Antiquity did not succeed in dividing labour and commodities so as to produce universal happiness, or even so as to escape the reproach of avoidable injustice. But it witnessed a many-sided mental development in which men sought to find the aim of life and the way to enjoy life worthily, and if minds had not derived much benefit from the educative effects of labour, yet on the other hand the developed taste of the liberal ancient culture had a stimulative effect upon labour by setting before it an abundance of interesting tasks. We see this effect in a pervading artistic grace and in the harmonious style of treatment to which it is owing that even in our view the numerous small remains of antique labour seem to represent a coherent wealth of ordered beauty in the surroundings of life. We see it also in the splendid works in which the organizing activity of political administration combined a multitude of subject powers. This condition of things was changed by the storms of national migration. The vague adventure-loving impulse of the heroic age again obtained ascendency over significant mental culture; slavery as a legally existing institution did indeed gradually disappear; but the labouring section of mankind, as contrasted with those who carry arms, sank into a state of dependence which in many respects was hardly different from slavery. Neither in detail, however, nor on the whole, did the newly dominant element afford to labour the stimulus of interesting tasks. For the requirements of private life were neither so varied nor so refined as before; the degeneration of political life into a multitude of territories loosely federated, and constantly at war with one another, prevented any of those great enterprises which had been the pride of antiquity. Yet ancient art and its productions lived on as well as they could; and these transmitted remains subsequently furnished an animating stimulus to renewed advance; but for a long time nothing new arose, and no age is so poor in progressive discoveries and inventions as the interval which divides the downfall of the classical world from the renascence of the sciences.

And it was just labour which by its peculiar development, especially in the more northern countries of Europe, was to change the whole aspect of life, and to give it a new and permanent direction. When the storms which stirred the nations had subsided, commerce which again began to traverse the different countries awoke new wants by the commodities

which it introduced, and new efforts to satisfy these wants at
the price of native productions. At the places where men
met to carry out these exchanges of commodities, settlements
were formed with which by degrees the native industries of
the surrounding country became permanently connected. Both
the absence of legal security in those times, and the imperfection and awkwardness of communication with distant
countries, necessitated close combination between related
industries, and at an equally early date made these combinations inclined to exclude any workman who had not, by
undertaking the duties of the brotherhood, also acquired its
rights. These noteworthy historical circumstances caused a
man's chosen work to become a fixed calling, which determined
for each individual his rank in the society; for in fact his work
was to him no longer a mere quantum of labour which he had
to get through, and by which an equally definite quantity of
enjoyment was to be purchased, but by his having on his
part voluntarily taken up this work he had become instead of
a mere specimen of the race an authorized constituent of
human society. The same articulation of society, which in
oriental caste had become as it were hardened into a natural
distinction, irremovable and extending from one generation
to another, was reproduced here, with the difference that it
was now an order in which the individual was entitled to
freely choose his own place; just as much a matter of course
as that each naturally belonged to one family was it that he
should not only do work or carry on business as a member of the
society, but that he should also follow a definite calling, sharing
its duties, rights, customs, and enjoyments. Thus all labour
was systematized into guilds; even beggars and vagabonds
were regarded as constituting a fellowship, having like the
others a right to exist, and having to establish this right by
the observance of certain customs. These combinations which
first arose from community of labour soon involved a community of all the interests of life; at social entertainments,
and in the administration of civic business, men took part not
simply as men and as citizens, but they felt both that it was

from the rank to which they belonged and from the guild
that their right arose to participation, and also that the same
source furnished them with the characteristic and expressive
forms of such participation.

Much in this constitution of society may now appear to us
as arbitrary restrictions; but that which makes us feel it
restricted then existed either not at all or only in a very
slight degree; and it is really doubtful whether our feelings
in the matter are quite justified. That remembrance of
differences of rank should be dragged into free social intercourse may easily seem to us preposterous; but there was
then no general culture which could make the interchange of
opinion interesting, and no generally accepted code of morality
capable of imposing fixed and beneficent forms of intercourse.
Still less active was the consciousness of a political order
representing social advantages of more than mere local interest;
on the contrary, those town communities which had arisen
from definite departments of labour were the only living wholes
which being united by reciprocal needs pursued common ends.
Thus it was natural that political importance should accrue to
individual trades in the localities where they flourished—an
importance by no means correspondent to the nature of the
labour in which they were engaged, but quite appropriate to
a society of men bound together among themselves by similar
habits of life and reciprocal duties and rights.

The results of this relation were of advantage to labour
itself as well as to public life. Consolidation of a trade into
guilds, beside which others exist, roused natural emulation and
made men desire to be esteemed for the sake of that condition in
life which they had chosen. There was developed that sturdy
temper which makes men seek to maintain before all the
world the honour of their handicraft, and makes them give
themselves to their work with heart and soul, in order that
they may increase its excellence; slowly and with difficulty,
not as yet helped and supported by any science, artistic fancy
once more gained a footing upon this path of thoughtful
labour. Public life gained in prosperity and beauty by the

humane institutions primarily founded by the brotherhoods
for the sake of their members and by the contributions
which they vied with each other in rendering for the
advancement of the common good; national codes of morals
having long ago fallen into disuse, family life, principally
under the influence of this industry, developed the new
growth of civic discipline, the strictness and steadiness of which
recall the golden age of Roman honour; and yet being pervaded
on the one hand by Christian thought which tends to freedom,
and on the other hand by the spirit of active industry, it shows
in not a few points an undoubted advance of the human race.

For a long time this form of life, in which work and enjoyment are blended as much as possible, was opposed to the
adventurous spirit of chivalry, which found that as society
became gradually consolidated, occasions of knightly deeds began
to fail, society having even to defend itself against the attacks
of the knightly order; but the new view of life made its way
notwithstanding, and if political independence or a recognition
amounting to the same thing were not very rapidly reached,
yet this philosophy soon began to determine the general forms
of society.   It is by it that the material wealth of modern
countries has been won; from it proceeded at a later period
the revival of learning and art; so to it was due nearly the
whole content of life; and it was but natural that it should
also influence the external character of life, even to costume
and the tone of conversation.   But it did not reach this
supremacy until influential circumstances of all kinds had
already begun to produce an essential alteration of its character.

§ 4. The great geographical discoveries with which the
Middle Ages closed, the rapid development of the physical
sciences which soon followed, the extraordinary effect which
the discovery of printing had in extending, accelerating, and
facilitating the communication of thought, and the similar
influence exercised by the development of navigation and
finally of steam power upon commerce—these things it is
that have chiefly given to modern life its distinctive character
as regards enjoyment, industry, and interchange of goods.

The outlines of land and water on the earth's surface have now been ascertained with a completeness which causes us to believe that we cannot look for any surprising discoveries in the future, and for the first time the various races that dwell upon the globe have come within sight of one another. The interior of great continents and their resources still remain in deep obscurity, and many nations are still seeking points of departure from which they may proceed to the formation of permanent social relations; but everywhere we find an investigating zeal which is no longer content to amuse the imagination with a description of distant wonders, but desires to bring all these unknown and distant regions into useful connection with our own civilisation. The explanation which science is now beginning to afford of the extensive connection between natural effects all over the surface of the earth already gives useful support to these attempts, hindering some adventurous undertakings by showing their economic uselessness, and encouraging others by pointing out their probable good results. Commerce, in equilibrating supply and demand in the most distant regions, and being able to effect desirable exchanges with increasing ease, is approaching the solution of its problem, which is to unite all parts of the earth into a single economic whole, to supplement the niggardliness of one climate by the fruitfulness of another, to guard against the dangerous fluctuations of society caused by famines in ancient times and in the Middle Ages, and to make the most inhospitable regions fit to be at least a temporary abode of human beings wherever Nature has not set limits to men's further advance by refusing the gifts which are absolutely indispensable to life. Political projects which have never been altogether independent of economic considerations are now obliged to be made with a more careful calculation of the much more complicated actions and reactions upon which the power and welfare of states depends. Perhaps an accurate judgment of what is here advantageous is in most respects still in its infancy; yet to some extent we clearly see the restraining power which is exercised upon the warlike instincts of mankind by the consciousness of this connection of complicated relations which men are bound to respect. Not indeed unfailingly, nor in all respects advantageously, is this influence exercised. For however desirable may be the restraint of coarse and merely destructive forces, it is by no means desirable that the whole of life should be fettered by material possessions and by that love of peace which would sometimes be willingly deaf to the call of honour from fear that such possessions should be endangered.

The opening of the boundless realms of the new world has in another respect had a favourable effect upon political life. Many institutions and conditions which had been handed down by long tradition, oppressed mankind as with the consciousness of a tedious and hopeless malady, and now an opportunity was afforded it of making vast new constructions ; it could now learn by its own fresh experience what strength and activity human life demands when men are forced to return to the most primitive labour, what benefits (perhaps too lightly esteemed) may be combined even with the evils of ancient civilisation, and finally what new and more vigorous institutions may be established when men are unhampered by tradition and are free to be guided by existing circumstances. It had hitherto been as impossible for history as it is for the physician to make the valuable experiment of trying how an existing condition, which has been treated in a definite way, would develop if subjected to quite different treatment. One of the most special advantages of modern times has been the possession of this new world alongside of the old world, and the being able without any sudden interruption of historical development to realize the events and life-experiences passed through by men in that great arena of aspiring powers.

To this extension of the scene of economic activity, with its important results, the growth of physical science furnished the means necessary for the complete conquest of the new territory. Useful discoveries have been made in all ages, but there has not existed in all ages that activity of imagination to which any success attained immediately becomes a starting-point for fresh undertakings; in ancient times and in the Middle Ages, the application of any newly discovered natural or artificial power was usually restricted to the immediate sphere of work which had given occasion for its discovery. It is different in our time. By experiment and calculation the principles and laws of action of forces have been arrived at in at least some departments of Nature; numerous observations have ascertained the various results produced by the action of these forces under arbitrarily established or altered conditions of their application; now every newly discovered material and every newly ascertained natural process is regarded from a variety of general standpoints and compared with a variety of recollections of what has been previously observed, and these not merely arouse but often forthwith give an answer to the question, What further advantage is to be gained by subjecting this new discovery to definite conditions or by combining it with known forces? Hence arise men's vigorous endeavours to follow out forthwith all the possible applications of a fresh discovery, and the frequent demand that definite instruments of progress (which are needed and from which are expected services which can be exactly specified) should be provided by searching out new chemical combinations, or new means for the composition of forces; and hence finally a knowledge of the hindrances which yet remain to be overcome in the accomplishment of a mechanical task, and of the direction which must be taken by any investigation which aims at removing these. These advantages depend upon the nature of our knowledge and the facility with which (thanks to the easy communication of thought) co-operative labour can be carried on; and they have not only conferred upon us an incomparably greater wealth of useful commodities than were possessed by men in ancient times and in the Middle Ages, but have also determined our mode of thought. Much which formerly seemed to us impossible we now regard as a mere matter of time; the combined energy of men applies itself to the most extensive undertakings with a calm prevision of success. This energy seeks not merely to transform the inanimate world but also regards the animal kingdom as a constituent part of a universe of usable commodities, modifying the physical formation of animals by careful breeding for arbitrarily chosen ends, and thus feeling ever more and more supreme over Nature and ever more and more losing the remains of that awe with which even as late as the Middle Ages the mysterious characteristics of natural elements were regarded; men anticipating more results from the wondrous developments of these (which they ventured only timidly to initiate) than from their own well-calculated interference.

These considerations extend to the coherence of society with reference both to its internal consistence and to its connection with physical conditions. The abundant and penetrating reflections of antiquity upon these questions were destitute on the one hand of a basis of observation wide as to both space and time, and on the other hand of the possibility of easily communicating the results attained. Statistical science with its characteristically developed methods of comparison is now able to utilize the rich material which the present owes to its greatly enlarged intellectual horizon, and the existing multifarious means of communication make its results the common property of much wider circles. Thus among the most characteristic features of modern times may be reckoned growing clearness and increasing extension of reflection concerning the foundations of the economic articulation of society, concerning the laws of exchange, and the connection of all human activity. If it were ever possible for the human mind to move on exclusively in a single direction, the injurious effects of the present preference for this region of thought would be developed still more plainly than they are. For taken alone it favours the disposition to regard all that happens as a mere example of general laws. It has a tendency to make man regard his own development, which had before seemed at least partly to be the work of his own free will, as the product of climate, of food, and of natural endowment, and the changes of these that take place according to natural law. In this connection of all things,

mechanically so clear, it is difficult to hold fast the thought
of higher ideals, ideals which are entitled to require something other than that which the natural concatenation of
causes and effects can of itself produce. In fact the flood of
materialistic views with which we are inundated bears witness
to this increasing disposition to leave to man no other destiny
than care for his physical nature, development of the capacities
of his kind, and the multiplication of those good things to the
enjoyment of which this part of his being leads him. Thoughtful reflection also, which does not take such a narrow view, has
succumbed to the temptation to regard social changes which
seem to be forced on by natural conditions, as being justifiable
simply because they are explicable; and to look on at the
stream of circumstances with tacit acceptance of events that
are accomplished, or are in course of being accomplished,
approving every turn and eddy of that stream.

§ 5. The greatest part of the peculiar form assumed by the
relations of labour in our times is due to the development of
machinery. The infinitely numerous possible functions of
the human hand in labour are found separated in machines,
each individual function being attached to a mechanism
which exists purposely for it, and each being on this account
endowed with greater strength, staying power, and exactness. Antiquity possessed but few of these advantages; it
had at best only tools, that is to say contrivances which do
indeed by their construction and manner of use afford to
human strength a more convenient hold of the objects upon
which men work, but yet find the spring of their movement
and action in the strength and skill of the human arm. It was
the utilization of steam which first substituted for them, and
that with ever increasing generality, machines the disposable
force of which is developed not indeed from nothing, and just
as little from a mere summation or transformation of human
activity, but from the efficiency of elemental forces, machines
merely providing for this efficiency the conditions of useful
action; and even this work is facilitated by the progress of
technical art. As from the beginning the earlier and coarser

tool helped to make a more delicate one, so it is machines themselves which make those parts of other machines that are difficult of construction ; and it is machines themselves which, in part at least, changing their action according to the changing requirements of the work, counterbalance the injurious incidental effects which that action would otherwise entail.

The costliness of machinery and of keeping it going, generally speaking makes its employment profitable only in uninterrupted production on a large scale.   As when the radius of a circle is increased, successive equal additions to its superficial extent are made with an ever decreasing proportional addition to its circumference : so with the same necessity in most kinds of labour, as the scale on which it is undertaken is increased, the increase of useful production exceeds in a growing ratio the increase of outlay ; when reduplications of similar functions are performed by one instrument there is hardly needed an increase of the activity which it would have to devote to a single performance of the same function ; most productions gain in perfection when their various separate parts are made by separate machinery which is devoted exclusively to them ; and finally this division of labour, advantageous in itself, is facilitated by the unvarying exactitude of mechanical action, the uniformity of its productions making possible their subsequent combination into a whole.

The advantages hence arising for the products of labour and for their distribution have been as often extolled as the disadvantages connected with them have been lamented.   It is without doubt due to the use of machinery in manufactures that there has been diffused among the people a great supply of the means of comfort and wellbeing which either were quite inaccessible to the civilisation of earlier times, or on account of the difficulty of procuring them were attainable only by a few.   But this industry has already absorbed much which used to belong to art, and though the artistic element may not have been wholly banished from its uniform productions, yet they are without the traces of that lively individual

imagination which is revealed in so many objects of ancient
or mediæval workmanship—objects which one hand had with
loving interest framed in every stage, from the raw material
to the final form. It is now more difficult than it used to be
to provide dwellings with harmonious furniture; it is the slight
interest which we can feel in furniture that has been purchased and brought together from a variety of places, that
makes us disregard the lack of coherent mental character in
our customary surroundings. On the other hand, the cheap-
ness of manufactures produced by machinery, as compared
with those produced by that human skill which has now lost
its value, is not so great as to allow of unpropertied persons
participating with any degree of completeness in these new
comforts and conveniences of life. In perfectly simple states
of society, the various dispositions which even there have place
appear side by side as if they all had an equal right to exist,
just as the different kinds of animals, for none of which is it
any reproof to be what it is; it is to a high degree of refinement,
that there is first opposed as its antitype that coarseness which
while it knows all the newly discovered and newly developed
moral relations despises or misuses all of them. Just in the
same way poverty of external appearance is no reproach, is
often even picturesque, at a stage of civilisation in which men
have but few needs and satisfy these in the most primitive
and simple manner. On the other hand, this same poverty
assumes the peculiar character of squalor when it appears in
the midst of a society the life of which is based upon a very
complicated and intricately branching system of satisfying
human wants. Poverty, taking isolated and disconnected fragments from this system, becomes subject to wants which it has
no assured permanent and adequate means of satisfying; and
substitutes for previous frugal needs and occasional inventive
sallies the awkward discomfort of surroundings which afford
adequate satisfaction of needs only by fits and starts, and of
an outward appearance of slovenliness. It is only in the south,
with its mild climate, that there still remains any charm about
the life of the majority; the vast and needy masses of the

civilised nations of the north pass their existence even now in such dwellings and under such conditions as to clothing and household furniture as must be hardly less repulsive than the hovels in which thousands of years ago oppressed Asiatics hid themselves away from their tyrants.

Still more unfavourable is the effect of the new forms of labour upon mental development.   What was so much feared in ancient times, the narrowing of men's intellectual horizon by unintellectual occupations, threatens the mass of the people more and more as the division of labour goes on getting greater.   Even in the division of manual labour in past times, many an employment constituted a fixed vocation which, if the matter had been settled by regard for untrammelled human development, must have been reckoned among the temporary occupations of household labour.   But independent handicrafts generally embraced a plurality of cognate operations; it was possible for the labourer to accompany the various stages of elaboration undergone by raw material before attaining its final form, with continuous activity and a satisfactory sense of the progress and results of the work.   The tool habitually used did indeed exercise an influence upon the bodily development, the demeanour, the character, and the sphere of thought of the workman; but yet he was not its slave: in every outline of the finished products he could, as it were, trace the strength and delicacy of his own formative touch.   On the other hand, man's share in the work that is done by machinery is limited to very uniform manual operations which do not directly shape anything, but merely communicate to some mechanism which is not understood an uncomprehended impulse to some invisible operation.   The completed product reaches the hands of the individual worker in a condition of which he did not witness the production, and passes out of his hands again to undergo further transformations which are brought about in a way equally obscure to him.   Hence arises the worst possible division of labour — the separation of the sagacious invention and guidance which, with the increasing complication of machinery, requires ever increasing circum-

spection, from the unintelligent manipulation which is able to
do without thought in proportion as all its difficulties are
solved by others.   For the only perfection which it is possible
for such workers to develop—the formal one of exactitude
without consciousness of the ends to be attained—is the very
same virtue which is required from machinery itself.   It is
only unusual talent that can succeed in raising itself, under
such unfavourable conditions and in entering the ranks of
invention; for moderate capacities labour is no longer either
enjoyment or a means of culture.   And this injurious result
cannot be counterbalanced by the compensation which intelligent benevolence seeks to provide for the labourer by giving
him a larger allowance of leisure and better means of occupying it.   He may have access given him to means of scientific
culture, to instructive lectures, to refined pleasures—he may
even be enabled to enjoy temporarily a luxury, which certainly
may possibly be made accessible to him by a system ot
industry that depends upon enormous consumption; but all
this does not alter the feeling which regards unintellectual
work as a mere means to enjoyment, and having no sympathy
or devotion for the work itself merely seeks to get it over in
order to obtain its fruits.   This lamentable division of life
into labour and leisure that are opposed to one another as day
and night, is at present undoubtedly progressing; when we
boast, as one of the advantages of our own time, that all kinds
of labour are now respected, this often means nothing more
than that the attainment of means of enjoyment by any kind
of effort is praised; it is not labour but its product that is
sought; men undertake to bear for a fixed term of years the
repulsive burden of this effort, which is destitute of mental
interest, in order that then the remainder of their life, sharply
marked off from this time of labour, may be spent in idle
enjoyment.

The social relations, too, which depend on the division
of labour, develop new and gloomy aspects.   As long as
production by hand-labour remains profitable, or in as far
as trade is concerned with simple products the indispensable-

ness of which insures their sale, honest endeavour may maintain a modest independence, without having any great superabundance of intellect and capital.   Wider knowledge of the connection which there is between the needs of extensive groups of countries, now makes it possible to anticipate demand to a much greater extent than formerly, the multiplied means of communication allow those products which can be cheaply supplied in large quantities to be easily got rid of, and the greatness of the resources employed makes it easier to weather the fluctuations of demand and exchange ; in many cases the greater excellence and uniformity of things produced by machinery contribute to drive out hand labour. There are not a few handicrafts which from an independent production of commodities have come down to the mere finishing off and fitting together of manufactured goods ; others no longer pursuing any trade of their own have to take a subordinate place as mere appendages of great businesses. The same conditions which in a general way make the combination of several different operations in one business more remunerative, have a specially powerful effect in concentrating mechanical industry in great manufactories, a system which, by its combination of mind and money, prevents mere faithful work from attaining independence.   It is true that within short periods the machine worker is more sure of his wages ; but whilst independent handicrafts depend upon the needs of a greater number of customers—a number which in a small trade is seldom altered suddenly—the existence of the machine worker depends partly upon the arbitrary choice and the insight of one person or of a few, and partly upon the fluctuations of universal demand and supply, which he can neither survey nor control.   This insecurity is by no means counterbalanced by the sense that he participates in a great whole, for he participates neither in the insight nor in the gain, but almost exclusively in the dangers.   Nor has he more cheering prospects as regards a gradual improvement of circumstances.   His wages are mostly insufficient for the attainment of ultimate independence; and a change of occupation is impracticable for him; since, generally speaking, a man becomes thoroughly competent for any definite work only by long habituation, which unfits him for any other. It therefore seems to the machine-worker that the best condition of life attainable for him is soon reached, and that striving after something more serves only to lessen the enjoyment of the present; the impulse to frugality is extinguished, and early marriages (contracted because there is no prospect of any advantage being derived from delay, and because the children's capacity of labour can soon be turned to account) rapidly increase the number of industrial proletarians, all doomed to the same prospectless and improvident life. The humanity of the masters, which is often present and often absent, cannot remove these evils without changing the principle of division of labour; even a patriarchial relation between them and their subordinates would not produce a complete solution of the problem, since this could only be found in the re-establishment of an independence based upon men's own activity.

In another direction labour has broken through earlier restrictions, with much advantage and not without some disadvantage. Historical relations had made it necessary that infant guilds in order to prosper should have strong internal coherence and external inaccessibility. But altogether rash was the view (which in course of time developed from these beginnings) that all human labour falls into a limited number of classes with a regularity like that of the animal or vegetable kingdom, each of these classes having an exclusive right to a definite circle of employments. The growing-up of new kinds of work, which could not be fitted into this system, led to the removal of such limitations, and this has certainly opened a free field of labour to struggling powers which were before confined; but the benefits of this improvement are abridged by the general condition of things. As there is scarcely any business which may not possibly be carried on in manufactories, the powers which have been thus set free may also divide into the two classes of employers of labour and dependent labourers. The possibility of going from one business to another may delay this result, but will also contribute to make men forget still more the idea of a calling and to dissolve the steadiness and security of ancient customs depending upon it ; life will become a succession of disconnected attempts to fight one's way through somehow or other.

The present age has met these wants by a resource which promises much though not everything, namely by voluntary combinations for definite objects. As Assurance Companies they distribute among a number the unavoidable damage produced by natural causes, effecting this distribution as a judicious economic measure ; as Joint-Stock Companies for carrying out undertakings which are beyond the power of individuals, they alone, combining self-interest with the common good, are able to succeed in works which can compare with the colossal undertakings of antiquity ; they appear in innumerable other forms in order to combine the separate resources of individuals whose wants are similar by buying the materials for work wholesale, saving the useless cost of retailing, and affording to small capitals, by co-operation in trade, the same rate of profit which large capitals can obtain. Cheering experiences already testify to the value of the further development which this principle is capable of.   Needy workmen combining their small savings into one capital stock, and thus being able to enter upon undertakings for the common benefit, have enlarged their modest associations into flourishing companies which afford to all participants the commercial advantages of business on a large scale. The united community of workers takes the place of the one employer, and the satisfaction of labour by wages regulated by the supply of unemployed labourers is transformed into a participation of the gain obtained by the industry of the society ; the oppressive and demoralizing effects of the relations between the sole lord and his " hands " give way to the animating and moralizing power of the sympathy which the individual feels for the prosperity of the whole to which he belongs. Without recourse being had to express prohibitions, vices

of excess, which are not congenial to the spirit of these
societies on the whole, seem to grow less of their own accord;
they have manifested a vigorous impulse towards further
cultivation by establishing educational institutions and
seeking means of instruction; without State support and
struggling against many obstacles, they have brought to their
members an amount of gain which secures and improves their
existence and their domestic life. It is hard to anticipate
experience and to determine what capacity of further development these associations may have; what they have hitherto
not afforded is the independence of individual callings, for all
they do for the individual is to guarantee him a competency.
The question is whether this ideal of family life, self-dependent,
economically self-supporting and constituting in itself a complete sphere of activity, is capable of general attainment in
our time, or whether it must not be sacrificed to the changed
conditions of labour. It still exists on landed properties
where the owner is the cultivator; but if the time of the
steam-plough should come and its superiority should make
necessary that cultivation on a large scale which alone is
suited to steam agriculture, then many fields will be thrown
into one, all the slight hollows will be filled up, all the slight
elevations will be levelled, and though individual rights of
property in the wide and fruitful plain thus created may continue, it will be handed over to the administration of select
committees, from whom after the harvesting the owners will
receive the produce or an account of it. The connection
between man on the one hand, and Nature and the labour
applied to natural objects on the other, will in this case as
in others become ever less perceptible; the earth also will
then be regarded as merely gain-producing, and not as the
object of an industry that is carried on with self-sacrificing
attachment.

The ties of neighbourhood already combine the inhabitants
of a village or town to a community of interest in most of
the affairs of life; and in the time when guilds flourished the
association between their members was even stronger, and

extended to the whole of life and not merely to work alone, all modern associations have hitherto had the disadvantage of being combinations for isolated objects, none of which captivates and occupies the whole man. As the implement which a man uses lays claim to him altogether as it were, but machinery, on the contrary, works for him, so formerly a man's calling encompassed him as it were on every side, while his present relation to work is like that of machinery to it, no devotion being required from him, but only the punctual fulfilment of a small number of conditions. Formal virtues are abundantly developed ; in the intercourse which is carried on in trade, by postal communication, by rail, in money exchanges, in credit, there is a stupendous reliance upon the trustworthiness of machinery that is withdrawn from all personal supervision and all individual influence, working for men as it were in the dark. What in ancient and mediæval times required a multiplicity of personal efforts, of emotional springs of action, of effectively calculated persuasion, of manifold manipulation, is now (with the least possible expenditure of excitement, with an economy that is sparing even of words) entrusted wholly to that machinery of communication which provides for all. But the more the real nature of business is understood and developed in conformity with its concept, the more are liking and personal devotion withdrawn from it. It is true that a great part of the good results which in earlier times resulted from this active participation is more advantageously obtained by the mode of business administration referred to; that by assurances, by a general system of poor-relief, and by the stimulation of intelligent self-interest, tasks which were formerly left to voluntary charity are to some extent lessened and to some extent more certainly fulfilled ; but after all these departments of human activity have been made as far as possible mechanical, the question becomes more and more prominent, Where, then, as a matter of fact does life itself begin if all which formerly filled it up is removed from the sphere of living interest and reckoned as merely among the preparations for life and instruments of living?

Enjoyment of the leisure which remains after all necessary labour has been accomplished is hardly on the whole estimated very highly in our own age. It is an age which is well acquainted with the bitterness of toil, but knows little of joyous festivals. With the disappearance to so large an extent of trade guilds and status, old manners and traditional customs with all the complex formalities of public festivals and entertainments and all significant ceremonies of social intercourse have declined; and amid the general formlessness, men are at a loss what to do with the leisure they have obtained unless they either turn again to the labour which was to have been got rid of, or seek that sensuous enjoyment which is always to be had. Exhibitions are the only peculiarly modern entertainments of a public kind, and public dinners for political or other purposes are the means used to strengthen enthusiasm. Neither Church nor State supplies the lack of popular inventive power; the latter neither favours the political activity natural to good fellowship, nor does it readily allow the use of social solemnities in even such political action as it approves ; and the Church by forbidding or disapproving natural impulses, abandons the imagination of the people to its own vacuity, without winning it to participation in the forms of worship and the enjoyment of genuine artistic beauty, by positive development of spiritual life.

Now if we take a comprehensive survey of these historical transformations, human life seems to be turned more and more into a struggle for existence ; the multiplication of small wants, which is not accompanied by a proportionate increase in the ease with which they are satisfied, consumes a large share of the strength which might have been devoted to more ultimate ends, while the kind of labour required does not contain in itself its own reward or even a part of its reward. The place of *Work*, which was once a self-animating exercise of activity, is taken more and more by *Business*, that wonderful creation of society, that with its complicated connections and its natural laws which are independent of our will in a certain sense leads a life of its own, and reduces individuals to the condition of its panting slaves.   Great advances in insight, in discoveries, in new social constructions of all kinds serve on the one hand to give new strength to this monster, and on the other hand to give some security, against the inexorable course of its development, to that humanity which it has itself created ; and we are accustomed to admire the one as well as the other.   We regard with amazement and not without satisfaction the growth of those giant cities in which the nature of business gradually concentrates the population, and often forget under what joyless and revolting conditions of existence a large part of humanity is thus placed ; we regard it as an advance when the tender strength of children is employed in useful labour, or there are opened to women spheres of work which secure to the increasing numbers of those who are unmarried the possibility of subsistence ; and we do not enough consider that at the best these arrangements are but forced and wholly unnatural attempts to counterbalance serious evils which owe their existence to the progressive development of all the relations of life.

That the sociological order when left to itself is necessarily such we do not deny, and we think that those are in the right who hold that it is unpractical sentimentality to wish for a condition which cannot be brought back.   But the remainder of the truth must also be told, which is that this course of things is not in itself a movement towards perfection.   The innumerable individual steps of progress in knowledge and capability which have unquestionably been made as regards this production and management of external goods, have as yet by no means become combined so as to form a general advance in the happiness of life.   For the growth of this happiness cannot be sought either in the mere multiplication and improvement of productions, or in the increasing bustle of industry, nor yet in the ingenuity that tries to maintain the same tolerable equilibrium between labour and wages under conditions that become ever more and more artificial and complicated.   For this maintenance is the utmost that is accomplished.   Each step of progress with the increase of strength which it brings, brings also a corresponding increase of pressure; the more varied the ways are in which the individual elements that form the social system touch one another—their connections being now more tense than formerly—the more do they both gain by the union of their forces, and suffer from the disturbances of others and the inner repulsions of all. Hence we find that never has there existed in such a striking degree the inconsistency of holding that the whole life with which men are anxiously occupied and which they eagerly participate in, is not at bottom the true life, and of dreaming that there is another and a fairer that might be lived and will be lived as soon as the lower life gives us time, and opens a way of entrance to it.

Let us see now whether in the midst of this noise of external progress, this better life has been preserved, and perchance by its own advance towards perfection provided a compensation for the deficiencies which we have indicated.

CHAPTER III.

BEAUTY AND ART.

Art as an "Organism," and as the Expression of Human Feeling—Eastern Vastness—Hebrew Sublimity—Greek Beauty—Roman Elegance and Dignity—The Individuality and Fantasticalness of the Middle Ages—Romance—Beauty, Art, and Æstheticism in Modern Life.

§ 1. IT is no longer our custom to personify (as myth-constructing imagination once did) the various forms of mental activity which in the course of history have been devoted to the same supreme aims, aided by ever new and perhaps ever more perfect expedients. But after thinking we had discovered in their historical changes an ordered and constant progress, we found, in the name and the notion of *spiritual organisms*, a means of ascribing to them greater independence of existence and development than really belongs to them. Philosophy and the history of philosophy have long been spoken of as if they embraced not only the ever-recurring efforts of human thought to grasp the truth which is always equally valid, not only the series of philosophic views by which the human heart seeks to rise above the doubts and difficulties and distresses of life; rather it seemed as though in them truth itself experienced a development of its own existence and content and validity, like the growth of a plant which is indeed tended and cultivated by our care and attention, but yet unfolds beneath our touch according to its own immutable law of development. Of the sphere of art, too, we are now accustomed to speak as if it were a mysterious region of enchantment, having indeed its place in our life, and yet separated from life, accessible to few, working in the service of eternal beauty according to laws and order of its own, holding together its various productions in a complete and isolated system, and governed, as to its history in time

by an innate law of development. We do not wholly dispute
the justice of such a conception, nor the good results which it
has had in deepening men's appreciation of all beauty; but the
few considerations which we are now about to offer are not
directed to this organism of art, for the development of which
according to its own laws the living passion of nations can serve
but as nutritive sap. On the contrary, our discussion is only
concerned with the varying attempts of men to make clear to
themselves the mood which governed them, and the peculiar
feeling awakened in them by existing conditions, by impressing
the image of that beauty which had most taken hold of their
minds upon everything that they did and experienced, both
upon the character of everyday intercourse and upon works
which were intended to remain as lasting monuments. As far
as posterity is concerned, it is commonly the constructions
of art which afford the most evident testimony with regard to
this æsthetic life of the past; to the men of any age the works
of art of that age are but one and that not always the most
expressive of its manifestations; for their production and their
greatness depend upon the number of creative and constructive
minds, and these, in consequence of some dispensation which
is to us inscrutable, are not distributed equally to all ages.
But even such minds cannot collect scattered rays if these are
as yet non-existent; and the appearance of such minds presupposes that men in general are in tune for that aspect of
beauty to which they are called upon to give form and
expression. Therefore where great artists are wanting, and
consequently the dreamy mood of appreciativeness is not
suddenly awakened to a clear consciousness of the ideal, there
the slow working of this less creative impulse produces
æsthetically expressive developments of life.

§ 2. The most ancient nations of the East found beauty chiefly
in what was vast. They may also, it is true, have been not
without appreciation of tenderness and grace, an appreciation
of which we have no testimony owing to the destruction of
their literature; but even Indian fancy, which exhibits this
feeling in a striking degree in such of its poetry as is still

extant, has an even greater preference for what is vast and
unmeasured.   This ancient world was pervaded by reverence
for what is colossal; tradition pointed to immeasurable distances of past time; its constructions towered to the skies,
and extended over the surface of the earth, or penetrated subterranean depths to an extent vastly beyond what might have
been expected from human powers, or what could be required for
human needs; sculptured figures of more than life size, and in
large groups, looked down from their pedestals in mart and
street upon busy commerce, which was struggling to assume
equally vast proportions; civilised countries were populated
by enormous multitudes; armies countless in number were at
the beck of conquerers, whose desires never stopped short of
universal monarchy; rulers, exalted above the rest of the
world by mysterious magnificence, became intoxicated with a
sense of their own divinity, and found nothing worthy of being
entrusted with the records of their conquests except the hard
and rocky tablets supplied by mountains that towered high
above the plains.

The impression of grandeur which the ruins of this bygone
world still make upon our mind, convinces us that its creations
were the result of genuinely æsthetic thought, which not only
covered its incapacity of estimating real beauty by an exaggeration of external proportions, but undoubtedly found in mere
magnitude a one-sided but true expression of beauty.   The
transitoriness of all that is human, and the swiftness with
which it passes out of sight, disappearing in the immeasurable
background of Nature, must have struck early civilisation more
sharply and hopelessly than it did a later age, which can look
back to a transmitted world of complex thought created by
human effort; it seems as though men's minds had sought to
alleviate this secret dissatisfaction by the greater boldness
with which they carried all images and monuments of human
life to such a magnitude as to entirely remove them from any
measurement by the standards hitherto accepted in different
stages of civilisation.   The colossal constructions of the
Egyptians seemed to force their way into the ranks of natural

objects of vast dimensions, as though they had been rivals of equal birth. As year after year they looked down undisturbed upon the inundations of the Nile, and the moving billows of desert sand, they inspired the beholder with a sense of the unending durableness with which the human race fills the ages ; religious worship—honouring the dead and ever mindful of the possible return of their souls to earth with a far-sighted-ness which was regardless of the flight of time, kept up this feeling—a feeling aroused by contemplating the native works of art, and by which these works of art had themselves been pro-duced. If one element in all beauty is an immediate certainty of the dominion of spiritual life over unconscious Nature, the manifestation of that life being inevitably connected with un-conscious natural instruments, those ancient nations have given to this thought its simplest expression ; they have sought above all things to represent the fact of the conquest of Nature by the living Mind ; and whilst they revelled in what was vast, and yet by no means always in what was without beauty of form, they made for themselves as it were space and breath-ing room in which, relieved from the pressure which all finite reality encounters, they might breathe freely with a sense of their own imperishableness. How much they attained in this way we know not ; for no tradition of their mental life has come down to us. It is only the writers of the Old Testament who tell us of the unbridled licence of the kingdoms of Western Asia, in which the life of pleasure flowed in fierce and mighty waves ; the monument of Sardanapalus, with its inscription—*Eat, drink, and love, for all else is but little worth*—seems to be the melancholy conclusion of this age, which in its struggles towards what was great was able indeed to assure itself of the strength and imperishableness of the race, but had failed to find for the individual any eternal content of life, and had, on the contrary, even minified that content by comparison with the colossal magnitude of works constructed by human hands.

It is only the Hebrew people who have left us speaking monuments of their early mental life. They must have

possessed an abundant literature besides the writings which
are now collected in the Old Testament; but judging by the
indications contained in these, those which are lost to us may
have been essentially similar to those which we still possess.
We know nothing about whether this nation had an inclination for scientific investigation; their language is not formed
so as to subserve this end, nor is it fitted to be the instrument
of a many-sided intercourse which makes it possible to
occupy a variety of points of view.   Not that there can be in
the original capacity of a language or in the principles of its
construction an insurmountable obstacle to the development of
any one side of mental life; but the condition of a language
at any time shows the direction which that mental life has
hitherto not taken, and in which consequently it has neglected
to develop the means of communication.   The Hebrew
language of the Old Testament, with its small number of
words for abstract ideas, and its great simplicity of construction, is favourable neither to scientific investigation nor to
intellectual conversation; but it is in an equal degree more
fitted for the most faithful delineation of the ever-recurring
fundamental characteristics of human life, and for the majestic
expression of divine sublimity.   A variety of points of view
which have been thought out and are well under command
generally diminish men's receptivity for both of these, or at
any rate their capacity for representing them; with regard to
both the Hebrew histories and hymns are imperishable models.
The treasures of classic culture are open to but few, but from
that Eastern fountain countless multitudes of men have for
centuries gone on drawing ennobling consolation in misery,
judicious doctrines of practical wisdom, and warm enthusiasm
for all that is exalted, so that mankind has become accustomed
to see in the characters of those most ancient stories and their
destinies, embodied exemplars of human life and of the different
characters which the variety of circumstances develops.

Here popular imagination is no longer directed to what is
vast, but strains after a *sublimity* that stands in need neither
of vastness nor of ornamentation.   Thus the descriptive

poetry of the Hebrews depicts characters and events with the greatest simplicity of expression, without the least artificial complication of motives, disclosing everywhere without reserve those natural springs of action which as long as the world lasts will be the real ultimate incentives of all that men do, however ingenious may be the mask thrown over their actions by the civilisation of any age. These representations do not employ even the figurative expressions with which Greek epic poetry incidentally adorns the objects of which it treats, in order to adapt them to the generally elevated tone of the description; on the contrary, their characters impress us with their sublimity by appearing before us without any adornment, in transparent naturalness, as though there were nothing in the world which could call in question man's right to be what he is, and to know that he, as he is, is the ultimate object of terrestrial creation. Their lyric poetry repeats the same sublimity, only after another fashion; that upon which this depended in their historical writings appears here still more obviously. Here the mind dwells upon its communion with God, and extols with all the power of the most passionate expression, as proof of divine omnipotence, every deeply-felt individual feature of cosmic beauty. For among the divine attributes it is certainly omnipotence which above all is felt, and gives a colouring to æsthetic imagination; we do indeed meet with innumerable pictures of Nature which taken separately have often that inimitable beauty and charm which civilisation, entangled by a thousand unessential accessories of thought, finds it so difficult to attain; but these pictures are not utilized for the development of a progressive course of thought, but merely juxtaposed as though to magnify from different but corresponding sides the omnipresent influence of that divine activity which they depict.

The earnestness of this religious bias of mind towards sublimity did certainly pervade life, but could not endow it with harmonious and many-sided beauty. The thousand petty cares to which notwithstanding their unimportance cheerful attention must be vouchsafed were too far below the soaring flight of this enthusiasm to be efficiently pervaded by it. The regulation of life continued to be left not to unfettered imagination but to instructive deductions from the great principle of religious belief; they filled it not with beauty but with ceremonies and deeds of the law which by connecting the smallest things directly with the greatest enabled the Hebrew people always to maintain, in their highest moods, the loftiness of character distinctive of them, but secured no uniform grace to existence as a whole, during the less exalted moments of relaxed tension.

§ 3. To what admirable richness and flexibility the mental life of the Greeks had developed at a very early period is most impressively shown by their language. In saying this I am referring neither to its wealth of grammatical forms, nor to its euphoniousness ; both make a language interesting, but do not show the greatness of those who use it. On the contrary, as at the period of greatest strength in animals, various parts of their bodies have been pushed out of place or have coalesced or wasted away—the body, which does not for a long while attain the fulness of living strength, having at an earlier period possessed these parts clearly marked out in significant symmetry and filled with vital activity—so in order to obtain a perfectly flexible instrument of mental life, the symmetrical body of language must have its bones to some extent displaced and its joints somewhat stretched ; and the influence of mental progress is shown in it chiefly by phænomena which concern the dissolution of its earlier structure. How many moods and cases may have continued in existence matters very little ; moods and cases cannot suffice for the expression of all possible relations ; but to increase them so as to cover most requirements is not in itself a nobler principle for the construction of language than the principle to which in the last resort recourse is always had when there is increasing demand for delicacy of expression—I refer to the independent indication of relations by separate words. That in this respect the Greek language reached a high degree of perfection is a trite remark ; its particles have always been admired. By their aid language could reproduce not only the essential content of thought but also the shades of the speaker's mood ; the sense of artificiality which perhaps in the dawn of civilisation accompanies every systematic recital and makes the more ceremonious form of verse seem most natural was, by help of these particles, replaced by a sense of easy communication ; just as in the sculptures of the Parthenon perfected art resolves the early stiffness of merely symbolic representation into the gracious ease of perfect beauty.

In all these respects the language of Homer holds a most happy medium between primitive unpliableness and later artificiality. In its copiously used conjunctions and prepositions we are made aware that the poet drew directly from a wealth of those temporal and spatial intuitions whence all languages derive their expressions for inner relations. Its structure of sentences connects thoughts paratactically without the hypotactic complications which later became customary, and continued to be intelligible to the quick ear of the classical nations without being in any striking degree a type of lucid discourse. If in this respect the language of Homer is language in its youth, yet its impression on the whole is decisively that of a language in which it was no new thing for human beings to be spoken of with human feeling. It was only after having been used for a considerable time in the intercourse of a people vividly awake to all the interests of life, that it could have attained such a degree of freedom in the expression of thought ; the metrical form itself must have been preceded by abundant practice in similar composition before its perfect harmony between the form of expression, the train of ideas, and the rhythm could have been produced.

But disregarding this merely lingual aspect, Homeric discourse, considered simply as discourse, bears witness to the early attainment of a high degree of human cultivation. The Homeric heroes speak much and willingly, and know nothing of the fierceness of dumb encounter with which barbaric energy does but hide its awkward incapacity of setting its own thoughts in order, and its still greater clumsiness in expressing and justifying them.   We see everywhere that habit of understanding things which makes men seek for reasons; Homeric men had long ago learnt how to converse with one another, and developed their natural reflections simply and fluently, not always confining themselves to the matter immediately in hand, but using comparisons and maxims which one feels to have proverbial weight, referring to a common social treasure of practical wisdom which had been for a considerable time in their possession.   In this respect the heroic poetry of the Germans produces a different impression; the spiritual depth which we admire in it lacks facility of expression.   The undeveloped structure of sentences; the meagre explanation of feelings and resolves, to the mere statement of which the discourse often confines itself; the occasional obscurity of the course of thought which yet seldom wanders from the immediate subject of discussion—all these indicate a stage of civilisation in which social intercourse is but little developed.   This un adorned conjunction of occurrences and actions between which we may in imagination interpolate unspoken mental agitation, is sometimes favourable to the loftiness of poetic representation; but since life does not consist of a continuous chain of adventures and great deeds, the cheerful interest shown by Greek writers in all intermediate circumstances testifies to greater progress in general tolerant regard for and treatment of the small and apparently insignificant elements of life.

And the Greeks knew what a treasure they had in their language.   When their poets glance at the history of human development, they do not omit to extol the endowment of speech as a great gift of the gods; to be able to express himself is the distinctive characteristic of man; to understand things by their causes, and to guide men's souls by eloquence, is a fundamental thought of their later development.   Homer can say nothing more bitter of the rude Cyclops than that they neither held markets nor had courts of justice, and that no man troubled himself about his fellows.   For the Greek all the real beauty of life arose from the most intense reciprocal action of mental powers in society; unburdened by transmitted science, and troubling themselves little about the knowledge of foreign nations, this dialectic people could attribute an importance to skill in the art of speaking which no later and dissimilar periods could honestly do, although even here unintelligent imitation has not been wanting.

The effect of this mental disposition, which so early turned to the observation and cultivation of human powers, expecting everything from their development, was shown even in the attitude of the Greek mind towards Nature. The penetrating glance of the Greeks could not fail to perceive either the beauty of their country or the significant characteristics of physical Nature, which in mysterious symbolism reflect spiritual life and its vicissitudes; even their mythology makes natural phænomena the background and source of religious thought in the broadest and fullest way; their poetry, by its wealth of clearly drawn comparisons, convinces us of the impression which the peculiarities of natural scenery made upon them, in an incidental sort of way; the very situation of their cities and places of assembly, theatres, and circuses, show how they felt the value of fine and beautiful natural surroundings, and wide prospects. But Nature affected them chiefly as the setting of their own lives, and they sought its beauty in the enjoyment of the mood which it produces in us, and regarded its productions as means of our refreshment and amusement rather than sought to live in sympathy with the mysterious life of Nature itself. It seemed to them, when all was said, that flowers had greater value as a wreath around some man's head than on the stalk where they bloom in solitude; and the saying that Plato puts into the mouth of Socrates—that men taught him, but that trees taught him not — certainly expresses the universal Greek feeling that the value of human society is far above any absorption in the beauty of Nature. Neither painting nor poetry showed much favour to the beauty of landscape; where the delineation of natural scenery can throw light upon men's feelings, we see all the poets, from Homer downwards, able to delineate it in a masterly way with a few impressive touches ; but it would have been nothing to them unless the enjoyment of some beholder had supplied the final life-giving condition. The words with which Homer concludes his description of a starry night in his wonderfully beautiful and striking way—*And from his heart the shepherd doth rejoice* — give the unchanging keynote of the Greek temper, which not only regarded all the glory of the heavens as merely revolving round the stationary earth, but also held that all the good things of earth were destined only for the adornment of human existence.

But all the more perfectly on this account did the Greeks make a real home of the earth, which was to them merely the stage on which was played the drama of human life. In this they were favoured by the situation of their country. If they had been buried in a primeval forest, without ever being able to take a comprehensive view of the situations of adjacent places, their sagacity would have developed in other directions ; it is probable that if they had never been able to take wide and comprehensive views in the visible world, they would never have been able to do so in the world of thought. But where, on the contrary, a bright, clear atmosphere reveals immeasurable distances, where the eye reaches from coast to coast, where the view from a mountain-top embraces seas and the straits (flowing between promontories) which unite them, and numerous human settlements along the shores—there alone does it seem as though the light of heaven really fulfilled its end, illuminating all parts of the world with the lucidity which can result only from showing the connection existing between them. A susceptible race of men could not dwell from their youth up amid such a breadth and wealth of bright and varied scenes without having the sense of spatial order sharpened, and with it the feeling for clearness and intelligibility of all kinds. Even in the Homeric songs we are surprised by the precision of geographic knowledge as long as the scene of the story is laid in regions which at that time we know to have been within the reach of navigation. There is

hardly a town which is not brought before us as a familiar
locality by some permanent characteristic of its situation—it is
on the sea, or in a valley watered by some river, or on a rocky
promontory; the routes of travellers are described with a
distinctness which teaches us that even then commerce had
established permanent paths, and that the sea-roads were
familiarly known. The world which presented itself to the
Greeks was different from the inland forest-covered regions
known to our forefathers; the Rhine and the Danube flow
through the world of the Nibelungenlied like two isolated
threads of silver, in the neighbourhood of which there is
light; but if any warlike expedition takes the heroes of the
song to a distance from these, indistinctness of geographical
knowledge closes like trackless night around them.

And finally, the Greeks were, from an intellectual point of
view, in full possession of this country with the physical
features of which they were so well acquainted. With every
locality that was marked out in any way, tradition had connected stories of the gods and heroes, and had made them
sacred; and to these their stirring historical life soon joined
the remembrance of great deeds performed by mortals. Thus
they were one with their country, and found satisfaction in
the soil itself; what lay beyond the limits of their native
land did indeed rouse a spirit of acquisitive enterprise, but
did not disturb their æsthetic imagination; the abode of the
gods was still within their reach upon Olympus, which was
not beyond the boundary of their horizon, and at the extreme
limit of which lay the entrance to the nether world; all
beyond might continue a chaos, peopled with fabulous beings
by which their native country was surrounded as by an
ornamental framework without order or significance. The
Hebrews were the only other nation that attained to anything
like a similar conception; the smallness of their country, the
never-forgotten connection of their tribes, the oneness of their
sacred traditions, shed upon Palestine too, that charm of an
historic light in which numerous coexistent points stand out
in the distinctness of their reciprocal relations.

A great part of the charm exercised upon us by pictures of ancient life depends upon the favour of Nature, which still endows the southern countries of our continent with a joyousness of life to which the north can never attain.   In their mild climate, which did not require that man should be shut off from Nature, the Greeks who, to begin with, were a finely-made race, learnt to regard nobility of form, dignity of carriage, and grace of movement as among the good things of life and the ends of education, in addition to that bodily strength and vigour the cultivation of which is common to all early civilisations.   It is superfluous to praise what is admirable in all this, and useless to investigate how far the reality corresponded to the pictures drawn by partial fancy when it peoples every rood of Greek soil with living forms of statuesque beauty.   The native poets with their love of satire have taken care to leave behind them testimonies of the frequent occurrence of ugliness and awkwardness.   But these do not alter our general impression; the Greeks present to all succeeding ages exemplars of human beauty; and probably as long as the world lasts the Spartans at Thermopylæ, the Athenians at Marathon and Salamis, the death of Socrates, and the kingly figure of Alexander the Great will continue to be celebrated as classic examples of self-sacrifice, of heroic courage, and of the spirit of enterprise.   Not that other times have not produced numerous examples of similar deeds performed to some extent from nobler motives; but nowhere, except in Greek life, has the intrinsic worth of the action been so perfectly manifested with a simple beauty which does not need that imagination should separate from it any perverse strangeness of exterior circumstance before enjoying the essence.

Such an artistic form had already been given to life when art, reaching the period of its greatest perfection almost simultaneously with the fulness of political maturity, gathered up as it were this living beauty, and reflected it back again upon life.   My intention is not to sketch here, even in outline, its magnificent development; it is sufficient to indicate what art was in relation to life.

Among the greatest and most attractive characteristics of the Greek mind was that mobility of fancy which can become absorbed in the intrinsic worth of any phænomenon, and which, while it did not bring with it any permanent bias of disposition, could sympathize with and accommodate itself to the changing nature of objects and of events. Yet this characteristic has a limit—not only that limit which is in itself a glory, the indefinable but perfectly distinct character which marks out the most varied productions of Greek art as having a common national stamp—but also another and different limit, which it would be idle to blame and perverse to imitate. That is, it was not really the intrinsic worth of things which the Greeks sought; everything was of value to them only in as far as it could be made instrumental to human development. Everything which could be utilized to produce a perfectly harmonious constitution of man's whole mental and physical nature, everything which could be permanently expressed in this constitution, or could through it receive some fresh manifestation, aroused their artistic imitative sympathy ; they were much less inclined to that which in its over-powering profundity and incalculableness left no alternative but contemplative subjection and submission.

We do not know their music, a fortunate circumstance which has left room for modern times to become great in this one art at least; but according to all that their authors have said on the subject, it was measure and harmony that they principally esteemed in music; they considered that those were the elements which one might expect to exercise a useful influence upon the temperament, disposition, and whole conscious life of man, the improved mental condition thus induced expressing itself in gesture, carriage, and action. Hence nothing was more natural than the close connection of ancient music with dancing ; the graceful and objectless movement of the limbs in the dance was the simplest and most sensuous expression and proof of the fact that the beauty felt in musical sound was not overpowering to human nature, but that on the contrary man could appropriate music as having special affinity with his own nature, and could reproduce it by the help of bodily organs. With regard to the development of any melody, this capacity does not count for much; the connection between successive phrases in a really beautiful musical composition carries us away from the well-known and familiar forms of our own existence into the wide ocean of a universal life in which all individual forms are dissolved; isolated turns and phrases may indeed charm by reminding as that even this beauty of sound is not wholly incapable of being reflected in human life; but taking it as a whole, we find that we have no choice but to give ourselves up to it with unreserved self-surrender; the agitation which it arouses may pass off in tears, but the content of this agitation cannot be presented in tangible form. Either this open sea of universality to which music leads us was avoided by the Greeks, or the error of venturing upon it was disapproved of by their æsthetes. The extremely meagre thoughts concerning music which are expressed with singular unanimity by their philosophers make it seem improbable that any striking degree of beauty had been developed in the actual practice of the art; on the contrary, the fashion in which (in the same matter-of-fact way in which one would draw up a catalogue of the most familiar objects) they set down definite mental conditions as effects which might always be expected to be produced by definite styles of melody, or hoped by State regulation of the kind of music to be cultivated, to establish a disposition favourable to the existing constitution—all this indicates that poverty of artistic content which commonly tries to make up for its deficiencies by doctrinaire over-estimation, analysis, and interpretation of that which has been attained.

Little has remained to us of all the wealth of song which Greece possessed. We have express testimony of that which we might have guessed—namely that among the ancient Greeks, as among all other nations, mothers sang lullabies to their little ones, and sailors lightened their toilsome rowing, and shepherd and peasant shortened the lingering hours with song; but this popular poesy has not been transmitted to us. The kind of Greek song which we know and which is framed according to the rules of art, presents two peculiar features. One is a predilection for the picturesque presentation of events which are set before us like a succession of living pictures, not with epic detail but effectively condensed; not so much related as brought into sudden relief by masterly delineation of the main outlines; not presented with the measured symmetry of epic verse, but seeking appropriate living expression in passionate rhythm. The inclination to make fable prominent may have a deeply-rooted cause in the fact that all human thought and action and life and suffering seemed incapable of being a worthy subject of poetry unless it had types and likenesses in the Olympian world and in mythology from which poetic imagery was ordinarily borrowed; on the other hand, it was no doubt a liking for plastic sensible phænomena which led Greek fancy not to linger in immediate contemplation of the content of feeling, but to illustrate it indirectly by looking at living examples. The other characteristic is the habit of storing up the outcome of poetic excitement in some general proposition or some proverb of practical wisdom—and thus in this way, too, taking refuge from the agitation of emotion in the definiteness and calm of a general conviction. It is difficult to estimate impartially this gnomic element, which in Pindar and in the choruses of the tragic poets continually alternates with graphic historic pictures. There is no doubt deep meaning in the trite expressions and commonplaces with which in practice we often try to brace ourselves in joy and sorrow; they could not have become commonplaces if they did not include something which, rightly understood, would suffice to completely calm our agitation. Now if the poet insensibly guides us in such a way that, as through a rift in a cloud, the content (still existent) of reflection which has thus grown into habit, suddenly appears to us in all its original heartfelt meaning, he will produce the finest possible effect by words and thoughts which in their insignificance seem to the uninitiated to be the

most commonplace on earth.   We not unfrequently meet this
lofty and earnest beauty in the songs of Pindar and the tragic
poets ; but sometimes only its external form is present, and
poesy hovers about the line beyond which what is really prose
becomes almost exalted into poetry by the solemnity with
which it gives itself out as such.   Greek lyric poetry moving
thus between the two poles of gorgeous historic painting and
impressive admonition, does not exhibit much of the true
spirit of song.   In the numerous remains of this lyric
poetry which we possess, we hear many a tone sweet or
beautiful or passionate or intense, but that which is expressed
in them is the mere human beauty of man's nature.   All the
charm and tenderness and graceful dignity exhibited by
favourable specimens of the race—especially in as far as all
this finds sensuous expression in gesture and demeanour—
exercised a strong influence upon the Greek mind, and was
apprehended and imitated by their artistic imagination.   But
this imagination does not reveal to us the unfathomable depths
of the individual soul, and the incalculable fashion in which
it apprehends the world.

To illustrate some universal truth of practical experience
by reference to great examples was the task undertaken by
the Greek drama also, and beside this task the full delineation
of human character and of the special justice which brings to
each his own peculiar and appropriate doom falls noticeably
into the background.   As mythology had once for all set out
the meaning of the heroic characters in the large firm outlines which the nature of the case demanded, the drama, without any great liking for mythology, borrowed from it, in order
to elaborate into characteristic individual forms these general
sketches of human dispositions and destinies.   This can
hardly be denied unless we apply different standards to old
and to new ; trying in the first case with microscopic acuteness
of vision to *prove* by instances the beauty of works of art, and
reserving for modern art an inexorable appeal to the immediate
impression produced, which alone is competent to decide how
far the beauty that has been proved to exist is æsthetically

effective. As regards the influence of art upon life, which is what we are here considering, this peculiarity of the Greek drama was an advantage. The subtle psychological analysis and delineation which in the masterpieces of modern dramatic art seeks to dive into the innermost recesses of the human heart, can never hope to be universally understood, nor even to meet, in narrower circles, with uniform and harmonious comprehension; but antiquity, ignoring those inexhaustible depths and taking characters that all could understand, depicted the destinies of mankind with broad firm strokes which found appreciative comprehension in the living sympathy of the people. And it did this all the more because both subject and mode of treatment were determined by ancient custom; the poet was not at liberty either to find his heroes in any obscure corner of the world, or to make any strangeness of his own humour the keynote of his representation. The fact that the persons of tragedy were always taken from the circle of native heroes; the repetition of the same story by various authors; the maintenance of the national philosophic views which yet allowed the special qualities of individual poets to make themselves felt—all this had a steady educative influence upon the people, and led it by a definite series of æsthetic presentations, without confusing multiplicity, to a capacity of judgment which has never since been so widely diffused as it was then at Athens.

Among the arts that deal with form, painting seems to have had least influence upon the national life, great as may have been the height of artistic development to which it had attained; of infinitely more importance was the constant sight of the noble and ideal forms which Greek sculpture, with a masterly perfection which has never since been reached, set before the eyes of the people. Having developed to this degree, the art of sculpture busied itself about the most insignificant as well as the most important tasks. To us, who admire the isolated remains, the thought expressed by many an ancient work of art seems to be too slight in comparison with the labour expended in presenting it in

sculpture; but such works were then intended to serve
as fitting adornments of edifices the most insignificant
details of which were pervaded by a coherent idea of harmonious beauty of form, and within the walls of which there
paced figures whose costume, ornaments, and gesture seemed
like the living embodiment of the same idea.   And from
what was finest and most beautiful in this world of art the
people were not excluded; traditional custom turned the
attention of creative art to the temples, the places of
public congregation; for the private dwellings of citizens
more modest adornment was thought sufficient.   In those
places which the nation regarded as sacred, in those festivals
in the arrangement of which no other nation has come up
to the Greeks, life was more thoroughly pervaded by all the
splendour of art than it has been in any other age; the
statues of the gods seemed to live among their worshippers;
music and dance appeared to be the natural expression of
the mood aroused by the words of the sacred songs, and in
looking on at theatrical representations the excitement of
feeling passed into a more calm contemplation of human
destinies, a mental condition permanently raised above the
commonplaces of daily life.   And the Greeks thus lived
and moved, and as it were had their being in beauty, without
that deification of art which is so common in our time;
they did indeed deify beauty, but not the human activity
by which it was produced.   They did not even possess any
word by which art might be essentially distinguished from
any handicraft skill; so self-evident did it seem to them
that every free-born soul is capable of appreciating beauty,
and needs for producing beauty no more mysterious endowment than that which in every kind of occupation distinguishes
productive from receptive talent.

§ 4. When the languages of the Greeks and Romans,
respectively, are compared, that of the latter seems the less
flexible.   If the Greek language forms its words in such a
way that each may be connected without break with those
that precede and follow, Latin seems to be animated by an

almost directly contrary endeavour. The vowel endings are
less numerous, the frequent inflexional terminations in *t*, *m*,
and *nt* necessitate slower enunciation owing to their incapacity of blending with most words that begin with mutes,
and give the impression of a sort of individual reserve with
which each word excludes its neighbour in self-contained
isolation. And the vowel changes have a more impressive
effect, since the phonic system of the Romans contains a
smaller number of differences and these more sharply contrasted, and there are lacking many intermediate sounds
which give gradations of light and shade to Greek speech.
The Romans gave up the article; each word appears as a solid
and independent whole without this prop; the conjugations
have fewer forms, and the declensions are only apparently
fuller because of their having retained the ablative. For as
compared with the Greek determination by prepositions, which
the Romans neither used so much nor possessed in such
abundance, the ablative hardly does more than indicate the
existence of some relation, leaving it to the hearer to guess,
within wide limits, the more definite nature of that relation.
The language is still poorer in those particles so frequently
used in Greek to indicate the subtle contrasts, connections,
limitations, and links between the different ideas of the
speaker, the expression of which contributes little towards
the communication of matter of fact, but helps greatly to
make clear the mood and the subjective view of the com-
municator. Hence, as compared with the soft drapery of
Greek speech which revealed the most trifling modifications of
thought, the Latin language has a sterner aspect; it groups
together more simply and concisely the items of fact, expecting the hearer to add that which is unexpressed. And
yet this mode of speech is not less expressive and impressive,
producing its effect by the position of the words, the peculiar
construction of sentences, and even by the omission of expressions which might have been expected. The gestures
which in other cases are an accompaniment of speech, and
can make clear the meaning of the most imperfect language,

are here contained, in a certain fashion, in the very structure
of the sentences; these characteristic forms of construction
supplement the meagre melody of speech as with a clear
harmonious accompaniment, and produce the impression
of that stern pomp and suppressed passion, which in the
Latin language always invite the reader to declaim, and give
the hearer the idea of a life full of power, and using its
splendid resources with calm mastery.

It is customary to estimate at a low value the artistic
endowment of the Romans as compared with that of the
Greeks.   Without disputing this judgment, which is well
founded, we must yet attribute to that which they accomplished (in harmony with the genius of their language) in
this department also, an historical significance which, though
different in kind from that which appertains to the art of
the Greek nation, is hardly less important.   The Greeks
made up in clearness of perception and in constructive power
what their imagination perhaps lacked in warmth and intensity of feeling.   As no living expression, no hidden
excellence of proportion in the human form, and no beauty
of attitude in the living subject was neglected by the sculptor's
art among the Greeks, so their poetry with lucid freshness
reflected all the habits of mental life, as well as external
occurrences.   It could enter into any circumstances with a
flexible sympathy which enabled it to represent how these
circumstances would affect the generality of men; it reproduced with the characteristic colouring every feeling of pain
or happiness commonly resulting from the experiences of
life in the human mind; it never lost itself amid those
obscure movements of distinctively individual emotion, which
as they are to one mind inevitable are to another unintelligible; it is nowhere disturbed by an intense longing to reach
beyond life as it is, to a higher peace—to a sacred joy in
life and an unforced equanimity in the contemplation of it.

The mind of the Romans seems to have been differently
constituted.   More phlegmatic, and with less airiness of
imagination, they could less easily be satisfied by the many-

hued brightness of life, behind which their religious belief
discerned a network of obscure connections between things—
enigmatical relations which were the more oppressive to
human life since no glory of redeeming beauty was shed
upon them (as it was in the case of the Greeks) by a circle of
divinities who were to them as living realities, and from
whose human-like customs these connections of things might
become intelligible. Also in social intercourse the Romans
exhibited a greater sense of their own individual personality
and of the mysteriousness of alien personality; the Greeks
felt themselves and regarded each other far more as mere
specimens of their kind, whose ambition might intelligibly be
directed to superior excellence in performances which might
be severally compared, but not to the attainment of something unique in the individual. Thus there arose among the
Romans that reflective turn which obtained for their poetry,
in the judgment of modern nations, a preference over the
colder and more objective repose of Greek poetry which it
did not quite deserve. For the greater warmth of their
reflective and contemplative imagination lacked that power of
artistic construction of which it required a specially large
measure. Now if to a soul that is passionately stirred it is
as unsatisfying to take things simply as they are, as it is
impossible to fashion the restless content of the mind to the
calm beauty of a nature not its own, there remains no alternative but voluntary renunciation—such as seeks to secure
to the soul that stands opposed to things a dignified com-
posure and an unchanging demeanour, by warding off all
disturbances from without and all outbreaks from within
that might interfere with the braced and steady calm of
manly firmness. This path of self-suppression was taken
by the Romans, and it led them to the development of a style
of æsthetic representation which has permanent historic
value.

Unceremonious communication is not generally carefully
precise in expression; the order in which we give utterance
to our thoughts concerning the connection of things is not

always in correspondence with those thoughts themselves
for sudden stirrings of emotion hurry on our words in advance
of the natural development of the subject, or force them back
to a point which they ought to have passed.   Greek speech
abounds in such incoherences and looseness, of which the
syntactic justification is often as difficult as the psychological
justification is easy, and which in facile superabundance and
in alternate sudden breaks and awkward additions reproduce
the natural and often charming irregularity of living speech.
The Latin style of expression is constructed with much
more conscious design, and even where it imitates Greek
models it does not simply follow the course of thought, but
(aiming at orderliness and a completeness which gives due
prominence to each essential relation, and omits what is
unessential) compresses the really important content in fixed
and regular structural forms.   Every other and perhaps every
higher æsthetic superiority may belong to the Greek style,
but the Roman style aims much more than it at an ideal
of Correctness.   It is pervaded by the sense of an intrinsic
order in all things which may be made the subject of
communication; without entering into the variety of their
nature with pliant imitative fancy, it seems under an obligation to observe with regard to them general forms of order,
which guarantee to their content, as it were, distant respect
without slavish submission, and at the same time secure this
respect from being violated by subjective caprice.

In the practice of art among the Romans, this characteristic
is repeated under a variety of aspects.   They copied all the
artistic forms of the Greeks, and always, even when they
borrowed matter as well as form, the copy in their hands
became something quite different from the exemplar.   Even
in the older imitations of Greek plays, of which there still
remain fragments, the sternness of the ancient Roman
character gives to the style a striking stamp of strength and
trustworthiness; as advancing civilisation permitted greater
refinement of form, Elegance appeared as the distinguishing
characteristic of Roman art.   The idea and name of elegance

occur here for the first time, and later culture has learnt afresh to value the quality by contemplating the specimens of Roman elegance which remain to us. There is no doubt that the Greeks possessed a gift of greater artistic value in their capacity of becoming absorbed in the full beauty of things without the intervention of reflection, and of reproducing that beauty with all the naturalness suggestive of having lived and moved in it; but in art, as in life, the higher does not so include the lower as to hinder the lower from developing to characteristic and irreplaceable worth if its evolution is allowed to proceed undisturbed. As the sharp-angled forms of crystals when compared with the unanalysable grace of flowers still retain their own inalienable charm, so the elegance of the Romans holds its own beside the beauty of the Greeks; and taking our civilisation as a whole, the former could not without loss be replaced by the latter.

The great master of elegance, Horace, has shown by precept and example what it is. When he requires the poet to say what is ordinary after an unordinary fashion, what he asks for is neither an idle play of enigmatical designations nor useless pomp of words, but a kind of justice towards things with regard to which we are in the habit of being unjust. The dust under our feet arouses neither our attention nor our admiration; yet the microscope finds in it crystalline and vegetable matters, the characteristic forms of which would captivate us if the confused intermixture in which it all appears to our eyes did not prevent our perceiving and distinguishing. In the same way the world and life are full of events, the frequent occurrence of which has diminished their value in our estimation, or to the characteristic significance of which we can only give an indifferent, distant, sidelong glance, because of the eagerness with which—and rightly—we press forward towards goals of more importance. It would only be a fresh injustice to bring forward and distinguish with special preference these things which have hitherto been unjustly neglected; what is just is, not to pass them over with the trite and well-worn phrases of everyday usage, but as we observe

them and then pass on, to suggest the forgotten value which
they conceal by some uncommon turn of expression prompted
by happy insight.   The appertaining of what is small and
insignificant and confused to the same world that holds what
is grand and beautiful and distinct, is brought into notice by
the careful and concise style to which we have referreo,
without offending against truth by artificial enhancing of
insignificant values.   This is what Horace calls the unordinary expression of what is ordinary, and with this
artistic intention which aims at elegance, the means which he
uses are connected.   So his poetic art—like that of others—
employs imagery not merely to give a twofold expression to
the same content, and also not merely—by help of the
palpable plainness of some simile—to give clearness to a
thought that is difficult to set forth; finally, it does not
merely reckon upon the probability that feelings which such
a simile may excite (and which attach themselves spontaneously
to it) may apply also to the object concerning which it is used,
without any express incentive—an incentive which, in fact, it
would not be possible for the poetry to convey in express terms:
on the contrary, by exhibiting the one event that it wishes to
emphasize by means of other and similar events, it abolishes the
isolation of the one, and shows it forth as entitled to constitute
part of a world in which the most essential features of its
character occur and are of value, in other places and under
other circumstances, forming part of the general plan of the
whole.   Roman fancy uses such similes with great precision;
by the perfect finish of its brief figurative expressions, a feeling
of certainty and assurance is awakened, and this feeling
is strengthened by the very strangeness of the construction which often essays to combine ideas from other
than the ordinary standpoint.   For the success of these
essays convinces us of the steady coherence between the
parts of the thinkable world; since this, being considered in a
variety of aspects, yet always appears as a self-contained
whole.   The same end is served by many analogous means—
the sparing use of ornamental predicates, the due proportion

in their distribution, and in the general grouping of ideas between which a musical or artistic play of connections and contrasts is plainly aimed at; and lastly, the predilection for working out a thought to that statuesque simplicity in which —all that is unnecessary having been got rid of, and all that is necessary having been brought into the sharpest relief—the thought is presented to us as the classical expression for all time, both of the nature of the object of thought and of the right way of regarding it.

Plenty of empty brilliancy of form has no doubt resulted from the following of these rules by poorly endowed poets; but this form of procedure furnishes a favourable testimony to the vitality and character of the people; it reveals even in the productions of depraved ages and unruly spirits the background of a grand discipline of thought which could never be wholly broken. And in other respects also Roman elegance is not to be despised in comparison with Greek beauty. Certainly its chief endeavour is to elevate and give weight to what is in itself small and slight and insignificant, in order to give to our temper and our philosophic views such equableness of tone as characterizes a good picture, and it is true that with this aim it minifies what is great; in place of the over-powering tones of living passion, it generally substitutes the colder reflection in which contemplative thought considers the gain and loss of a struggle which has already come to a conclusion. But when such procedure cannot attain the highest poetry, it may yet give an air of grandeur to the prose of life. Society, as well as intercourse with Nature, produces innumerable situations from which all really striking beauty has wholly disappeared; means to an end which are in themselves indifferent, and the attention which they require place keenly felt obstacles in the way of mental activity; a world of worthless externalities bars the way to that for which our soul longs. Where any occurrence of domestic or public life may be transfigured, either by its own content or by immediate connection with a world of æsthetic or religious thought, the Greeks have not failed to consecrate it thus in

a striking manner; but to give interest and an air of
stateliness to that prose of life which obstinately refuses to
be transformed into anything but prose, and to do this by the
mere mode of treatment adopted, was a task the merit of
accomplishing which belonged to the Romans.   Their mode of
thought, which in art created the special notion of elegance,
introduced into life a not less special dignity in the formal
treatment of all kinds of subjects.   With the declining
vitality of the nation, reverence for the sacredness of legal
institutions (once the fairest flower of Roman thought) became
weakened, and only ceremonial and the external regulation of
splendid ostentation continued to receive further development; and these themselves were elements which, after the
fall of the empire, helped (amid the chaos which characterized
the beginning of a new order of international life) to preserve the thought that everything has some particular mode
which, and which only, is right for it.   From this legacy left
by the Romans the men of succeeding centuries derived a
large part of that which gave beauty to their life; and that
portion of this legacy of which we are the historic heirs, still
works more powerfully within us than the artistically more
important heritage that we have received from the Greeks,
which affects us by rousing us to conscious imitation.
Numerous forms of expression which have been transplanted
into modern languages, the character of our public solemnities
and the difficulty on all such occasions (and for inscriptions
on monuments, records of solemn ceremonies, or brief and
pregnant sayings) of replacing the statuesque style of
Roman speech and custom by substitutes of home growth—
all this still bears witness to the lasting influence of Roman
civilisation—an influence from which, even now, we have
scarcely begun to try and emancipate ourselves, and for the
advantages of which we do not as yet know any adequate
substitute.

§ 5. Between the fall of the ancient world and our own
times, the temper, morality, and æsthetic feelings of mankind
have experienced many changes, which must be passed over

In silence by our brief survey, which is concerned only with the lasting results of these developments. There were set before imagination increasingly difficult tasks, which roused it to passionate agitation; but there was an absence of those favouring conditions which in the age of classic antiquity made it possible to impress upon life a stamp of harmonious beauty.

To the ancients the starting-point and goal of all human endeavours were, as a whole, plain. Nature lay before them as the only reality; in unceasing creation, which is its very essence, and without pursuing ends situated beyond the sphere of its phænomena, it brought forth even the human race, as the fairest among its perishable blossoms; that man should live in harmony with Nature was the common conclusion at which the ancients, setting out from the most various premises, had arrived. Excellence of national disposition and the intellectual candour of an active spirit of investigation prevented this adherence to Nature from being carried out by obedience to every rude and blind impulse, and every noble and attractive quality of the race was cherished as a distinctive endowment by which Nature prescribed to man a path which leads beyond the limits of the animal world; to the fair ideal of humanity thus formed, a rich and harmonious development of characteristic morality and custom was insured by an almost undisturbed national evolution. But no recognised aims lay beyond; the course of events might pursue the same round for ever and ever; Nature might go on to eternity producing fresh relays of short-lived mortals, each generation of whom, after having exhausted the good things which its organization enabled it to develop and to enjoy, would be reabsorbed into the same universal Nature. Now doubtless there will always be a secret contradiction between this sacrifice of self to Nature and its transitoriness, and a civilisation which, the more noble the aims which it recognises, only presupposes the more an eternal preservation of all that is good; the impetus of eager and exuberant activity easily carries men past unsolved problems which press upon those

who have leisure.   So that antiquity did not in theory overcome the discrepancy in its philosophic view, but neither did
it allow this discrepancy to influence its temper.   It neither
sought nor found that higher world, into the eternity of which
the transitoriness of this debouches; yet it did not, like
oriental pantheism, take pleasure in extolling the frailty of the
individual.   A happy talent for making the most of mundane
existence, and pleasure in the increasing success of efforts in
that direction, helped to compensate for the great deprivation
of not recognising any significance beyond that of a mere
passing natural occurrence in even the very highest of its
works, that is, in the cultured development of human life.
As long as the creative activity of antiquity traversed an
ascending path, and as art and political life were fruitful in
the production of new forms expressive of the ideal of the age,
while historical circumstances were favourable to attempts at
their realization, so long the still impetuous general movement
of civilisation carried men safely over the weak place in their
philosophy, and the fits of doubt and despair which appeared
in isolated minds and at isolated moments, had little influence
upon the general temper.   In course of time such favouring
conditions failed, and antiquity, having exhausted its creative
strength, developed uncertain, dissatisfied, contradictory tempers
which attacked the hitherto received philosophy on all sides.

Another foundation had from the beginning been given
by Christianity to the new civilisation which was to grow up
upon the ruins of that which was passing away.   Christianity
had demolished the calm self-sufficingness of the secular
world; the life of humanity which to the ancients seemed like
a never-ending uniform stream, was by it compressed into
a course of stern dramatic development between the two events
of the Creation and the Last Judgment, and (as compared
with the Kingdom of Heaven) depressed to a mere brief stage
of transition; that to which man was destined no longer
appeared as the goal at which our being naturally aims, but
was regarded as attainable only by conflict with innate impulses, of which the noblest seemed to be hardly more than

splendid vices; great Nature herself was no longer considered as the sole cause of things or the mighty Mother of all, but merely as an instrument in the hand of Providence; and even to this vocation she was thought to have been untrue— the intrusion of sin had distorted her features, and there was in her a mingling of memories of what was divine with inexplicable self-will and the seductive charm of evil powers. These richly coloured pictures of a vast cosmic history entered perhaps more generally and deeply into the imagination of the people of the Middle Ages than the spiritual content of Christianity did into their heart; and they did not have merely the same effect as other similar oriental pictures which afford us glimpses of the beginning and end of the world hovering in mythic obscurity at inconceivable distances of time. In times of historic light—times of which the detailed outlines were recognisable—there had happened the greatest marvel in the providential guidance of the world; bringing with it into its own dazzling reality, all connected circumstances whether past or future, making them look as if they had either just happened or were just about to happen. Men did not see symbols, with regard to which they were uncertain as to how much was figurative and how much real and serious, but they actually stood in the current of universal history and felt themselves carried forward by it.

Thus, whilst antiquity only cared to see with the eye of intuition what things were, and whither their development was tending, the imagination of the new age developed a taste for subtle inquiry; it distinguished everywhere between what things appear and what they signify or what they are a means to ; life was to be ordered after a pattern, the sole content of which had first to be discovered by interpreting an ideal that soared high above all reality ; but resigned obedience to the ordering of this life had at the same time to struggle with the discouragement constantly arising from a consciousness of the merely conditional value and temporariness of all earthly existence ; finally, this difficult task had fallen to the lot of nations which were not supported by any heritage of

long-accustomed civilisation. Christianity did not immediately supply this want; it had indeed ennobled from within
the developed forms of ancient life as long as these lasted;
but systematizing ideas capable of furnishing a foundation for
new constructions, could not be easily obtained from its
simple ideal content. Perhaps it is rather the case that all
the characteristic contrasts of the Middle Ages were held
together by the fact, that the vigour with which they
grasped a high ideal lacked all thoroughly developed insight
into the articulation of the instruments necessary to prepare
a place for it in the world of reality. With the aim of
antiquity—to develop what Nature prescribes—was given
also the way by which that aim might be reached; but
the new ideal of sanctification towards the attainment of
which all Nature affords no aid, left the question, What shall
we do to be saved? without any such definite answer.
Proximate ends, the earthly vocation of men, admitted of
various interpretations; salvation might be sought in various
ways. Yet neither in penitential aversion from all the
interests of earthly life nor in the excitement of knightly
combat was full satisfaction found; both these modes of
life were at the best conflicts with threatening evils; but
they were not productive of any material gain which
could be cherished and guarded; just as little was labour
capable of setting all longing at rest; occupied by the
pressing needs of life which were regarded as being necessary
only on account of earthly imperfection, labour for a long
while felt a sense of its own meanness and could not
regard itself as direct service in the work of sanctification.
Thus human life attained to no clear views concerning its
earthly tasks; it was the reconstruction of society which
gradually, at first, toned down the excitement of the prejudice
which made men think that they must do once for all in
this life work which had an inalienable place in the universal
order; instead of feeling themselves called upon to be conscious participants in the construction of the great universal
fabric, men learnt afresh the lesson of valuing every unim-

portant situation resulting from human intercourse as affording scope for the exercise of moral strength, and learnt not to seek in life anything more lofty than it is capable of affording.

Thus there had not been developed a generally received type of human culture ; but every rank and condition had its own code of morals, and sought in the exact observance of transmitted ordinances an historical justification of its mode of life, in place of that ideal justification which it lacked. There was never a greater multiplicity of forms and observances than in the intercourse of the society built up out of all these multifarious distinct elements ; but this very state of things corresponded to the theoretic philosophy developed by the Middle Ages in contrast with antiquity. The eye of antiquity was captivated by that which is general and homogeneous in human life and in Nature, and which is ever recurring in inexhaustible variety of manifestation ; it made no great effort to comprehend the world as a whole. It was not possible for Christian imagination to have so much sympathy with this generality ; what it regarded as the really efficient agency in the world was not that nature of things which works homogeneously in a variety of subjects, but that divine Providence which has a special purpose with regard to every individual, and assigns to each his share in the building up of the whole. Minds were very earnestly directed to this unity of the world, which consisted in the congruence in one plan of innumerable individuals ; speculative philosophy as well as practical life neglected the region intervening between the Whole and the Individual— those generalities of homogeneous activities and simple laws by means of which alone the materials of any edifice can be combined into one whole. The knowledge handed down by tradition having become meagre, the educational curriculum of the Middle Ages sought to compress encyclopædically the sum of all that was knowable into one great whole, in which the sciences were arranged in an order that corresponded to the place which the subject of each seemed to occupy in the

divine plan of the universe.   What was accomplished was
far from being equal to what was designed; but even the
external forced and far - fetched concatenation which was
brought about shows how vivid was the belief that all
things are closely connected parts of a divine cosmic order
—the unsubstantial truths of mathematics as well as human
history and the rich variety of Nature in products and
events.   In this cosmic construction, which was regarded
not as the simultaneous production of a manifold from a
homogeneous cause but as the combination into one whole
of the most heterogeneous members, a social system comprising many varying codes and callings naturally found a
place.

This mode of thought which regarded nothing as self-
contained, but considered everything as either significant of
or connected with something else, could not favour impulses
to æsthetic construction.   An exaggerated leaning towards
symbolism caused a disproportionate value to be set upon the
significance of phænomena, and weakened men's susceptibility
to beauty of form, which depends more upon general laws of
the reciprocal relation of several elements than upon the
intellectual significance of the whole which these constitute.
Delight in the splendid profuseness of life itself was foreign
to this philosophic view, and would have remained foreign to
the age also but that it is not possible for any philosophic
view, however deeply rooted, to wholly alter the unvarying
natural tendencies of the human race.   So that the men of
the Middle Ages, notwithstanding the oppressive solemnity of
their idea of cosmic connection, had also a liking for fun and
enjoyment; and notwithstanding their mania for symbolic
distortion, took pleasure in self-sufficing beauty of form.   But
even in the imitative arts they did not attain to any originality in the reproduction of beauty; for a long time sculpture
and painting were mere vehicles for the expression of actual
thoughts, feelings, or situations—aiming at first at mere conventional indication of their meaning, but afterwards at
natural and powerful expression.   At last art bethought

itself that its productions ought not to be of merely commercial value, but should be developed to creations having a full, beautiful, and characteristic reality of their own. In architecture alone—the activity of which does not, to so great an extent, presuppose unfettered and original skill—it was possible for works of great and special merit to be produced by imitation of existing models, and a sense of the complex beauty of proportion (a beauty susceptible of realization) both in the whole of an architectural production and in its details. Such works sometimes combine into clearly expressed unity a multitude of members differing from one another; and sometimes by adopting a principle of construction which seems rather suited to a picture or a landscape than to architecture, they recall that characteristic manifoldness of human life which it is difficult to take in at a glance. Poetry, as an art of words, needed for its full evolution a considerable development of language, and this during a large portion of the Middle Ages was lacking; for not only were the languages of some of the nations slow in becoming fixed, Latin remaining for a long time the instrument of communication among the learned, but the undeveloped state of society had still more influence in hindering the advance of language as the instrument of social intercourse. There lacked that cultured language which thinks and poetizes for us, and the thorough development of which, up to a certain point, is undoubtedly a necessary condition of complete perfection in poetic form. Profound feelings did unquestionably find powerful expression in national songs; but even narrations which conformed to the rules of poetic art did not succeed in giving a perfect representation of the rich poetic content of ancient legends; form remained inferior to content.

And this was the general fate of the age. It lived a life full of poetic impulses from the strength of which it suffered; but it was only in the mind of posterity that there was developed a comprehensive consciousness of what that age might have been to itself, if it had not been hindered by so many obstacles from recognising and realizing its ideal.

As life began to take in a high degree an intelligible form, imagination, which always seeks to find its way by a short cut from the pursuit of common aims to the secret of the Eternal, turned back with a feeling of preference to the picture presented by the Middle Ages—or rather to the ideal antitype of this which it had constructed for itself. For indeed as a matter of historical fact this romantic temper in looking back could nowhere find such an age as that which it thus preferred ; the actual Middle Ages were richer in good and in ill than the dreamy temper of romance, which everywhere sought the infinite in the finite, and turned away from intelligible ends—richer in real interests, the obstinate individuality of which was not wholly exhausted in symbolism ; and likewise richer in natural barbarism and eccentric cruelty --that are the heritage of primitive savagery (which it took Christianity a long time to tame thoroughly) and of those fanatical wanderings to which a misunderstanding of great ideals commonly leads. But still this age has left to us a very important legacy, namely that dissatisfaction with what is merely phænomenal and that longing for the infinite which give the keynote to the æsthetic temper of modern times and to its poetry ; although the age itself, mistaking the noblest sources of its life, not unfrequently imagines that it may become greater by imitating other ideals than by developing its own.

§ 6. If we glance at the monuments of Romanesque and Gothic architecture, at the flourishing condition of painting in the fifteenth and sixteenth centuries, at the progressive development of music and the treasures of poesy which the Romance and Germanic nations of Europe, vying with one another, successively produced in rich abundance, we are convinced that the human race was not lacking either in susceptibility to beauty or in power of artistic construction at the period of transition from the Middle Ages to the modern era. A decision concerning the comparative greatness of these two endowments at this period and in antiquity finds equal hindrances in the difficulty of the subject itself and in the

many prejudices that have been produced both naturally
and artificially ; there will be more unanimity in the complaint that the echo which even the best of the more modern
art found in real life appears to have been incomparably less
than in antiquity, and even where considerable to have been
of a less satisfying kind.   For the Greeks at any rate appreciation and enjoyment of beauty were a substantial part of
life ; and though no doubt the culture which makes men
capable of both was unequally distributed among them, yet
the less intelligent were surrounded, as by the atmosphere
which they breathed, by a kind of artistic rhythm which
had impressed its stamp even upon the customs of ordinary
life.   The gulf which separated the life of more modern
nations from their art, was wider ; men became accustomed
to contemplate an ideal kingdom, far removed from living
reality—a region which it was both possible and delightful
to look up to, yet the contemplation of which could not be
regarded as part of the proper business of life, but rather as
a relaxation from it.   It seemed to them that among the
innumerable wonders which the universe contains, and in
which men (incapable of examining more than a part of the
whole) take a spontaneous interest, art is one—that it grows
and blooms like an exotic plant, the marvellous productive
impulses of which, deviating from all indigenous models, from
time to time captivate and interest the fancy.

We here find art not as yet detached from all connection
with religious, public, and social life, though the nature of the
reciprocal contact shows its superficiality.   In antiquity,
religious worship was the living act of the national mind, to a
great extent supplying poetry with its *raison d'être*, its content,
and its form ; what art furnishes to us is formal powers (which
it attributes to its own nature) that may be used to embellish
religious worship ; even now, in moments of peril which
rouse passionate feeling, it may rise to adequate expression of
the national consciousness ; but in times of rest it finds no
fixed popular ideal of morality and life from which to borrow
the form and content of its productions, and there is put at

the service of its formal means of expression only one-sided party tendencies, or petty private interests, or capricious individual views of life; it does not penetrate social life in such a way as to become as it were its very rhythm, but among the many dishes which society serves up to help while away the time, art also brings its contribution, which makes a change and is an assistance.   It would be a misapprehension of these remarks to take them for a denial of the real worth of modern art or of the powerful effects which it produces even under such unfavourable circumstances; but we think it desirable to bring these circumstances into prominence.   It would, however, be just as great a misapprehension to take what we have said as applicable only to the dull multitude which has always been without appreciation of beauty; in order to understand all the barbarism of our attitude towards art, we must call to mind arrangements which, by their commonness, have already wholly ceased to affect us unpleasantly.   We crowd pictures together, one above another, in galleries, so that the impressions received from them are mutually destructive; the resolution to erect any great architectural work is followed regularly and as a matter of course by a discussion as to the style to be adopted, that point being regarded as an open question; at concerts, which are given in places and at hours the choice of which is determined by causes known only to the person who provides them, the hearer's soul is carried compendiously through a whole series of masterpieces; occasionally some quiet valley, invaded by a troop of singers, without knowing why, suddenly hears chanted by a hundred voices the praises of its modest violets which bloomed so long unseen; the theatres are opened almost nightly, and it would be hard to say whether the sentiments or the taste of the spectators are most cultivated by their rich variety of material and style; fortunately there is a less frequent recurrence of the pleasures of the Carnival, which is as incoherent in itself as it is devoid of any living connection with life, and which has long ago forgotten what originally gave rise to it.   All these exhibi-

tions of varied beauty and artistic skill take place for their
own sake, and do not mark any important epochs of human
life; they connect the enjoyment of art with fixed times, in
the same way as, at any rate, Protestant worship does divine
service; as in the one case the world is left to itself for six
days, but on the seventh men "go to church," so in the other
case the prose of life is sharply marked off from moments of
poetic exaltation.

Of all this we can alter nothing. The modern spirit,
which analyses and investigates critically, has begun in all
departments of life to seek for rational foundations; with
conscious calculation it aims at constructing society according
to principles which do not leave to the once characteristically
various multiplicity of social conditions either a *raison d'être*
or any significant task to accomplish; the very course of
events, by inevitably procuring recognition of the human
rights of every kind of labour and of every labourer, has contributed to the levelling of society, even to uniformity of
costume, and has fixed a moderate temper as giving the tone
to social intercourse—a temper which has to be on its guard
against the intrusion of elements of intense dulness, and
which will scarcely allow that the external forms of life
should be informed with beauty. The tendency of the general
instinct seems rather to be towards entirely purging social
intercourse from all poetic elements, which would appear as mere
fantastic inequalities in its measured sobriety, and to reserve
all excitement and enthusiasm for the retirement and solitude
of the private life of individuals. Here the best part of our
mental development is accustomed to take refuge more now
than formerly, fearing all publicity as almost a profanation.

I have already remarked that this characteristic of our time
does not in itself make it impossible for art to exercise great
influence upon men's minds, nor for its productions to have a
high degree of perfection; yet in both these respects the
characteristic referred to is not without effect. The less the
thought and style of art are the direct expression of popular
philosophy, and the more its works seem to be the arbitrary

constructions of an imagination that is merely making unrestrained trials of its strength, the more easily does art evoke
critical estimation of the merit of its representation, instead of
sympathy with its content. There has been plenty of criticism in all ages; and on the other hand I do not maintain
that single-minded devotion to and enthusiasm for beauty are
things unknown in our day, but a careful comparison of
the productions of art (the business of which is to embody
beauty) is more frequently met with; the peculiar pleasure of
connoisseurship, the satisfaction arising from intelligent knowledge of the instruments and tricks of art, their historical
development and their application in particular cases, and a
half critical, half literary interest in the procedure of creative
imagination—all this lessens our susceptibility to the immediate impression which it is yet the sole final aim of such
imagination to produce. As the collector shuts up in port-
folios the works of art which he has brought together, content
to possess them and to know what æsthetic impressions they
are capable of producing, if ever the hour of unrestrained
enjoyment should come, so all of us are in a general way
satisfied to possess an intelligent consciousness of the latent
power which beauty has to stir our souls; æstheticism con-
gratulates itself on increasing sympathy, in proportion as the
living emotion produced in the soul by the objects which it
judges becomes rarer.

Art itself has also suffered from the causes which have produced these conditions. Mankind have not, indeed, wanted
for great geniuses since (from the end of the Middle Ages
onwards) the increasing enlightenment and many-sidedness of
social culture have afforded opportunities of evolution to such
minds. With the exception of sculpture and epic poetry
(essential conditions for the prosperous growth of which were
lacking), there is no art which has not in this period reached
the highest point of development. A long series of the most
illustrious names, versatile minds equal to the greatest of
antiquity, adorn the annals of Italian art; more solitary
indeed, but in the same degree more great, is the lofty genius

of Shakespeare, whom Northern Europe can boast. Yet there is a frequent complaint that the productions of these powerful minds, together with those of the illustrious men of later times, are (notwithstanding their greatness) lacking in that classic perfection of form which has made antiquity the one epoch that can be regarded as affording models to the art of succeeding times. I hold that neither this praise of the ancients nor this blame of the moderns is just, if taken in the careless generality with which both are commonly expressed. The ancients seldom failed from individual caprice; their world of artistic thought and their favourite methods of treatment grew so directly from their popular philosophy, and were so generally established by tradition and constant practice, that even the less highly endowed minds attained to the harmonious use of artistic forms as easily as in our time they do to irreproachable social behaviour; and this very harmony of treatment occurring in an immense number of works of art causes us to regard as among the essentially necessary conditions of beauty, much which even in the antique works themselves is mere conventional manner. Modern art lacks the advantage above referred to. It grew out of passionate needs of the soul, the satisfaction of which men had to search for since they did not find it ready to their hand, either in science as it then existed, or in social intercourse, which was in a state of disruption, or in the political constitution of public life. Modern art therefore had not the simple task of giving an artistic reproduction of beauty of which it had had living experience, but it had the double task of finding first an ideal which should satisfy its longings, and then the forms in which to embody this ideal. The revival of antique art could only partially further these ends; much could be learnt from its forms, but as far as its content was concerned, this did not come up to the demands made by the spirit of the later age. When for some time men's dominant endeavour was to reanimate literature, art, and politics with the spirit of antiquity, what took place was not an historically necessary development, but a conscious movement, which, choosing freely among various directions that stood open as possible paths of further development, selected a particular one in preference to the rest. The want of a generally accepted ideal, and the necessity which there is that every age, every nation, and every individual genius should fix once for all its own highest aims and its own forms of expression, introduced into modern art its varied and rapid alternations of style, and gave to its works as compared with those of antiquity a predominant stamp of intellectual wealth. For we may very well describe by this phrase the impression which we receive when imagination, instead of being borne along by the general current of the age, and reflecting without effort some representation of the universe which has become a kind of second nature, undertakes independent investigation and analysis, in order to arrive at some interpretation of reality, of which reality itself cannot refuse to recognise the truth.   Incontestably this free action of imagination is oftener exposed to æsthetic failure than imaginative activity which works in subordination to a fixed ideal; modern art was not satisfied by representations of universal, typical, generic beauty, but became absorbed in profound depths of human existence which had been previously untouched, and sought to investigate the mighty coherence of the universe with many a passionate question concerning its significance—thus it was in danger, on the one hand, of arriving at fanciful conclusions, not recognised by reality as justifiable, and, on the other hand, of neglecting formal beauty of representation on account of the predominance of reflective activity.   In many works of wit and sarcasm and insolent caricature, capricious fancy has no doubt overstepped the limits of beauty; but, on the other hand, if poetry attempts to portray the secret development of human character, if painting is only satisfied when it can succeed in presenting a reflection of the story of such development compressed into the action of a moment, if music, stripping off from our feelings all remembrance of their earthly occasions, so enlarges and exalts them that their movement becomes the interaction (not describable in words) of those univer[~]al forms of the connection of elements upon which all the joy and all the pain of reality depends—if all this is so, and if we take as our model abstractions derived from a far simpler age, it is easy to disapprove a large part of the wealth of modern art, but difficult to be impartially just to the lofty beauty which has assumed new and unique forms under these more complex manifestations; finally, it is in any case impossible to give up what we now possess, and to return to that greater simplicity which can no longer satisfy our hearts.

In spite of its slight connection with the higher aims of art, modern life is not wanting in a special æsthetic element, that has, in course of time, made itself felt in many and various ways: The modern spirit of criticism and of self-conscious reflection first showed itself in Italy; the cultivation of knowledge of all kinds and formal excellence in all the dexterities and refinements of style, both in language and in the intercourse of life, were the ends at which it aimed, and which in many brilliant instances it attained; the large and significant views which constructive art inherited from the Middle Ages, views by which it held fast and which it was able to embody with a technical perfection which made rapid progress, afforded a wholesome counterpoise to the unrestrainedness of this subjective spirit. Political disasters interrupted the progress of this development, and Italy abdicated to France that living dominion over the rising modern world which it had for a time possessed unquestioned. In France the gradually perfected centralization of governmental power had caused the formation of a coherent and exclusive society of aristocrats, who, being compelled to keep comparative peace among themselves, and being furnished with abundant means, but destitute of any great aims in life, were forced to employ their intellectual strength upon problems of social intercourse. The condition of the people, which furnished the necessary basis of such a society, was miserable to a degree; indeed, the epoch taken as a whole was by no means a Golden Age, that men need wish back again; but it was undoubtedly this isolated concentration of action and reaction between the most favoured constituents of a great State, which first gave to the spirit of modern times a characteristic æsthetic expression.

It was language which above all experienced the influence of these favourable conditions. It was developed as it had been in Greece, by means of living conversation, though not as there in the publicity of a great political life. Such conversation dealt with all imaginable subjects of reflection from all possible points of view; and being thus compelled both to use brief and clear expression, and to clothe opinion in an agreeable dress, the French style became formed into the most perfect prose that up to that time the human mind had succeeded in constructing. There is but little of the aroma of poetry about it, as might be expected in the instrument of expression used and formed by a society not accustomed to manifest its deepest emotions; but it has the well-defined, lucid, orderly movement and the conscious respect for generally recognised conventional rules which were likewise necessary in such a society; it does not show the interesting but awkward originality with which, in the prose of the ancients, we often see the thought that is to be communicated unfold from its germ, and as it were seek its fitting form, but as becomes the heir of an old and reflective civilisation it skilfully lays hold of the most diverse among familiar points of view, and accomplishes its object by means of abstractions and modes of combination applicable to them all; in these respects it corresponds to the character of the modern spirit, the strength of which consists not in flights of artistic enthusiasm in which it rushes upon its objects, attracting attention and betraying its own inward excitement, but in the unobtrusive business-like way in which it gets rid of difficulties, being conscious of knowing ways of solving them which are of general application. It is not surprising that through this spirit of clearness and precision the French language obtained dominion all over the world—a prerogative which it has only gradually lost. In Germany the rise of a higher kind of art to which the genius of the French language was not adapted, caused its supremacy to be set aside, but a substitute for its prose has hardly yet been found in that country. The living unity of society was lacking there; the too great predominance of learned culture thence arising, and the inherited error of not only learning from antiquity, but also of imitating it, caused German prose to be for a long time awkward and confused, and the language itself and its resources to be more unfamiliar to the people than in other countries. For let the Germans not deceive themselves— though the whole nation can read and write, he is a happy man who need not hear the reading nor see the writing; the gulf that still exists between the perfection of the language in the masterpieces of German poetry and the style of ordinary life is wide indeed. It will only be gradually filled up as the education of the circles which do not go to antiquity for guidance increases to such an extent that they can give to the modern modes of expression which they use for modern views and interests the established character and fixed form which it is quite in vain to expect from ancient models.

The peculiar character of the time found more whimsical but not less animated expression in the much abused Rococo style which became dominant in the ceremonies of social intercourse in costume, buildings, furniture, and even in the laying out of gardens and of pleasure-grounds. It is easy for us, guided by the teachings of historic periods which were more favoured in an artistic point of view, to reproach this style, because, being destitute of feeling for the characteristic truth of things, it distorted the real nature of everything without exception that it attempted to beautify, and with odd caprice imposed arbitrary forms and laws upon every department of life into which it intruded; yet it cannot be said that this caprice was incoherent and inconsequent. Certainly it had no other principle than that of the sovereign and unrestrained will with which the subjective mind moulds all given material into a creation that is according to its own fancy; but it did not merely apply this principle with rare

consistency to things, but with stern discipline brought even
human life under self-imposed laws of etiquette. Certainly
the forms which it forced upon all objects and all relations
cannot be understood by reference to any artistically justifiable
principle of form ; but the very end aimed at was to be
invariably graceful even amid all the complete arbitrariness
of this procedure, and where there is a cessation of all rule
dependent upon the nature of the thing, to find by the power
of the mind itself a definite law of the production of pleasure.
It would be mere scholastic pedantry to deny that in many
cases this was accomplished; not only do we trace with
pleasure in countless individual utensils, buildings, and
fashions of the time the bright and graceful flight of
this arbitrary fancy, but among all the styles which have ever
pervaded life in all directions this as a whole seems to be
quite the most in harmony with natural receptivity. Who
would not admit that Classic and Gothic art unfold a
refined and lofty beauty that is more to be reverenced than
this ? But at the same time we may admit that they are
alien to us, and that especially every renewal of the antique
in our life looks like a learned pretension to the possession
of superior understanding, whilst in defiance of all æsthetic
systems, we always sympathize with the Rococo style.

But this too has passed away ; and the æsthetic elements
which life in the present day still retains appear much
more insignificant. We often hear quoted the saying that
architecture is *frozen music :* hence I have some hope of
gaining a modicum of undying fame by taking a step further
and calling mathematics *desiccated music.* For what element
of music does mathematics lack except the living sound ? All
its other elements and resources are common to it and to
music, or, more correctly speaking, music borrows them all
from mathematics. Now it seems to me that what has
remained to us as the good genius of our age, is just a mathematical element of exactness, neatness, concise clearness and
simplicity, supple versatility and pruning away of all superfluities. As compared with the roundabout procedure and

awkwardness of innumerable regulations of earlier times, what a preference do we now see for that elegance which characterises the most concise solutions of difficulties! What brief and severe simplicity do we see in the structure of machines! what vast effects produced by the ingenious combination of simple means!

Undoubtedly there is beauty even in this, and we may rejoice heartily in that genius of modern times, which no longer wearing antique draperies, or dreaming through life with flowing hair, goes with shorn locks and close-fitting garments; and we may hope that it will raise from this small germ a mighty tree filling life with fresh beauty.

CHAPTER IV.

THE RELIGIOUS LIFE.

Comparison of Eastern with Western Life and Thought—Nature and Social
Life as Sources of Religious Ideas—Preponderance of the Cosmological
Element in Heathendom, and of the Moral Element in Judaism and
Christianity—Christianity and the Church—Returning Preponderance of
Cosmology in the New Philosophical Dogmatism—Life and the Church.

§ 1. THE East has been the birthplace of all those
religions which have had a decisive influence on
the destinies of mankind. And not only has it (as the fatherland of all nations historically important) forestalled future
ages by giving birth to the germs of all religion—religion being
one of the things earliest developed by the human race—but
also even in later times the religious life of the West is
distinguished from that of the East by a permanent difference
of disposition and of the course of development. In the latter
the imagination of men became early susceptible to the
numerous analogies by which visible reality points to something beyond itself, and drew in grand outlines pictures of
a supersensuous world, which contained the beginning and
the end, the completion and the explanation of the world we
know. And the manifold content of this faith was no mere
impotent dream of enthusiastic moments; the thought of it
pervaded the insignificant customs of everyday life, the rules
of commerce, and the ordinances of morality; the obligatory
commands, which seemed to flow from it, received unquestioning obedience, whether they demanded the long self-denial
of a life of penance, or some one supreme sacrifice; even
general social and political arrangements were (without
separating between divine and human law) governed by an
ever-present thought of the great universe, of which all
earthly things make but a dependent part. This broad and

widely-comprehensive view remained in many respects peculiar to the East, and still has an imposing effect upon us, but the blood of the Western nations cannot endure for a continuance that repose of cosmic contemplation in which this view causes men to become absorbed.

The more exclusively imagination aims at combining the manifold of reality into a whole in the unity of one plan, the more is every particular arranged and fixed in its own proper place, and cared for and subordinated within the clearly-marked outlines of this whole—supposing, of course, that the attempt at unification appears successful. We are stimulated to advance by the unknown reaches of the path that lies before us; to have an early view of all attainable goals only makes men wish to continue undisturbed in the position in which they happen to be, and beyond the horizon of which there lies nothing essentially new. To such an early survey and to such quiescence did the nations of the East attain; the universe as a whole seemed to lie finished and complete before them; it had been such from eternity, and the future could add nothing to it. Many things in it seemed uncertain, but nothing really was so; there was no such thing as a merely probable development of cosmic history capable of being determined by some exercise of human freedom; there was no field for the exercise of inventive activity which might enrich life by new productions, or accomplish by purposive struggles anything more than that which, being preordained, would come to pass without human effort; according to the immutable ordering of the whole, man can choose nothing except what he cannot avoid, namely to live that part of the life of the universe which falls to his share, and to suffer and rejoice therewith. It is true that even within these limits human nature (which is never wholly brought into subjection to its own philosophic views) finds room for untamed passions; but the only goals which these can have are visions of pride and sensual delights—visions which fall to pieces when the passions that gave rise to them are burnt out, and which do not affect the old order of things, which goes on unchanging and undeveloping. Therefore, however agitated the course of oriental life may be in detail, looked at as a whole all its activities appear to be enclosed by a broad framework of resigned quietism.

The West developed a contrary bias, and this the more vigorously in proportion as it freed itself the more thoroughly from oriental traditions. Its imagination was never directed so eagerly to a comprehensive view of the world as a whole, but all the more eagerly to those universal laws upon which the reality and movement of the world itself in every particular depends. The oriental representation of the complete and finished condition of the world and of the circle of its phæno-mena exhibited a universe that had been perfected once for all, which no one could add to or take away from; but to gain a knowledge of these universal laws the world had to be regarded as something imperfect, to the perfecting of which it was possible to contribute; for these laws taught men to comprehend not only the condition of what actually existed, but also the possibility of much that as yet did not actually exist; and opened to the mind that was struggling onwards a prospect of reconstructing—for its own ends and by the help of these laws—both external Nature within narrow limits and human life within much wider limits. For such a mind there was possible a history in which human action should determine the as yet formless future to new and hitherto indefinite developments of reality.

It is said of philosophy that if the cup is merely tasted it leads man away from God, but that if it is deeply drained it brings him back again. Perhaps this saying is equally applicable to the whole mode of thought which in occidental civilisation gave rise to its characteristic restlessness—to that spirit of progress which must be for ever bringing change into every department of life, and to the investigating and analysing spirit of philosophy itself. For certainly that which first makes a distinct impression upon the mind is the alienation from God and from what is divine to which on the whole the course of this period of civilisation has unremittingly tended.

In as far as imagination influences life, the horizon of human imagination has undergone progressive contraction in proportion as there has been an ever-progressive increase of clearness in the diminishing field of vision to which it restricted itself. With growing knowledge of natural products, and increased skill in making use of them, men's insight into the connection between them and the supersensuous world has not become clearer, but attention has been weaned from dwelling upon the connection as one of the problems which have to be considered ; and life and morality have become more and more separated from the content of religious belief, regarded as the source of obligation, and have become more and more established upon secular principles of their own. Æsthetic sensibility, averse to ideals of vast and eternal significance, has turned from what was great and exalted to what was elegant and correct and to the activity of intellectual resource exhibited in it. Art is hardly able to cope even with what is merely historically great, but in genre-painting it gives characteristic reproductions of fragments of life. In science, dependence on experience has taken the place of speculation, and elements and general laws of action have supplanted the predetermining oneness of creative and formative Ideas as means of explanation. In a similar way in the department of practice individual rights are being brought into ever greater prominence as compared with the duties demanded by consideration for the whole ; and finally, we see that increasingly general acceptance is accorded to the principle of letting every individual power act unhindered, and of expecting the most satisfactory condition of human affairs from the equilibrium which the various forces will reach of their own accord through the reciprocal action of all.

All these features cause Western civilisation as compared with Eastern to have the aspect of a wholly profane or secular life which does indeed willingly submit to the general conditions and laws which govern the course of things, and skilfully contrives that these forces should work for it ; but is little conscious of any necessary connection between its thought and

action as a whole and a supersensuous world, and is of opinion
that it only needs, and need only regard, as much of what is
divine as may be expressed in the form of general laws for the
regulation of moral conduct.   Undoubtedly the entrance of
Christianity into the Western world was like a mighty inflow-
ing wave which interrupted this ebb, but it has not prevented
it from resuming its course.   Dogma and worship are equally
poverty-stricken, and efforts which aim at their rehabilitation
have to encounter increasing aversion ; religiousness disappears from morality even while morality increases in
humanity and refinement ; not only does the articulation of
secular society avoid all ecclesiastical control, but even the
coherence of church communities becomes loosened by the
growing demands for independence made by individual
opinion.   Are these conditions signs of a general retrogression
of humanity, or do they conceal an advance which appears to
us to be primarily occupied in breaking up the old forms of
religious life, but which does not leave us without hope that
in the future those old forms will be replaced by new ones ?

§ 2. Nature is commonly our earliest guide to religious
contemplation.   Observation of Nature leads in various ways
to attempts to supplement the perceptible content of reality
by continuations which are visible only to the eye of faith.
Imagination looks to the past, seeking in histories of the
origin of the world an explanation of the wonders of the
existing universe—wonders which could not, it seems, have
owed their birth to such an order of Nature as now obtains—
and it looks also to the future, seeking to find some continuation of Nature into which the swift-flowing stream of
earthly life may empty itself and find continued existence ;
the two lines of fancy are connected together by a more or
less comprehensive knowledge of reality so as to constitute a
whole of greater or less completeness.   If no other interest
than the merely theoretic one of explanation were involved
in this cosmological construction, it would attract no greater
sympathy and attention than the geological opinions of the
Neptunists and the Vulcanists, or than the equally divergent

conjectures which imperfect astronomical science once put forth concerning the structure of the starry heavens. But in those cosmological views there is always contained some expression of men's conclusions concerning the worth of the world, and the amount of satisfaction which the order of Nature affords or refuses to the irrepressible needs of the human soul; and with the pictures which are drawn of the powers which create, preserve, and guide the universe there is always connected a more or less developed view of the position which man occupies with regard to them, or the attitude which in action he should hold towards them. It is only for these reasons that we can with any justice seek the germs of religion in such complementing of natural phænomena and such combinations and explanations of them; we can attribute all the significance implied in the name *Cosmology*, to those systems only in which that is made the standard of worth and the point of departure which in existing theories has but a secret influence—I mean a conscious recognition of the unconditioned validity and truth of Morality and Holiness as compared with all that is, or seems to be, matter of fact.

Now if the whole of Nature lay before us we should see its manifoldness combined in a unity which—being the perfect reflection of what ought to be—would teach us what the significance of Nature itself is, what our place in it is, and what the aims of our existence are. But such insight as this is reserved for the end of time. To every nation that has entered on the path of civilisation, Nature has displayed but a small section of its whole content; different in different zones and climates and unintelligible in its connection without the enlightenment to be supplied by investigations which have not yet been carried out; unfit to form the basis of a comprehensive view of the world, because the condition of that which has been observed seems to leave diverse modes of completion equally admissible. Imagination always finds in the course of Nature traces of harmonious and beneficent wisdom; besides these it always finds also traces of discord, harshness, and cruelty; it finds much which leads it to believe in a righteous Providence and much of which the Nature is such that this belief can only be held in defiance of it. Different nations have become absorbed in the confusing complication of these facts—men with different degrees of mental activity, with different temperaments, and under the influence of very divergent modes of life; and according to the measure of their endowments in these respects they have attained to philosophic views of greater or less fulness and lucidity. But even the greatest fulness, with the keen eye for Nature which belongs to developed cosmologic insight (such as characterize the mythologies of the classic nations), can scarcely be regarded as having ever been a blessing in themselves. To the distant observer the richly coloured and realistic circumstantiality of those mythologies appears as an enviable filling of man's whole life with thoughts which unceasingly connect all its trivialities with the grandeur of the supersensuous world, and it exalts, in our view, the æsthetic importance of those nations with whom it is found; but these nations themselves were hardly ever led by the natural-philosophic element of their religion to any useful progress in life and humanity, but often enough to great errors and to a useless squandering of human powers.

Observation of Nature easily leads to a conviction that there is some supersensuous power which rules events, but no observation of Nature teaches moral truths. It can teach that the destruction of every individual may have its significance in the plan of the whole; that from every life that is trampled out another life may spring; that all the powers of Nature in an unceasing cycle may combine in the continual production, destruction, and reproduction of phænomena in never-failing regularity; but with all this it leaves wholly undecided whether indulgence towards others and sacrifice of oneself, or conversely trampling upon others and asserting oneself, is that to which we are morally called; as a conscious prolongation of the course that Nature unconsciously takes, the one mode of action has as good a claim to consideration as the other. That which is, does not enlighten us concerning that which we ought to do, unless we know beforehand what meaning we ought to attach to that which is. But how this ambiguous world of phænomena is to be taken and understood by men, whether the way in which it is interpreted and used will be a blessing or a curse, is determined by the mind which man brings to it—by the degree of civilisation which the moral influences of society have enabled him to attain, and upon the development of which Nature herself (not as instructress but as the sum of conditions promotive or obstructive) undoubtedly has an important effect.

If social conditions have provided but meagrely for the cultivation of the moral consciousness, men must be destitute of standpoints and conditions necessary for taking a coherent and comprehensive view of Nature and of the order of events .—a view in which there is room for the accommodation of individual contradictions. And being thus destitute they must lack also that wholesome ballast which is capable of preserving imagination from yielding unresistingly to the impressions produced by individual striking phænomena. In such a case the unstable mind is driven by the incalculable influence of fortuitous combinations of ideas, first of all to this or that interpretation of phænomena, and then to such or such maxims of conduct—perhaps to maxims of foolish soft-heartedness or perhaps to others of barbarous cruelty. And this danger is a permanent one; it reappears under some fresh form at every stage of civilisation. It is a danger that threatens even when a vigorous and developed intelligence that has long been in possession of many-sided experience and of various standpoints from which to estimate things, can no longer be imposed upon and led into narrow-minded mistakes by isolated phænomena, being able to rise above many individual contradictions to a consciousness of the all-pervading and eternal harmony of the universal order. For even supposing that it does thus rise, yet a just perception of facts does not of necessity involve a just estimation of their worth. On the contrary, the higher our trains of thought soar in their progress to ever wider generalizations, the more unstable does their equilibrium become ; it needs but a slight alteration of mood and at once our mobile imagination beholds the same facts in a light which altogether transforms them, without their having themselves undergone any change. When this happens, nothing but a thorough and established moralization of life can furnish a counterpoise of sufficient weight to withstand the effect on conduct of the wild theories into which speculation is only too easily drawn, in its attempts to take a comprehensive view of the universe. And finally, even when reverence for the content of moral Ideas, undisturbed by any doubt, rules the general mind and is the point from which by common consent all attempts set out which aim at following by faith the course of the world into regions which no experience can reach : even in these times of religious culture in the strict sense, the old danger will always lurk in men's preference for a cosmological construction of philosophy. With the voice of conscience and with that which we venerate as revelation, we build up but very tottering bridges, which are none the more secure because we use them with presumptuous confidence as a means of obtaining untrustworthy glimpses of the construction and articulation of the universe as a whole. Still more untrustworthy will be the conclusions as to practical life which men deduce from cosmologic philosophy, as though it afforded a representation of reality which might be relied on. The aim of such an application of these conclusions would be to deduce from a supramundane metaphysic of the universe holier precepts and aims for human guidance ; while perhaps on their account silence would be imposed upon the simple absolute commands of conscience which have no pretensions to universal knowledge.

If therefore the name of religion is to be exclusively reserved for that form of spiritual activity which regards a recognition of the divine order of the world and the subordination of our life to it, as conditions of salvation (and it is in this sense that religion is commonly opposed to unbelieving morality), we should be expressing but a part of the truth in lauding the improvement of the human race as attributable to the influence of religion ; we should have equally to admit that the progress of humanity due to the action and reaction of society and to the development proper to secular life, on the one hand has supplied religious belief with new questions and subjects of consideration, and on the other hand by its quiet, obstinate, and ever present resistance has blunted the edge of those injurious extravagances into which the world-interpreting, world - creating flights of devoutly inspired speculation were apt to run.

§ 3. By what thread of connected tradition or by what recognisable law of progressive development those successive forms of religion may have been determined which have gradually arisen among the civilised nations of our hemisphere, are matters which I leave undecided, considering that they cannot be exhaustively discussed in this place. And even the hasty survey which I propose taking for the confirmation of the foregoing remarks, must be curtailed.

Where social life is very little developed and reflection lacks the breadth of view which can be given to it only by a stirring life and constant intercourse between one's own thoughts and those of others, the foreshadowings of a supersensuous world which may be called into existence by even the most everyday occurrences, remain chaotic and incoherent. Fetich-worship, with very natural confusion, while it reverences the mysterious power residing in every object which happens to strike the senses, neither identifies this power with that in which it inheres nor clearly distinguishes it therefrom. It is not this lack of conceptual clearness which causes Fetichism to take such a low place among the different forms of religion, but the absolute indefiniteness of its ideas concerning the nature of the supersensuous power which it venerates. It regards this as nothing but a certain degree of mysterious indeterminate capacity, not any fixed kind of volition or activity, susceptible of specification. Such power is to be found in every object, but any one object may possess it in a higher degree than any other ; for men to try, by prayer and sacrifice, to make it favourable to them is but a transference of natural human action in reference to human wills ; in the nature of the incalculable demon itself there is no intelligible ground for even this most simple worship. The same poverty of thought makes it difficult to estimate the gain to life of presentiments of immortality. The idea of the absolute annihilation of anything which has once been observed in the vigorous exercise of perceptible activity is as incomprehensible to undeveloped thought as the idea of anything's arising from absolute nothingness ; belief in the continuance of the soul after death is more natural and more ancient than the belief in its annihilation, which is among the earliest mental products of a somewhat advanced civilisation. But the poor philosophy of the early stage is equally unable to assign a content to the continued existence in which it believes and to its notion of a supersensuous power in things ; where future existence is not conceived of as a copy of earthly life, the soul is supposed to join the ranks of the obscure powers of Nature; it continues to exist as a ghost, that is, with the general attributes of the spiritual life of man, but without humanly intelligible ends. Such unsatisfying ideas neither can become sources of moral convictions nor do they readily admit of being connected with such convictions ; but the ideas themselves would have taken a very different turn if a greater degree of moral cultivation had led men to seek beneath the surface of phænomena something other than vague forms of life and powers different from our own. What is taught by fear and sympathy can at any rate, as contrasted with such a faith, be developed to practical precepts and the rudiments of worship ; but what such precepts and such rudiments shall be is decided by the purely accidental course of unbridled imagination and the bias of temperament; they are apt to run into superstitions deformed by witless sorceries and bloody abominations of sacrifices to the dead.

One of the errors that seem to us most strange is the paying of divine honours to animals, and yet there is an intelligible cause for it in dawning religious feeling. Social intercourse teaches men to know one another in a wholly secular aspect; they find each other busied with small and changing and contradictory interests which are perfectly intelligible and have nothing of the obscure grandeur which imagination admires in those natural forces which work unconsciously. When man has once begun to contrast himself and his fellows and all his human interests with the world and that strange power residing in it which constitute the first object of his confused reverence, he can find nothing in which this power appears more expressively than in the activity of the animal kingdom, which in all its manifestations impresses us the more on account of its voicelessness and our inability to understand the extraordinary instincts which it displays. It is true that without some flights of imagination this contemplation cannot give any definite content to our notion of the supersensuous, but at any rate it views this under the exalted notion of a spirit-life that broods over strange ends, unintelligible to us. We can see that while men lived a life in which attention had not as yet been attracted from physical existence by a multiplicity of peculiarly human interests, such considerations might easily give rise to the idea of transmigration of souls, an idea which afforded an abundant field for the exercise of ingenious comparison and constructive imagination. There is no doubt that at one time men's minds were seriously possessed by this idea, and that in consequence a vast amount of human activity and attention were squandered on wholly unmeaning and fictitious objects. The belief was not refuted by science, but died out from its own lack of interest, as there grew up around it a civilisation which has its centre of attraction in the worth of social and moral relations. At present we hardly think of animals except as objects of domestic economy, or of natural history, or as ornaments in a landscape; that they have a multiform mental life allied to our own, is a proposition which we sometimes timidly advance as a probable conjecture. And just as indifferently do we turn away from all the unremembered past which preceded our earthly existence ; as to what lies beyond this we refuse material analogies in as far as our abiding need for some sensuous representation of the supersensuous will permit.

In every case in which fully developed civilisations have culminated in comprehensive religious systems, in Egypt, in India and in Western Asia, investigation takes us back to the grand all-encompassing phænomena of the heavens as the point of departure from which religious ideas have set out. Far removed beyond the reach of earthly contact, the heavenly bodies for that reason stirred imaginative forebodings with their far-away brilliancy, but they attracted attention still more by the regularity of their movements; the reverence paid to them applied not only to their gladdening light, but it was also the first homage that was offered to the notion of truth, and law, and order, as the genuine content of the supersensuous. But this germ which promised so much, seems to have come to nothing as far as the development of religion was concerned.

Egypt owed to it noteworthy beginnings of astronomic science, and an attempt to construct cosmic order by connecting it systematically with natural forces that were personified as divine beings. From the cultivation of this wisdom (on which the ingenuity of the priesthood was exercised) no gain accrued to life—nothing but the burden of a ceremonial worship, which at best could only serve to keep up a general feeling that it was being offered to supersensuous beings, but the symbolic significance of which was unknown to the people. On the other hand, the wonderful phænomena of the Nile valley, connected as they seemed with the course of the heavenly bodies, must have directed general attention to the regular activity of the natural forces which in steady rotation alternately call forth and destroy life. The contrast between generative and destructive power not only aroused mystic speculative reflection, but was also the subject of popular mythology and of many solemn rites. Still the whole sphere of religious thought does not seem to have been dominated by it to such an extent as in Babylonia, where imagination was carried away by similar incentives to the most extravagant worship of the universal generative power of Nature. In Egypt alongside these cosmological myths, and connected with them in a way that to us appears merely external, there was developed a religious view of human life. This view was characterized by a conviction of the immortality of the personal soul; combined with the idea of a judgment which should summon the spirits of the good to a life of blessedness, and condemn the wicked to infernal punishments and the purifying penance of passing through earthly life again under the forms of men or beasts, this system of doctrine most happily succeeded in keeping itself from being overgrown by the speculations of natural philosophy, and brought together those elements of moral conviction which the full and various life of the oldest civilised nation had developed.

This was a comparatively healthy realism, which, though it attached human existence to an all-embracing cosmic order, left the determination of the ends of human life to the development of life itself, and not to cosmological speculation. The excess of such speculation in India led, on the other hand, to an idealism which, while it took away all meaning from the world, took away also all meaning from human life. Here imagination turned from the primitive worship of the heavenly bodies, not to bring into prominence their order and regularity, but to lay one-sided stress upon their changeableness and transitoriness, and emphasized with fatal ingenuity the necessity of one eternal primal being, which we should conceive of wrongly if we imagined it to have any definite content, and most wrongly if we imagined such content to be continuous eternal rest. Indian speculation found it as difficult as later philosophy has done to get back from this indefinite being to the world of reality. It avoided those mythical genealogies of divine beings which in other cases fix the successive steps of the creation of the world, while at the same time the failure to explain how and by whom this progression was accomplished is hidden

by the imagery. Thus it came to pass that our want of
insight into the cause of the origin of the world was taken
to indicate an origin which had no cause; the primal being,
misunderstanding its own yearnings, is represented by this line
of speculation as developing into a world which is illusive,
and which only seems real to its own individual members.
An appearance which arises without cause, and which appears
in orderly fashion to its own constituent parts, is but another
name for a reality which is as yet unexplained; hence this
mode of representation is metaphysically inadequate.   On
the other hand, it contains a decided expression of opinion as
to the *worth* of the world; the world is a mere appearance,
not because it is not real, but because it is not what ought to
be.   As regards that which ought not to be, man's only duty
is the effort to remove it; in the universal nothingness of the
world, the condemnation of which is unceasingly expressed by
the primal being itself in the constant destruction of all
created things, human life has no worth and no special ends;
salvation lies only in turning away from it, in withdrawing
oneself from the influence of that world of appearances, which
is what it ought not to be, by annihilating all passion, and
finally all ideas and all thought, and returning to the painless
condition of the unconscious primal being.   This despair of
life is not to be regarded as resulting from speculative
error in interpreting the universe; it must have proceeded
from psychological causes, from the general tone of mood and
feeling which we can no longer analyse, for it pervaded all
Indian thought and even practical life with a power which
belongs to no doctrine that is not in harmony with the
popular mind.   Even Buddhism, after it had sought to free
men's minds from the fetters of Brahmanism, of ceremonial
service, of distinctions of caste, of the horrors of transmigration of souls which threatened ever renewed tortures of
existence, ended with the same thought and aimed only at
facilitating the return to nothingness.   The power which
this belief exercised over men's souls is shown by that
inclination for an ascetic life which inspired such countless

numbers with an enthusiasm for penance and unheard-of self-torture. The great mental endowments of the people were expended uselessly under the guidance of such views. The development of knowledge was insignificant; notwithstanding great refinement of feeling, morality did not recognise the unconditional sacredness of goodness; strictly speaking, it knew nothing of sin, but only of ill, which is the cause of mental disquiet; hence all virtue consisted in cultivating skilfulness in escaping from this ill. Finally, in course of time, like all other similar extravagances which, becoming unable to maintain their original elevation, produce some mechanism of custom as a residuum of enthusiasm, Brahmanism and Buddhism (and the latter in the end to a greater extent than the former) became secularized into the utter aimlessness of monastic life and ceremonial pomp.

Thanks to a more robust mental constitution, the cognate Iranian races obtained better fruits from the germs of religion which were common to them and to the Indians. Zoroaster's teaching added a dark shadow to the light which men worshipped; here, the delusion by which the primal being was supposed to have been confused, and misled to create the world, was replaced by the darkness of an evil principle which limits, but only apparently, the just and true development of the good principle of light; at the end of that conflict between the two which fills the world, the evil will succumb to the kingdom of light, and then nothing will be except what ought to be. In this conflict man has to take part. The natural symbolism, which in all times has made Light the image of the Good, and Darkness the symbol of Evil, allowed of this hurtful, equivocal, ill-favoured, natural phænomenon being assigned to the realm of Ahriman, and (while the final victory of Ormuzd in the future was held to be certain) also allowed a multitude of practical precepts, which prescribed intelligible ends of daily action and reasonable moral obligations, to be connected with the clear dualism of principles which was adopted. But neither did this form of religion escape the fate of having its great

thoughts buried under a superfluity of external forms by the
ceremonial pedantry of a growing priesthood.

§ 4. We encounter other phænomena on European soil.
The Greeks as well as the nations above referred to felt something divine in natural phænomena before they recognised it
in the law of conscience. But their thoughts were absorbed
neither in the abyss of universal being in which all form disappears, nor in considering the intelligible secrets which each
particular in its own place was called upon to indicate ; what
they took hold of and clung to was the beauty of the whole
and of each of its parts ; the more their civilisation advanced,
the more did that didactic part of the content of their myths,
which at one time was common to them and the Eastern races
with which they were allied, fall into the background beside
the characteristic beauty with which they endowed their
divinities and the world they inhabited. Calm, steady
development, the domination of motley multiplicity by the
unity of one ever-repeated rhythm and all the fair proportion,
clearness, and purity which the world of the senses presents
to us—these are not in themselves moral concepts, but they
are modes in which things exist and comport themselves, which
we strive first to realize in ourselves as conditions or results
of morality and afterwards to find again in the external
world. Hence favourable natural surroundings from which
such impressions may be obtained, may contribute their part
to the taming of wild impulses and to mildness and beauty of
disposition, but the larger share is undoubtedly contributed by
a successful development of moral life in society ; it is this
which first gives susceptibility to and interest in the beauty
of the external world. And this it was which early withdrew the attention of the Greeks from the significance of their
deities in Nature, a subject the consideration of which has
always proved unfruitful as regards religious development ;
their imagination substituted for the vanishing mysteries of
this secret meaning the obvious and expressive beauty of ideal
forms, the characteristic variety of which reflected the
infinitely higher secret of the manifoldness of mental life.

This representation of the world of gods (which was not accomplished without frequent misuse of poetic imagination) in making them human made them at the same time moral. As often as the popular conscience recognised the beauty and urgency of some new moral obligation or some new ethical Idea, men tried on the one hand (from the natural desire to understand that which is greatest in the world as being also the most perfect) to assure to the divine world the possession of this beauty as a side of its wealth that had hitherto remained unknown, and on the other hand they tried to raise recognised duty above the fluctuations of individual judgment and of variable moods by deducing it from the will of the gods. Thus the Greeks improved their faith by the results of living culture ; their most profound poets struggled to infuse into the transmitted content of this faith their consciousness of sacred truths and precepts, thereby deepening that content. And it was just on this account that at last the feeling became overpowering that the original basis which men sought thus to ennoble was inadequate ; they found that all which gives worth to human life may indeed be externally connected with the names of the mythic gods, but has not any essential dependence upon them. Then there came into honour the simple name of God or of the Divine, used to indicate the true source of what is worthy, to which source the living longing of the nobler minds turned back in anxious search.

It was the religion of individuals and not of the people that came to this conclusion ; the popular religion which at last fell wholly into ruins, never attained the coherent unity of the religious systems of the East. Mythology arose neither from a single impulse, nor from impulses that worked on uninterruptedly. Notions that had diverged somewhat even in the Asiatic home where they had their birth, had become still more different in the European settlements in which the various tribes lived on for a long time in isolation from one another ; migration and intercourse with other peoples had introduced foreign ideas concerning God ; local circumstances had reduced many an image of some divinity which had formerly been the same for all, to various different embodiments; and finally, all such notions had early fallen into the transforming hands of poetry. All this collection of characteristic ideal figures, consisting of symbolic personages from ancient national legends and from the poetry of untrammelled imagination, had grown to such vast dimensions that perfect agreement about them had become unthinkable, and dogmatic instruction as the foundation of a settled confession of faith impossible. The world of the gods in its boundlessness stood over against consciousness as physical Nature had stood over against it from the beginning; the latter, too, is not known in all its parts by any man, but its main outlines are known to all; each has a limited region within which he lives, and the peculiar worth of which he understands from actual experience. So in the wide world of mythological divinities each had a special circle of tribal gods ; and to honour these with traditional forms of worship was enjoined by the state, the family, or some ancient religious guild, on all who wished to be reckoned as belonging to it. But there was no church to guard pure doctrine or to see that it was followed, no established priesthood with any power over consciences. The priest was the expert who knew the secrets of the particular sanctuary in which he served and lent his aid as mediator to the pious worshipper who came with offerings. Wherever there was any censorship of religious opinions, it was exercised by the political community ; the national worship of the gods, upon which, as upon a primitive sacred treaty, the welfare of the state was supposed to rest, was defended by the state itself, on the one hand against the intrusion of immoral foreign worship, and on the other hand against the disintegrating enlightenment of home-born philosophy.

Before the moral deepening of the idea of divinity had made it possible for men to pay unceasing reverence to this idea by their mode of life, prayers and sacrifice and songs of praise continued here, as in all cases, to be the only expressions of gratitude, of spontaneous admiration, and of awful

fear called forth by the gods, whom men regarded as beneficent, or exaltedly beautiful, or finally as threatening powers
of Nature. A mixture of these feelings was the frame of
mind which the Greek conscience continued to require as
piety towards the gods. It is a long step from this frame
of mind to the definite actions by which it manifests
itself in men's lives. The will of the gods men did not
know ; to reverence it while yet unknown, and also to regard
the scattered revelations in which it now and then made itself
known ; not to be in any way haughty or presumptuous, but
to maintain a moderate frame of mind, being conscious that
the guidance of all things is in higher and mysterious hands
—such was the sole further development that the Greek conscience was able to give to this εὐσέβεια. Mythology could
not teach any more pregnant connection between human life
and divine decrees ; it had too entirely lost all remembrance of
the comprehensive world-history with which human history
had been interwoven by oriental imagination ; for it everything was but a radiant present, the echoes of whose past
lived only in a few obscure legends, and which saw before it
no unfathomable future, nothing but its own steady uniform
continuance. Under however glorified an aspect men might
regard the gods, they yet never regarded them as the creators
of the world ; they continued to look upon them as conditioned beings, the fortunate firstlings of a hidden creative
power ; as ideal men and powerful helpers of their weaker
brethren in difficulties which yet even for themselves were
still difficulties. And for this very reason the moral deficiencies which were blots in their representations of their gods,
when the natural symbolism of the early legends had been
transformed into histories of personal beings, did not disturb
the sincerity of their reverence to the extent which might
otherwise have been expected. These pictures of the gods
lived in men's consciousness as expressive and characteristic
representations of natures, some of which were noble and some
ignoble, but all having the freshness and reality of life about
them ; and the gods themselves were regarded as superhuman

combatants who had been our forerunners in the battle of life, forerunners for whom men felt the same kind of devoted and confident attachment that soldiers do for their leader.

In the external forms of worship the Greek mind preferred the solemn beauty of mystic elevation, and avoided, except in a few points, the sensuous enthusiastic passion of the worship of God as practised by the Asiatics. Many of the customs handed down from antiquity had become unintelligible to the people. Although every divinity might be called upon in any locality, yet the more solemn worship of each was connected with special places where help had been vouchsafed to men on particularly memorable occasions, the recollection of which was intended to be preserved by significant ceremonies, yet which notwithstanding did not escape oblivion. Thus sacred ceremonies remained attached to particular places of worship, as being of traditional obligation ; almost like the peculiar feudal obligations which vassals of the Middle Ages owed to their feudal lords ever after the occurrence of some forgotten adventure. Yet the Greeks were impelled to maintain conscientiously the integrity of these ceremonies by that piety with which they believed that they ought in all cases to honour the uncomprehended will of the gods.

And uncomprehended as to its final secrets did this will ever remain to the Greeks. There is a mild, pleasing, unaffected naturalness in their religious views ; they do not, however, set up a kingdom of heaven in opposition to the world, but exhibit the beauty of a moderate, serene, peaceful enjoyment of life springing from a judicious and intelligent appropriation and improvement of earthly conditions, in contrast to the splendour of oriental despotism and unmeaning luxury. It was only this which Solon set before Crœsus when he declared the peaceful life of Tellos, or the happy end of Kleobis and Biton cut off in their youth by a blessed death, to be preferable to the renowned good fortune of the Lydian king. There is no reference in his words to a happiness which is not of this world, or to a peace of conscience which can outweigh external misfortune. Solon urgently admonishes the king to think of the end, not as though he were then to be judged according to the worth or worthlessness of his life, but because no man is truly happy who is not happy to the end. According to Greek ideas, a disaster quite late in life mars all a man's previous happiness, just as in art the beauty of a whole work is spoiled by failure in the smallest detail. These remarkable people even tried to make the end — which they regarded as the final end — artistically satisfactory; any connection of the whole life with a future beyond it was never a dominant thought with them. It may be that in the religious mysteries of the Greeks there were handed down some ancient Eastern teachings as to immortality, and certainly cultivated Greeks were not unacquainted with the idea of a continued existence, such as lightens the hard life of so many rude tribes. But if this belief had had any deep-reaching influence, we should know of it, without any special proof, through the immediate impression produced by Greek national life as a whole. This impression, however, testifies decidedly to complete satisfaction with the present world. The wide gulf between the Greek view of life and that of Christianity cannot be filled up by bringing together isolated expressions of which we can never be sure whether they gave voice to a fixed and hearty belief or whether they were mere poetic images without serious meaning, which served the æsthetically cultured people who used them as mere ornamentations of life.

§ 5. The noblest representatives of Greek speculation had learnt to know God as the first and unmoved mover of all things, as the operative essence of the Ideas of the True, the Beautiful, and the Good; but to the Hellenic mind (of which the one-sided reverence for knowledge was kept up by its consciousness of scientific achievements, and to which sin was only intelligible as error) the Supreme Good was without any content of its own, and melted away again into Beauty and Truth. However great the interest with which we may

continue to regard this final religious outcome of the classical
world, which is great regarded as the fruit of human investigation, yet it is but as a modest rivulet compared to that
rushing river of consciousness of God which, from a long
previous period, had swept through the life of the Hebrew
people and overflowed in their sacred poetry with a power
compared to the assured reality of which the highest flights
of Greek enthusiasm seem but as mere problematic conjecture.

Learned investigation may discover traces of foreign influence in individual features of legend and custom, and in the
artistic and ceremonial development of Hebrew worship, but
the essence of their religious philosophy was wholly withdrawn
by the Israelites from the influence of heathen culture, with
some aspects of which they were in long-continued contact.
Those principles of natural philosophy which smothered the
religions of the East with their rank and injurious growth are
almost entirely absent from the religion of the Hebrews ; here
the motive-power of development is to be found in ethical
Ideas, which, though not indeed alien to the life of other
nations, were not the source from which their religious notions
were derived.   With what ingenuity must the Egyptians have
determined the succession of the cosmic powers to which the
order of the universe is due—if, that is, we can trust the
equal ingenuity of their interpreters.   But for religious life it
has all about as much worth as the infinitely more trustworthy
teachings of modern geology concerning the stratification of
the earth's crust.   The Mosaic history of the creation (to
which only a strange misunderstanding can seek to attribute
natural-historical significance) is distinguished by its contempt
for such cosmological speculation.   It does not make any one
phænomenon a basis for the development of any other ; with
the greatest uniformity it repeats in the case of every creature
that God made it, and in describing the series of creative acts
it hardly thinks enough even about observing an order
corresponding to the interdependence existing between
different parts of the material world.   It was sufficient that
God made everything, and that everything as He made it was

good; sufficient that man was regarded as the crown of this creation, and the creation itself as the garden in which he was destined to live after the likeness of God. Nor was any higher place assigned to Nature later; as regarded the one living God, natural phænomena had no meaning but as signs of His goodness, His almighty power, or His wrath, and as such, poetry depicted them in the most striking colours; but except in hasty sketches, imagination never busied itself in attempts to see God's being symbolized in the order of Nature, as though such a manifestation were necessary to Him, or could suffice Him. But this God who had no serious ends in Nature itself, but used it as the scenery of a magnificent drama, had special designs for the human race; while the cosmographic horizon of the Hebrews was narrowed to almost idyllic dimensions, and all interest in Nature as a whole was relinquished with indifference, the promised land was raised to the sacredness of a special sphere of divine influence, and became the stage on which a course of action and reaction between God and man was played out.

Attention being turned away from the structure of Nature itself, the danger was avoided which had misled those religions that had a cosmological foundation—the danger, that is, of regarding first natural ill and then moral evil as necessary constituents of the cosmic order, and as metaphysical consequences of the Divine Nature. According to the Hebrew faith God was wholly good, and neither in Him nor in the creation as it came from His hand was there any seed of ill; it was human freedom which, perfectly unfettered and unconstrained by any metaphysical fatality, brought sin into the world, and, as its punishment, death and the ills of life. This kingdom of evil which had now arisen was not something which must be necessarily thought as a part of the world; it was something which need not have been and which ought not to have been; the command to be holy as God is holy applied to man, and applied to him as one which it was possible to fulfil in the fear of God and of His law. The doubts to which the human mind must always be led by the consideration of these most important matters, were not theoretically solved by the Hebrew faith; but their suppression gave to life for the first time a thoroughly religious foundation. Moral obligations, conscience of which is everywhere developed by social action and reaction, appear here consolidated into a Will of God, which has to be fulfilled and glorified, not only by the individual in inward disposition and outward works, but also by the whole nation in a theocratically regulated life of the community; the national history is the account of a continuous intercourse with the God of righteousness, who has attached promises of favour to the sanctification of His will, and who punishes obduracy towards it.

Neither did the external destiny of the nation bring the fulfilment of what had been promised, nor did the people find in conscience the evidence of its own uprightness; the end of the struggle carried on in the attempt at self-justification towards God lay yet in the future, and was anticipated as the temporal glory of the whole race, which, with somewhat obscure hopes as to the eternal significance of the individual soul, felt itself called to constitute a kingdom of God upon earth. Christianity regarded itself as the realization of the predictions which seemed to point to this, but it was not recognised as such by the Jews; in the view of Christianity all which men had hoped with regard to the Messiah was found realised in deepened significance in the person of Christ —the crowning prophecy of a final revelation, the high priest's office as mediator by means of sacrifice (sacrifice and mediator being one), and the sovereign power of Him who is to be King over the Church in all ages.

§ 6. If we separate for a moment that which the doctrine of the Christian Church does not allow to be separated, namely that which is revealed through Christ from faith in the historical fact of His revelation, we shall see that the former contains exclusively religious truth conveyed in a form of expression which is also exclusively religious. The order of visible Nature is not a subject of its interpretation and explanation; pervaded as a whole and in all its details by the foreseeing and preserving will of God, it does indeed in its totality form the background of our life, and to it the mind may appeal when seeking some witness to the truth of its belief; but to know its construction and its articulation does not belong to the one thing needful. In the ordinances of the Law even Judaism had given to natural reality a significance not its own; although it insisted on holiness of mind, it still saw in the performance of actions a service which was in itself of some significance, and without the doing of which the world would lack that which human action was intended to contribute to it. From this sometimes outspoken and sometimes hesitating reverence for works Christianity turned away, caring exclusively for man's spiritual temper and the sanctification of this; what is primarily to be aimed at is not any particular state of things, nor even any particular state of mankind, revealing the kingdom of God in *external* ordinances by the harmony between men in different orders of society—but it is the new birth and the transformation of the individual human being, whose immortal spirit is to become the temple of God. It was only the chosen people as a theocratically regulated whole which Judaism had regarded as worthy to be such a temple. Hence Christianity developed directly social theories as little as it did cosmological wisdom; but in the new inner life that it demanded and made possible was the essential germ from which might be developed not indeed knowledge, but the renovation of man's nature, not a definite form of social relations, but a capacity of using and modifying any existing state of things in the right way.

If the thought of the merely conditioned worth of all earthly life lay at the foundation of this peculiarity of the Christian revelation, whilst the earlier religions of the East regarded this life not as a preparation and a school, but as occupying the place of real existence in the plan of the world, it might have been expected that at any rate the connection of earthly reality with the secret of the divinely ordered universe, of that which is with that which ought to be, would be all the more clearly developed by Christianity. This expectation would be deceived supposing its object to be an enlightening knowledge of the construction of the super-sensuous world; but, as the history of centuries shows, completely fulfilled if it ask nothing more than certainty as regards the blessed significance of the connection which that world (whatever its definite form may be) has both in itself and with earthly life. Revelation speaks of a Personal Spirit who is Almighty Love, but it is not absorbed in answering those questions concerning the metaphysical form of His existence which human knowledge raises in order that it may understand this after its own fashion; it describes that aspect of God which He shows to men, but it merely indicates without attempting to analyse that glory which only the angels in heaven see. It regards the world as the creation of this God, but with regard to its beginning and its end, it makes no essential addition to the knowledge possessed by the ancient faith; it is pervaded by belief in the immortality of the individual spirit, but intrusive questions concerning the nature of future existence it declines to answer; there is much to be told which we cannot bear yet. For just in proportion as this future existence is more certain, the less necessary is it to try and mature beforehand upon earth the fruits of the higher knowledge which it will afford us, and the more exclusively necessary is it to prepare ourselves for this great future. Thus it may appear as though revelation really revealed but very little, and in truth in a doctrinal point of view it is neither extensive nor circumstantial; it does not enrich science by an abundance of individual truths, but establishes a new life upon a foundation of truth, which is not considered to be possessed if it is merely known, but only when it pervades the whole man as the prevailing tone and temper of his life.

To characterize this essential germ of Christianity more in detail than we have attempted to do in our short survey of the course of history, is not our present business, but we may recall some aspects of its relation to other philosophic views.

Human nature is so similar everywhere, that wherever there is a sufficient amount of the social intercourse necessary to develop its capacities, the moral convictions that are evolved are similar in all essential points. But at the same time, the faculty of drawing from our own premises all the conclusions which they involve, and the effort to attain complete harmony of character, are so deficient in us, and (being the result of growing reflection) appear so late, that nearly everywhere in human civilisation, that has grown up spontaneously as national culture in practical life, we find between coexistent moral principles obstinate discrepancies, to which men are blinded by habit. Therefore, on the one hand, it may easily seem as though Christianity had brought no other moral ideals into the world than those which mankind had already discovered for themselves ; but, on the other hand, it will be found that its efficacy was not expended in introducing coherence and completeness into the contradictory convictions of heathen ethics.

The ground of all moral obligation is understood differently by it and by the heathen world, which in its rough beginnings was led to moral habits, partly by natural good dispositions, and partly by experience of their usefulness ; and when it had reached a higher stage of civilisation felt bound by the obligation of moral commands for their own sake, just as unconditionally as it found itself subjected unconditionally to natural laws. For Christianity the command to do God's will was not merely a comprehensive expression for the content of all individual moral ideals, but it also supplied at the same time a reason which justified, or at any rate explained, their binding power. The ordinary opinion of more or less scientific reflection is that there is here a retrogression as compared with the philosophic view of heathendom, to which the Beautiful and the Good seemed to be obligatory, in virtue of its own power and dignity and not as a law, even though it might be a law laid down by the Supreme Will. The faithful Christian will judge differently. He will admit that he learns the interpretation of the divine will only from the deliverances of conscience, and will shun the frightful consequences which have always arisen from the admission of any other source of enlightenment; he will not conceal from himself that his conviction lays upon thought new difficulties which are hard to overcome; but yet he will maintain that through it alone is he able to understand the phænomenon of conscience.   For it will seem to him simply incomprehensible that through some original and primary necessity there should be laws which have binding power over our actions but yet serve no purpose—serve no purpose because their whole business is to insist upon their own fulfilment and realization, the fulfilment when it has come about being the end of the matter, as though it were some new fact, without any good having been produced that did not previously exist.   The Christian seeks to escape this labour in the service of impersonal laws, this mere bringing about of facts; it is only in the pleasure which God has in what he has done, that he finds that ultimate good for the sake of which all moral action has worth in his eyes.   If love is the great commandment, then that that great commandment must be carried out for love's sake is a necessary corollary; neither the realization of any Idea for its own sake, merely in order that it, devoid of sensibility as it is, should be put into act, nor the residence of all excellences within ourselves, the egoistic glorification of self, but only love to the living God, only the longing to be approved not by our own hearts but by Him—this, and this only, is the basis of Christian morality, and science will never find one that is plainer, nor life one that is surer.

There is a close connection between this foundation and the fact that with the Christian precepts promises are always conjoined; this, too, is a rock of offence for that quixotism of pure reason which regards its efforts as almost disgraced if the kingdom of heaven and eternal blessedness are offered as their reward.   It would be wicked to deny that the human heart is capable of the greatest self-sacrifice, even without admitting to itself that it cherishes a hope of such reward;

for we have no right to doubt the instances which we find in history and in life, or to attribute motives by which these instances would be made more intelligible to us. But while we recognise the merit of that virtue which, in sincere devotion to the moral ideal, prefers destruction to defilement, we regard as incomplete any philosophy which holds that good may vanish out of the universe unrequited, and which lets the joyousness of action be damped by this conviction which can never be in itself a motive to action. Yet indeed it is not merely in order that the universe should be in itself harmonious and perfect that Christianity connects blessedness with moral fidelity as its result; it does undoubtedly also hold forth the crown of life which it promises as the motive which is to confirm and uphold that fidelity even unto death. Can we then contest with those who denounce all Eudæmonism the right to apply this name of reproach to Christian doctrine also, and to prefer to it as more exalted their own teaching which commands virtue and self-sacrifice without any reward? This latter requirement may indeed seem more exalted; but from the sublime there is but a step not only to the ridiculous but also to the inane and the preposterous. And this pedantry of reason runs the risk of taking not indeed the first but the second of those steps, if it is really in earnest. For without a supreme good to which the lesser good would be sacrificed, and as mere continuous labour for the establishment of a definite external condition of things, or of some definite condition of the inner man, in what would our moral struggles differ from any blind activity of natural forces, except in the accompanying but inexplicable feeling that one ought to do something which, when it is done, is of no use to any one? But in fact this step is not taken by that quixotic virtue to which we have referred; it is conscious that at bottom it, too, aspires after a Supreme Good—namely, Self-esteem; and it would certainly give up all moral effort if it were not rewarded by this result. Perhaps it would even be much less inclined than the more open Eudæmonism of Christianity to labour in the service of moral commands if obedience to them did not enable it to reach by the shortest road, and with the greatest possible directness, that which it regards as the Supreme Good.   The distinction, then, is between the proud inflexible Eudæmonism of self-esteem, which is self-sufficing, and the Eudæmonism of humility, which is not self-sufficing, and seeks its Highest Good in standing well not with self but with God, and in being beloved by Him.   The sacrifices which Christianity imposes on men in order to the attainment of salvation are not less than those required from them by the more self-sufficing doctrine; but while the latter sets out with efforts to reach that which is sublime, and finds little opportunity of returning thence to what is meek and lowly, the former begins with what is joyous and attractive, and yet mighty enough to produce also what is most sublime.   And that this way alone is the true one is an opinion confirmed by a consideration even of those æsthetic ideas upon which our moral judgment is only too dependent.   Such a consideration would show us how hollow is all sublimity that aims only at being sublime, and how imperfectly it is conceived when, being carried beyond its necessary relation to an Absolute Good to the power of which it testifies, it is set up as independent. Christianity does not see this good in the mere existence of a world of being and action, regulated according to moral Ideas, but only in the blessedness produced by the enjoyment of this world; and the gospel is glad tidings just because it carries out to this its final logical result the abolition of all reverence for mere blind factual existence, and reveals the hidden priceless jewel of salvation as the final secret for the sake of which all the vast expenditure of creation and human life has been made.   It never aimed at being sublime or magnificent; and yet because it is "*glad* tidings" it is also sublime and grand.

By Judaism too and by all heathen religions, the moral commands which life has taught have been interpreted as the requirements of God or of the gods, and they have promised happiness as the result of fulfilling these commands. But the gods of heathendom were too much occupied with

Nature; their care for the spiritual world and for mankind seemed to disappear beside the splendour of their manifestations in Nature, for the significance of which, alien as it was from human life, they demanded reverence; in this world which had no special definite aim, man must strive to win, by careful piety towards the easily offended unknown powers, mere toleration for mere transitory happiness. By the Jews too God was regarded as the Almighty whose acts, whatever they might be, were always righteous, because they were not measured by any higher standard of right; by this Almighty Being blessings were promised as a reward for the submission of mortals, which He eagerly desired although they were as nothing before Him. Not only did Christianity bring into prominence the spiritual world as the only true world and that in which God specially works, but moreover man is no longer as nothing before Him. It is true that the hope of attaining happiness in his own strength is taken from him; but as a child of God even the meanest knows himself to be an object of unceasing care to the Almighty, to whom the manifestation of the glory of external Nature is now regarded as being but a secondary consideration. Traces of all this are visible throughout history. Men had heretofore felt themselves to be individuals of a species, members of the nation to which they belonged, and they had sought in the external order of political society to realize those higher goods of life which the individual could share in only as the joint production of the race. Christianity gave to this characteristic on the one hand cosmopolitan breadth, and on the other hand individual depth. All distinctions of earthly rank and calling disappeared as unimportant in the sight of the one God; the immediate relation to God, which is possible for every faithful soul, gave to each individual a worth of which he could not be deprived, a worth that did not arise primarily from his position in human society, and that was the work not of Nature but of himself. Each man was to his fellow now no longer a mere specimen of the race, whose whole nature was transparent and familiar, but in each individual there was something hidden and sacred that forbade intrusion. It is of course the fact that under favourable social conditions men had always developed varieties of disposition and indeed wholly distinct types of character; it was Christianity that first supplied a deeper reason than this for demanding respect for the individual by rousing a sensitive regard for personal honour, through the ascription of eternal significance to the soul of the individual man.

§ 7. The full joyous assurance of the truth of these doctrines, the subjection in lowly humility of all one's own strength to the grace of God, the consciousness not only of that imperfection which has a meaning in the cosmic order but also of the sinfulness which always is but never ought to be, the confession of the inadequacy of all one's own deserts, and the hope of redemption from all evil through the love of God which no one can deserve but every one can win— all this is characteristic of a temper of mind which has been regarded by many in all times as that which entitles men to call themselves by the name of Christ. The Christian Church has judged otherwise. It has attached the right to this name to a faith which believes not only in Christian doctrine but also in the whole historical account of how this came to be revealed to the world. The Church holds that Christian doctrine alone does not contain the seed of a redemption which through faith can take root and spring up afresh in every soul in every age ; on the contrary, it holds that once and by one act, which belongs not to earthly but to divine history, the work of redemption was accomplished ; and that its benefits are to be obtained, not indeed without the living appropriation of the doctrine, yet also not by this alone, but only by this in conjunction with faith in Christ as the mediator of future generations. The moral doctrines of Christianity have encountered no other hostility than that which wickedness and folly have always opposed to all religion, and the best civilisation of the modern world is built upon these doctrines, whether consciously or unconsciously or against its will. But, on the other hand, the demand that the strengthening and blessing which they give should be earned by faith in the Bible history, has met with a growing opposition which has called down upon the present age the reproach of increasing irreligiousness.

The most essential point, the recognition of providential foresight as an historical fact, is regarded by this civilisation not with aversion but rather as answering a secret need. Only one-sided habituation to observation of Nature could prefer the thought of a cosmic order established once for all, and according to the unchanging conditions of which nothing is possible except a brief and continually repeated cycle of phænomena, to the idea of a cosmic history, at the different stages of which God does not work uniformly but is constantly adding to the world in genuine action, something new, something which was not there before. A simple natural religious temper will be inclined to conceal from itself the difficulties which this idea of a history of the world involves, or to hope for a subsequent solution of them. If it is once admitted that in the changing destinies of mankind there is a temporal succession of things which cannot be regarded as the mere repetition of previous cycles, it is hardly likely that serious opposition will be offered to the demand that the relation of God to the world should be conceived as one which changes as history goes on—that men should believe God to be nearer to the world at some periods of history than at others, and His influence to have been imparted in a manner wholly unique in some periods of which the temporal limits are clearly defined. But readiness to admit this much is not held to be enough; and when orthodoxy demands either the acceptance of the whole content of the Biblical history, or the recognition of those doctrines which the dogmatic theology of the Church has connected with them, a difference begins which it is impossible to adjust.

The sacred writings will always captivate men's minds by their majesty of content and their grand beauty of expression, the simplicity of which is more effective than any conscious art. But that which primarily hinders us from taking them

quite literally is not the incredibility of that which they
report, but the figurative form of their teaching which must
be interpreted in order to be understood.   And then (since
we were bound to the Scriptures only by our reverence for
the doctrine which they teach), in the second place, doubts
arise concerning the history of those wonderful events the
credibility of which cannot be the same to us as it was to the
age from which we have received the account of them.   It
was natural that that age should demand to see the presence
of God confirmed by signs and wonders which yet could not
have as much significance for them as they would for us.
For the thought of an order in Nature connecting natural
phænomena according to universal laws, was alien to antiquity,
which regarded *every* force that works in Nature as being
directly guided by the end at which it aims, and as having
the power to realize that end.   Hence miracles did not lie as
contradictions *outside* the order of Nature, but were actually
the natural exercise of a superior power, which, under
unwonted conditions of time and space, made its appearance
within the sphere where lesser powers were used to work.
In this sense the order of Nature was not independent even
as regarded the heathen gods; each petty deity could violate
that order; even men had at their command enchantments
by which they could alter its course; and for this very
reason miracles could not be received in those times as convincing proofs of the presence and working of the supreme
and the one true God.   It is only to the modern conception
of Nature that a miracle could seem really miraculous, for
this conception recognises no impulse of which the result does
not follow necessarily and according to general laws, from a
pre-existing collocation of conditions.   At the same time,
those who hold this view of Nature are in a position to admit
the general possibility of miracles in as far as the idea
corresponds to a mental need, although they may lack faith
to believe in them as recorded in Scripture.   For to them too
the whole course of Nature becomes intelligible only by
supposing the continual concourse of God, who alone mediates

the action and reaction going on between different parts of the world. It is only as long as this concourse takes place in similar ways that it (being then a constant condition in the course of events) does not appear as a condition of change; and as long as this is so the course of Nature seems to be a self-contained whole, that does not need, nor experience, nor admit, interference from without. But any view which admits a divine life that is not fixed in rigid immutability, will also be able to understand the eternal divine concourse as a variable quantity, the transforming influence of which becomes prominent at particular times, showing that the course of Nature is not independent. And this being the case, the completely conditioning causes of miracles will be found in God and Nature together, and in that eternal action and reaction between them, which is not without governing rules, although perhaps it is not simply ordered according to general laws; it is this idea only, and not the idea of complete fortuitousness and arbitrariness, which the mind frames of a miracle when it would see in it an object of reverence. But the recognition of this general thought does not suffice to lead Natural Science to a recognition of the reality of miracles in the form in which religion generally demands it. So immeasurably preponderant is the weight of all experience in favour of a steady development of all natural occurrences, each step preparing the way for that which succeeds it, that even this general admission prepares the mind to believe only in a noiseless, ceaseless working of God in Nature, not in sudden interruptions of the established order by occasional interferences of divine power. Such a belief could only arise if the ideal significance of miracles in the system of the universe were sufficiently clear and important to cause us to regard them as a turning-point in history, for which the efficient forces of the universe had always been preparing unperceived.

And the wonderful events which glorify the life of Christ in the sacred writings would certainly in themselves give rise to this thought if their physical reality were not made dubious

to us partly by the change in men's conception of Nature
which has occurred since Christ's time, and partly by the way
in which we take the spiritual meaning which the record of
these events is intended to convey.   While the earth was
regarded as a flat disk, and the visible heaven above it as the
abode of God, it was possible for the ascension into heaven to
appear to men's minds as a real return of the Divine to God ;
but since astronomy has taught us that the earth is a sphere
surrounded by immeasurable realms of homogeneous space, we
fail to see what intelligible goal the upward ascent of Christ
could have.   In an age that could hardly distinguish between
the sensuous and the supersensuous, men might regard the
bodily resurrection of the Saviour with reverence as a
guarantee of their own immortality ; but to us this reanimation
of the body is not an object of hope; if it were really to
happen, it would only secure to us the continuance of this
life during the existence of the body which it animates ; what
would really give us comfort would be some proof of a continued life of the spirit after its return to that invisible world
by which the visible world which we inhabit is mysteriously
surrounded.   Rationalism in interpreting these circumstances,
which are described to us as external facts, as visions of those
who describe them, has overlooked the point which can here
give more worth to visions than to actual external facts.
Rationalism supposed that out of mere psychological trains of
ideas, there arose in excited minds fancies due to memory and
subjective conditions, which had nothing objective corresponding to them ; the very thing that it had to take account
of was this spiritual world which though unseen is everywhere, and in which that which has no actual corporeal
existence is present and none the less real.   Between this
world and the world of sense, actions and reactions might
take place which are foreign to the ordinary course of Nature ;
and from these, which are true, real, living impressions upon
the soul of something divine and actually present, those
visions might arise, being apparitions not of the non-existent
but of something really existent, and (as the direct inward

action of the deity) not mediated by help of the course of physical Nature, which has no independent worth, or by disturbances of that course which are incomprehensible to us. The significance of the resurrection lies not in this, that the soul of the risen person now as heretofore inhabits a body which is visible to the eyes of men, but in this, that without any such mediation, his real living presence, and not the mere remembrance of him, takes hold of men's souls, and appears to them in a form which has greater strength and efficacy of influence than the restoration of the actual bodily presence would have.

But to the religious frame of mind from which such attempts at explanation arise, the prosecution of them to any great length is naturally repugnant; it seems impious to make that the subject of theorizing ingenuity which, when received uncritically, never fails to produce a deep impression, but which critical analysis can never bring to certainty in detail. Such awe is not aroused by the dogmas in which in the course of history the content of Christian faith has come to be expressed. The human mind will continually be forced to renew its attempts to grasp and retain in scientific form the truth which it has believingly appropriated, in order that it may maintain this truth against unbelieving civilisation, and that it may satisfy its own cravings after unity and clearness of philosophic view; we see this work of human speculation in Dogmatic Theology, which is respectable on account of the earnestness of its efforts and the connection it establishes between all earthly life on the one hand, and the kingdom of heaven and the divine order of the universe on the other. Yet this dogmatic theology, as being the antiquated ecclesiastical philosophy, is subject to criticism, as is also every fresh attempt at a philosophical explanation of the universe. The content of this dogmatic system has become alienated from modern civilisation (which, owing to its great advances in secular matters, has grown careless of religious interests), and is frequently regarded by it as a fabric built up out of traditions, having no root in reality and no significance for human life; a less hostile consideration of the matter would speedily show that, on the contrary, dogmatic theology is concerned with but few merely subtle inquiries; it deals principally with serious and weighty questions, which our civilisation may indeed seem to get rid of, but to which we are led back by every searching reflection on the destiny of man and his relation to God. But with equal plainness we may say that dogmatic theology has neither succeeded in giving, nor indeed attempted to give, to these questions any answer which cognition can accept as satisfactory; it formulates in its tenets the burning and inextinguishable interest which we take in these great problems, and expresses without satisfying our craving for enlightenment.

It would be a misinterpretation of this avowal of dissatisfaction to consider that its cause is to be found in the demand for an explanation of the possibility and process of realization of something which in itself surpasses the powers of human reason to elucidate, and to require in place of this presumptuous demand the faith which is lacking; for faith where it exists does not find that its own content can be embraced by dogma. Faith does not require explanations, impossible to be given, of how things come about, but it must require the clearest determination of *what* it is which dogmatism presents as the fixed and central truths towards which the vague yearnings of faith itself gravitate. And this is just what is not given; that of which, as the right and true, we are fully conscious in the dim impulse of faith, almost always receives from dogmatism a mere figurative expression, which, instead of immediately determining what we believe, itself requires a fresh exercise of interpretation, the admissible limits of which, again, can only be fixed by that same dim impulse. When Christian theology calls Christ the Son of God, it gives expression no doubt to the most distinctive article of its belief; but it does this in a figure the exact signification of which it can by no means precisely determine; what that phrase expresses and is meant to express is clearer to the believing soul without than with the dogmatic determinations which have been attached to it, for the figure taken simply merely indicates the intimate nature of that relation between God and Christ which is clear to feeling; it contains no explanation as to the mode of that relation, all adequate knowledge of which is impossible for us. Direct religious feeling meets the Church's teaching concerning the redeeming power of Christ's mediatorial death with ready faith, but this faith is not rewarded by any increase of knowledge. For that idea of a sacrifice to which dim emotion first betakes itself, no other idea is substituted which makes the redeeming power more comprehensible without at the same time diminishing the value of the mediatorial death. We all feel that evil has taken hold of us, and that sin, like some inheritance inexplicably entailed, runs through the whole race; but the thoughts which arise from this consciousness, and have not been worked out to any clear conclusion, cannot be led to such a conclusion by way of dogmatism; ideas which go so far astray as belief in the complete solidarity of mankind, and in the actual inheritance by the whole race (as by legal representatives) of the sin of our first ancestor, cannot by reason of their own obscurity afford any illumination to our minds; they merely give an incisive statement of the problem at which we unsuccessfully labour.

Besides those harmonious and early-developed teachings which the Church adopted as part of its confession of faith, men's speculative impulse has driven them to make innumerable attempts to find an explanation of the world which should be in agreement with Christian doctrine, but the greater divergence of these explanations from accepted teaching has prevented their being similarly accepted. The Protestant theology of our own time is more active in this direction than it has been for a long period previously, believing on the one hand that it possesses in the results of modern philosophy new and previously unknown levers of religious truth, and on the other hand being animated by a courage of conviction for the assurance of which I do not know the grounds. The self-imposed limitation which led philosophy at the end of the last century to give up all claim to a knowledge of the super-sensuous, caused the prominence of a rationalistic system of ethics which, since it lacked any views concerning the place of the moral world in the plan of the whole, came at last to be without any religious colouring whatever.   But our highest wisdom cannot consist in following general rules of duty without caring in the least what benefit may or may not ultimately result from their fulfilment; we need to be con-vinced of some intelligible cosmic connection in which we can trace the destiny of human life and the eternal significance of all moral effort.   The suppression of this impulse to a cosmo-logical development of philosophic views has by a natural process of reaction been followed by its reappearance in a prominent form, and it has now, as it seems to me, far exceeded the limits within which it could hope for success and for salutary influence on Christian life.

For not only do we doubt whether the methods of modern philosophy can make possible that which has always hitherto been impossible, but we also lament that dogmatic investiga-tions seldom make a conscientious use of even the modest results which this philosophy has perhaps obtained.   Christi-anity does not furnish any immediate revelation concerning the structure of the world; the essence of its ethical teaching and scriptural sayings which only incidentally involve cosmo-logical notions, are the sole materials which Christians can use for making a construction of the universe.   But from moral Ideas the most careful investigation can never develop anything more than the universal conditions to which the cosmic construction must conform in order to avoid coming into collision with the Supreme Principle of Good; and only a very undisciplined fancy will imagine that it can learn from this source those definite concrete forms of the cosmic order by which the conditions indicated are satisfied; we cannot even use these Ideas to carry on the world of experi-ence, which lies before us, beyond what is actually given, or to find with any certainty that continuation and completion of it which is hidden from our observation. Therefore such attempts run great risk of ceasing to ask what *must* be, or even what *may* be, and of asking instead what it is that would be most delightful *supposing* that it were the actual condition of things ; and this matter is decided by the prejudices of individual character, which are insusceptible of discipline. Yet the inclination which we here blame is supported by a philosophy which expressly regards the meaning of things, their Ideas, as being (and that without any limitation) their active essence ; and which, in seeking out and determining these Ideas, requires no strict and formal proof, but regards poetic justice in the coherent development of thought as a sufficient warranty of truth. This being the case, the dogmatic investigation of our own time has, with great expenditure of philosophic profundity, and with little method and much self-satisfaction, plunged into inquiries which the modern spirit of general culture refuses to enter upon at all, not only from a consciousness of its probable ill-success, but also from fear lest, by presumptuously insisting upon trying to know all things, it should intrude upon those divine secrets which it respects. And the divergent results of these attempts do not promise unanimity of knowledge on questions concerning which believing minds have been always at one ; they only give to modern dogmatic theology as a whole a character of anarchy tempered by sterility.

For no gain accrues to life from all these attempts either to set forth in detail by uncertain interpretation of uncertain texts the whole story of Creation—and that after a fashion which is in conflict with the results of scientific investigation of Nature —or to make out what will be the end of the world and the exact nature of man's future life, without taking into consideration our progressive knowledge of the physical world, which (though it can indeed never solve such problems) may furnish our thoughts concerning them with a background that sets limits to too great extravagances. And finally, we blame, as being both unfruitful and little in conformity with the spirit of Christianity, a predilection for speculations concerning the divine Trinity in Unity, in which many declare, to the profound astonishment of their hearers, that they have found the key to all knowledge, sacred and pro- fane—though they have not hitherto done anything to make men hope for the fulfilment of their promises.   In the living Christ, faithful souls beheld, not indeed God, for Christ Him- self said, *The Father is greater than I*, but the Son of God who is one with God in a way that we do not understand, and who came into the world, not because His coming had from the beginning been the necessary consequence of some natural law of cosmic order, but because the love of God, which is greater than all the mechanism of necessary develop- ment, though it need not have sent Him, yet did send Him to the world.   To this dualism of the Divine Personality faith might also add, as an object of veneration, that Holy Ghost, the Comforter, whom Christ promised to send; but neither had this Spirit appeared in the course of history in personal form, nor was there any need to understand it as other than some divine activity.   Dogmatic theology, with but a weak foundation in passages of Holy Scripture which indicate the dawn of speculation in Christian thought, has endeavoured to develop from such material a Metaphysics of the divine nature which the further it advances gets further away from that to which simple faith would cling as the blessing of Christianity.

And yet it is a natural need which leads men to make these attempts.   It seemed that the divine revelation was not estimated at its full value if it were regarded as an historically incalculable addition to an intrinsically independent cosmic order (the content of the revelation being indeed at first taken hold of by men's minds for its own worth, without any inquiry as to the process by which this content was made known)— it seemed as though this revelation must be inwoven both in the past and in the future with the whole economy of the universe, so that there might be nothing in that economy which was (as to either the nature or reality of its existence) independent of the revelation.   Thus it was that the image of the historical Christ grew into the thought of a power that worked in God before the world began; the same purpose of the love of God which was made manifest in the historical act of redemption, came to be regarded as having been from the beginning that regulative will through which things are what they are. Now this spiritual need of finding unity in the nature and acts of God could be satisfied by the belief that that which moved God to redeem the world should be conceived as a thought which had been from everlasting, and had not been called into existence by any temporal occasion; it was not necessary that the unity thus reached should be endangered by the impracticable demand to make two persons into one; still less was there in the content of faith itself any cogent reason for a similar personification of the Holy Ghost. On the other hand, as we shall see later, the secular speculations of philosophy lead to a trinity in the beginnings of the cosmos—that is, to laws *according to* which things are, to powers *by* which they are, and to ends *for the sake of* which they are what they are. The recognition of this trinity is no triumph of philosophy, for it is in reality a confession of human incapacity to identify as one in cognition that which according to the demand of cognition itself must necessarily be one; and for the rest, however those three may be conceived, they can never be anything other than forms of divine activity which are incapable of being derived one from another. This trinity—a fateful gift—has been offered by philosophy to theology, and has been accepted, although its several members correspond neither with the historical Christ and the promised Holy Ghost, nor with the three persons of the divine Trinity in Unity as confessed by the Church. Now it may be that theology in the narrowest sense—the dogmatic determination of our notions concerning the nature of God—cannot be made complete without reference to that philosophic trinity of essentially different principles; but all the assistance that philosophy can give will never apply to more than the first article of our confession of faith: Christology gains nothing by it in a scientific point of view, and loses as regards its significance for living faith. For what faithful souls cling to is the living Christ, the complete personality of the Saviour, not taken figuratively or in any symbolic sense; if this personality is interpreted as some necessary phase of the Divine Nature, as some secondary potentiality of the concept of Divinity, as an antithesis within the Deity, as a world-creating λόγος, our faith is only disturbed.   For we do not see why we should separate from God energies which we are accustomed to regard as among His attributes, and we cannot discover that any metaphysical glory of Christ as a superlatively supernatural God of Nature, is greater than the moral majesty of the Redeemer.   It seems to us that such speculations transfer us from the place in which Christianity has set us, from faith in the sole and final reality of what is good and holy, back to the old heathen cosmology which regarded God as manifesting Himself, not in unfathomable deeds of love, but in those emanations of His being which take place according to natural laws.   It *seems* to us so; for we do not in the least wish to conceal from ourselves, nor to withhold here our acknowledgment, that the attempts which have been made in this direction have been determined by the need which men feel of making the world and all things in it subordinate to the ethical plan of salvation; neither the Christian temper in which these attempts have been undertaken nor the earnestness with which they have been carried out seem to us to admit of doubt; all we affirm is that the impression produced on many minds by the results at which they have arrived is the very opposite of that at which they aim.   But we pass over with silent contempt those essays which simply trifle with the notion of trinity in unity, after the fashion of that numerical mysticism indulged in by the Pythagoreans; and which almost seem as though they set great value on the Trinity merely because of its involving the number three.   It would be just as reasonable to include in our confession of faith veneration of the prime numbers, or of square roots.

§ 8. We have said that these speculations were for the

most part unfruitful ; that we are able to confine ourselves to this reproach is due to the opposition which secular civilisation has for so long offered to the power of the Church. The vagaries of millenarian dreamers have now come to an end ; if they were still in fashion, other and more important consequences would be entailed by the rococo of belief in a devil and other similar doctrines to which the dogmatic renascence of our time is inclined to return ; the Humanism which has had a salutary and pervading influence on theology, as science has revived and a sense of practical justice has received increased development, will, we hope, in the future prevent speculative errors from being carried out in practice. But this greater security of personal faith is connected with increasing insecurity of the ecclesiastical edifice, the pulling down or re-establishing of which is at present a subject of dispute.

The fact that their gods were chiefly important because of their significance in Nature, prevented the heathen world from regarding the whole life of man as a continual service of worship towards the divine splendour of these deities ; the plurality of divinities, to particular individuals among whom particular tribes attached themselves, made difficult the combination of all mankind or even of one nation in that close communion which unites the members of a community drawn together by common spiritual interests ; where, in consequence of the greater unity of mythological teaching, this hindrance did not exist, still religious communion did not exist independently beside political communion, but men's confession of faith was itself national ; nothing was required beyond civic virtue and ceremonial acts, and the national religion had no power to bring individuals into communion as the subjects of a higher and spiritual kingdom ; in India, where more than in any other part of the heathen world, religious feeling had entered most deeply into all mental needs and distresses, the despair of life at which men arrived was no bond of any community of life. The Church is an institution peculiar to Christianity. Disregarding distinctions of nationality, sex, rank, and education, it aims at uniting all mankind in a service towards God which consists in the subjection of one's whole life to Him.

The Church began as a free community, without any other bond of union than love and a common faith; like every growing society, it developed forms of administration and of internal intercourse that were binding on its members, but it did not claim any authority over the rest of mankind, although even then it felt itself raised above all temporal combinations of men by the consciousness of being a union entered into for purposes of eternal import.   When the Roman persecution of Christianity had given place to recognition, there grew up in the Church the consciousness of being an institution to the ordinances of which secular national life was bound to conform, and departure from which was no longer regarded as a step which men might take of their own free choice but as an act of desertion to be judicially punished.   With a still bolder flight it finally rose from the position of an earthly institution to the importance of a cosmic power which not only has given to it on earth all supremacy over the consciences of men, over the authority of magistrates, and over the lands of the heathen, but which is able also, through those means of grace which it alone administers and distributes, to reach beyond this life, and not only teaches men how to find or avoid the paths to salvation and to damnation, but actually opens or shuts the entrance to these.   Thus the Church became the grandest and most noteworthy constituent of that great department of cosmic order which the human mind has added to the existing order of physical Nature.   Even the constitution of States depends upon objects of the physical world, upon the land and its boundaries, the produce of the soil and men's right to it, and the distribution of the wealth which is produced, and nowhere do its pretensions to power extend beyond the earth itself; the Church alone binds the spirits of men and fills the whole of life with a pervading consciousness of its connection with the other world.   Hence it is easy to comprehend the admiration which the dazzling impression produced by this mighty phænomenon ever calls forth afresh in receptive minds, and the longing which men feel to be received into the steady shelter of its mighty order, and thus escape the fragmentariness of a life which pursues its ends with vacillating purpose.

But the more completely the plan of any organization corresponds to an ideal, the more injurious is the effect which this organization has if it is forced upon any life as a form that must be complied with, when that life is not adapted to realize it voluntarily. The most fatal error of human efforts consists in prematurely attempting to realize *ectypes* of perfection in cases where what ought to be considered is the organization of *means* for approaching *in practice* as near to perfection as circumstances will allow. Such an error was involved in the constitution of the Church; it sought to reach in this life a condition which is only possible in another life, and suppressed the free activity of powers which cannot reach this goal here below though they may prepare the way for it. It believed that it possessed complete truth, and endeavoured to hinder any search for truth; and believing itself to be in enjoyment of this possession, it undertook cares which belong only to providence itself; it interfered, commanding and forbidding, with the general secular concerns of mankind and the consciences of individuals, as though it had been the immediate plenipotentiary of God and the guardian of those laws according to which eternal wisdom chose to regulate mundane affairs; it assumed a right of punishing and persecuting all who resisted any part of the extensive ramifications of its doctrine and its regulations, and all this universal dominion which it arrogated in the name of the Holy Ghost, it could only carry on by means of human personages whose incurable frailties were in innumerable particulars in contradiction with the sacredness of their office. It is the spirit of orientalism which culminates in this colossal attempt not only to teach but also to found and establish a cosmic order, and to assign to human life, with all its multifarious interests, a place in that order. But

as it was the West and not the East which reached this
highest summit of religious cosmology, so from the time when
it was attained all the powers of Western civilisation have
been actively engaged in an unceasing struggle against this
vision of an earthly anticipation of divine order, which at a
distance promises happiness but disappoints those who have
drawn near.

The Protestant mode of thought has given up the cosmical
significance of the Church; according to it, the visible Church
at least is once again regarded as a mundane institution of
which the business is to minister to the religious life of man.
But this being so, the course of events has brought the
Church into a connection with the State which abounds in
anomalies that are difficult to remove, and that have caused
her members to withdraw their sympathy from her in increasing measure.   The Roman Catholic Church, having one
supreme head, an established doctrine, and extremely homogeneous forms of worship, is spread abroad among the nations,
and may be regarded by those who belong to it as a great
objective and independent organization.   If Protestant Christianity had been able to maintain a similar unity of doctrine,
of worship, and of Church government, the various national
Churches into which it has split up would be less prejudicial
to the vigour of religious feeling; they would appear as the
locally diverse secular organizations which guard sanctities
that are everywhere equally hallowed.   And in fact this is
the part which the secular power professes to assume in
religious matters; but the unity to which we have referred
never existed in any completeness.   Hence although the times
will not return in which governments could forbid their
subjects to make profession of any religion or to change their
religion, yet there is still much room in the interpretation of
the established faith for the exercise of political power, and of
the favour capriciously shown to divergent points of view
between which Protestant freedom permits a choice.   The
absence of uniform doctrine; men's feeling that its place and
name are taken by the subjective convictions of individual

ecclesiastics ; a perception that the character of these convictions changes considerably within brief periods; the not always
just yet still not always unjust suspicion that these changes
are to some extent influenced by the pressure of political
motives—all these circumstances cause the Church to be
regarded as a political institution, the pressure of which
arouses aversion, because it intrudes into a region in which
obedience to it ought not to come into conflict with men's
spiritual convictions.

We cannot prophesy what the future will be, we can only
prepare for it.   It is not to be hoped nor is it to be wished
that the Protestant freedom of religious conviction and investigation should be suppressed or voluntarily surrendered ; it is
to be hoped and wished that dogmatic theology, becoming less
confident in its assurance of knowledge, should diminish the
number of arbitrary interpretations of things which do not
admit of interpretation ; and should by greater unanimity in
matters that are essential, and by abandoning useless disputes,
strengthen in the members of its communion a sense of trust
in Christian faith ; it is to be hoped and wished that thoughtful sensitiveness of conscience in treating all the concerns of
life (that most wholesome fruit which living Christianity has
produced in many souls) should be recognised as greater than
the temper of mind which, turning away from all that is best
and fairest in modern secular civilisation, affects matters that
are inscrutable and useless, and archaisms which offend taste
without strengthening faith ; finally, it is to be hoped and
wished that a greater share in the management of Church
matters should be given to the laity, and that thus they should
regain that interest in these matters which they have lost
through being so much excluded from them.   But though it
is certain that among the things most to be desired in the
future we must reckon the continued existence of the Church
as an objective reality in which the religious life of the
individual issues, finding therein both a guarantee that its
efforts are well directed, and spiritual comfort and edification ;
yet still if those changes which we have indicated as desirable

are not carried out, we should hold that the renewed attempt to maintain the external integrity of the Church, while it lacked the internal conditions of truth, would be less salutary than its ruin——a ruin which our opponents point out to us as an inevitable consequence of the Protestant principle. It is certain that the time immediately succeeding such a catastrophe would be neither desirable nor agreeable; but we may confidently hope that not only would living religion grow when relieved from conflict with unsuitable external ordinances, but that also the ineradicable need which men feel of not standing alone in religion and of having their faith recognised, would lead to the voluntary establishment of great ecclesiastical communities that would be free from impracticable claims to authority over men not belonging to them.

CHAPTER V.

POLITICAL LIFE, AND SOCIETY.

The Family, and Tribal States—The Kingdoms of the East—Paternal Despotism
—The Political Constructions of the Greeks—Civic Life and Law in Rome
—Political Life and Society in the Middle Ages—The Autonomy of Society
—National and Historical Law—Practicable and Impracticable Postulates :
Duty of Society as regards its Members ; State and Society ; Constitutional
Government ; Socialism ; International Relations.

§ 1. THE Family, as being most directly founded upon
natural relations, has always been regarded as the
indispensable basis of Society, and often as the root from
which this has grown ; and its constitution has always furnished
the model to be imitated by all social order. Unless ennobled
by the civilising influences of a life rich in manifold interests,
natural family relations in themselves and exclusive regard
for them, have not produced either " the white flower of a
blameless life," or social arrangements conducive to progress,
or just towards the just claims of individual human beings.
And this is not surprising ; for Nature does indeed lead us to
form connections which, understood in a right sense and used
in a right spirit, afford abundant occasion for the development
of moral beauty, but we cannot have the right understanding
or the right spirit except as the result of many-sided reflection to which we are forced by the multiplicity of the tasks
and conflicts of life.

The world, with all those complicated relations of existence
produced by the historical course of human civilisation, is now
spread before us as an immeasurable field in which there lie
concealed a thousand sources of happiness and of evil ; to go
out together into this dim distance (into which our anticipatory dreams have long ago ventured) purposing to share each
other's joy and sorrow, and with the hope that agreement in
estimating that which the future may bring will strengthen

mutual fidelity—such a resolution (when such it is that leads
to the establishment of family relationships) does undoubtedly
ennoble the natural impulse from which it springs.   On
the other hand, the poorer life is, and the more monotonous
men's anticipation of the future, the less worthy will family
happiness be, and the less removed from that which Nature
affords even to the beasts ; and the more plainly will there
appear those immoral results, of which (in barbarous minds)
natural relations are actually the occasion.   For the superiority
of the man's strength over the woman's need of help, and of
the fully developed vigour of adults over the tenderness of
childhood, are indications of Nature which have been always
understood and followed by the barbarous men of uncivilised
times.   And the less the security of life and the activity of
trade, the more does the woman, who is dependent and obliged
to seek the protection of the man, have to do for the support
of the family, and so there arises polygamy, not as the result
of a direct indication of Nature, but as a proximate consequence of natural relations ; and polygamy entails a general
degradation of women, degrees of importance among the wives
of one man, and differences in the hereditary rights which
descend to their children.   The relation between parents and
children is in the same way deformed by this incapacity of
ennobling natural bonds.   That profound secret of cosmic
order by which each generation of men springs from that
which preceded it, and by which parents are endowed with
the wonderful power of bringing into the world immortal
souls like themselves, appears to the untutored mind to be
nothing but a most commonplace example of causation, and it
seems to it that all the power which a maker has over the
work of his own hands belongs as a matter of course to parents
—or rather to the father, since maternal rights were very
early ignored.   This paternal power had as regarded the child
a right of life and death just as unconditional as is the right
of a possessor to dispose of his lifeless chattels ; it knew no
distinction between immature youth and the dawning of
manly independence ; it was without respect for the ripening

individuality of human souls, and made no attempt to renew the bonds of relationship in a spiritual sense, by learning to enter into fresh views of life, but was ever harking back upon one past fact—the fact of physical generation. This paternal power was the direct result of straining to the utmost limit those natural relations upon which the family is founded ; we trace it clearly in the beginnings of every civilisation, and see that it disappears from practice in proportion as the growing complexity of human relations leads to a more refined estimation of the rights of individual men.

Even apart from such crude misinterpretations family life does not teach social morality. Special and unique relations bind the members of a family together by feelings which do not flow from general duties of men towards their fellows ; these feelings do indeed incidentally enrich life with a passionate intensity of affection, which is no doubt an element of the best human happiness, but so far from illuminating men's consciousness of general moral duties, they only obscure it. Through forgiving lenity and precautionary discipline they hinder justice ; in the education of children they often abridge freedom which should be permitted, and permit them much to which they have no claim ; even where their demands and permissions agree with the general commands of morality, there is in the mixture of piety and love which prompts them a combination alien to the obligatory power of moral laws. For what we do from piety, that is from a devout feeling, which is not clearly conscious of the grounds and limits of reciprocal duties, seems to us (being indeed, as it is, only the result of our temper of mind) as the mere efflux of our own devout individual character; and even where all vanity of self-exaltation is absent, it appears to be something which is by no means necessarily present in the world, and in fact would not be so if it were not for our good disposition ; we by no means think of it as something which others have a right to receive from us, which right would be eternally valid even though no one should regard it. Every one acknowledges the advantage of this foundation of piety in the domestic life of families, but public morality is not based upon it. A man only comprehends what he owes to his fellows when he comes into contact with those who are nothing to him ; it is only when all the claims to consideration, friendship, love, and reverence founded upon those special natural conditions have fallen away, that general duties and their necessary general motives become clear. Hence as long as the social conditions of a growing nation are regulated after the pattern of family relations, we do indeed find many beautiful and poetic traits of character, but scarcely any advance towards justice—rather, on the contrary, many traces of its opposite. For instance, it is quite common to find in early civilisation, even among people of otherwise mild temperament, extreme harshness in the punishment of crime ; without weighing the degrees of heinousness in different offences, and still more without taking into consideration those extenuating circumstances which lessen guilt in particular cases, the piety of national morality, when once wounded, proceeds with indiscriminating pitilessness. This is quite natural to a temper which is accustomed to be guided in its demands not by recognised rights of others, but only by its own general feeling, and which therefore when it is offended is conscious of nothing but the offence to itself, and in unconditionally repulsing the insult is not moderated by any consideration of different circumstances.

When a numerous people arises from the multiplication of families, the feeling of being bound together by ties of kindred disappears, and is replaced by the feeling of a community founded on similarity of language, custom, and thought. The more self - centred and exclusive any such people, starting from a basis of very special conditions, can make its life, the further will its condition be from corresponding to the ideal of human society. To æsthetic feeling it may seem that in comparison with the vacillating half-heartedness so abundantly produced by every complex civilisation, that unwavering stability of national character is much to be preferred which is easily and homogeneously developed in all individuals when the whole circumstances of their life are fixed and never subjected to doubt; but this advantage is not in itself to be reckoned higher than the beauty which belongs to some species of animals, and likewise always reappears under certain conditions. There is in it no germ of progress; its morality, which has only grown up through custom, has not the flexibility which can only be given by general principles; it presses upon individuals with the force of rigid prejudice, and condemns all those individual impulses running counter to the narrow-mindedness of tradition, which now and then arise from the inextinguishable diversities of human nature. Hence all such thoroughly national civilisations of past times are characterized by unintelligent intolerance, and this only disappears when, having been forced into contact with the morality of other nations, men's illusion as to the universal validity of their own maxims is destroyed, and they are constrained to learn in their most comprehensive form those universal moral obligations without the recognition of which no human society can subsist.

It was by nomad tribes, whose unity depends predominantly upon the remembrance of their past history, that the bond of consanguinity was held in highest esteem; and in the early ages it often happened that when they changed their nomad habits for a stationary life, this fact had a great influence upon the political arrangements which sprang from their connection with the soil which was to be henceforth the permanent object of their activity. The different tribes distributed conquered and unowned land among fathers of families and heads of houses, and sought by many ingenuities of legislation to make this distribution permanent; in so doing they gave to the constitution of the State a distinct genealogical stamp, assimilating it even in reference to its physical basis to the internal order of a family sprung from one ancestor. When they made such attempts they had little knowledge of the real tasks of life, and did not foresee that the new connections with territorial possessions, into which they were entering, were at variance with the sentiments and plans by which they were still swayed.   While the Hebrews were yet wandering shepherds, they regarded the preservation of their race as the most sacred duty, and believed that their God had promised to them, the chosen people, the multiplication of their seed as His primary blessing.   In fact, if the historic life of wandering tribes were not carried on in the ever - renewed traditions of never - failing generations, such tribes would leave behind no signs of their existence and activity, for they produce nothing that is physically durable ; they would vanish from the earth and from reality altogether, leaving no trace behind, and would be as though they had never been.   The Greeks as well as the Hebrews were not without a longing to live for ever in their descendants ; but history did not afford them any such uninterrupted retrospective view of their ancestors.   And yet when the Dorians founded the kingdom of Sparta, they seemed just as eagerly anxious to establish by artificial regulation of property an immutable complement of families of equal fortune, by which the Spartan nation should be represented through infinite future ages.   Both the Greek and the Semitic races sought to strengthen their national fabric by hindering free self-determination in various ways, and to secure the continued existence of every family even by the help of legal fictions ; and thus both greatly retarded their own social development, and their political constructions were eventually swept away by the natural current of events.   For stationary life brings men into such manifold contact with the nature of things, and awakens in them such strong ideas of the rights which accrue to them from the activity which they expend upon objects, that any family morality which does not recognise the independence founded upon such personal rights is sure to be at last broken through ; and Nature itself forbids that the number of families should always continue the same ; some families multiplying greatly while others become extinct, laws which aim at the maintenance of family interests are likely to promote the advent of intolerable extremes of wealth and poverty.

A strong feeling of unity animated the tribes whose members were bound together by ties of blood; this feeling ceased to be possible when need and the spirit of adventure had caused nations to attack one another, and through the subjection of many by one, had formed communities which indeed hardly deserve to be called communities, and still less states, but were simply kingdoms. For it was only the authority of government and not a desire on the part of individuals for such association that held together these political conglomerates which were produced in greatest number, though not exclusively, by the East. That a victorious tribe should regard the vanquished as destitute of rights and should arbitrarily dispose of their lives, is a thing that the general characteristics of human nature make easily intelligible; and from a consideration of actual circumstances and of the better aspects of that nature, we can also understand how it was that what befell the conquered was not unmitigated slavery, but that the details of their life were left to be determined by their own codes of morality, absolute submission being required in only a few particulars. But that within the dominant tribe there could be developed the authority of one individual ruler is a fact which can be explained only by the co-operation of many conditions. In time of peace patriarchal authority might be established in a tribe, and the leader of successful expeditions might win ardent attachment; the exaltation of the authority thus obtained to sovereign majesty seems to be made permanently possible only by the transition to stationary life, and to be facilitated by the subjection of alien communities. For however slight political insight may be in other respects, the claim of sovereignty over wide territories, the inhabitants of which differ in their mode of life, must teach that some system of order and administration is necessary; care for the general security recommends that in the government of conquered races there should be no divided mind; and finally, the greater complexity of conditions makes it

possible for the ruler to withdraw from daily intercourse into
exalted unapproachableness.   This last circumstance seems
always to have been serviceable to oriental governments in
establishing and exalting men's reverence for their rulers, and
impressing upon the minds both of the dominant race and also
(and with less difficulty) of the conquered that this sovereignty
of one over all was an irrevocable decree of Nature.   Obedience of the multitude towards a power to which it feels
bound by ties neither of morality nor of affection, and which
at the same time would be incapable of opposing any adequate
physical resistance to the united will of all its individual
subjects, rests chiefly upon the uncertainty which each individual feels as to the sentiments and interests of the rest.
There have been but few governments which could have
outlasted the moment (if it had ever come) of a general
revelation of the secrets of all hearts ; men would have seen
how little the law corresponded with the real will of all ; and
with such a discovery there would have been a general revolt
of will against it.   But such knowledge of a possibly existent
unanimity could only be in very simple conditions of society,
where the circumstances of all are thoroughly homogeneous,
or where there are great facilities for the exchange of thought,
and a highly developed public opinion.   Yet in comparatively
recent times the leaders of nomad nations have been able to
put the world in dread, thanks to the enthusiastic and unconditional obedience of their followers ; the will of the
leaders being nothing more than the concentrated and unified
expression of desires which they both found pre-existing and
also helped to intensify in their uncivilised and hardy tribes ;
but nations at this stage of civilisation universally reject
despotism in times of peace.   Where such unanimity has
become a thing of the past, and community of public opinion
has not yet arisen, the ambition of rulers derives its strength
from the paralysing uncertainty of each man concerning the
views of others.   For submission to an express law addressed
to all, must ever promise most security to him who does not
know (because they do not manifest themselves) those counter-

forces in the society which are able and willing to offer
resistance—who can never know what interests beyond his
own intellectual horizon may alter the sentiments of men
whom otherwise he would naturally conclude to be like-
minded with himself—and who finally, if he knew all this,
would still not be able to call into combined action at the
right moment those forces which he knows to exist. In this
lies the great superiority which any established order, whatever it may be, generally has over all attempts at innovation
—the certain evil, to which men have learnt to accommodate
themselves, is preferred to the uncertain, of which they
cannot see all the bearings.

The sentiments cherished towards Asiatic despots by their
subjects could hardly have been other than these. They
were reinforced only by the strength of habit, which confirmed
patience in the one case and confidence in the other, causing
him who was ruled to regard being ruled as a fate which
could not be even thought away, and him who ruled to
consider that he had a natural right to rule. The material
with which this framework of society was filled in, differed
according to the temperament of nations and of their gover-
nors. In the East the giddy height to which the position of
ruler had been raised brought to powerful minds little more
than dreams of universal sovereignty which did not lead to
the purposive accomplishment of any social organization, but
yet, with unconscious historic efficacy, enlarged the intellectual
horizon of the nations and the bonds between them, and
aroused in men a general idea of vast and comprehensive
order. And since these dreams could only be carried out by
means of the strength of subjects, the resources of subjects
had to be spared, and protected by a regular administration;
such administration could be carried out in detail only by
the conquered nations themselves, since they alone were
acquainted with their own circumstances, and for this reason
despots left national institutions uninterfered with for the
most part, only reserving to themselves the power of disposing
absolutely of the resources produced by means of these.

Hence the fall of kingdoms and the transference of dominion to other tribes altered but little the general features of society; it was only organized within limited circles, not being to any extent systematized as a whole.

The ancient political communities of China and of the American Indians deserve the name of states much more than these Asiatic kingdoms; in them, in place of empty arbitrariness, we find the thought of an ordered administration of human affairs which the ruler is empowered to carry out. Asiatic despotism left the life of the people to its own luck; it ruled indeed, but did not govern; but China, Mexico, and Peru lacked neither an administration regulated in detail, nor generally received laws and traditions which sought to bring the tenor of individual life into harmony with the well-being of the whole; the rights and duties of subjects and morality and education were determined with provident wisdom and sometimes with much refinement of feeling, and connected with rules founded at the same time upon natural equity and judicious policy. Peru especially had in many respects realized the Platonic ideal of a state, though presenting that interesting superabundance of characteristic practical arrangements which always distinguishes social institutions that have resulted from actual circumstances as compared with logical deductions from general principles. Yet none of these states were promotive of progress for long; China has retained its isolation up to the present time, Mexico was on the verge of dissolution when destruction fell upon it from without; Peru, notwithstanding the devotion of its people to their native government, could not long withstand the pressure of the Europeans. For all these states were founded, not upon any basis of justice, but on well-meaning administration and consecrated tradition. They had laws, and these were not merely arbitrary ordinances; but a sense of equity, attainment of definite ends, and traditional usage, were the sole grounds from which they proceeded, for they were based on no recognition of universal principles of right. They had an ideal of social life, which they regarded as the concern of the state, and sought to realize by complex organization and strict centralization; but for them society did not rest on individual personal rights, which always demand recognition even where their exercise has to be renounced for the sake of the general welfare; they rather set their political ideal before themselves as an immutable goal, and deduced from it all individual rights and the comparative cogency of every claim. Hence when there came a dissolution of this form of political constitution, in which (as is commonly said of organisms) the whole was actually prior to the parts, the parts had no vital strength of their own, which could enable them to attempt new political constructions. Any structure that arises from the inherent powers of its constituent parts is, by the ever-active reciprocity of these, renewed under some fresh form, whenever the old form disappears; that more organic construction of society, in which every detail has reference to the one informing Idea of the whole, may have a more imposing appearance as long as it lasts, but if its integrity is once broken up, it falls into a condition of corruption incapable of producing fresh life. The European nations, who had a strong consciousness of personal rights, due partly to their own natural character, and partly to Roman influence, have been able to escape without political dissolution from conditions of great social confusion; for the Peruvian, the possibility of social life depended upon the existence of his Incas, and upon the continuance of a thousand historically transmitted institutions; accustomed to a definite form of the *whole*, he knew nothing of that power of the *universal* which makes the formation of new wholes always possible. Under the dominion of their well-meaning princes, who were prudent in policy and not unskilled in economics, these Indians may have felt very much happier than they would probably have been under the dominion of the philosophic expert whom Plato would have called to the throne; but their edifice of protective despotism, when called upon to resist unforeseen disturbances, did not stand the test.

§ 2. A settlement had been formed on the banks of **the**

Eurotas by a warlike nomad tribe called Dorians. It was natural that this community of foreigners, surrounded on all sides by enemies, should retain those habits of constant readiness for combat, strict fidelity to one another, and stern discipline, which the obligation of self-preservation had taught to them during their wanderings, and which besides were ancient habits of their race. Hence is explicable a great part of the political constitution of Sparta—both of what it commanded and of what it forbade; it established as the permanent order of the commonwealth institutions which had been adapted to the temporary needs of the infant state, and the position in which it at first found itself.

In modern times the State is not expected to teach society what are the important aims of life, and to make regulations by which individuals may be guided to the attainment of such ends; it is sufficient and seems most desirable that public institutions should do no more than protect all free and lawful personal activity, affording merely the possibility of general human culture, which every individual may use in the particular way that suits his own talents. By the Greek mind generally, the unceasing discipline and guidance of individual life was regarded as both the right and the duty of the state; it was carried out in Sparta in such a way that all individual powers were forced to exhaust themselves in the work of keeping up the whole, efforts for private ends being neither justified nor encouraged. What was demanded was not the blind obedience of slaves but the conscious self-devotion of citizens to the common weal, the laws and traditions of which were impressed upon all by a careful course of education; but the individual had no freedom either with reference to this genius of the state, or even in other respects; every exercise of human powers which was left to unfettered self-determination was held to threaten the security of the whole. There was no choice of callings, the possible differences of which all disappeared before the one task which the state set itself, that of ensuring constant readiness for war; the behaviour of individuals, family relations, and in addition to these the social enjoyments of life, were even in unimportant details subjected to state regulation.

Yet it would be an entire mistake to suppose that on this account Spartan life was destitute of all the mental wealth and all the happiness which can rejoice the human soul. That stern discipline itself produced so many and such admirable virtues of manliness, constancy, moderation, discretion, and fidelity, that the very consciousness of this strong and splendid development was in itself a source of exalted pleasure, as it became for contemporaries and posterity an object of genuine admiration. Yet a question arises as to the independent and intrinsically worthy good, which this state (since it took away from individuals the liberty of choosing their aim in life) seemed the more bound to set before all as that which every one should strive after. For all those virtues which we have enumerated are yet but formal excellences, preparatory discipline of efficient powers, which strain towards some ideal in the service of which they may receive the consecration of humanity; they do not in themselves set man much higher than many favoured races of animals which walk the earth in native beauty and with all the grace of consummate strength. The Spartan state lacked the content of mental life to which we refer. It was not animated by any unbounded impulse towards the enlargement of knowledge; on the contrary, it regarded such impulse with suspicion; for the innumerable small interests with which cultivated minds often amuse and occupy themselves, generally winning by the way some fragment of eternal good fraught with delight, the Dorian mind felt no sympathy or indulgence—felt nothing but the contemptuous superiority of which a hardy nature is conscious towards those which are more finely organized; it even seemed as though the moral perfections which it inculcated were required less as a result of devotion to that which was in itself fair and noble than as formal conditions the fulfilment of which were a guarantee to gods and men of the safety of the commonwealth; at any rate Sparta seemed to regard intellectual and artistic culture with suspicion, and to refuse them room for further development as soon as the stage was reached which from this point of view was desirable.

This strange round of political life—of universality that tolerates no divergencies, of a whole the parts of which have no task but to constitute that whole—is very clearly expressed by Plato when in describing that ideal State of his which reminds one of the Dorian reality, he makes the candid remark, " We are concerned here not with any wellbeing of the parts, but with securing to the whole, to the State as such, the greatest possible power of self - preservation."   Both Sparta and Plato leave us asking the question, " What good is it for any such State to exist in the world at all, and what interest can one take in a machine which expends all its strength in self-conservation and turns out no useful product ? "

We owe it to history that we need not leave the first question so entirely without an answer as the second.   We can easily conceive that a tribe of Indians might have a Spartan form of government, many of the Spartan institutions, and much of the Spartan virtue, and yet that with all this if it lived surrounded by allied tribes, it might not far surpass the average civilisation of the race.   But the Spartans were Greeks and lived in Greece.   Their constitution did not favour mental progress, but the more it came into contact with the advanced development of the rest of Greece, the less did it suppress in its own subjects the natural capacities of the Hellenic race.   The necessity of combating harmful excesses of opposed political tendencies had caused the nation to have an inspiring remembrance of great deeds in the accomplishment of which all had taken part, to have its pride in the national formal virtues confirmed, and its intellectual horizon enlarged by acquaintance with that civilisation against the political consequences of which it fought. Continuous peace or permanent isolation would have undermined the political life of Sparta by increasing unintellectuality and a growing consciousness of its aimlessness; but the external relations we have referred to—the necessity of undertaking the *rôle* of political opposition—provided it for some time with a vocation for the accomplishment of which subjection to its stern discipline might with some reason be required and was willingly rendered. Gradually the causes of which we have spoken had a disintegrating effect, at first slowly and afterwards more rapidly; the irrepressible desires of human nature were roused by an acquaintance, which crept in and grew, with luxury which the old constitution had taught men to lack with dignity, but had not taught them to enjoy with dignity.

In the parts inhabited by tribes of Ionic tongue, the common evils of unequal distribution of goods, and misuse of inherited authority, were the primary cause of attempts at innovation which, however, did not stop short at the attainment of their proximate ends. The mobile nature of these more social people whom trade and industry had early made familiar with various civilisations, impelled them generally to wish to take a personal part in the administration and guidance of public affairs. The nature of the country seemed to harmonize with this inclination; it favoured the independent development of small communities, the mental powers of which, exercised in constant and concentrated action and reaction, connected with a circumscribed district the remembrance of many famous deeds in which the community as a whole had participated; their native city, adorned with monuments of artistic labour, appeared to all as the visible embodiment of mental wealth, to preserve, protect, and increase which was a debt of honour transmitted from generation to generation. They consciously held fast to this principle of political development; they required that the state should embrace a territory large enough to render it independent of foreign supplies as far as essential necessaries were concerned, but small enough to allow of the personal intercourse of all the citizens. An enlargement of the state which while all the population enjoyed equal rights would have withdrawn the conduct of affairs from the general view and handed it over to a government which could not be inspected, they would have regarded as the beginning of a suppression of freedom.

For them the co-operation of more extensive powers could only be attained by means of confederations which, however, often sacrificed the freedom of the less powerful allies to the interest of the principal one.

The smallness of the stage upon which the actions and reactions of these exceedingly active societies were carried on, accelerated their maturity and decay. The participation of the people in the course of public affairs is free from danger only at times when political development is just beginning, or when it is fully accomplished; in the first case when established national custom is still an effective check upon individual caprice, and at the same time the political course of all is guided into predetermined paths by simple and unvarying tasks; in the second case when long experience (producing respect for necessary restrictions of which men have at last become conscious) prevents even those who disapprove from inconsiderate interference with the course of events. In the first period men will submit without envy to the guidance and authority of a few; in the second it will seem to them necessary that the State as a living historical whole in which past and future as well as present generations have a part, should in some form or other be contrasted with Society, with the aggregate of living men, as an organism which does not altogether coincide with that aggregate. Athens lived over the first period; it was not destined to reach the second; the complete removal of all popular restraints led to a political dissolution, and any reconstruction from the ruins was hindered by the inroads of misfortune from without. We find that even in time of calamity Athens produced some splendid examples of self-sacrifice and enthusiasm, but these—alternating as they do with instances of fatal rashness—seem but as an echo from better times that have passed away; certainly a large number of the most gifted minds appeared in this age of decay, but all withdrew their interest from the present, and looked back with longing eyes to the superior simplicity of the past; unbridled freedom had brought no advance, but it was only gradually that it could destroy all the good that had been developed in that highly endowed people, by wise legislation, the rule of gifted tyrants, and the thoughtful enthusiasm of a less self-seeking generation.

By this double example of developments in opposite directions, the merits and errors of which it exhibits with inspiring and warning effect to later times, Greece became a decisive turning-point for the political development of the West. To it belongs the glory of having led the human mind from stupid acquiescence in traditional order to conscious participation in the good and ill of a commonwealth; of having transformed the child of a tribe into the member of a nation, and the mere subject of a ruler into the citizen of a state. That which gave stability and order to other nations was not without influence among the Greeks also; they, like others, had begun with obedience towards historical tradition, but at a later stage they held fast (not with the blindness of mere habit, but with conscious piety) all that changed circumstances made it possible for them to retain; they, too, knew well what a magic bond of union between the members of a family or of a race is the retrospective contemplation of a long line of forefathers; but long-continued participation in one common weal was regarded by them as a more powerful bond than the natural tie of race or blood; and when they contrasted their much-divided nation under the common name of Hellenes with the world of Barbarians, they felt themselves connected not as descendants of any one ancestor, but as being the only branch of mankind capable of true political life; finally, they were very ready to trace back their constitutions to the authority of lawgivers and political founders, and to consecrate them by the idea of divine co-operation; they did not, however, receive the ordinances which they ascribed to this source as alien statutes, but recognised in them (as though they had been the expression of a covenant between gods and men) that which had caused their recognition and adoption in the first place. Thus the State seemed to them neither an ordinance of

Nature, nor of directly divine institution, but a construction of human reason, which, with conscious reflection upon existing circumstances, endeavours to order things according to that which is good in the eyes of both gods and men—the national conscience affording the revelation of this good.

To return once more to a consideration of the splendid results of these new political views would be superfluous; scarcely less obvious than those results is the danger which they involve, and involved in an extreme degree when for the first time in history it was attempted to establish political life on its own principles, detached from theocratic grounds and from the constraining influence of instinctive obedience towards traditional authority.   Whether what is just exists by Nature or depends upon human institution was a disputed point much handled by Greek sophists.   With this question were connected the inferences that if right exists of itself, it is binding upon all, but that if it is the product of human institution, it is not binding for any power which is able to break it.   The question when put in the form above given did not admit of any plain and simple answer.   Eternal Ideas, valid in themselves, might or might not determine those simplest principles of sentiment and action which must be exemplified in individual actions, in a world of objects that is conditioned by circumstances; but as regarded the *obligation* of these moral Ideas in as far as the Greek national conscience was acquainted with them, no doubt was felt, or at least no doubt but such as was raised by the most idle sophistry— scholastic not practical doubt.   The dispute as to what is just related to those definite rights and duties, laws and institutions of social life, which were based upon existing circumstances But with regard to these, Dialectic, in its attempts to prove their bindingness by showing that they proceed directly from the majesty of the supreme ethical Ideas, always fell short. Speaking generally, there are several arrangements by which these Ideas may be introduced into life with almost equal perfection; whatever such arrangements there were or are, have always resulted from human institution, for in this dispute concerning what is just, gradual growth from the unconscious action and reaction of felt needs is included under the notion of a condition of things produced by the free action of human wills. But this origin of justice in the concrete seemed to diminish its binding force, and the more the Greeks felt that they were in advance of other nations, because of their social order being established on maxims the worth of which they consciously recognised, the greater was their danger of falling into the error of regarding that which they thus recognised as resulting from their own will and choice and always revocable, and themselves as not bound by it. This error, which henceforward has never disappeared from the history of political life, confounds the departments of science and practice. Truths can never be decreed; they can only have their validity recognised; and their validity, as regards reality, is always complete and full, never partial and merely approximate. But on the other hand that which ought to be is determined only by universal Ideas which, as Ideas, form no part of the real world, and always have to wait until human wills give them some special definite form under which they become part of the world of reality. In this sense all justice is the work of men, and can only exist as such, and undoubtedly the sacredness which belongs to the supreme Ideas themselves does not befit it; it has a claim to respect only in as far as it reflects them; but it does not lose its binding force and become of no account merely because it is mediated by human action by which alone reality can be given to it. He who reverences only the supreme Ideas and despises all positive law and justice because of its human ingredient, entirely mistakes the work and destiny of man in history. Our institutions do not exist in order to arouse the admiration of the angels in heaven by their ideal perfection; but their business is, while partaking of that mundane defectiveness which attaches to all human existence, society, and history, to serve as testimonies and results of human reason, which, working by the best light that science and conscience give, tries to make the ideal (as

far as understood) the rule of its action within the sphere of
existing circumstances.   For this work it is entitled to
demand respect, for its worth is not reduced to nullity
because it is not the highest conceivable.   The attempt to
give greater stability to human institutions by tracing them
directly to divine revelation, or regarding them as the
mysterious consequences of some metaphysical cosmic order,
shows imperfection or retrogression of political development;
it is an attempt to perpetuate that which, when all is said, is
but the work of human creatures; here again what is
demanded of men is to be faithful in that which is least, and
to feel bound by the relative validity of that for which
absolute validity is impossible—bound, that is, as far as is
required by the destiny to which they are called—that of
going through a course of rational development with the
steady continuity of historical progression.

This true political instinct was by no means lacking in the
fair infancy of Greek state-construction ; in fact, there was a
period in which the people regarded with religious awe and
scrupulousness the laws which they had imposed upon themselves.   It was the sophistry of a corrupt time which first
raised the question that we have been considering.   But yet
before this time there existed motives for raising it.   As long
as the traditions of unchanging custom were powerful enough,
reverence for law was upheld by habit ; and to this reverence
there was not opposed in men's minds any strong consciousness of having themselves created law.   According to the
legend, special personal obligations of the people to their lawgivers, ensured to the first great legislations of Lycurgus and
Solon a continuance sufficiently long to reproduce the same
habit of respect ; when subsequently social evils and ever-
recurring passions had repeatedly changed the aspect of
public order, great statesmen did indeed insist more emphatically than ever upon the sacredness of law—they insisted upon
this notwithstanding (rather indeed because of) the fact that
they based law upon the free and unanimous consent of the
community, but they no longer succeeded in convincing the

popular mind.   As in every period which has experienced
the misfortune of numerous constitutional changes, so in the
later ages of Greek power, political life seemed to be a mere
stage upon which arbitrary ordinances and experiments,
unsupported by any authoritative force, might clash and
struggle.

With regard to the actual order which they established,
the views of the Greeks were different from our own.   Among
the civilised nations of modern times many circumstances
(among which the influence of Christianity is most prominent)
have contributed to develop a sensitive consciousness of the
significance of human personality.   Not only does the nobler
spirit find true life in those relations to the supersensuous
world which are the result of its own mental labour, and
ward off from this inner sanctuary all intrusive curiosity
or inspection ; a similar sense of individual personality has
become natural even to simpler minds, which without
being conscious of the foundation of their claim, feel that
there is something in them which no power in the world
is entitled to pry into ; every one requires that at least in
his family life, his work, his favourite tastes and hobbies
he should be left unmolested, and the restrictions for the
general good which interfere with him within this sphere
he feels to be restrictions indeed.   Hence we regard the
State as the sum total of ordinances and institutions
necessary for securing permanently the free development
of individuality, having due regard to the needs of human
life and the means which material Nature presents for their
satisfaction ; and we all along make the tacit assumption
that this security must be effected with no more constraint
than is involved in limiting the freedom of each individual
member of the society so far as to secure the equal freedom
of all.   The Greeks did not share this high estimation (which
is in some respects an over-estimation) of human personality.
They regarded men chiefly as products of Nature, and character
as dependent upon degrees of intelligence ; it was not in their
thoughts that there is in us a third power, the Will, which in

good and in evil can fight against insight or natural inclination ; as in thought they were little addicted to pondering the
problem of Free Will (to which our time loves to refer the
very inner sanctuary of personality), so in life they were not
averse to being regarded as homogeneous examples of the
human race.   Absorption in work, in the supersensuous
world of belief, and in the heterogeneous circles of
thought familiar to those who laboriously investigate the
extant fragments of past civilisations, contributes to favour
capricious peculiarity of personal development among us.
These sources of interest did not count for much among
the Greeks, and so there was but little which they could
have felt impelled to withdraw from the observation of
public life as a sacred private interest.   They did not therefore oppose to political order that sensitive consciousness of
the respect due to every man's individuality, which demands
that each several person should be judged by an unique
standard ; the State appeared to them as a system of social
ordinances by which alone man is originally raised above
mere animal existence, is made acquainted with the work of
his life, is educated to fitness for this work, and has determined for him the aggregate of his rights and duties towards
other men.   Not that Greek consciousness lacked either
universal moral Ideas or notions of equity and justice in
matters of private right ; both of these were inevitably evolved
by life itself ; but neither reached a development corresponding to the perfection of political theory, and neither was
independent of this.   The Greeks always held fast to the
distinction between Greeks and barbarians, bond and free,
strangers and guests, friends whom one should benefit and
enemies whom one should hate ; and this shows that they
did not look for justice (the specially moral perfection
among the four which they extolled) in the general disposition
of man towards man, but in the performance of the mutual
obligations imposed by social position.   But State regulations
interfered in such a way with private right as to diminish
many natural privileges, and elevate many others into duties,

seeming in all cases to be rather the source whence rights proceeded, than to find its business in the recognition of those which already existed. Even when the actual condition of things no longer allowed the rein of law to be so tightly drawn, we still see a disposition even in the most enlightened minds to make the disposal of property, the choice of a calling, marriage and the production and education of children the object of State regulation—both these and a multitude of other matters, all of which modern feeling would not even permit to be brought under public consideration. Variety of mental development was not hindered everywhere as it was in Sparta; but even in Athens it was not unfettered until the time of political retrogression; at an earlier period this development itself was in harmony with public opinion; when it was not so, it was, like many religious opinions, suppressed — not as being a sin against a Divine Spirit, but as being an offence against one of the securities for political order; and as a last resource the individual whose existence, even without his own fault seemed to threaten this order, might be removed by banishment.

This complete subjection of individual life to a general rule is not peculiar to antiquity. It lives again not only in religious societies and orders, in which it has its special and easily recognisable motives in feelings of contrition; even where ordinary political society cannot content itself with remedying evil in detail, but thinks that its whole order must be reconstructed, we see both in the carrying out of this purpose (which is rare) and in the plans for doing so (which are frequent) an inclination towards this excessive regulation of life by law. In this case the source of the impulse is not so obvious. Man jealously guards his personal independence in most respects, and yet there is in his mind some mysterious attraction towards renouncing it again, and trying to live as a mere exemplar of his species; the constant exertion of strength which is necessary in order to carry on his individual plan of life is relaxed, and is exchanged for refreshing ease, when he swims with the stream which flows in an accustomed channel. The want of courage which lurks in this impulse is veiled by the æsthetically elevating impression made by the thought of a strict universal order in human affairs; and that which was partly customary submission, partly exhaustion, takes on the more pleasing aspect of self-sacrifice. If even the more favoured are dragged down by these two motives to the liking for an uniform mechanism of life, the oppressed find in it their only hope of relief; it will at least let them have some weight as individuals in the crowd, as examples of their kind, and assure to them a position in life which they could not have won by their own strength. All these impulses were influential in Greece; there was powerful pressure from below caused by envious desire for equality; it was met from above by a self-sacrificing appreciation of the value of law and order on the part of the more noble spirits; thus it happened that freedom came to the people as a whole only in the form of autonomy, that is, the power to make their own laws; the only freedom left for individuals consisted in the consciousness that all which they did and all which they left undone was determined by rational ordinances of the commonwealth.

Thus Society and the State were almost wholly coincident, and both suffered from the admixture. If there had been realized in society any permanent order, always corresponding to social needs, or if it had been possible to enlighten it to such a degree that it would have made every necessary transition in the quickest and most direct way, then the State would have been of but little importance compared to it. But when the development of society proceeds naturally, it is a struggle of selfish interests, which in seeking their own satisfaction violate the rights of others, and thereby disturb the conditions of general prosperity, and finally damage their own welfare. To society in this stage, the State is as it were a conscience. As the guardian of universal justice which is superior to all individual interests, it protects the existing condition of things from all encroachments which would overturn and disregard it as being of no account, at the same time allowing any new development to set it aside in a lawful way; being keeper of the maxims by which the commonwealth is guided in its external behaviour, it is deaf to those promptings of eccentric fancy which would impose upon the nation tasks that are unsuitable and do not historically devolve upon it. Now it is difficult for this conscience to become articulate and to give judgment if it really resides only in the various individual consciences of conflicting parties, and is not opposed to them as a third and higher power, having a definite embodiment. The present age enjoys a superabundance of this privilege; antiquity had not enough of it. Not only do monarchies embody the impartial justice of the State in the one person of the ruler, to whom the base envy of private interests is unknown, not only do those officers whose connected activity constitutes the government oppose to individual wills a plain systematization of the general will, but also very frequently the authority of the State encounters the mobility of society with superfluous and vexatious constraint, in the form of an excessive number of subordinate officials; finally the large size of states, the enormous extent and complication of State business, and the great development of the science of jurisprudence are all conditions which make it necessary to suppose that the assumption of governmental office should be preceded by special technical preparation by which government as an embodiment of the State is marked off from the rest of society.

Whatever disadvantages this sharp limitation may have, the Greek states which were without it suffered from the deficiency. A small group of reverend officials consisting of men of whom some belonged to the natural aristocracy of age, some to the still more respected aristocracy of noble birth, and some again to that of the rich landed proprietors, were originally contrasted with the nation as its guides and rulers, representing the ancient traditions of justice and civilisation. The progress of democratic sentiment and the increasing power of uncertain riches deprived them of all these advantages.

Want of respect for work as work prevented the formation of any regular circle of occupations which would have divided society into ranks and classes, and have made men desire that the various great interests of human life should be represented; hence it came to pass more and more that every individual felt himself a Citizen of the State pure and simple, and that the National Assembly felt itself identical with the State; growing envy and the struggle of all for equal rights caused an increase in the number of governmental officers, and these degenerated into mere business managers in a society the decisions of which were guided by no respect for any developed system of universal law, but merely by traditions of the past in as far as temporary interests allowed them to prevail, and which was turned to good or ill by the eloquence of individual leaders. The battles which society under the supervision of the State had to fight out, were thus transferred to the domain of politics, and since each party tried to get possession of the helm of State, these battles continually endangered the stability both of the constitution and of those individual rights which were too dependent upon it. Indeed the strife of parties assumed a more monotonous aspect than might have been expected after so much splendour of mental development; it became at last nothing but a struggle between poverty and wealth, and ended in Sparta in an intolerable ascendency of some few rich families, in whose hands was accumulated the possession of all the land, and in Athens in the supremacy of the unpropertied majority, who thrust upon the diminishing class of the well-to-do all those State burdens which resulted chiefly from their own measures, and were intended to satisfy their greed and their political vanity.

§ 3. Between Greece and the present there lies Rome; and to it cultivated minds have often looked, hoping to be taught and elevated; it is with Rome only that the political development of the modern world stands in real causal connection, partly by means of many special historical bonds, which there is no need to mention in this place, and partly by a great intellectual heritage which has been transmitted by her to us.

The development of Law, of Jurisprudence, and of a general sense of Right has given to modern society a foundation by which, even in its aberrations, it is essentially distinguished from the states of the early ages of antiquity; and this foundation is a legacy from Rome.

The Greeks had been animated by a strong impulse to sociality, and an inclination to devote themselves to speculative knowledge. The first led them to seek above all in both theory and practice a perfect plan of social order which should secure the most complete and permanent satisfaction possible of the need they felt for communication, for human intercourse, for consideration among their equals; the other characteristic led them to recognise and disentangle moral and æsthetic Ideas which as supreme exemplars determined the content of a beautiful and worthy life, which they regarded as the goal of human development. Neither of these two spheres of thought favoured the development of a strong sense of right.

Special emotions accompany the approving or disapproving verdict of conscience, being different for different classes of the objects which we judge of, and similar for individuals of the same class. Our approval of what is beautiful is not merely an affirmative judgment that differs from a judgment expressing approval of what is good only in this, that it concerns a different object; on the contrary, in both cases there is an affection of the whole mind differing in kind in each case; and in the same way there is a difference between the recognition of what is just, and of what is benevolent and kind. This subjective impression which the thing judged of—or if we look at it objectively, the nature and degree of that worth which we ascribe to it—makes upon us, is expressed by the general names of good, beautiful, or just, but these names contain no answer to the question, What must anything be in itself in order to produce this impression, and hence to merit this ascription of value? Hence from the Idea of the Beautiful, no theory of æsthetics can show what kind of individual thing it is that beauty appertains to;

and yet it is only individual things that are beautiful, and not the general concept of Beauty.   The Idea of Justice does not lead us to know the kind of action that corresponds to it, any more than the concept of Usefulness (to which in a logical point of view it is wholly similar) enlightens us as to what things are useful, and for what.   Hence a predilection for these universal concepts which are without content, and for systematic deduction from them, leads men on the one hand (in order that they may have something to deduce from) to put into them some content more or less suitable, and supplied perhaps by cultivated taste, perhaps by a happy inspiration, but not warranted as certain and exhaustive by any full and careful preliminary investigation of particulars. On the other hand, it forces men to take those individual cases in which unsophisticated feeling must recognise the validity of the determinations of value referred to, and with logical art to fit them into a previously constructed scheme.   Both these procedures are likely to interfere with the just estimation of particulars, in which alone, all the while, the universal can be realized.

The Romans were protected from this danger by their lack of speculative impulse.   They were just as firmly persuaded as the Greeks that there is one single eternal universal Cause which, directly or indirectly, makes everything right that is right; but it did not occur to them to take this Cause and under the form of an Idea of Right to make it in itself the objective source from which the particulars of what is just and right are to be derived; it was known to them only as the agreement of the Practical Reason with itself, this reason never being able to express its whole thought fully at once, but giving, when consulted on special cases, approving or disapproving judgments, all of which are consistent with one another.   They made use of this organon for the discovery of right, and thus, by the same path by which hitherto every science has collected its material, attained possession of a multitude of truths relating to right conduct, which referred primarily to very special circumstances, but were in this isolation much more evident to men's natural sense of right without any mediation, than they could have been as known mediately by deduction from an universal. When the accumulation of the material thus obtained began to make it worth while, and when changed habits of life seemed to require it, there was developed great ingenuity in the discovery of the next higher general principles which lay at the foundation of individual groups of maxims—of analogies by which fresh objects of ethical consideration might be brought under the rules of cases already treated—and finally, in the adjustment of the reciprocal limitations required in cases where different principles came into conflict. But when—stirred partly by the systematizing spirit of Greek philosophy—they finally attempted to express those ultimate principles upon which the abundant store of their ethical wealth rested, they succeeded as little as all later philosophy has done, in finding anything that was at the same time fruitful and conclusive.

This inductive temper which, if need be, can content itself with secure possession of the particular if it cannot find the universal for which it seeks, but cares nothing for any universal from which the particular cannot be obtained, confirmed the peculiarity which marked the political bent of the Romans. Intercourse with one's fellow-men was not a prime necessity of life to the Romans as it was to the Greeks, who could not conceive of human life except in society; least of all did the Romans look to society to bestow or to establish personal rights. A lively consciousness of these—of all to which the individual man lays claim as naturally coming under his power, both as regards the family of which he is the head and the goods which belong to him—was before all other considerations with them; in their view these rights could be bestowed by none, but must be recognised by all. Now life taught the impossibility of carrying out these claims without any modifications, and obliged men to form social ties; but social order did not bestow rights on subjects previously destitute of them, but resulted from the renunciation by individuals of a part of the rights which they already possessed.   Hence it depended on practical limitation of rights recognised in theory, and not upon the establishment of fresh rights.   I need only note briefly that these remarks are not intended to describe the actual origin of the Roman state, in which (as in all great historic events) many causes co-operated; they merely serve to indicate a predominant sentiment, by which, as we think, the Roman world was animated; it is the sentiment which led them to a splendid development of the Law of Private Right, and to a development of Public Law that was by no means narrow and merely national.

The changing relations to one another into which the course of life brings individual persons, form the most natural school for the development of a sense of right.   The claims of different men daily come into conflict, whether as regards the use of material objects, or with regard to those return services and compensations which the actions of some impose upon others.   The frequent occurrence of cases that are similar, though seldom exactly alike, does of itself to some extent secure just judgment; the speedily-felt ill effect of a false judgment helps to bring about its correction; any selfish inclination which a man may feel to maintain such a decision for his own advantage will be in every case suppressed by the apprehension that he may be the next person to suffer from it; from the great multitude of particulars men naturally arrive at general points of view, analogies drawn from which may serve for their guidance in fresh cases; and at the same time the frequent recurrence of individual cases makes clear the errors that may have been committed in incorrectly setting down as similar that which is dissimilar, thus sharpening the distinction between things that are only superficially the same.   And further, the course of life brings into circumstances that are the same, or at any rate similar, the most different classes of people—some bound together by strict ties of love and reverence, others unconnected even by the merest acquaintanceship, and having no cause to reckon upon any definite reciprocity either of benevolence or of ill-will.   Thus it becomes so much the easier to separate from all considerations of sentiment the regulation of any special relation, and the just determination of that which, under the circumstances, is due on both sides; and to look at the matter with reference to that which the nature of the relation itself (in as far as it actually occurs among men) imposes upon those between whom it obtains, whatever other ties may connect them together. Thus custom and right become gradually disjoined, and by degrees it grows clearer how much of that which custom enjoins is required by the essential nature of the case, and what modifications of these requirements are a spontaneous contribution of personal feeling. And not only is the multiplicity of persons important, between whom there may arise relations involving private rights, but also the infinite variety of objects with which they may be connected, is of consequence. Superstition may easily attach to particular objects and arrangements in Nature which are permanent, or grand in their way, a mystic significance, which interferes with a just practical treatment of them; but the vast multitude of things which, in the highest degree various, prosaic, and in themselves unimportant, may yet at any moment become the objects of conflicting claims, do not admit of this false illumination; in dealing with them, men become accustomed to regard things as what they are, not as symbols of something else, and look to find the right treatment of them in such a procedure as their nature requires, in order that they may satisfy as completely and permanently as possible all existing claims.

Now the organization of political society, which is to accomplish such a limitation of the rights of individuals as will make them compatible with one another, and is to afford them efficient external protection, is (if we look at its nature) the furthest goal to which our search after right can approximate, and at the same time (if we look at the need of it) one of the first which our search is bound to attain. The difficulties which oppose its establishment are quite different from those which are met with in the establishment of the individual relations which belong to the sphere of Private Right.

Men cannot, as in the latter case, learn from observation of innumerable examples; the injurious effects of any established error only become apparent after a long time, and cannot be easily traced to their source; the organization has to deal with permanent differences of status by taking account of permanent relations, and hence finds it difficult to avoid laying down fixed rules that favour permanent but unjust interests of individual classes of society; it finds it very difficult to escape the influence of those general prejudices with regard to men's different positions in life and reciprocal obligations, which have been produced by custom as time went on; finally, it has to guarantee not only all private rights, but also the welfare of the whole, both of which partly depend upon external circumstances, and also to afford by its institutions, positive satisfaction to the desire for reputation and the impulses to action which stir individual men. These tasks have to be accomplished under conditions which are continually, though slowly, changing; just judgment concerning them is being continually disturbed by party interests, which are not (as is the case in questions between man and man) held back from persistence in injustice by the fear that they themselves may be the next to suffer the evil consequences of an unjust decision. Hence it early came to seem as though Private Right were a kind of immutable justice, founded in the very nature of things and of relations, and inherent in them; and just as naturally Public Law seemed to be the result of human convention and incapable of being made definitive. Indeed the former was not established in Rome by governmental action, but discovered by the sagacity of experts, as instruments of the natural sense of right, whilst many of the ordinances of Public Law have the character of a treaty between conflicting parties, the content of which is binding not in Nature but through the combined wills of contracting parties, until they agree to revoke the treaty.

Notwithstanding this difference of origin, devotion to the commonwealth was not less in Rome than in Greece. When social order had once been constructed by the submission of all to limitation of their rights, the individual did not cling to it merely because it represented his interests among others; a sense of the grandeur and power which could only be attained by the community as a whole, pride in the great deeds achieved, and a consciousness of the manly virtues which through that order had become elements of everyday life—these won for the State the self-sacrificing attachment of the citizens, and that habit of uncompromising obedience, which caused them to put up with many governmental deficiencies, and more than once to drop complaints about pressing grievances without having obtained the redress they demanded, when government used legal forms as an instrument against them, calling upon them for services the rendering of which prevented their following up their grievances. In later times political storms did indeed by violent and illegal measures disregard all law, yet still even the Empire was far from being a return to Asiatic despotism. In truth the fact is that from this time forward the life of man was based upon a consciousness of inalienable rights which might indeed in many individual cases be disregarded by the temporary representatives of political power, but the theoretical validity of which was no longer a matter of dispute.

We have supposed that personal rights should be recognised and limited by society, but are not bestowed by it. In itself this view of the origin of personal rights is by no means absolutely just; it is with capacities only that Nature endows us; a man's right is something which he first feels as a duty towards others, and hence regards as being also a duty owed by others to himself. This second aspect of the case leads more easily than the first to the conception of right as something universal in which all mankind have part. We only agree to a limitation of original rights with regard to those who profess a willingness to make similar renunciations, that is with regard to members of the same political community; and then an outsider is admitted to participation in political rights only by being received into the community,

and to procure or to permit this reception is left to the discretion of custom and prejudice. The political development
of Rome was in harmony with all this without being altogether
determined by it. Its original town-community was indeed
obliged, by the course of events, to construct legal forms for
the regulation of intercourse with those who were not citizens,
in addition to the native law according to which they themselves lived ; but for a long time this original community continued sole sovereign over the growing multitudes of conquered
subjects, in as far as Public Law was concerned, and it was
but slowly that the rights of Roman citizenship spread
to the provinces. Previously these had been simply rifled
for the benefit of the metropolis, and given over as a prey to
the covetousness and tyranny of her officials; and even when
the imperial government abolished this metropolitan privilege,
it still did not loose the bonds of slavery in which a great part
of the population languished.

After slow historic changes, Roman Law at a later period
again began to restrict the national legal customs of more
modern nations. Not only did it at the time encounter
suspicious aversion, but even at the present day it is re-
proached with having caused the loss of much legal insight
which was of the very essence of the national spirit of
different peoples. It does not lie within the scope of our
brief survey to determine, in reference to this reproach, the
limits of its validity ; we are more concerned to remember the
beneficial effects which resulted not so much from the introduction of Roman Law, as from the way in which all relations of
life became pervaded by the spirit of Roman Jurisprudence. It
is owing to it that there have disappeared the poetic, significant,
and spirit-stirring forms of legal administration and carrying
out of sentences, but also simultaneously with them the barbarous justice that was exercised with so much fantastic
pageantry ; to its cool clear logic it is due that completed
actions, incipient attempts, remote intentions, and obvious or
merely supposed inclinations, not yet put into act, came to be
no longer indiscriminately regarded as deserving of one and the

same sentence; that different offences were no longer visited
with the same frightful and unvarying measure of punishment,
which customary morality (in such cases always too severe) or
examples of Biblical history seemed—to an offended sense of
justice not much used to draw distinctions—to demand.

§ 4. All through the storms which disturbed the beginning
of the Middle Ages, the thought of the solidarity of mankind
had been kept alive only in the Church; and it had regard
rather to the heavenly goal which men had in common than
to any ordered action and reaction on earth. Afterwards the
Empire sought, but with very imperfect success, to bind together
at least civilised Christendom in political union; any consciousness on the part of society of being an universal human
community had been lost amid the multitudinous fragments
of nations which struggled on with difficulty, in conflict with
one another, and without being able to take any comprehensive view of their mutual relations; there were indeed families
and tribes, corporations and communities, nationalities and
kingdoms, but no political construction deserving of the name
of a State. This kind of dispersed and fragmentary social life,
notwithstanding that it produced here and there some splendid
fruit, was not favourable to the growth of civilisation; society
was first delivered from it by the growing absolutism of
kingly power, which (at first with the help of the towns)
broke the independence of the feudal lords, just as these had
already destroyed the freedom of the common people. Where
this subjugation of the vassals took place after a long struggle
and over a wide domain, the prince might not unjustly
identify himself with the State, for he in his own person
represented the political unity of the whole. And this not
merely formally, in as far as his sole will was supreme; in
addition to this, a considerable part of the intellectual store
which was common to all, and by which the national consciousness was nourished, was to be traced to him; wars,
although carried on without reference to the real needs of
the whole and merely for the sake of dynastic interests,
yet accustomed nations to internal solidarity, and to that

jealous national hatred without which no young state becomes great; and many undertakings in art and science, although due to a liking for useless ostentation and other misguided impulses, yet furthered civilisation by the abundant means that were placed at their disposal.

The condition of society certainly changed according to the disposition and insight of the rulers, but this Absolutism was very far from being a return to oriental Despotism; and however strange the forms which sovereign power took here and there, the idea which both rulers and subjects had of it was founded upon entirely different principles. The sovereign was neither the possessor of the whole country nor the sole source of all private rights, which if due to such a source are not rights but only favours bestowed; and powerful and fierce as were some of the attacks made upon these rights, they were either regarded as violent and illegal measures, or were based upon previously established ordinances, and though the content of these might be arbitrary, yet the reference to them showed that it was not a retrospective and baseless decision, given after the fact, and applying only to the individual case, that was taken as the rule of procedure, but a general precept affecting the future. But not only did the power of rulers find a limit in this recognition of general rights which it could not evade, but also their claims to majesty could not quite do without some substratum of respect for the people over which they were supreme. It was not merely distinction of descent which ennobled the kingly office—or perhaps this very distinction itself consisted in the transmitted heritage of kingly rule, the worth and dignity of which depended on the worth of the people ruled. Hence it happened that though the resources of a people might not be always used for their own best advantage, they were not employed exclusively for the personal benefit of the ruler; it was felt necessary to glorify the name of the country from which a prince took his title; under this name was veiled the thought of the State, which now again began to come into prominence; in all external relations the prince felt himself to be the representative of the State, but in relation to his subjects he was more apt to lose this consciousness. Hence we find that this absolutism has a paternal character, and there are very numerous examples of princes who sought to employ the whole strength of their people for objects in which they thought they discerned, not their own personal advantage, but the good of the whole ; and we can easily understand the subsequent transition to a much-governing bureaucracy, the activity of which was not particularly useful to the welfare either of the prince or of the people, but seemed to be advantageous to the orderly maintenance of the State, the notion of which had not yet found its right place with regard either to the notion of Society or to that of Government.

Respect for the kingly power in the minds of subjects was only to a very small extent based upon general convictions respecting the necessary order of human affairs, and not exclusively upon that personal feeling which results from long-continued intercourse. Like almost all the institutions of the Middle Ages, this sovereign power was based upon historical tradition, and its justification, limitation, and extension upon treaties and concessions, which, though they may have been brought about by force, became in their turn a source of law, and grew to be consecrated by prescription. In this way conflicting legal rights of individuals had long been maintained; when finally the power which became supreme was victorious, it—in as far as it was victorious— became another case of acquired right, which now had a place in history, and continued to influence its course. Even the Church, when it alternately confirmed and attacked secular power, did not act in the name of universal principles of right, but proceeded upon isolated historic facts of confirmation and remission. This general bias of the time towards deriving the binding force of an existing condition not from one general source of all law, but from its establishment upon the factual validity of earlier conditions, favoured the development of the notion of legitimacy, that is, of a kind of legality that rests not upon natural universal right, but upon the historic accumulation of acquired rights.   Hence, in point of fact, the beginning of all legitimacy is illegitimate, although it need not be at the same time illegal; even where the rise of any power is due to moral impulses of personal feeling or common consent, the character of legitimacy properly belongs to it only at a later stage of its existence.

In proportion as absolutism consolidated the connection between different parts of the State and removed the restrictions which hindered their reciprocal action, it taught society to feel itself a community, and roused it to further efforts which became dangerous to the stability of absolutism itself. This supreme power did not fully accomplish its natural task; although exerting every effort to make all the forces of the nation directly serviceable to the State, and on this account hostile to all subordinate legal power which in any degree withdrew power from itself, it yet did not succeed in breaking down all those barriers to its own authority which were at the same time hindrances to the free movement of society, and which, like it, were based upon traditional custom, but were not, like it, capable of justifying their existence by rendering great services to general progress.   When the struggle between these powers began, men became more conscious than ever before of the contrast between absolute natural right and historic and legitimate right, as a ground of dispute; even in our own day attachment to the one or the other distinguishes men's convictions as regards the political constitution of the State, as regards international relations, and finally as regards the educative, controlling, and punitive power over its members which society ascribes to itself.   We will now for a brief space devote our attention to these still influential questions, leaving undiscussed the immense abundance of various social and political constructions which have filled up the period intervening between antiquity and the present time.

§ 5. When the Roman looked beyond the boundaries of his own kingdom and noted the similarity of capacities with which Nature has endowed all nations, and seemingly predestined them to unity, he recognised that all men belong to one Human Race. For Christian thought the place of this notion was taken by that of Humanity, men being regarded not specially as called by likeness of natural gifts to likeness of joy and sorrow, but by likeness of supernatural appointment to help make up the composite whole of a complex system of life. Finally, in the present day the expression Society of Human Beings is preferred, and it indicates a new change in the way in which the matter is regarded. In the notion of the Human Race, the prominent thought was that of an Universal, existing in Nature, and exemplified in every individual; in the notion of Humanity, the prominent thought was the idea of a Whole, that makes the Individual the means of its realization; in the notion of Society, the Individual plainly stands out as both goal and point of departure. Society does not exist for its own sake, and its ordinances are not ends in themselves; society is formed and its internal relations developed, partly in order to compensate the needs and deficiencies of individuals, partly in order to make use of the capacities of different individuals for the mutual benefit of all; but the general order which results from this systematization is only valued in proportion as it produces some good result which, coming back to the individual, is consciously enjoyed by him.

In the unconcealed expression of this conviction even well-meaning persons often see a tacit threat of opposition to nearly all the forms under which human life has always hitherto gone on, and still continues to go on—to the institutions of morality which, in family relations and social intercourse, restrain the caprice of self-will; to traditional respect for the rights of property, and at the same time to everything that hinders the free exercise of this right; to the grouping and dividing of nations by political boundaries which have arisen without reference to social needs; to that self-sacrificing obedience which the State imposes as an inherited obligation upon its citizens, generation after generation; to the general duty of respecting obligations of historic growth which happen to conflict with temporary needs; finally, to all that could call in question the sovereign power of society to rearrange its construction at any moment.   It is believed that if this mode of thought were practically accepted, it could only become the source of an instability and lack of rule which would cause the speedy disappearance of all the most treasured possessions of humanity, and that there ought to be upheld in opposition to it, the absolute authority of those intrinsically binding rules of life to which all human striving after happiness has to submit as to a divine order.

And certainly all that we have here referred to is called in question by this mode of thought ; but not in order to be denied—on the contrary, in order that it may be reaffirmed, and for better reasons than before.   The modern idea of Society and its imprescriptible right of autonomous legislation is not new in itself—it is new only as the final and consciously formulated expression of a presupposition which has at all periods of history driven men to attack existing relations, and which too is almost inevitably supreme for some time in the life of the individual.   For all of us are earlier conscious of the restraints which the condition of society imposes on our activity in many directions, than we are of the grounds of their justification, and of those return services for our benefit which that society renders and which are as omnipresent, and hence as unnoticed, as the atmosphere, the pressure of which holds our bodies together ; with that well-known inclination to neglect all middle terms which characterizes Idealism of all kinds, youth is accustomed to demand *empty* space in which to exercise the free wings of its soul. It learns by degrees to understand the value of resistance and friction, and then recognises in the restraints of human relations the unavoidable modification which every ideal must put up with when it is realized in a community of finite beings.   The same revolt against the existing order, which is to some extent a reasonable result of unjustifiable evils, and is to some extent carried beyond its due limits by the confusion due to passion, has repeatedly in the course of history shaken the whole fabric of society ; but how-

ever often this storm of revolt may have threatened to destroy
all those established forms of human relation which we regard
as sacred, and may for a brief period have actually destroyed
them, the waves have always at last sunk down again leaving
these same forms, as a plain indication that it is only the
misunderstanding of passion which fails to recognise them as
what they are—that is, as parts of an organization which
society itself (just in order to partake of the good which
it seeks) would have to assume consciously, if it had not done
so unconsciously, from some obscure impulse, far back in the
course of history.   Now what distinguishes our own time
from earlier times is chiefly the extraordinary facilitation
which has taken place in the exchange of opinions, views, and
experiences, and the proportionally high degree of clearness
with which we are able to survey, in long periods of past
time, like movements of human society with their motives,
their degree of justice, their mistakes, and their issues.
Therefore if the present age again takes up the supposition
that society ought to be self-ruling, it will not want for
warnings against errors which experience has long since shown
the destructiveness of; if it is able to develop the principle
in peace and without being roused to passionate convulsions,
we may hope that the new interpretation which it will give
of the foundation of human duties will not endanger the
continuance of any of those forms of order upon which, from
the beginning of time, the value of life has depended.

But in fact it is not the mere actual continuance of these
moral forms that will content the opposers of the modern
view ; they ask to be given another reason for respecting
these forms.   It is demanded that the validity of great social
institutions should be based not upon proof of their usefulness
or even of their indispensableness for the preservation of
society, but on some inherent and absolute right in them to
fashion human existence, however much its needs may change
with changing time ; it is required that they should not have
merely the significance of *axiomata media* of order proved by
experience and found to be conducive to the greatest good of the

whole, but that by their own intrinsic majesty they should be ideals which men are bound to accept, and to follow which gives worth to life. Those who hold these views meanwhile impute to the effort of society to develop itself, a one-sided desire for material welfare as its source. It is certainly true that the majority of mankind are always inclined to this, and certainly particular periods, the industry of which has to make up for the deficiencies caused by previous ignorance or idleness, are especially exposed to this danger. But neither does the general principle of social self-development in itself exclude the satisfaction of the noblest mental needs from the list of our aims, nor have they been always excluded by practical efforts in this direction. Men have taken upon themselves many sacrifices in the name of freedom, and on the other hand many and great mental advantages have been sacrificed for the sake of rules of human life supposed to be of absolute validity. Whatever errors may be committed in practice by ill-regulated passions, the theory of the autonomy of society is free from the reproach of base egoism; it can reckon the unconditionally binding dictate of conscience, as readily as it can any existing natural need, among the actual conditions which must be considered in any attempts of society to determine what its order ought to be. This theory, too, has at heart not merely material prosperity, and the unconstrainedness of individual wills; it too, since it wishes to satisfy at once all moral æsthetic and sensuous needs, seeks the kingdom of heaven here upon earth, or at least such an approximation to it as is possible upon earth; but it does indeed seek all this in a way different from that which is sometimes assigned to it.

We here again renew the old war which we have so often waged against the worship of empty forms. It is lamentable when Science degrades the rich warmth of reality to mere representations, wholly devoid of interest, of reciprocal action between Unity and Plurality, Finity and Infinity, Centre and Periphery; it is still more deplorable when Art and Religion, instead of gathering enthusiasm from that which warms all hearts, seeks what is highest in dogmas and symbols, the signification of which, if it is ever grasped, can produce nothing but empty astonishment; but it is wholly unbearable when social and political life are attempted to be forced into forms which signify something or other, but help men not at all. And yet how much has been required of us in this way by the profundity of our own time; how often has the attempt been made to deduce from comparisons analogies and symbols, of which we can understand neither the justification nor the evidential force, that which can only be derived from practical needs which are actually felt! Following the comparison which had already proved a failure in the hands of Plato, the different ranks of human society have been obliged to submit to be regarded as an imitation of bodily functions; at times when astronomy has impressed men's minds, the different gradations of relationship between the central body of a system and its planet and between the planet and its satellite, and the complicated regularity of their orbits, have seemed to exhibit a mysterious type of political order; less arbitrary is the procedure of those who seek the exemplar according to which the articulation of the social and political organism is to be carried out, not in any one isolated case in Nature to which another and contradictory case may always be opposed, but in the very ground and base of all things, in the nature of God, in the Trinity in Unity, in the reciprocal action of the divine attributes. All these attempts forget that what is right for one thing is by no means also appropriate for another thing, which is really dissimilar, or which perhaps does not even appear similar; what is just in these comparisons is not valid for us in virtue of the analogy—but it is because that which is just is quite independently and originally valid for the relations with which we are concerned, that the analogy may be conveniently used as an ornament of our discourse upon such subjects, without however having any further evidential force. More deceptive than these arbitrary fancies and equally baseless are views which would regulate human relations according to notions which have a wider application, which through their own power constitute themselves supreme principles, and the expression of which in phænomena would seem to be the necessary business of all reality.   As logically the particular is subordinated to the universal; as rest is physically the result of equilibrium, and movement the result of the reciprocal action of unequal forces; as we derive æsthetic satisfaction only from a plurality which may be apprehended as a clearly discerned unity: so also, it is imagined, is society bound—in the separation and subordination of classes, in the division of labour and distribution of rights, and in the connection of the whole under the unity of government—to exhibit those fundamental ideas of reality in actual life.   I say *to exhibit*—for certainly what these views regard as of primary importance is not that such social arrangements should be useful or necessary or unavoidable; they are required, not to meet a want, but to exist in order that those formal notions of order should be reflected in them.   But the end of human society is not to act proverbs or to present *tableaux vivants,* the symbolic meaning of which may delight spectators dwelling in other planets; human life, with the infinity of struggles, passions, pains, and cares which it includes, is far too earnest to be used for such a purpose.   The only order that can be obligatory for us is that which in some actual and legitimate causal connection is indispensable or helpful for the accomplishment of our human destiny.   I do not mean by this that the organization of society should be limited to some slight and rough systematization corresponding to the most pressing needs, and should despise every arrangement the ideal significance of which might adorn life; in as far as this significance is vividly felt it is rather itself to be reckoned among the conditions which effectively promote our improvement; but it must be felt in order to be justified. Every form of which the symbolic or speculative meaning is plain only to the erudite or in isolated moments of reflection, without rousing or restraining any activity in real life, is an artificial thing that has no binding force.

A general over-estimation of human affairs, of which our

philosophy is not altogether innocent, produces or favours
these errors. For a long time historical existence was regarded
as a confused stream in contrast to the immutable order of
Nature; subsequently reflection, finding in it no less than in
Nature, traces of intelligible development and system, grew
accustomed to regard the forms in which this intelligibility
was expressed as being ends in themselves, in much the same
way as those which are pointed out by concepts of kind in
Nature. As Nature brings forth animals and plants, in order
that there may be animals and plants, and not for any other
reason, so the political construction of the State came to be
looked upon as the evolutional end predetermined and prefigured by its eternal Idea. The State itself, it was thought,
should exist for its own sake alone—to produce a State
in order that there should be a State being part of the business of mankind, who are called upon to realize this among
other forms of their organization — the concepts of such
forms (as ends in themselves and eternal) being regarded
as the goal of human development. This view naturally
has a dangerous bias towards doctrinaire deduction of political
principles; it believes that there is an eternal Idea of the
State not only in the sense of a permanent task which it has
to accomplish, but also in the sense of a type which in all the
detail of its permanently binding systematization is a form
that ought to be realized for its own sake, independent of any
other purpose. We cannot agree with this view either in
regard to the State or in regard to all the other forms of human
life which with the State have been concatenated into a series
of stages in the development of the World-soul—stages which
that soul (in the part of its path in which it wears human
form) has to go through, in order that in each successive stage
it may realize its own being with increased perfection. All these
developments seem to us to be not individual phases and
forms of heavenly light, whose outline and configuration are
actually filled by the Supreme Good, but forms of human effort in
which men struggle to reach that Good. There is no real subject,
no substance, no place in which anything worthy or sacred can

be realized except the individual Ego, the personal soul; beyond the inner life of the subjective spirit with its consciousness of Ideas, its enthusiasm for them, its efforts to realize them, there is no superior region of a so-called objective spirit the forms and articulation of which are in their mere existence more worthy than the subjective soul. It is imagined that the objective spirit reveals itself in the mere forms of social life—but all relations between individuals are of worth only in as far as they are, and because they are, not only *between* those individuals but also *in* them, being felt and enjoyed in living souls according to their worth. There is nothing gained in the existence of family relations, if by family we understand merely the formal connection between parents and children; in this sense animals and many of the plants in a garden are parents and brothers and sisters, but it is nothing to them; that which ought to be realized is the sum of the feelings which such formal relations produce in the minds of those who belong to one family, their minds being like foci in which alone the rays that elsewhere are without meaning concentre to form a bright and living picture. And of just as little consequence is it that Political Society or State or Church should simply *exist*, or be developed in this way or that; if all these have necessary forms which men are bound to keep to, the binding force of such forms always depends upon the degree in which they correspond to permanent or temporary human needs, and are capable of being brought nearer to perfection both in themselves and in those external states in which they can be realized.

§ 6. Radicalism is accustomed in atomistic fashion to oppose the individual to society, and absolute and inalienable personal rights to social privileges and restrictions. But it does not succeed in showing how an isolated human being can be a subject of rights. Against the powers of Nature, the ravages of disease, the fierceness of wild beasts, we can establish no right to the security of our existence; we feel that what Nature has endowed us with is only more or less extended capacities and the wish to exercise them; but our natural claims only become rights when there is some one else who can recognise them. It is certainly true that they then become rights not merely to the extent to which recognition is actually accorded to them, for the recognition may fail where it ought to be forthcoming; but it is equally true that this recognition, when it is accorded, does not consist in a bare perception of rights attaching in finished completeness to individuals as such without any reference to reciprocal inter-course. Not only to fear the claims of others as a power that may possibly be turned against ourselves, but also to respect them as rights, is a thing to which we can be compelled by nothing but the feeling that we are morally bound to help forward the accomplishment of the task to which mankind are destined (and which society alone makes possible) by renouncing absolute freedom of arbitrary individual will. Our right is something which another feels to be obligatory, expecting in return that we shall be similarly bound towards him. Therefore if we speak of the original rights of human persons, we here regard each man not as a solitary individual, but think of him, under the concept of a person, as one who is in intercourse with others, as member of a society of which all the constituents are not indeed always acting and reacting upon one another, but still only have rights, as regards one another, in as far and for as long as this reciprocal action goes on.

It may here be regarded as sufficient (though the thought is not in such a case apprehended in quite the same way) to admit that the opportunity of making rights effective first occurs in society, but that their actual content remains fixed, as a series of requirements to the fulfilment of which man's destiny carries him, in anticipation of all special relations. When the transition is made from theory to practice, it is soon seen that these anticipatory demands consist only in extremely general claims; and that when we come to the question of how a number of men are to live together in the same world, how they are to make a common use of their resources, and how to manage their sources of enjoyment, these claims need much more close and definite limitation before they can be carried into effect. Here we may make a distinction between two things that can never be separated in reality. When a number of men live together in one community, there daily arise a multitude of ever-recurring similar cases of collision between conflicting claims; and hence to make any kind of rational existence possible, it becomes necessary to force individual wills to give up their freedom in some definite degree. The rules of such renunciation (concerning chiefly relations dealt with by the Law of Private Right) are thought to have binding force because they are, in general at least, the dictates of an ever-present reason, founded in the nature of men and the nature of things, and hence receive fresh confirmation at every moment. The case is supposed to be quite different as regards those legal determinations which have grown up as time went on, and which, embracing and enclosing the whole life of man, set to his arbitrary will bounds for which no justification can be found either in the notion of human destiny or in the nature of things. It is thought that to allow these customary arrangements to continue in force is to resist that eternal law of reason which requires that all human affairs should be continually guided directly by its own immutable laws.

It is a matter of course that among the conditions, which are only historically explicable, there are primarily reckoned political relations and the division of society into different ranks, but the line of demarcation has not always been steadily fixed; communism shows that even essential parts of the Law of Private Right may be reckoned among those laws and legal institutions which seem to drag on their existence like a slow disease. This very fact shows the untenableness of the whole distinction. If man could live his destined life in solitude, and if he entered only incidentally into social relations, then indeed no form of society which had grown up historically would be binding on him without his individual consent. But man has no power over the place and time of his birth, both of which involve his life from the first in a network of conditions that have grown up historically; he does not rise to the independence of which his nature allows without the assistance of others, who in this very work are protected only by an historically established reign of law in his society; his mental development would be a nullity if the same condition of society did not bring to him in countless ways the material of mental growth, and aid him in making use of it. Thus then before he becomes a person having rights concerning which he can dispute, he is profoundly indebted to the institutions of society for the very development of his personality. And the same thing holds of society as a whole. If a number of beings endowed with mind, without ancestors or previous history, were suddenly to arise in space, having similar natures and being at a similar stage of development, they would be at liberty to reconstruct their social order at any moment by arbitrary convention. But any human society embraces countless gradations of age, with just as numerous gradations of rights and obligations, of rational insight, and of helpless nonage; hence it can never, as a whole, constitute one subject, able in real truth to exhibit and realize a homogeneous general will; it must always regard the resolutions which it takes as binding even upon those of its members who were incapable of taking part in them, who will therefore on the other hand be unable to refuse to recognise, as having a right to exist, those transmitted conditions which it had no hand in establishing. We find nowhere realized the assumption of Radicalism, that in the construction of human society an altogether new departure might be taken, or the past be treated as of no account.

This, however, is in fact only one aspect of the matter. History continues its course, and the conditions by which one age seeks to order human life can neither furnish irrevocable rules for the future, nor have, as means for the attainment of human ends, the unconditioned majesty of the moral commands themselves. Only it would be an error, destructive of the security of human existence, to treat obsolescent institutions as being in themselves invalid, laws that

are growing unfair as becoming naturally extinct, and inno-
vations of which the intrinsic justice is indubitable as though
they were legally established claims. It is just the historical
connection of all things which makes that which is growing
old remain a power still to be recognised and to be got rid of
in a legal manner ; and new impulses to development cannot
grow up unrestrained as in empty space, but must come to
terms with that which already exists. Not even a condition
that owed its origin wholly to unlawful force can, when it
has subsisted for some time, be summarily set aside as invalid
with all its consequences ; the life of society could not pause
from the time of its establishment, nor hold back from all
connection with it—engagements laudable in themselves will
have been entered into, legal agreements of unquestionable
validity concluded, and the prosperity of society advanced, all
in formal recognition or acceptance of the illegal condition ;
and if this unjust foundation were done away, it would be
impossible to give up along with it the covenants and gains
in negotiating which men had had to make use of it. Still
less can that which was once law disappear of itself simply
because the spirit of the time has changed ; the consequences
of such law will have pervaded society in all directions with
personal rights and duties which can only be sacrificed to the
new condition that it is desired to establish, by means of
compensation and voluntary renunciation on the part of those
concerned. To accept this view is the moral duty of all
parties ; the generation that is going out cannot bind all
future ages by its own conceptions of life, and that which is
growing up cannot lay exclusive claim to all the world, seeing
that it has been brought into possession of it by those who
went before.

§ 7. We see that men of the present day led by such
convictions are in a variety of ways occupied in seeking for
legal forms which may admit of the necessary progress being
made without breach of legal continuity. These efforts can
only be successful in particular directions ; the historical
work of humanity cannot once for all be brought up to a

point from which onwards all further development may proceed without struggle as naturally resulting from the final
adjustment of the social mechanism.   It must suffice if men
will accept the guidance of general principles which are
favourable to this view; difficulties there will always be,
either old or new, which at the moment when they are most
strongly felt can be obviated only by temporary expedients,
and not fundamentally solved for all future time.

The individual will submits more easily to any limitation,
if this appears to be in fact an unavoidable prerequisite of
social life; but it is irritated and offended if the same demand
is enforced upon it as an original right of society, without
regard to this practical signification.   As a matter of fact,
society will always exercise an educating, guiding, and pro-
tecting power over its individual members; but among the
first of those general principles referred to above must be
reckoned the maxim that society should not formally use
this influence as a right belonging to it, nor (as is always
very likely to happen) systematize this in permanent institutions to a much greater extent than the nature of the
matter requires.   Of all that is demanded by moral custom
and the spirit of the age, by habit and fashion, no more should
be made law than is indispensably necessary in order to preserve social life from the encroachments of rude and arbitrary
caprice; and these laws will have to take the form of
prohibitions, not of commands.   Without doubt also it is
the interest of society that the culture of its members should
reach a certain stage, and should take some definite direction
in preference to others; and we do not in the least oppose those
who look upon guardianship of this interest as an exalted part of
the historical work of society; but desirable as it is that this
conviction should be powerful in the minds of all individuals,
and should strengthen their readiness to acquire such a degree
of culture, yet it should by no means be regarded as the
source of an authority entitling society to claim obedience for
any educational system which it may see fit to prescribe.
The moral spirit which should animate humanity will every

where be more perfect in proportion as it is more immediately guided by the loftiest views; but the mechanism of social arrangements has to be based upon proximate and unquestionable grounds. The historical work of any age and the proximate goal of its culture is not written visibly in the heavens, that all who run may read, but is interpreted by individuals according to their intelligence; if the uncertain content of this interpretation is made a legal basis of social institutions the result is apt to be a guardianship of the many by the few which, though acquiesced in quite contentedly in as far as it grows up spontaneously, always offends when it appears as legal ordinance. Hence society has not only to refrain from fixing its requirements in too great detail, but even that which it demands it must (in as far as it demands it) regard as being the condition of a return service which it is itself able to offer. Upon this sober ground of reciprocal and general interest, public institutions will rest more securely than upon pretensions to insight into the eternal cosmic order, of which no one has been appointed the exclusive interpreter.

And these return services of society consist far more in the natural reactions of the interests which it embraces, and which it has to make men respect, than in advantages afforded by it expressly and of set purpose. For society does not establish and confer individual rights, but recognises them and guarantees the possibility of their exercise, on condition that individuals in some respects renounce their unrestricted use. It only confers privileges which arise from its own constitution — governmental posts which cannot naturally belong to any individual, because they are themselves a result of the voluntary renunciation of other men; for the rest its power is limitative, and its activity (in as far as legally determined) is exercised chiefly in affording security and protection.

The amount of limitation which it can impose upon individual wills must itself be but limited. No system of human regulations can claim authority to dispose irrevocably of a man's whole life; every one must be allowed liberty to leave the State to which he belongs, his social rank and calling, and his Church, and to break national ties; every one must be free to throw off those conditions of dependence into which he has been born—not indeed unconditionally, and not without having paid the obligations which he owed, yet still as a right and not a favour it is due to him, being a free person, that he should at his own choice give or withhold at least a supplementary agreement to a condition of things into which he came at first without any agreement on his part. And indeed freedom cannot always be restricted to this possibility of separation when existing relations are felt to be irksome; nor can society everywhere require that if any man disapprove the laws that rule in a given region, he should quit that region; but while entitled to treat in this fashion the wilfulness and insubordination of individuals, it cannot take such an attitude towards any wider current of change in the general mind. It is not the duty of society to accommodate itself unresistingly to new claims; but where it loses its own unity and is divided on essential points, a conservative minority cannot permanently exclude a dissenting majority from participation in the benefits of social order to which it has a multitude of traditional claims—claims that cannot be extinguished by its divergence from rules which are not immutable. It is easy to scoff, and to say that in this way the majority of votes (so often irrational) would decide human destiny; and to demand that votes should be weighed and not merely counted; those who make such a demand do not perceive the tremendous assumption involved in taking for granted that there anywhere exists an infallible organ able to weigh votes in the manner required. Even to count the majority of votes is hard enough; and we must content ourselves for the most part with this imperfect means of decision; taking care, however—with due recognition of its imperfection—that it should, whenever possible, supply its own corrective. And practically this can only be done through the delays which the existing laws and constitution oppose to the realization of the new claims.

We can only hope from the influence of time that (as a result of that free exchange of opinions which must be allowed and fostered) beliefs may be amended, rash haste moderated, misunderstandings cleared up, vague dreams developed to practicable projects, and the abiding heart and seed of fluctuating efforts held fast and cherished— the greater weight of just opinion being thus in truth assured of victory over the mere majority of votes. It is indeed not impossible, and to some it may seem very probable, that notwithstanding everything, mankind will still go on permanently wandering in error ; but this would be an evil fate that could not be remedied by any legal measures. It is not to be expected that every one should patiently acquiesce in this ill fate ; we can praise the heroes of history who in doing battle against it either conquered or were destroyed — though in delivering this judgment we altogether transgress the boundaries of the consideration with which we are here occupied. For historical development, which remains ever superior to our poor political and social art, will not in the future any more than in the past proceed without much disturbance from breaches of law, *coups d'état*, and violent subversions of existing relations. Such historical events, which simply show that the guidance of affairs has temporarily escaped from governmental control, may be regarded as blessings or curses according to their results ; but as long as it still remains a question whether human reason can guide the course of history, they can never be taken into account beforehand as admissible factors in its development. All coherent interest in the public affairs of mankind is injured at the root as soon as notions which are opposed to law are regarded as entitled to the practical guidance of these affairs ; to build hopes upon arbitrary caprice which puts itself in the place of Providence, is like expecting the cure of some bodily ailment from the doubtful issue of a frightful disease artificially induced. Let us appropriate the golden saying of Kant—*If Law ceases, all worth of human life on earth ceases too.*

§ 8. We have spoken of human society as existent and as being actually occupied in giving itself the organization corresponding to its destiny and its needs. But when the notion of Society first arose, men had lived for thousands of years divided into various states, between which there had constantly been hostile contact; and for a long time before, each of these states had by an organization of its corporate life proceeding from quite other causes, anticipated the work which social theories are supposed to begin. Hence it may seem useless to distinguish the notion of Society from that of the State—the State being the only form in which hitherto communities embracing all the interests of human life have been able to subsist. Yet it is not quite useless to consider which of the two notions presupposes the other — whether the State is to be regarded as the basis of all possibility of human solidarity and the source of all rights and duties ; or whether the destiny of Society should be considered as the goal for the attainment of which Society itself requires state - organization as a necessary condition. In the latter case Society may be entirely hidden beneath the State, as the root which is the source of growth and nourishment below the spreading ramage of the developed tree ; but amid all the storms that might hurt the latter, hope and help would be derivable only from a knowledge of the vital impulse that flows forth from the former. The foregoing remarks will have made it plain in what way we ourselves should answer this question ; but the characteristic development of modern times gives it another more practical signification. It makes it seem at least possible to hopeful minds that the numerous states which still divide the world may finally be replaced by one Universal Society, which just on account of its universality would no longer have altogether the form of those states the work of which it would undertake—or that at any rate, without the demolition of existing political structures, Society may be called to exercise over them a power which hitherto it has only possessed by their means.

The increasing relations between the different divisions of mankind have indeed in a great measure changed the signification of political boundaries, and have given new stimulus to the thought of cosmopolitanism. Similar forms of social intercourse and way of life, and like notions of honour, duty, and good manners, are diffused over countries and among the various ranks of their populations in proportion as the intercourse between these becomes general; not only are arts and sciences to a very large extent cultivated in similar ways, but their most splendid productions are being more and more brought into connection, so as to form one store of universal literature accessible to all; the Church, not indeed embracing all the interests of life but cherishing those which are noblest and highest, has for ages spread its arms abroad, regardless of the boundaries of states, and distinguished chiefly by a complex organization, bound neither to territories nor to nationalities; innumerable associations for the pursuit of economic ends have long reckoned among their members men belonging to different states, such associations being made possible by respect for commercial obligations, kept up by mutual interest; thus, over a great part of the earth the individual finds himself supported and restrained by a spirit of social order which is not directly due to any political connection between those whom it affects. It will be objected that in the fortunate case of voluntary performance on both sides of what is equitable, this international intercourse could only dispense with state - help in the same way as under similar conditions it could be dispensed with in the intercourse between individuals within a country; while its general possibility depends upon the further possibility of calling upon the governments of distant countries to aid, in virtue of treaties, in compelling the rendering of reciprocal services which have been refused. And this is certainly the case at present; but the cosmopolitan theory will reply that it is so only because hitherto individual states which have grown up historically have regarded themselves as constituting divisions within society; and because, each recognising as binding on itself its own special legal customs (likewise of historical growth), none has been ready and willing to undertake all the obligations of such international intercourse, each requiring special treaties as a basis of its procedure. An Universal Society would also not be able to dispense with administrative, legislative, judicial, and executive organs; but it would establish organs that would not have their action interfered with by the intricacy and diffuseness which the present plurality of states causes.

We will not discuss such projects in detail; to sketch out plans for the organization of a community is always hazardous. To deduce from the notion of a State its necessary functions, and to set up a special organ for each of these, and definite rules for the co-operation of these organs, is all quite worthless if it cannot be shown that men will give themselves up to carrying out and enduring whatever this logically developed organization of life may require of them. And it is a fact that careful consideration of men's nature and habits, such as can only result from many-sided knowledge of life and history, furnishes no ideal pictures of assured practicability; for however well-authorized and however respectable may be men's efforts to attain some yet unenjoyed good, their being so is never any guarantee that the use made of it when attained will be either respectable or admissible. The general conscience of mankind may slowly grow in insight into our duties and destiny; but the successive generations of living men who are to fulfil this destiny grow up each afresh with all the imperfections and faults of the breed; and it is seldom that those who come into power show themselves capable of establishing that better condition for which, when in opposition, they fought with good right against existing defects. Hence detailed plans for the future organization of society seem to us worthless, but very important the general thoughts and sentiments which are expressed in them; for these will be capable of giving a definite character to our treatment of actual historic conditions, even when the attainment of those general ideals has to be given up. From the remotest times struggles between different states have filled the world with their noise, and the general temper of the present age—intent upon the development of all mental powers and material prosperity—very naturally doubts the authority of political constructions, which on the one hand claim that their organization shall embrace the whole of human life, and on the other hand are continually exposing the treasures which men have won to the most destructive shocks.   And hence it is worth our while to ask, what the State is and must remain for Modern Society, and in what sense that form of it which has grown up historically can be transformed so as to harmonize better with the growing need which men feel for freedom of development.

Similarity of language is an indispensable condition of the civilisation of even the smallest communities; for without direct mutual understanding, extending even to the small things of daily life and equally possible for every member, we cannot conceive a society of which the members have all the interests of life in common.   But within such a small circle the requirements of civilisation do not find full satisfaction; even for the supply of material resources foreign intercourse is necessary; and not less is there an effort of mind to supplement the one-sided stimuli received in daily intercourse by manifold contact with foreign but intelligible spheres of life.   Hence where geographical conditions do not hinder, it is communities whose languages are similar that first draw near to each other; and the higher the development attained by art and science the more closely are such communities united in reciprocal sympathy and praiseworthy emulation, not only by similarity in their views of right and in institutions promotive of mutual intercourse, but also by the consciousness of common intellectual possessions.   But as the individual does not know until he is in a foreign land all the worth of home, so also national culture and the feeling of kinship between those who belong to the same country only receive the finishing touch from contrast with that which is extra-native. This takes place in a less degree as long as foreign surroundings oppose to native culture nothing but barbarism, and in a greater degree when amid general civilisation the contrast is no longer between humanity and brutishness, but between the most refined and subtle peculiarities of national character and custom. Hence in the modern world the wide diffusion of many similar elements of culture has not caused the disappearance of contrasts between different nations, but has produced generally and in an intensified form those struggles towards unity which aim at combining as intimately as possible all the material and mental forces of races having one language—the object of such combination being partly that the intensified action and reaction thence resulting may ensure to the nation an honourable share in the human work of civilisation, partly in order to protect it from the arrogance which disposes every developed nationality to be oppressive towards others.

But mere community of origin, language, and custom does not suffice to build up the form of a State ; it is not nomadic but only stationary peoples that have been able to develop their national life in this form. And, indeed, the territory inhabited by any people is not the mere locality in which the nation dwells and may be found, as plants and animals in their habitats and haunts—it has much rather to be considered as a permanent object of joint labour which first weaves into a strong and lasting fabric those elements, akin in language and origin, which before had had as it were a merely parallel existence. For the division of this labour causes the separation from one another of various branches of employment, and the indispensable action and reaction between these makes men feel the necessity of a fixed, comprehensive and complex administration ; the transmission of the same work from generation to generation makes the history of the nation, and gives it the consciousness of an historical task, in the accomplishment of which are done the most splendid of those deeds which give exalted worth to human life ; even those delicate shades of national thought and temper which constitute the spiritual possession of a people are more or less connected with its modes of labour. A territory large enough to provide within its own limits a great variety of occupations, and rich enough to be able to do without foreign help except for adornments of existence which are not indispensable— belonging as an inherited possession to a people speaking one language, and bound to its native land by a wealth of historical associations, and exerting all its economic and mental forces under a strong and united government in order to fill its own particular place in the movement of civilisation—such a territory presents a complete picture of Human Society, and it can neither become enormously enlarged without losing its distinctive character, nor very much contracted without losing its importance.

In as far as the historical condition of things provides the material for such social constructions, and the possibility of introducing them without breach of law, attempts to realize them are justified; and it is neither to be expected nor to be wished that in the future these many-coloured contrasts should disappear. It is not to be wished; for even the desire that moral commands should rule all mankind with equal power, if it give rise to a demand for the extirpation of that variety, only betrays afresh the oft combated prejudice which would make reality a mere example of those universal laws, in obedience to which, alone, it is able to develop its living content. The whole of morality for the individual does not consist in this—that each has simply in a general way to fulfil the moral commands, and therefore each to be just the same as others; but within the limits of this obedience to the universal, it is the duty of each to develop his own individuality, and, by the good which he and none other can thus accomplish, help to exhibit and to realize the glorious results which the moral Ideas are capable of producing. The task of the nations is no other. They too are not meant to be mere general colourless examples of human communities which might, without suffering loss, melt into the uniformity of an Universal Society, but each has its own special forms of life to develop, without prejudice to the general validity of the moral principles by which all its reciprocal relations must be regulated. Such a blending, however, is as little to be expected as it is to be wished. All the facilitations of intercourse to which we may yet look forward may suffice to compensate economic deficiencies, and to cause a salutary enlargement of the intellectual horizon of the nations beyond the narrowness of native prejudice ; but they will never bring the great masses of mankind into such thoroughgoing contact with one another, that a growing consciousness of the duties of cosmopolitan intercourse will abolish all national peculiarities of character. As far as the last result has actually taken place—as, for instance, in the disappearance of many national characteristics of manners, costume, and speech—we must regard it as being for the most part pure loss, and only to a small extent as a sacrifice indispensable for the attainment of the advantages referred to. Hence we hope that through an ever-deepening conviction of the moral and economic unity of mankind, the progress of civilisation will thus more and more realize an Universal Society in such a way that its existence and the stability of its organization may be beneficially felt by every one who seeks after them ; but on the other hand, we do not doubt that this great universal whole will always seem so vast and so impossible to take in at one view that each individual will still find indispensable for the feeling, thought, and action which fill his life, the narrower home which he can find nowhere but among his own people, in his own fatherland, under his own government.

Only a nation to which there have been left or given in the course of history those favourable conditions of existence which we have hitherto presupposed, can have at the same time means for the complete development of a State, and a natural impulse to such development. Wrecks of nations speaking different languages, which by the course of events have become attached to territories, the unfavourable position of which prevents their making up for the smallness of the population, may be constrained by strong economic interests to reciprocal ties ; but their union, although it may embrace a widely extended community of administration and of organization for defence, has rather the character of a Federation than of a State. The individual members are lacking in independence, and the whole in a permanent and natural unity; for even the economic conditions of countries vary in course of time, and parts which at one time seemed to belong to each other may at a later period show a tendency to disintegration. Regard to advantages which international intercourse, properly arranged, would of itself bring, has here become the determining cause for the construction of a political unity, all the essential conditions of which are not given; and the lack of these cannot be altogether supplied by the unifying force of a long history common to all.

The State is developed from Society by the recognition of an historical obligation on successive generations to maintain and increase a store common to all of material and mental wealth, which each living generation of men has to regard as a trust from past ages, and for the use and development of which it is answerable to posterity. Every association freely formed is entitled to complete liberty in the choice of its ends and means, and in every temporary alteration of both; the only thing it is bound to do is to make allowance for the minority who will not follow the new path; and as no one can have an indissoluble right to participation in an arbitrarily formed combination, the minority will have to content itself with such allowance. Life as a whole is no matter of free contract, but individuals are born into it; society at any moment is not entitled to ignore the legal constructions of past times or to fetter all future ages by its decisions; it has neither the right nor the power to make allowance for those who will not follow its changes of opinion. Prosperous human life is only possible when society endeavours to protect itself against itself, and secures all the essential foundations of its existence—not only the general principles of right accepted by the national conscience, but also those maxims of life and administration which its position makes necessary—against the influence of its own changeable moods, by the formation of a strong and stable government that keeps watch over all the traditions of justice. The four elements which go to form the State are (1) a people speaking one language and having a natural interest in its own unity; (2) an inherited territory furnishing it with means for the maintenance of its independence; (3) a government which represents the historical continuity of the national mind; and finally (4) the general conviction that all freedom of individual development, all its struggle and its progress, must result from the possibility of legal harmony between the people and the government.

This too is an ideal the realization of which may be sought in various forms of the relationship between the people and the government. Certainly every constitution which is intentionally adapted to certain circumstances is, under such circumstances, to be preferred to any other which is not adapted to them; but in saying this we do not mean to deny generally that different forms of political constitution have different degrees of value. We are far from wishing to show from any doctrinaire grounds, which have no significance in practical life, that Hereditary Monarchy is a necessary institution; but we agree with the modern view that practically in it alone has been found the form of government which in itself and under present conditions offers the greatest security for steady development. We cannot hope by any possible discovery to hinder all disease and all evil and all unhappiness; there is no social institution which can prevent the possibility of its own abuse or imperfection in its accomplishment; and finally, there is no constitution which can secure full satisfaction to the restless and envious desires of folly—a satisfaction to which those who cherish such desires can have no claim whatever except as simply being members of the society. With regard to every form of government, it is necessary that no more should be demanded from it than is possible; but hereditary monarchy seems to afford more possibilities than other forms.

What is necessary above all is that the natural struggle between different classes of society, of which each pursues its special interests as far as possible, should not become a struggle for political power; and this one condition it is that prevents Republican Constitutions from being salutary except under certain conditions.   Where in a small state, the peculiarity of its soil and position or special historical circumstances seem to point out that all the members should have a similar occupation and similar sources of gain, this homogeneity of interests may allow a great number of individuals to take part in their representation.   There have been agricultural republics, pastoral republics, commercial republics; but none of them have cultivated all branches of human civilisation until, having become rich, they have produced an aristocracy which left to the majority no political influence worth mentioning.   The assumption of a republican form of government by a larger society with a multiplicity of occupations working into one another will always be detrimental to the many-sidedness of civilised existence; and wealth, being in the ascendant, and restrained by no counterbalancing power or individual currents of life which influence the whole, will make use of political institutions in a one-sided way for private advantage, or by stirring up opposition expose them to continual unsteadiness.   It is fair that the different classes of society with their wishes and demands should be heard and considered by government, but it is not desirable that they themselves should constitute government.   In every form of constitution in which either the supreme head or the governing body is elected, whether for life or for a fixed period, the jealousy of the different ranks and callings and beliefs have an undue influence, disturbing to the steady continuity of national development; it is essential that the highest power in the State should be raised above all competition, that the governing will should belong to no class of society; that it should not be forced to seek maintenance and gain and all that adorns life by the one-sided pursuit of any particular interest; but that rather the exceptional position which it enjoys should afford to it from the beginning all the good things of life, and leave no other goal for its ambition but the

glory of employing its power conscientiously for the preservation and increase of that which has been entrusted to it.

That society consists, as some say, of concentric or, as others
say, of intersecting circles, or that it rises, pyramid-like, from
a broad foundation, and that all these constructions require
some indivisible central or terminal point—all this is indeed
no argument for hereditary monarchy, which is neither a
geometrical object nor a matter in which we are concerned to
produce a picturesque effect. It seems to us to be of more
consequence that a nation desires to see itself, its interests
and its genius, embodied in some Representative Person ; for
this embodiment is in fact of psychological efficacy, because
it is not mere abstract symbolism, but establishes a relation
between persons, capable of pervading even the minutiæ of
life with its influence—with living feelings of fidelity, reverence,
admiration, and love on the one side, of justice, benevolence,
and favour on the other. But this embodiment of the State
is not necessarily presented in the unity of one supreme head ;
it may also be imagined to exist in an aristocracy, a patrician
order of leading families ; and so great is the natural inclination of the multitude to let itself be led, that if its interests
and feelings are but considered to some extent, it will reverence even in this plurality of persons the embodied
representation of the whole. But generally speaking a
favoured class is more narrow in its prejudices (which are
nourished by continual echoes from within), more arrogant in
temper and more harsh in its conclusions than an individual
whose position, raised above all comparison with others, needs
no defence against the intrusion of claims similar to his own.
We should sooner expect forbearance and the removal of
unfair pressure of laws and conditions from an individual
monarch than from aristocratic and democratic majorities,
to whom on the one hand the greatest harshness of doctrinaire
consistency and on the other hand cruel passion are made
easy by division of responsibility and the impersonality of
their resolves. But what history warns us above all things
not to do is, by dividing monarchical power among a plurality

of individuals, to make that power an object of competition, each striving to get undivided possession of it by splitting up the nation into parties and unjustly favouring the more powerful interests.

The primitive leadership of tribal princes sprang from a sense of these reasons, but later developed monarchy did not; they, however, are the motives which permanently keep up in nations a readiness to regard the guidance of public affairs as attaching of private right, as a heritable possession, to the family to which in the course of history royalty has come to belong. The more definitely this ground of legitimacy has been recognised, the more necessary has it seemed to the nation to oppose to the possible abuse of this power representatives of popular rights; and thus there has been founded Constitutional Monarchy—the favourite political product of the last century.

§ 9. It would be interesting to know beforehand what later times will think of the Constitutionalism of the present day. They will certainly be wrong if they wish to give up its fundamental thought, which is that there should be an ever-renewed understanding between the living men of any generation and the government—which, to whomever entrusted, represents the historical Idea of the whole as opposed to the changing interests of the hour. They will perhaps have more reason if they doubt whether the system of constitutional forms which at present prevail are, taken as a whole, the happiest possible arrangement considering the circumstances of the time; and they will certainly be right if they hold that this system is not an ideal in the sense in which it has been lauded by well-meaning *doctrinaires*, as though it were the product of perfect political insight. The extremely unjust oppression of the third estate, which being overloaded with burdens was not able to command any regular representation of its interests, had aroused at the time of the Revolution passionate hatred towards all legal, political, and social distinctions; the not less oppressive hindrances which the antiquated forms of companies, guilds, and corporations opposed to the free movement of labour caused even the more peaceably disposed of the reforming party to prefer the complete abolition of these institutions to the transformation which they so urgently required. Thus there arose the notion of a *Citizen of the State*—a strange theoretic invention, superfluous if merely intended to indicate those who without being under some other guardianship are directly subject to laws the same for all and directly bound by general obligations—but very dangerous if intended also to indicate that these legally equal constituents are also equal in political importance. I am indeed in complete disagreement with the prevailing opinions of the time, in that I regard this low estimation of the corporate element as our most essential fault. Of course we do not want to go back to corporations for the subsistence of which we can find no even plausible reason, in order to accumulate privileges for which there is still less any conceivable rightful claim ; but on the one hand a living bond between those who are really connected would maintain the discipline which we so greatly need, but which yet we cannot enforce by means of general laws ; on the other hand such combinations— representing partly the most important callings (agriculture, manufactures, commerce, art and science), partly the special local interests of different districts—would form the true unities, the representatives of which, by equilibration of the interests of each, would cover the wants of the whole.

I will not weary my readers by attempting to set forth the rank and number of these unities, but will merely remark that they cannot be of equal importance either among themselves or for every state ; a more detailed determination of their co-operation is the business not of Formal but of Material Politics. Here, too, again, I am in conflict with modes of thought which are received with favour at the present day. As corresponding to the notion of *Citizen of a State*, these lay stress also upon the notion of the State itself in such a way as to imply that instead of political forms being developed by nations to further the ends of their existence, we should rather regard the State as a rigid framework to which all national life must accommodate itself.   It is certainly true that owing to the homogeneity of the needs of all human societies, the formal outlines of different constitutions are to a great extent analogous ; and on the other hand, it is equally certain that some conclusions are valid only within the limits of the individual state from the representatives of which they proceed, and are commonly caused only by such circumstances as exist in these cases.   But though it may hence be true that in particular instances this deification of the notion of the State may not have an injurious effect, yet it gives a false colouring to our endeavours ; it gives rise to a superfluous abundance of doctrinaire wisdom, which seeks to centralize and establish as political functions much which should be alterable with altering conditions of time and place ; while conversely it is disposed to approve and enact only temporarily measures which belong to the irrevocable necessities of national life.   We owe the unification of Germany to the military resources created by political powers ; but that readiness of self-sacrifice which secured its success resulted from love for the German fatherland, and not from enthusiasm for " The State "—the most various examples of which general notion Europe offers for our choice.   In fact the more we regard this abstraction as the highest source of our rights and the recipient of our services, the more doubtful becomes the ground of our obligation to render these services to this one state and to shun as treason the lending of our support to foreign states.   Other nations are not influenced by such ideas.   When a supreme moment comes we hear, *England expects every man to do his duty*—or *France demands. it*—or *Holy Russia calls her children, and the Starry Banner summons its followers ;* it is only the strong hand to which our external affairs are entrusted that can raise aloft the national flag as a rallying point for our internal life too, and this national flag it is that we shall follow ; the mere general State flag, without colour or device, which is waved by theorists and party leaders, is hardly likely to attract an enthusiastic following. What we lament is not that great branches of commerce should be brought under public administration ; for the needs of the nation itself may justify this measure ; but we think that much danger will be incurred if the nation choose to regard its whole life as being held in fee of the self-created state, and to look upon the existence which is common to all, as a legal relation—hence seeking, with logical consistency, to replace the government, which is always an essentially political activity, by mere state-administration.

These are but theories. If I could specify measures that would enable them to be easily put in practice, I should believe that I had solved one of the great problems which exercise statesmen. Only this is clear—that existing institutions can never be adapted to the accomplishment of our wishes. To ensure to the votes of experts on every question that weight which is their due, by means of a representation based upon incorporations of the different classes, would no doubt be regarded as a punishable attack upon the rights enjoyed by the citizens of the State—rights which men have become accustomed to think can only be exercised by means of popular representation. And yet a simple calculation teaches us that though any mechanism of direct or indirect election by the body of the people may indeed afford to the individual the small formal satisfaction of having taken part in it, yet the further result slips out of his hands altogether —and it is party leaders and their supporters who take all decisive resolutions, without reference to the wishes and expectations of the individual elector. It is also clear that the proceedings of great assemblies are, at any rate, no longer the only possible form of such reciprocal action between the factors concerned as we should wish to see substituted for the present condition of things. On this point, indeed, we are not altogether without experience ; and I suppose I shall not meet with overwhelming opposition if I say that it would be worth while to offer a prize for an answer to the question whether it is not possible to obtain the real advantages of

constitutional government without the form of popular parlia-
mentary representation?

And from another quarter the existing order is pressed
upon by Socialism; and Socialism, too, demands that the life
of the community shall be established upon quite new foundations. But it has not yet been able to show that the introduction of the new order could be accomplished without the
pressure which now falls upon one part of the people being
transferred in all its burdensomeness to another; and just as
little has it been able to show that the desired institutions
have so much adaptation to practical life as would justify the
attempt to establish them. The theory of the abstract State
equally fails to supply a remedy; a system of privileges and
obligations may be developed from it, but not any information
as to the means of using the one and fulfilling the other.
Here, if anywhere, do we need a pliant and active imagination that will be guided by circumstances and not by immutable principles; it is such alone that will not only be able to
meet the ills which Nature sends, but also and above all to
fulfil the task of alleviating the misery that springs just from
the existing organization of social relations. No machine has
yet been discovered which will work without any friction whatever, and Mechanics aims at no such impossible achievement;
but it has means of diminishing the friction which does arise,
or of rendering it innocuous. We would sooner look for the
spring of such serviceable action, as well as for the capacity
of counteracting immediate evils by known and attainable
means, in corporate bodies, than in those unorganized assemblies where, for the most part, all that takes place is, struggles
between some few wide and well-known party questions which
leave difficulties of detail undecided.

§ 10. A great part of political evils is due to the international relations of states, and to the complete absence of a
developed and recognised system of International Law. The
permanence and obligation of treaties; the rightness or wrong-
ness of intervention in foreign affairs; the difficulties caused
in the case of hereditary dynasties by the political union or

separation of foreign or allied peoples—these have always
been and still are points which are usually decided in individual cases upon grounds of interest, well or ill understood,
and with reference to which fixed rules of justice have
scarcely yet begun to be formed, far less have obtained general
recognition.

To regard as irrevocable every treaty which has once been
agreed to by two contracting parties would involve the
assumption that these were possessed of superhuman wisdom;
for only such could foresee that circumstances would never
arise to make them change their mind, or to make the carrying out of the treaty senseless, or to make it turn out excess-
ively injurious to the one side.   Treaties between nations and
their governments must be not more but less irreversible than
these; since it is impossible that the will of one generation
should bind irrevocably the generation which succeeds it.
Not because centuries ago some old treaty established the
eternal union of two countries, ought that union still to be
held indissoluble; but only because the present mind of
living men declares in favour of the agreement, and freely
consents to it.   If this free consent is wanting, all force of any
treaty is wanting too, and the only obligation remaining is that
of carefully discharging the legal claims, based upon its previous
validity, which have grown up in course of time.   It must
be required that nations should respect a public treaty during
the time that it is received as valid; treaties that have not
been made public, inevitably succumb to the logic of facts;
and when at last they are broken, no one can complain—since
in fact they were concluded without due authorization.

The moral duty of trying to settle any strife of wills is
limited in private life by the respect due to the personal
independence of others; this respect preventing the interference of a third as long as the strife does not imperil his
own interests.   But no one denies to his opponent a right to
accept the active partisanship of a third person, or to this
third person a right to make the quarrel his own; it is only
the privilege of coming between the combatants as arbiter,

and of not taking either side, that will be unwillingly
acknowledged by him who is favoured and always decisively
disowned by the other party.   The attitude of nations
towards combats between nations is just the same.   They
have never complained of injustice when the number of their
declared foes has been increased by alliances; they have
never doubted that others had a right to become their open
enemies; but the notion of intervention, which involves the
assumption of arbitrative power over their internal affairs on
the part of some foreign nation, has always provoked irritation and revolt.   Nevertheless it is just this doctine of the
right of arbitration that the science of modern politics delights
to develop.   As individual vengeance has been replaced by
the public administration of justice in societies, so it is
desired that at least in the circle of European nations bloody
outbreaks of self-defence on the part of individual nations
should be averted or suppressed by the verdict of the whole
body, which now (in consequence of the intimate connection
between the nations) sees its common interests threatened by
every struggle.   At present there is nothing wanting to this
theory, excellent in itself, except the conditions necessary
in order to make it practicable, and uprightness of intention
in its advocates.   The national Areopagus, the incorruptible
integrity of which this scheme of international arbitration
presupposes, does not exist; and it is hardly likely that it
ever will exist except in the shape of a few great powers,
which will adjust the claims of the less powerful in accordance with their own special interests.   But even if in practice
the egoism of these motives could be paralysed by a general
representation of states in the arbitrative congresses, the
analogy between this jurisdiction and that which an individual
state exercises over its subjects would still be incomplete in
many essential points.   There would be lacking, for instance
(1) the possibility of making the matters in dispute perfectly
clear to a court of foreign delegates, who (being influenced by
different national feelings and different historical memories)
would be neither able nor willing to appreciate the value and

urgency of the claims and counter-claims that would be made; (2) the established rule of universally valid International Law, in place of which an untenable regard to passing expediency would be substituted by a narrow diplomacy, having regard only to the immediate future; (3) the absence of any personal interest on the part of the judges in the matter in dispute (an absence which could not be compensated for by mutual jealousy, which instead of making people anxious to decide, generally makes them only anxious to delay); and finally (4) the possibility of carrying out with certainty, against the resistance of the non-contents, any sentence that might be pronounced. Not *one* of the great international problems which have, so far, been taken up by European Congresses, has received a satisfactory solution; not *one* of the political constructions which they have brought into existence has showed that it is endowed with a capacity of permanent vitality; not *one* of those which they have suppressed has been so broken as to be prevented from subsequently disturbing the general tranquillity with ever-recurring convulsions.

No one can say what the course of history would have been if such or such conditions had been different; yet one of the follies of our time consists in this, that so often in historical discussions (which can regard only the actual, and not any merely possible, course of events) we allow ourselves to consider some brief series of good results (which is all that has so far come under our observation) as a justification of preceding perversities—not thinking that the very morrow may bring tardy retribution. And just because we do not know the future, it must be the business of politics, under all circumstances, to respect the Right so long as it is in any way cognisable; but where human wisdom is no longer able to recognise it with certainty, it is perhaps best to defend at all hazards that which one honestly believes to be right, and to commit the result to Providence; better than to act Providence oneself, and by maintaining a hollow truce to increase for society in future times, difficulties of which it will make quite enough for itself without any such addition.

BOOK IX.

THE UNITY OF THINGS.

CHAPTER I.

OF THE BEING OF THINGS.

Introduction—Three Elemental Forms of Knowledge and the Problem of their Connection—The Being of Things a State of Relatedness—Comparability of the Natures of Things—Necessity of the Substantial Connection of Finite Multiplicity in the Unity of the Infinite—Summary.

§ 1. THERE are certain problems concerning our origin and destiny—questions as to the significance of the world which surrounds us and our own position in it ; as to the ends which are set before us in the great whole of cosmic order and the good things that await us in the future —which men have reflected upon in all ages, sometimes with passionate zeal that hopes to find at last a solution which yet has never hitherto been found, sometimes with the moderation of conscious weakness which, giving up all hope of complete success, contents itself with credible opinions concerning what seems so far beyond us.   We behold the results of this intellectual effort in the cosmic theories of religions, in philosophic speculations, in the widespread inspirations which animate art in different ages, in the convictions which have impressed their character on many forms of national life and morality— and all this lies before us in its many-hued attractiveness as the most worthy of all objects of consideration in which we can become absorbed.   But it is hardly possible for any such attentive consideration to attain results different from those which have been afforded by our hasty survey ; men have never succeeded in forgetting the old doubts, except during brief and historically favoured periods, animated by some freshly roused enthusiastic practical activity, or some hopeful inspiration of new ideas that seemed full of promise ; but this temporary lulling of doubt has never been transformed to a

permanent solution at any time, either by the spirit of the
age or by the discoveries of individuals.

And do we yet, notwithstanding all this, desire once more
to attempt the impossible ?　Do we desire, in this concluding
Book, to try and lay the foundations of a true philosophy which
shall for ever dispel the doubts of preceding centuries ?

It would indeed seem so in a certain sense—yet what we
desire is not quite this, and we shall not be justly open to the
reproach of outrageous boastfulness even if most of the
expectations which we have unintentionally aroused should
meet with disappointment.　For the reader himself is an
accomplice in our attempts ; as long as the world lasts, the
human mind will go on wearying itself out in labouring at
this impossible task, and perhaps in doing so find greater
enjoyment than in the initiation and prosecution of labours
which experience has taught us are capable of completion and
lead to indubitable results.　And how could the leisure of life
be worthily filled up if there were permanently excluded from
the occupations of men all reflection which, sometimes more and
sometimes less near and perhaps never reaching its goal, yet
in unceasing movement, circles about those problems ?　We are
with respect to them much more helpless still, when at isolated
moments, stirred and shocked by the events of life, we are
forced to think of them, but with thoughts that are hasty,
unsteady, and fragmentary.　I make no higher claim for the
remainder of my book than this (which it will perchance
justify)—that it may present to the reader the coherent
results of long reflection which have grown dear to me, with
the candour that every one ought to use in communicating his
best thoughts in any earnest converse, so that moments of
leisure may be exalted to moments of mental concentration
the effects of which will not pass away.　This living personal
relation to the mind of the reader, if I should succeed in
establishing it, would be worth more to me than the happiness of seeing a place in the development of philosophy
accorded to the philosophic view of which I am now about
to summarize the outlines.　For nowadays all of us certainly

doubt to some extent the convincingness of a faith, accepted
not so long ago, according to which the very essence of cosmic
history was to be found in the progress of philosophy, and in
every change of speculative systems the dawn of a new phase
in the life of the Unconditioned Cause of the universe. And
even if we had no reason for such doubt, the consideration
whether any philosophic theory which one had to propound
fitted into the rhythm of an evolutional history already begun
—whether it were not late or premature, or altogether out of
course and to be banished from the regular succession of
systems—these and all other similar questions of etiquette
would seem to me unimportant in comparison of the serious
doubt whether that which I wished to communicate would be
capable of comforting or relieving or refreshing any oppressed
soul, by clearing up some obscurity, by solving some doubt, or
by revealing some fresh point of view. Not in playing at
development, but in such services from one living man to
another, is to be found the worth even of those speculations
which are concerned about the highest truths.

No other and no higher has been the aim of all the preceding portion of this work ; and the sympathy with which it
has been received encourages me to press on to the conclusion.
My aim would indeed remain unattained if I did not try to
weave together the loose threads which I have spun, in a
pattern that presents the results which, as it seems to me, may
be reached in estimating the subjects of which I have treated.
I feel the need of such a synoptical conclusion all the more
because I have not felt that I was entitled in the foregoing
portion of my work to make explicit use of a philosophic
view from which its parts taken separately might seem to be
logically developed. I held that it would be more fit, and
thought, moreover, that I should best earn the reader's thanks
(supposing I could earn them at all) if I entered fully into the
doubts which life calls forth with reference to those several
questions which have been in turn the object of our consideration. I have everywhere endeavoured to trace the prejudices,
partly tacit, partly appearing only in isolated indications,

which (springing from æsthetic interests of the feelings, and other mental needs) are the roots that really give to the most different opinions their hold upon our minds.   Hence but little use could be made of philosophic notions and principles, which for the most part are furbished and sharpened only at a later and dialectical stage, for the establishment, defence, or refutation of such prejudices, permitting but little recognition of the real and living worth which these have for the human heart.

My work taking this course I have not been able adequately to present the connection that exists between the views which I hold; I shall now have to show that, looked at in the light of that connection, many apparently conspicuous contradictions do not exist; that many later turns of thought were at the foundation of earlier ones with which they seem to be in conflict, and that taken as a whole, the convictions which I have here sought to communicate are connected with that which I pointed out at the beginning as the aim of my whole work.   The unusual nature of my task must be an excuse for the imperfection of this concluding portion—both for the repetitions which I shall not be able altogether to avoid, and for the references which solicit the attention of the reader to earlier sections, in order that the repetitions may not be altogether too numerous.

§ 2.   Various philosophic systems, setting out from stand-points not wholly similar, and having the course of their investigation governed by the special interest of some particular mode of putting the question, have believed that they have found more than one exhaustive expression for the ultimate source of those difficulties in which our view of the cosmos is involved.   Recalling in a synoptic view the points at which the lines of our previous consideration have come into contact, it would seem as though this supreme problem were to be found in the reciprocal relation of three elemental forms of our knowledge, forms upon which we must base all our judgment of things, without, however, being able to embrace all three in one comprehensive notion, or from any one to obtain the other two by logical deduction. All our analysis of the cosmic order ends in leading our thought back to a consciousness of necessarily valid *truths*, our perception to the intuition of immediately given *facts* of reality, our conscience to the recognition of an absolute standard of all *determinations of worth*.

But none of those necessary truths reveals to us what is; as universal laws they speak only of that which must be if something else is; they show us what inevitably follows from conditions the occurrence of which they leave wholly doubtful. On the other hand, none of those intuitions which present to us the actual features of reality, exhibit those features to us as necessary; however difficult it may be for our imagination to free itself from the impression of those forms in which experience as a whole has accustomed us to see things be and happen, yet we do not find in them any reason why we should regard them as indispensable; they might either not be, or be different from what they are. Finally, none of our Ideas of what has worth, of what is holy or good or beautiful, can of itself give rise to a definite world of forms as its own proper consequence; even where reality clearly reflects the content of any such Idea, this realization still remains in form and colouring but one out of many possible realizations, conditioned by existing facts, while other and different facts are quite conceivable which would have caused the same content to take an embodiment wholly different in form and colour. Still more obscure than this connection of the necessary laws of thought on the one hand to the worth-determining Ideas, and on the other to the factual condition of reality, is the bond—wholly concealed from us—which connects together those Ideas of what is holy and good and beautiful with the indifferent but immutable content of mathematical and metaphysical truth.

This incoherence not only hinders our knowledge from becoming complete, but is also the source of the doubts which oppress our life. As long as we cannot help thinking that the world as by an unfathomable fate follows the fiat of necessary laws; that then from some fresh and independent quarter there comes in the reality which is required for the carrying out of these laws; that finally, there are added Ideas of that which ought to be, which have to be realized so far as on the one hand the limitations of those *à priori* laws and on the other hand the inertia and resistance of this underived reality permit—so long as we cannot help thinking all this, our cosmic theory has not the unity necessary for knowledge, and our hopes lack that confirmation which would make them strong and vigorous. That it is not so—that there is but one origin of the world from which flow, as from a common source, its laws, its realities, and its worth—that this origin is not to be sought in that which in itself is unmeaning though necessary, but that that which is most worthy is at once the Alpha and the Omega of all—by this conviction (which has animated all our considerations hitherto) we abide; and we now seek for it both a more exact expression which may take the place of that just used, and such verification as it is capable of. Neither that expression nor this verification will in all respects come up to what in other cases we have—and with justice—required from scientific statements; we must to a large extent content ourselves with making clear what it is that we mean and that we require, without being able to show how that which we require and mean can be; we shall not be able to prove throughout the necessity of that which we are seeking, and to develop its whole content with the certainty of a strict logical deduction from undeniable premises, but must be content to remove the difficulties that hinder a living faith in its existence, and to exhibit it as the ultimate goal to which we have to approximate, although we may not reach it.

These preliminary remarks, which anticipate the special questions in the prosecution of which only their real meaning can be made clear, are offered merely in order that impracticable demands may be met beforehand with a confession of general inability on our part. I will add to them but one other observation, which may serve to indicate our proximate task or the nearest way to the accomplishment of our task. Convinced of the formal incorrectness of views which teach that for an explanation of the world nothing need be considered except the animating breath of a creative Idea, our considerations have hitherto, and for quite long enough, been occupied in taking the part of the Finite against the Infinite, of the blind necessity of the mechanism of Nature against the freedom of Spirit-life, of Plurality against Unity; to many it will seem as though they had, in all essential points, taken the side of the small against the great, which they have provisionally neglected. Now that we take the opposite standpoint (being convinced of the perfect legitimacy of claims which, in the form in which they are generally put, we felt bound to reject), we cannot consider that we should gain anything by setting the view which this standpoint opens to us in opposition to the one we formerly took, as being *also* existent and having *also* a foundation in the condition of things. We should only gain if the earlier mode of thought, traced back to its real principles, itself constrained us to enter upon the path which leads inevitably to this other view of the world. It may make one happy to exhibit the world, in fresh enthusiasm of feeling, as presenting an unspeakably lofty and beautiful content, which rather possesses our mind than is possessed by it; but the nearer we come to particulars the more are they felt as hindrances which compel the lowering of this lofty flight; we should be really raised higher if from a right handling of these particulars we could derive an upward impulse, which would give us hope that at the end of our way (now secure because all hindrances have been overcome) we should reach that which is highest.

The way that we would indicate is familiar to the natural course of men's thought. Almost every attempt to justify or to communicate the fundamental truths of religious conviction is introduced by the assertions that if there is a conditioned then there must be also an unconditioned, if something that passes away, then something that is eternal, if plurality and something that may change, then also something that is necessary, some being that is one and immutable. It would not be easy to show a valid connection between antecedent and consequent in any of these concise maxims; as commonly used, they juxtapose the beginning and the end of a long train of thought, suppressing the intermediate links by which they must be connected. They meet with approval and seem evident to the hearer, because he too has been long accustomed, by a combination of ideas the justification for which he has perhaps never been clearly conscious of, to strive to pass from the thought of the Finite to that of the Infinite, from that of the Many to that of the One; whilst whatever essential ground there may be in the content of the one idea on account of which its reality guarantees also the reality of the content of the contrasted idea, this recovery of the links which justify the maxims and connect their members, may be regarded as the task which first demands our attention.

§ 3. Things—each of which is an harmonious group of properties—seem, when we first look at the world, to be, in all essential respects, immovable wholes, untouched at bottom by the alterations which some of their less important characteristics undergo. But when investigation begins, it soon appears that the disturbance which seemed only lightly to graze the surface of things, really penetrates far deeper into them, and finally affects everything in them which we had thought to be permanent and unchanging. Each of their properties appears to be ultimately dependent upon conditions, and to change when these conditions change; and all these conditions consist of variable reciprocal relations subsisting between many things—that is, of activities which they exercise and are affected by. Thus in the most favourable case we learn the relation and behaviour of things under definite circumstances, but not what they must be in order that they may be able to exhibit such relation and behaviour. But it is not only the content of what is that remains enigmatical to us, the significance of its being is so too—as far as we are concerned it resolves itself into mere action. For even the most stable properties of things, properties which in their permanence our imagination may use as the very image of a changeless being, appear upon close investigation to be undergoing continuous growth and decay; their existence at every moment is the transitory result of reciprocal action between many elements, unceasing renewal of this action being required in order that their apparently steady continuance may extend for even but a small space. Where then can we find, and wherein con-sists, that uniform undisturbed permanent being which we have been used to regard as comparable to the unchanging channel in which the current of events flows on? Even now we do not find that we can do without it; for the content of one moment conditions that of the next, so that there must be some stable reality, embracing equally all phases of Be-coming, and assuring to each its power to condition the next; what then is, and in what consists, this Being?

Let us, in order to be brief, follow the path which has been taken by an ingenious modern system. Our first question concerning the τὸ τί of things has regard, not to that nature by which each is differenced from others, but to that in virtue of which all are similar and all are things. But this name, *thing*, indicates—as far as known to us—nothing other than the performances which we expect from what we call things as evidence of their reality; they are things in as far as *they* are at least participant in immutable independent being, and present the fixed points to which is attached, in whatever way, the varying course of events. Now, having once become doubtful of the correctness of the ideas which we formerly applied with unquestioning confidence, we must first consider and make clear to ourselves what that being is which we require in things in order that our theory of the world may find in them a firm foundation; in the second place, we must ask how and what things may and must be, in order that they may participate in this being, of which we have found the meaning.

The content of the simplest notions does not admit of being built up out of constituent parts, but only of being detached from the examples in which it occurs. Therefore

we are justified in starting from the fact that existence is first
present and intelligible to every one in sense-perception ; that
is, which is seen or heard or in any way perceived, and at
this first stage nothing indicates the existence of things except
their being perceived by us.   But even at this first stage we
recognise too that this illustration of existence does not suffice
to express that which we mean by it.   For, as we think, the
existence of things remains, even when our attention is turned
from them ; they were when we did not perceive them, and
for that reason, when our senses are again applied to them,
they may afresh become objects of sense-perception.   Consequently their existence, which at first consisted for us only in
their being perceived, must belong to them without reference
to our sense-perception ; but in what, then, does it consist ?
This question, too, is readily answered by ordinary thought,
according to which, whilst things are not perceived by us,
and perhaps when they have never been perceived by any
one, they still continue to stand in relations of various kinds
to one another; it was these relations that formerly gave to
them a firm hold upon reality ; and these constituted their
existence up to the moment of their being again perceived by
us.   But this being perceived is itself nothing but a new
relation which is added to, or dissolves, the old ones ; while of
greater importance for us, because it is only through it that
we come to have cognisance of existence, it is to the existent
thing itself not more indispensable for its existence than those
relations which subsist or subsisted between it and other
things.

Ordinary thought generally keeps to this standpoint; a
state of relatedness is regarded by it as being the existence
which it has in view; it is only philosophic reflection that
tries to reach beyond this and in a reality devoid of relations,
in a wholly self-sufficing self-dependence, to find the true and
pure existence which belongs to things in themselves, and first
makes them capable of serving as points from which relations
may start.   And in fact to stand in no relation to anything
else—neither to be known nor to know—not to be brought

into connection with any other thing, either as having position in space or order in time—neither to be affected by anything nor to have any perceptible effect—is exactly what in ordinary thought is regarded as the fate of the non-existent, but not as the nature of the existent; and it asks, and with reason, In what respects then is this pure existence distinguished from non-existence, if not in the fact that we *choose* to understand by it the opposite of non-existence? Now this question would undoubtedly be foolish if it were the expression of a curiosity to know the process or the inner structure by which such existence is endowed with the reality that differences it from non-existence; but the impossibility which we here find of separating that which we mean from that which we do not mean, even by determinations of thought, points to some error of commission which we will now endeavour to discover.

From the total content of any idea by which we think some fact of reality, analytic abstraction easily separates individual ideas which are admissible and just as long as they, conjoined with others in the further course of thought, lead to conclusions that are again coincident with real facts, whilst they have not the kind of validity which would enable them, of themselves and out of such combination, to denote any reality whatever. From the idea of the movement of a body, an idea which in its completeness denotes a fact of observation, we drop all reference to the body, and lay stress upon the idea of movement alone; we analyse further this idea itself and thus get the notions of velocity and direction— pure abstractions, which are just and useful because their content is capable of an elaboration in thought the results of which when again applied to the complete idea of the movement of a body, gives us enlarged insight into the nature of this movement; but neither velocity in itself without direction, nor direction without some movement in some direction, can denote anything which could of itself actually exist. That the notion of pure existence is to be reckoned among these notions, which are in themselves valid but can have no real independent existence, is most easily shown by the other names which are given to it—the names of absolute Positing or Affirmation.

Affirmation would be a most inadequate mark of that which we mean by existence ; it is only our habit of virtually thinking of the real and not of the unreal as that which our affirmation concerns that can mislead us into thinking that we exhaust the notion of existence by the notion of affirmation. It is plain that existence is denoted not by affirmation simply, but by affirmation of existence ; and both the meaning of this and its difference from non-existence remain wholly untouched by appeal to an affirmation which may apply to its opposite just as much as to itself.   The notion of Positing gives us nothing better.   It is readily conceded that something or other must be thought which the positing posits, or the affirmation affirms ; but even this addition does not give to either of the two notions such completeness as would make it possible to accept them in the sense here assigned.   The affirmation of a single notion has no meaning which we can specify ; we can affirm nothing but a proposition in which the content of one notion is brought into relation with that of another ; and just as unmeaning is it to speak of any positing in general without at the same time thinking of and naming those relations, the being brought into which constitutes the very positing of that which is posited.   Nothing can be simply posited without being posited in some way or other, in some specified circumstances or connection—and the assertions that characterize the true existence of things as *unconditioned irrevocable absolute positing* cannot compensate for the failure to state in what this positing consists and what its effect is, by predicates that emphasize its importance.   Undoubtedly the general notion of Affirmation may be separated from affirmations of various propositions, and the notion of Positing from the manifold positings by which we bring various things into various relations, and both will furnish serviceable abstractions ; but neither can we in thought make an affirmation thus without content, or a mere bare positing, nor can there be, beyond our thought, any reality corresponding to either.

The failure of a definition does not do away with the validity of the notion intended to be defined. Hence we shall have no difficulty in admitting the fruitlessness of these attempts to cover the true meaning of existence by the notions of empty Affirmation or Positing, while yet there may be very plausible considerations leading men to think themselves justified in seeking it nevertheless in an absence of relatedness which we thought to be equivalent to non-existence. For no view can bind the reality of things to a definite number of definite and immutable relations; but things are *things* just because their existence lasts on undisturbed throughout the ceaseless change of all their relations. If we now take away, all at once, all those relations that we are undoubtedly justified in taking away one after the other and separately—if we deny all relations—this denial will not concern that which was independent of what we denied; there will remain, it is supposed, as the object of a distinct and assured opinion, Pure Existence—which now without relations is the same reality that it formerly was with relations; less easily described indeed in its simplicity than it would be in any of its relations, which would give us an opportunity of telling something about it, but not the less something certain and positive in itself, because of our inability to characterize its self-dependence in any other way than by denial of that which it excludes. Thus, it is thought, do we reach a confirmation of that which we vainly attempted to call in question —things must *be, before* they can stand in the relations in which indeed alone their reality can become perceptible to us; and it is thought that this hidden existence is permanently distinguished from non existence by the capacity of that which exists to enter at any moment into that network of relationships in which its reality becomes apparent.

What I object to in this train of thought is this insignificant *before*. When we recall those individual ideas by the joining of which we make clear the simple meaning of existence, it is very natural that this idea of reality which cannot be further analysed, should, just because it is contained in the notion of *every* existent thing, take the favoured position of something precedent to the various and changeable determinations by which one existent thing is distinguished from another—these determinations being only subsequently added to the pre-existent reality. If this were expressed thus :—In order to think the existence of things one must first grasp that reality or affirmation by which all existence is distinguished from non-existence, and then understand as that to which this affirmation relates, all those determinations and relations by which one definite existence is differenced from another—we should have no objection to make to this logical arrangement of the notions referred to. But this succession of ideas, which always arises in a similar way when we compare numerous examples of some universal, does not always correspond to a uniform actual process in the compared objects themselves ; and even in the case which we have taken it may be shown that the priority of unrelated to related existence is merely this logical priority, not the metaphysical priority which would be expressed in the assertion that there is real unrelated existence, taking *real* in the same meaning in which we apply it to related existence.

We only speak of things and of their existence because these ideas are indispensable for the intelligibility of the changeable phænomenal world. Now it may seem quite allow-able to assume that a thing may emerge at any moment out of the complete unrelatedness in which it reposes, secure of itself but not as yet contributing anything to the play of events in the world, and may enter into those relations to others in which it is capable of making an efficient con-tribution to the sum of what is going on in the universe. But nothing can enter into relations at all without enter-ing into some definite relation to the exclusion of all others. Yet wherein can lie the grounds of decision for the choice of this relation, if not in other relations, which, however unobserved, have long subsisted between that solitary element and the rest of the world, with which it appears—but plainly only appears—to enter now for the first time into conjunction? If, allowing ourselves to use a spatial image, we represent the whole of reality as a sphere of infinite diameter, but that solitary element, which as yet has no relation to it, as actually having a non-spatial existence from which it will pass into spatial reality, its entrance into space must take place at a definite point to the exclusion of all others; it is impossible to find any reason for the choice of this place except in the direction of some movement which the solitary element already had towards it, even when it seemed to us in its spacelessness to be devoid of all definite relation to spatial reality. Therefore if we leave undecided whether there is admissible at all the notion of unrelated existence separated from related existence (in the idea of which indeed the idea of unrelated existence is contained), yet we must maintain that anything which actually *was* so unrelated could never enter into those relations through which it would assert itself in reality as a real thing among other real things. And just as little is it possible that any existing thing which had once been in relation to others, should get rid of all relation to the rest of the world; it could only get to a greater distance from it, this, however, being just as much a relation as its former proximity. Hence there was an error in concluding that because it may be possible to deny an individual relation therefore some real existence would remain when all relations have been denied. In the same way consciousness remains when any individual idea is removed, but it disappears if all are removed simultaneously.

Then—it will be finally objected—there is nothing stable at all, since the existence of everything presupposes the existence of some other to which it must be related, and thus neither can be without the other which is its foundation. Without doubt this is the most erroneous of all objections; it wholly mistakes the business of philosophy. For that business is not to state a mode of procedure according to which the world might be created if unfortunately it did not yet exist, but only (and especially here) to understand the connection of the world which already exists. We do not inquire what difficulties a world-creating power might have in producing this reciprocal tension of all those constituent parts of the whole arch which mutually presuppose one another; perhaps it had no difficulty at all, for why should this power have been so one-armed that it could only fix one element at a time? It is just this continuous arch of mutually related things which is the primal reality that constitutes the object of all our investigations—the object which is given and which alone we can recognise; to seek to discover the laws according to which the course of changes takes place in it now that it exists seems to us a possible aim of these investigations; but to ask by what device it has been made, or how it has been brought about that there is any coherent world whatever, instead of none at all, we hold to be a wandering flight of fancy that shoots beyond the mark. Hence, we may observe incidentally, it would be an advantage to banish from the consideration of existence those expressions *Affirmation* and *Positing* which have already troubled us. Being, in form, designations of actions, their use keeps up the prejudiced belief that some process may be stated by which the existence of that which exists is produced—as if in existence itself there took place that succession which obtains among our ideas when we try to comprehend it.

And so perhaps we may most quickly come to terms with the mode of thought opposed to us, in the following way. When it maintains that each Real being in its own pure existence floats in a condition of complete unrelatedness, but not only *can* enter into relations with others, but does so in reality with infinite frequency, we only ask to be allowed to regard this relatedness—which is admitted in fact, and therefore recognised as possible—as the only kind of real existence; but that pure existence to which reference has been made, as something that does not occur in any place or at any time. If according to that view existing things need no external relation in order to exist, we would add that at any rate existing things now *have* enough and to spare of such relations, and have them everywhere and from all eternity, and that in point of fact reality contains nothing that is or could be isolated in its own pure existence, and out of all relation. If then there *is* nothing that is unrelated, we are entitled to say that it belongs to the notion and nature of *existence* to be related. For he who holds that existing things devoid of relation are conceivable but admits that none such do actually exist, plainly does not speak metaphysically of existent things, but logically of what is possible but not actual, and hence certainly not existent.

§ 4. We have already had repeated occasion to distinguish between the relations which seemed to belong to things themselves and others into which they are merely brought arbitrarily by our thought. It is only in the first class that we shall now seek to find those relations, to be in which constitutes the existence of things ; and yet those of the other class are not less important, and it is only apparently that they are foreign to the nature of things. For to establish by arbitrary conjunction relations which have no foundation in the content of the things conjoined would be not thought but mental aberration ; even a relation of comparison must, in as far as it is correct, have its root in the actual condition of that which is compared. If we compare things as contrary or greater or smaller, it is not our comparison that makes them contrary, greater or smaller, but the things compared actually had these relations to one another before we came to consider them, and the relations are found, not invented, by our thought. Yet their remains a difference. Sometimes the contrasts are brought together and exhibit their opposition only in our thought ; sometimes they encounter in reality and cancel one another ; sometimes in our thoughts the greater is opposed to the less without affecting it, sometimes in conflict it makes its superior power felt. It will easily appear from a generalization of these examples that the former relations afford definite grounds for the form and content of *future* action, and that the latter are the effective

conditions of *actual* action, in which the related elements do
and suffer as the former indicate.   It is to this last relation
that we now wish to devote some attention.

That all which takes place in the world takes place in
obedience to laws will be readily admitted by every one; but
there is less agreement as to *what* that is which the most
general laws of existence and action impose upon the
demeanour of all things.   Yet it is not this variety of views
as to the last point which stirs us up at present, but just that
assumption in which they all agree.   But whilst we raise a
special question as to the conceivability of this assumption—
that there are universal laws—we must be careful to guard
the meaning of the question itself against misapprehensions.
Of course we cannot demand to know how it is that a *primitive* truth, which is not derived from any other, can be true,
nor how it comes to pass that an *universal* truth should be
valid in all the cases of its application which we think of;
the only thing that we wish to make clear to ourselves is,
how a law can be not only a valid truth in the realm of
thoughts, but also a determining power in the world of things.
And this question we do not ask in the hope of obtaining a
graphic picture of the arrangements by which the subordination of things to the law is brought about; all we want is an
explanation of the several thoughts which always accompany
the notion which we have when we think of any general law
as a valid truth; for when we desire to transfer the law to
reality as a governing power, we must, as a condition of the
possibility of that transference of the law, carry with us in
our thinking the content of those thoughts, into the sphere of
reality.

Now every law regarded as a valid truth attaches some
definite consequence to a relation that either always exists or
can be established between some two factors.   But in order
that it may be general and not the expression of an individual
case, it not only assumes that that consequence and the conditioning relation to which it is attached belong each to a
series of which the members are connected in some definite

way; but it also has to make the same assumption as regards the factors between which there is this connection of condition and consequent. And indeed it needs but little consideration to show that not only must each one of these factors be of a kind belonging to some genus, but that also both the genera themselves (to which the two factors severally belong) though not indeed necessarily kinds belonging to a higher common universal, must be at any rate members of some relation in which they occupy definite positions. Under these conditions the law expresses the general mode of dependence by which in each individual case the kind and magnitude of its consequent is determined, in accordance with the given kind and magnitude of an assumed connection which may be variable, and with the special nature of the factors between which this connection has place. The general axioms of mechanics would furnish a multitude of illustrative examples for the further consideration of these briefly indicated relations ; here I would only once more emphasize the point (which is, in this place, of importance for us) that on the one hand the relations between things, and on the other hand the effects resulting from them, must be comparable though different cases of general events, and not only so but also the natures of the things from the relation between which an effect is to arise, cannot differ to an immeasurable and incomparable degree, as long as the part which those natures are to contribute to the formation of the result in any case is to be determinable by some general law. And indeed it is not sufficient to allow to the things such homogeneity as causes them to be co-ordinated under the general notion of *thing;* it is further necessary that the qualities by which one of them is distinguished from the other should be comparable values of general qualities.

In developing this demand, more laboriously than perhaps might seem necessary, what we have been insisting upon is not some assumption that is not naturally made everywhere in attempts at philosophic explanations of the world, but certainly one the significance of which men, when they make it, do not adequately perceive. For we only give a different expression to the meaning of this assumption when we maintain that it makes impossible any thought of the independence of things which allows an individual thing to be what it is without reference to others ; that on the contrary it constrains us to regard the specific nature of everything as being a definite member of an all-embracing series in the existible world—a series of which the equally special natures of other things constitute the remaining members.   That this assumption is everywhere tacitly made we learn from the procedure of those who formally deny it.   Wholly un-determined—they say—are the qualities of Real beings ; each may be what it will, if only what it is is something simple and positive, and if in order to be what it is, it needs no relation to anything else.   But as soon as the explanation of phænomena makes it necessary to give an account of the consequences which arise from the relation that (not-withstanding their independence) comes to exist between two Real beings, this assumption has to be supplemented by a correction which makes it useless in itself.   For the simple natures which make it impossible to divine how the conjunc-tion of the beings to which they belong could produce any result, it is thought that there may be substituted combina-tions of several qualities as equivalent expressions of their content, and this without detriment to the supposed simplicity ; and because all opposites tend to cancel one another, these substitutory expressions, by analysing the simple qualities into similar or opposite or otherwise contrasted constituents, give us some insight into the way in which the reciprocally acting natures of things work into each other so as to produce a new condition of reality.   And thus contingent aspects skilfully reach—by a roundabout path, and in a form which has dangers of its own—the same assertion which we held from the begin-ning, namely that the natures of things do not differ in an incomparable degree, but that they are members of a series (or a system of series) susceptible of comparison.   Each indeed has a value of its own, by which independent of others it is what it is, but all these values are in a condition of relatedness through which it first becomes possible that the conjunction of several of them should furnish adequate ground for a definite result. Realism has an intelligible but unfair interest in preferring this roundabout path ; it desires that the independence of each individual Real being should not be endangered by the thought that the commensurability of its nature with the natures of the others belongs to the very notion of it ; only as a completed fact, and a fact which might have been otherwise, does it admit, as something supplementary, this comparability of things. But here again it does not escape confusion between an effort of logical thought and metaphysical knowledge. For to assume that in itself the notion of an existent thing does not require that it should be comparable with others, and then to admit that in point of fact what exists is comparable, can only signify that just that being which is insusceptible of comparison belongs not to what exists but to those possibilities of thought with which abstraction plays when it takes to pieces the notion of reality and for its own ends substantiates parts which in reality only occur in combination.

Now whilst in practice explanations of the world admit the comparability of things, they misunderstand (as we remarked above) the significance of this admission, for they regard the content of it as far more natural and self-evident than it is. According to them just because nothing keeps the differences between the natures of things within certain definite limits, it is equally possible that they should all be comparable and that each should be immeasurably different from the others. Now if in reality only the first and not the second alternative is met with, there need not on that account be any more intimate relationship, any bond between individual things that could in any way detract from their independence ; each might be wholly independent of others, the reality of *its* content (as in every instance of a general notion) might, in spite of its similarity to some other, yet be wholly independent of that other. Hence if things should happen to be partly similar and partly contrary, there would be no reason for regarding such a state of things with suspicion, and it would be self-evident that the reciprocal action of things would have the same result which the opposition or the similarity of natures brought together in some definite relation must always have.

But is it a fact that all this is self-evident ? Or if it seems so, does not the self-evidence result from long custom, which dulls our apprehension of what is wonderful in familiar things? The combined impression of all experience early taught us to know the world as a coherent whole, within which each several content, every state, every quality, every nature of anything comes into conjunction with other contents, states, qualities, and natures in such a way that from the combination there may arise the complete cause of a new result. At present, *after* having this experience, it does indeed seem to us self-evident that each individual, however isolated and independent it may at first seem, is yet included in the web of this universal world, embracing truth and correspondence of all existence; but considered in itself this fact is calculated to excite inexhaustible wonder. And this wonder is by no means allayed by the cool reflections which we have just cited; the equal or unequal probability of various cases can be discussed in such a manner only when those cases themselves are regarded as being already constituents of a world with reference to which there has already been established from the beginning the universal validity of certain laws—laws that enable us to distinguish the possible from the impossible, and to estimate the different or equal probability of different cases. Only when we have already assumed that there is One Truth which is valid amid the multiplicity of reality, and have once for all resolved to regard the signification of this validity as clear, and not to ask further in what it is precisely that this dominion of truth and subordination of existence to it consist—only then is it that everything in reality which is in accordance herewith seems to us to be self-evident, and to have its validity guaranteed by that universal truth which is past comprehension. But what exactly is it that we do in making this assumption? How do we reach the assurance that truth, if only it be true, will achieve dominion over all things, whatever the nature of these may consist in?

There is a perverse way of representing these things which I have already found frequent occasion to criticise. We are accustomed to speak of the laws of Nature and of the cosmic order as though each were something independent, and were between or outside of or above things, and ready to enforce their obedience to its commands. A glance at social relations showed us a case of this error. Where would be the law of a state if all its citizens slept, or if the plague had swept them all off, or if all willed something different from the law? In the last case it would at any rate have an efficient existence as causing reproaches of conscience in the minds of the disobedient; in the first two it would exist only in the form of a temporary continuance of that order of material conditions that it had created; in general it has controlling efficacy only as it lives in the citizens, as conscious idea, as disposition, as personal conviction, and as conforming will; it never exists between or outside of or above them. And with the laws of things the case is no otherwise. It is not that they constrain things to act as they do; but things themselves act, and act in such a fashion that in reflecting upon their action we are able to find a law guided by which, in predicting a consequence from given conditions, we reach a conclusion that coincides with reality. But after we have developed this thought of a law which at bottom is nothing more than the unvarying nature of real things and of their action, this creation of our thought grows under our hands and easily comes to wear the appearance of a truth valid in itself and preceding reality; and it then seems to us self-evident that even existent things should obey that which is in itself true and necessary. The self-evidence we may now admit; but not for the reason given, which represents it erroneously. It exists for us in so far as we regard it as belonging to the innermost being of

things—the being in virtue of which they are things—that
their natures are not incommensurably different, but are comparable; that none of them is simply unique of its kind; that
even a thing which had no equals would be distinguished
only by the special position which it occupied in the cosmic
system, or by some peculiar combination of qualities which
are also found out of that combination as constituents of
" contingent aspects"—thus having definite relations among
themselves and to other things.   It is only if these presuppositions be made that it is, in our view, conceivable that
One Truth should control the Manifold of Reality, and that
changing relations should produce a system of causes from
which springs an ordered sequence of results.

We do not ascribe to this assertion—that there is a correspondence between all things which is a necessity of thought—
greater significance than it can possess; it contains no reason
for understanding the connection of things as being yet closer
—of intensifying it, for instance, to a common origin from
one source or to continuous inherence in one substance.   Yet
it abolishes the supposed self-dependence, in unconstrained
freedom and isolation, of each thing, and draws attention to a
connectedness between the contents of things which is everywhere assumed in attempts to explain the world without its
being made quite clear how much the assumption admits.

§ 5. We have been hitherto speaking only of the comparability of things or of the relations between them which
contain the ground of some future event; we have not as
yet spoken of those connections the introduction of which
constrains that nature in things to which we have referred
actually to produce these possible results by their reciprocal
action.   In turning to this subject we shall for the present
disregard some questions which at this point are beginning
to force themselves upon us, but would divide our attention detrimentally.   It may remain undecided under what
form of intuition we are accustomed to imagine, or have to
imagine, those connections between things which constrain
them to reciprocal actions; and we may likewise leave

undisturbed the question whether we should regard them generally as sometimes being present and sometimes absent; or on the other hand as always subsisting, but as being sometimes forced upon our observation and sometimes withdrawn from it by an infinitely varied gradation of their intimacy or closeness, and the correspondingly varying magnitude of the effects that depend upon them. We will set out from the fact that they have hitherto appeared to us as relations between things; connecting with this fact the question how they can exist thus *between* things; and how—supposing that they do thus exist—they could act upon things as conditioning forces.

When in thought we compare two things of which one is greater and the other less, and recognise a difference between them, the dividing and connecting *between* that arises here consists in the consciousness of a change of our inward condition which we experienced when our ideation of the greater passed into ideation of the less. This third idea, which is a state of our mind in the same sense as the two previous ones which are compared in it, partakes of the same kind of reality as they do. Now what is it that can give to the *between* of things themselves—to the connection which joins them and not their ideational images—a reality similar to that possessed by the things? Besides what exists there is nothing except what is non-existent; that which neither is the things themselves, nor is in them, must sink unsupported into a complete vacuum in which it neither can simply exist, nor exist with various definite values; and can least of all subsist as a unifying and connecting power superior to things. It is easy to imagine a connecting background of all things, on which, as a firm support, connections may run from one thing to another; but as long as things themselves do not constitute this background, on further consideration the question will always recur, How can the connection, being a state of this bond between things, have any power over things which are themselves other than the bond? Between this bond and things there must be another Between, which the connections *must* include—and yet *could* not include, either if the Between were empty nothing or if it were filled up by some reality foreign to the things themselves. This ever-recurring difficulty will later force us to recognise that the thought of an objective connection between things is altogether impossible, and that what we use to call by this name is in all cases some state or action *in* things themselves. But at the present stage, when further elucidation of this assertion would lead us astray from the proximate topic of discussion, we will content ourselves with the recognition that at all events connections which exist between beings would be without significance as long as they existed only *between* them, and had not produced any internal state in the things themselves. As long as things feel and know nothing of the connections that hold between them, these cannot contain the cause of a change in things, and just as little the cause of their reciprocal action upon each other. Any being can be caused to change its state only by something that is actually in itself, by some passion of its own; only in as far as two beings cause this passion in one another, can they be reciprocally acting causes. But since they cannot produce this passion in each other by means of connections between them, the change which we assume in one must *be* a direct passion in the other, and the question arises, Upon what assumption is the fulfilment of this requirement thinkable ?

We may escape the tediousness of the explanation here required by a reference. We have repeatedly had occasion (cf. i., pp. 357–8) to consider the possibility of reciprocal action between things and the suppositions that have been made in the hope of explaining it. We convinced ourselves that all ideas of some influence passing from one thing to another, ended in impossibilities and contradictions. It could hardly be made clear what exactly that should be that did thus (as was supposed) pass between them—if it was some third real element, that detached itself from a first in order to pass over to a second, its movement between the two might indeed be capable of presentation in idea, but the problem of reciprocal action would remain unsolved; and indeed doubled—for we should need to ask how this third element could be sent out from the first, and how its reaching the second could be the cause of any passion in that second; if the third something were a force, an effect, or a state, neither of our two obscure points would be made clearer, but even what before was plain would become obscure—namely, how all these which can exist only as attributes of a being could detach themselves from any one element, float for an instant in the vacuum between the two, and then taking a definite direction, arrive at the second as the goal of their movement; this second being at the same time the point of their return to the sphere of existence.   All these difficulties have frequently led to the attempt to deny altogether this inexplicable reciprocal action, and to put in its place a predetermined harmony of cosmic order, according to which the states of the different things accompany and correspond to one another, without having to be produced by reciprocal action.   But it was quite idle to imagine an order separated from the things in the changes of which alone it could have any reality.   Only if the course of all, even of the most trivial, events were fixed by immutable predestination, could the assumption of a Pre-established Harmony—not indeed *explain* anything, but — tolerably well *describe* the facts.   But it is impossible that there could be such a harmony which as a general law should predetermine the necessary consequences of contingent events; for if a change of some constituent of the universe (and it is of such that all these consequences must finally consist) has to follow and correspond to any event that may or may not happen whenever it does happen, then that constituent must be able to distinguish the occurrence from the non-occurrence of the event by some passion which the event produces in it, and the action and reaction which it was desired to banish would thus be necessary for the comprehension of that harmony which is intended to replace it.   And the most desperate efforts to find in the continual mediating activity of God the bond to which it is due that the states of one thing become the efficient causes of change in another, cannot obviate our speculative scruples, as long as they separate God and things from one another in the same way as individual things used to be separated from one another. For these views, too, only double the unsolved problem—they suppose an action of things upon God, and a reaction of God upon them, and explain neither the action nor the reaction. It has seemed to us indispensable to remove this separation, and in a substantial community of being between all things to find the possibility of the states of one becoming efficient causes of the changes of another. It is only if individual things do not float independent or left to themselves in a vacuum across which no connection can reach—only if all of them, being finite individuals, are at the same time only parts of one single Infinite Substance, which embraces them all and cherishes them all within itself, that their reciprocal action, or what we call such, is possible. For only then can the change which any one of them experiences *be* at the same time a state of the Infinite, so that it is not necessary for its influence to extend across a gulf which can never be filled up, in order to produce this state; only then can the result which this state produces in the Infinite, in accordance with the truth of its own nature, appear at the same time as a change of other individual things without there being any need of some fresh process by which it may be produced in them.

Now how this itself is thinkable—under what form that one all-embracing Being may be represented in idea, and how in its unity the plurality of finite things may be contained— is a question which we reserve for later consideration, being quite conscious that what we have hitherto done has been to make a demand which was unavoidable, without having as yet shown that it was capable of being satisfied. But there is another question which we do not reserve for later consideration, and the repetition of which would only convince us of the fruitlessness of all previous considerations—the question how within any one being that action could take place which we must presuppose in order to understand how any fresh state of the being in question could result from its preceding states ? I should be glad to hope that I had succeeded in making clear the self-contradictory circle which this curiosity involves. For whatever process it may devise to fill up the apparent chasm between reason and consequent this process would always consist of a longer or shorter chain of events of which every two consecutive ones would be connected by the same uncomprehended action, the very possibility of which it had been attempted to explain by means of their collocation. It is not at this impossible explanation that we have aimed ; how a cause begins to produce its *immediate* effect, how a condition is the foundation of its *direct* result it will never be possible to say ; yet that cause and condition *do* thus act must be reckoned among those simple facts that compose the reality which is the object of all our investigations. But there was an intolerable contradiction in the assumption that though two beings may be wholly independent the one of the other, yet that which takes place in one can be a cause of change in the other ; things that do not affect each other at all, cannot at the same time affect each other in such a manner that the one is guided by the other. It was necessary to remove this contradiction, in order to make room for recognising the fact of that ever-incomprehensible connection of states ; but we have never held that by removing a hindrance that stood in the way of acknowledging its occurrence, one could make more intelligible the actual way in which this connection is brought to pass (if I may for once make use of this self-contradictory expression).

§ 6. The detail with which on a previous occasion (cf. i., p. 365) the last part of this train of thought was elucidated, may justify the comparative brevity of the above repetition, which is only intended to call to mind the results which we now wish to combine with the preceding results of our reflection.

I. To our minds all intelligibility of the cosmic course depends upon universal relations, which connect all things

together.   Of course things must *be* in order that they may
be connected with one another, but the being which we think
as yet unrelated and which we represent to ourselves in idea
as the ground of the possibility of related being, is not a
reality that occurs independently, beginning from which things
enter later into reciprocal relations, and to which (getting rid
of all relations) they can return again; the truth rather is
that it is latent in the forms of related being, and inseparable
from these, and is in truth only the affirmation, positing, or
reality of these relations themselves.

II. Also that τὸ τί of things—namely their nature—by
which each individual one is distinguished from every other,
is at least so far similar or comparable for all things, that
there can be one universal truth, valid throughout the world,
according to which, from certain definite relations of things,
there flow definite results, and from other relations other
results.   The possibility that any combinations whatever of
things should become adequate grounds of a consequence
definite in itself or capable of being specified, forbids the
assumption that anything whatever can have a content
absolutely unconditioned or unique; at most it could only
be the sole actual example of one content which, whether
simple or compound, may be thought as an universal occurring
in various examples, and being combined in thought with
other contents, according to the universal laws of truth,
may be regarded as the adequate ground of any third
content.

III. Not only are the natures of things actually so adapted
that they can supplement one another so as to become the
causes of results, but also this fact of their correspondence
must be understood by reference to a continuous and substantial unity of all.   That correspondence is not a lucky hit
which alone has been realized among many equally possible
but actually unrealized cases of non-adaptation of beings
independent of one another and perfectly self-dependent as
regards their content—but it depends upon this, that all
which exists is but One Infinite Being which stamps upon

individual things in fitting forms its own ever-similar and self-identical nature. Only on the assumption of this substantial unity is that intelligible which we call the reciprocal action of different things, and which in truth is always the reciprocal action of the different states of one and the same thing.

CHAPTER II.

THE SPATIAL AND SUPERSENSUOUS WORLDS.

The Doctrine of the Ideality of Space—The Correspondence of the Real Intellectual and of the Apparent Spatial Places of Things—Removal of even the Intellectual Relations between Things ; Sole Reality of Reciprocal Action— Notion of Action—Summary.

§ 1. IN considering the connection between bodily and mental life we have had frequent occasion to ask what is to be regarded as the real τὸ τί and nature of things. The numerous transformations which in the course of our reflection are undergone by the answers which we at first put forth with undoubting confidence, and the gradual removal of the prejudices by which at the beginning we commonly allow ourselves to be ruled, will have been sufficiently traced in our previous considerations to allow of our now connecting further reflections with the results which we had provisionally reached, without the need of repetition. The things of the sensuous world will first occupy our attention.

We have long ago left behind us the standpoints from which it appeared at first that things consisted directly in a combined multiplicity of sensuous qualities, and then that the matter which was at the foundation of them all constantly occupied space; even the atoms into which the need of explaining Nature necessarily drove us to resolve that which is efficient in the world of sense, we could not regard as homogeneous but minute particles of that ever-extended universal matter ; spatial extension, form, and magnitude could not belong to their being, much less constitute the whole and exhaustive content of that being. It seemed to us that these spatial properties belonged only to what was composite, not to the simple elements from the repetition of which the composite arises; that unextended beings sending forth their effects from different points of space, and by their forces reciprocally prescribing positions to one another and maintaining these positions, produce images of extended substances which we intuit, and which with more or less intensity of coherence and impenetrability seem always under different conditions to occupy different parts of space. The nature of those simple beings themselves we left undecided; we only characterized them—in expressions chiefly of negative signification—as supersensuous, intellectual, and intensive, in contrast to that which we, in accordance with common opinion, had up to this point regarded as the τὸ τί of things; we could only point to the nature of souls as furnishing an illustration of what was meant by these words (cf. i., pp. 326 *scq.*). But throughout all these considerations we have in one point held fast to common opinion—we have retained the idea of an infinite space stretching beyond us and between things, in order to serve as a place for things; as the theatre of their actions and reactions and an ever-present background, making possible the existence of connections between them; and finally (by the alternations of remoteness and nearness which it allows) conditioning the exercise of these effects sometimes as hindrance and sometimes as furtherance. We have now come to the point at which we must reject this assumption— the temporary acceptance of which was necessary for the simplification of the problems with which we have had to deal, and was possible because the changed view which we must now substitute for it will allow without detriment of our using in all the details of our investigation of Nature, the modes of expression founded upon it.

Although obscured by the last stage of modern philosophy (which regards its retrogression in this point as a particularly successful step in advance), Kant's doctrine of the Ideality of Space is still so fresh in the remembrance of modern culture, that I shall most simply express what is essential in my own view by briefly agreeing with it. I hold that space and all spatial connections are merely forms of our subjective intui-

tion, not applicable to those things and those relations of
things which are the efficient causes of all particular sensuous
intuitions — this kernel of Kant's doctrine I accept unreservedly.　I should be happy if I could accept with equal
unreserve the arguments by which he supports it, or the way
in which he uses it for the construction of his philosophic
theory.　But I can do neither; and being unable to refer to
any accepted doctrine, I am constrained to attempt a very
brief outline of my own view, which could only be
demonstrated by a special scientific investigation, since it
would necessarily involve laborious examination of countless
objections.

Our ideas are not what they signify—the idea of sweet is
not sweet, the idea of half is not half.　And our intuitions of
extended things do not themselves possess those properties
which make up the content intuited, and there do not exist
between them those spatial connections the existence of which
between the objects intuited are indicated by them.　Our
idea of the greater is not itself greater than that of the less,
our idea of a triangle is not triangular, and our idea of something which is to the left is not itself situated to the left of
the idea of something else which is to the right, having the
same position and distance with respect to it which are, *by*
the two ideas, ascribed to some two particular points of an
object.　Therefore, however certainly it may appear to our
senses that endless space extends around and beyond us—
however self-evident it may seem that the definite local
relations of things which we perceive do exist outside of
us and between the things themselves — yet our intuition
of this space and our perception of these relations proceed
from the reciprocal action of impressions, or inner states of
our being, of which none has in itself spatial form, and
the mutual relations of which are like anything rather than
relations of position in space.　Hence space and extension are
not forms of *intuiting*, that is, not forms in which those mental
activities work which produce our ideas of the extended
universe ; but they are forms of *intuition*, if by this name we

mean to indicate the result of those activities, the finished picture itself, the vision of endless extension which floats before our consciousness as contrasted with the non-spatial and merely intensive activities of ideation to which that vision is due.

If now the very means by which this space-world that we intuit takes hold of our own mind are so wholly unlike that world, we may easily be tempted to conclude that there will therefore be as little, or if possible still less, likeness between it and the outer world by the influence of which upon our mental states the space-intuiting activity of our soul is aroused. Things without form, not therefore unsubstantial but characteristically differenced by the variety of their supersensuous content, arranged in a multiplicity of relations not spatial but intellectual, would then by the direct reciprocal action subsisting between them and men's souls constrain those souls to make the various impressions communicated by things the objects of intuiting consciousness. But in concluding thus, we should have arrived at a right conclusion by a wrong road; for the way in which spatial intuition arises from the reciprocal action of non-spatial impressions in us, decides nothing concerning the spatial or non-spatial character of the external world from which these impressions come. We have long ago reached the conviction that it is in this way that our intuition of space must arise, whether an extended universe exists outside us or not. For even if it existed, our mind, which is not extended, could never be entered by extended images of things, with their relations of magnitude and position; and even if such did enter the mind, their actual existence in the soul would have a different significance from their being intuited. Even the impressions of a real space-world must, in order to exist for us, be transformed into an ordered multiplicity of non-spatial excitations of our soul; and in any case it is only from these that our intuition of the world of space could be built up. And hence psychological investigations as to the way in which the intuition of what

is extended arises in us, or does not arise, but is, may be,
innate, cannot decide the question with which we are concerned.   Only a metaphysical discussion as to the kind of
reality that space, after it has been thought and as it has
been thought, could have on account of that which it then is
or signifies, can establish its ideality or its Realness.

Now apparently it is not quite easy to say what we really
hold space to be, when we think of it as empty and
infinite extension; the attempt to do so soon makes us feel
the uniqueness of this idea, for the elucidation of which we
can find no homogeneous analogies, and hardly any images
which are not borrowed from the wholly peculiar nature of
just that which is extended.   To regard space as something
infinite or as a property of things, is to entertain thoughts
which no one in the present day will think it necessary to
turn back and refute, for even in the pre-Christian era it
was plainly seen in what contradictions we should be involved by this assumption, partly as to the existence of
things in space, and partly as to their movement through
space. ˌAnd the habit which modern culture has of calling it
a form, a relation, or an order of things, is little more satisfactory ; for all this is just what it plainly is not—formless in
itself, it could serve but as a background, lending itself to the
purposes of form, relation, or order, and being in its nature
capable of having an endless variety of forms inscribed in it,
countless relations subsisting in it, and the most varied
imaginable arrangements of a plurality presented in it.   But
it will scarcely escape observation that this name *background*
is but another denomination of space itself ; and hence from
this correction of the ordinary view we learn not so much
what space is as what services it can render to our combining and discriminating intuition of a conceived manifold—
it appears as the possibility of the juxtaposition of a plurality ;
but what it is in itself that makes it capable of affording
this possibility remains unexplained.   To this we may add
that even this last expression involves a circle, for juxtaposition is a kind of simultaneity which is distinguished from

other kinds of simultaneity only by its thoroughly spatial character.

And this very remark may put us into the right way of fixing our attention not primarily on space but on the general laws of extension, letting space itself arise at a later stage from the application of these laws—as without doubt psychologically the intuition of space as an infinite whole comes later. For we have given to us originally, whether as innate gift, or as the first result of the reciprocal action of our impressions, the certainty that any point can be reached from any other point by one, and only one, straight line, that all the points in this line are perfectly homogeneous and equal in value when considered with reference to the two terminal points, and that they hold a similar relation to every other point. We have, I say, the certainty expressed in these statements, or in any others—we need not here inquire what—by which the nature of juxtaposition is so completely expressed that the first principles of geometry may be based upon them. In logical form, our expression is an universal law, as must be also any more exact expression by which its place may be supplied; but yet the peculiarity of its content essentially distinguishes it, even formally, from the formative law which every universal concept imposes on its particular examples. The concept only requires that each instance of it, regarded in itself, should contain a definite group of characteristics combined in a definite manner; hence it does indeed subordinate to itself its individual examples, but it does not establish between them any significant connection, by which they may reciprocally work upon and affect each other. For what we logically call their *co*-ordination, only indicates the complete similarity of the way in which they are *sub*-ordinated to the universal, and beyond the similarity which must of course belong to them on account of this subordination it has no effect upon their reciprocal behaviour, this remaining wholly undetermined by it. The same thing holds of all other general laws which comprise under them a variety of individual cases; they are valid in *every one* of these cases taken separately, but do not bring the different cases into any mutual connection.

It is quite the reverse with the law of juxtaposition. When it declares that between any two points one and only one straight line is possible and necessary, it not only asserts in a general way that every second or third pair of points is subject to the same law of connection, but it requires at the same time that the second pair should be regarded as connected with the first—in short, every pair with every other pair—in the same way and after the same fashion as the members of every pair with one another. Thus it combines all the different instances of its application into one *whole* which coheres together according to the same rule by which any two of its parts are connected, and does not allow us to think of any single case of its application existing as it were isolated in a world of its own, without attaching itself to this whole as a part of it. Hence it is that here for the first time co-ordination has a special meaning; particular spaces are not only subordinated to a general notion of extension as examples of it, but they are at the same time joined to and co-ordinated with one another according to the general laws of space-construction as parts of one space. Thus it is that space is as it were a picture, and this is the reason why we prefer, and see preferred, for it the name of *intuition*, which denotes something essentially different from that which is denoted by *notion*. In the same peculiarity of the law of extension or juxtaposition' which we have emphasized, there is at the same time contained (as in the nature of every series) the possibility of endless progression by which to the members already given new ones may constantly be added, according to the same formula by which the old ones are connected; thus it is that space extends to infinity. By an arbitrarily chosen expression that has a tinge of contempt, we may, it is true, describe this by saying that space is unending because of its inherent incapacity of self-limitation. Without going into scholastic controversies we may cheerfully accept even this interpretation; we are not aware of needing any other infinity of space than that which is here asserted, and we can hardly regard as the mere lack of some better property its characteristic of not only not resisting any advance beyond temporarily assumed limits, but of moreover pointing out a definite path for such progression. If we put together what we have said, space appears to us as a kind of integral by which that whole is given which proceeds from the summation of all the infinitely numerous applications of the law of juxtaposition, when we abstract wholly from the nature of the reality that stands in those relations, and substitute for it the mere empty framework of the related points. Now when we have once got hold of the intuition of space, space appears to us as the all-embracing whole, in which and through which is possible the multiplicity of all those relations from the summation of which it has itself really originated.

Now if this is the signification of space, the question as to the nature of its reality scarcely needs a special answer Even those who regarded space as mere empty form fitted for the reception of things, must have acknowledged that empty forms could be thought existing previous to real things only as formed material, and therefore themselves something real and capable of receiving the other real things; as unreal forms, unsupported by matter of which they are the form, they can of course exist only in thought that has abstracted from matter. Just as little could relations and arrangements have an independent existence previous to the things which are to enter into them; they too, if separated from those things, could have a place of existence only in that activity of mind by which they are thought. It is scarcely necessary to add that still less can space as comprising the collected results of an infinite number of possible relations have its existence anywhere else than in the activity of intuition which is conscious of this result of its relating movement, manifested in combination, division, and systematization. Space does not exist *between* things and preceding them in such a way that things are in it, but it diffuses itself in things, at least in souls, as the extension which can exist only for thought, in which we assign their places to impressions which we receive through reciprocal action between our minds and the outer world, that is the things which are not ourselves. It is only misplaced respect for a venerable error which would rejoin that even the non-existent may exist, and that even relations which have nothing real in them may have an objective existence independent of our thought. There is but cheap wisdom in asserting that even mere appearance and nothingness and error do exist after a fashion; that the sense in which the past and the future are non-existent is different from the sense in which what has never been and what is for ever impossible are non-existent; it is just this fashion and the meaning of this existence which we have above, in treating of our present subject, space, endeavoured to fix in a somewhat better way than by means of these indefinite expressions. We did it by trying to ascertain the kind of reality that can be attributed to space, instead of consigning space offhand to the region of non-existence. And this reality is to be found in its existence as intuition in ideating beings, not in existence as a vacuum independent of them. By such determination its reality is not diminished, but its nature is fixed. As events really *happen* although they never *are ;* as light really *shines* although only to the eye that perceives it; as the power of money and the truth of mathematical laws really *exert their influence,* though the first *exists* nowhere except in the estimation of men, and the second nowhere except in the actual things to which they relate—so space has reality although it does not exist but only *appears.* For reality is like a sun that rises upon the just and upon the unjust; it embraces not only the existence of that which exists, but also the process of that which happens and the validity of relations and the appearance of phænomena; the mistake is in attributing to any one of these the kind of reality which can belong only to one of the others, and in complaining when there is assigned to each of them the place and the particular kind of existence possible for it.

§ 2. It was other reasons than those here advanced that led Kant to his doctrine of the Ideality of Space, and caused a further development of that doctrine to which we cannot give our adhesion. Among the thoughts that belong to this subject we will only briefly mention the practice of regarding space as being a subjective form of *human* intuition alone, and of considering it possible that other knowing beings may make use of other forms of intuition which we cannot even guess at. If the chief stress is laid upon reckoning space as an innate *à priori* possession of the mind, as contrasted with that content of knowledge which is brought to us by experience, it is natural to bring into prominence the thought that its peculiarity depends on the nature of the intuiting mind, and that in differently constituted minds different forms of intuition may take its place. Herbart's recent attempt to exhibit all the *à priori* forms of our knowledge as results which must necessarily be produced by the reciprocal action of different ideas in every ideating being, has led to the opposite presupposition with regard to space itself—namely, that to every being whose mode of cognition depends upon a mechanism of reciprocally acting individual ideas, the plurality of his impressions must appear in space-relations. I do not consider it possible to choose decisively between these two opposed views. It seems to me that all the deductions have failed which have attempted to show the necessity that space must be, or must be intuited, either from assumptions concerning the necessary development of the cosmic content or from self-evident laws of the reciprocal action in all ideation. Such attempts, when they took the first way, have only deduced certain abstract postulates from the notion of a self-developing Absolute—postulates which did not even posit space or show how space could be deduced from them; postulates such that he and he only who was already acquainted with space, could guess that by it they would be satisfied. Where they took the second way they have only succeeded in producing space by taking certain figurative expressions borrowed from it (expressions which, as they at first maintained, they used only in an abstract non-spatial sense), and reintroducing into them somewhere in the course of the deduction their proper spatial meaning. Hence it does not seem to me to be proved that in every ideating being, whose mode of cognition may be compared with ours, the intuition of a manifold must everywhere take place under the form of space; I would not precisely assert, but still I conjecture that this demonstration, which has always hitherto failed, is impossible.

It was natural and right to oppose this space-world as phænomenal to the world of real existence, but erroneous to exaggerate the distinction between the two to such an extent as to make it appear that they were insusceptible of comparison, and especially (after the fashion adopted by popular culture when penetrated by the doctrine of the ideality of space) to revel expressly in the thought of this incomparability, as though it were a guarantee of everything that is best. It was erroneous to regard space as a form of our intuition, by which things were received, extension being quite alien to things as they are in themselves; for it is certain after all that nothing can be received by a form to which it is not in some way suited. Equally inexact was the expression used by Kant himself—that cognition having for so long accepted from experience the laws by which it judged of things, it was time to see whether conversely cognition could not prescribe laws to things, if only the laws of their appearance—for it is obvious that the cognizing mind itself may determine the general colouring of the reality that appears to it, but in order that it may cognize at all it must receive from the nature of that which appears, at least the special outlines of the phænomenon. More generally expressed, the inadequacy of this view lay in this—that while it attributed the intuition of space to the mind as an innate possession, it did not attempt to explain the application of this possession. We have not only an intuition of empty space, but also a spatial intuition of the full content of the world; and it remained to show how in those empty forms with which, as it was said, we encounter the reality of experience, this reality can take its appointed position, and assume its appointed form. The solution of this problem was impossible without the assumption that there exist between things themselves manifold connections the special distinctions and meanings of which are reflected in corresponding forms of spatial relation, or may be transposed into the language of space ; however unknown and inscrutable in other respects the nature of things may be considered to be, the view in question cannot, without cancelling itself, disclaim this much knowledge concerning them.

In order that this standpoint which we have taken up in the above discussion, and which we do not wish to keep always to ourselves, may become sufficiently familiar to be intelligible, we need only call to mind how, when we draw comparisons, whatever the object of our thought may be, spatial images always press in spontaneously to give the greatest attainable degree of clearness by making it, as it were, visible to the mind. We may indeed think of a non-spatial plurality ; but we never represent it to ourselves without distributing the plurality in different parts of co-presented space ; we illustrate any unity by spatial boundary lines by which it is shut off from others and shut into itself ; there is no abstract idea of variety, contrast, or degree of relationship that we present to ourselves in idea without mentally endowing the content of these notions with visible form, by images of various spatial situation, form, direction, and distance. And even these words (content, contrast, presentation, and so forth), as well as innumerable words which indicate relations (to which the progress of civilisation has gradually attached the abstract signification which they now possess), plainly show, when etymologically considered, that they owe their origin to spatial intuitions. Therefore we scarcely need to exhibit further this capacity of space to give sensible form to the most multifarious variety and gradation of intellectual relations by the unbounded multiplicity of the possible relations between its points ; we rather need to convince imagination, accustomed as it is to this symbolism, that those very relations which it loves to represent spatially have a special meaning of their own, that is merely reflected in this spatial form without being bound to it. The structure of the world of sound or of mathematical truth may serve as illustrations of such relation. Without the spatial images of height and depth and intervals, the relations between tones would not be clear to us in thought, although in sensation we are conscious of their simply qualitative nature; as regards mathematical truths or the relations of pure number, since they have no sensuous images, we more easily comprehend them as what they really are—as systems of members the reciprocal dependence of which, varying extremely in degree, is of a wholly abstract nature, neither standing in need of spatial symbolism for its subsistence nor even, in some cases, admitting of it. These examples will suffice to illustrate provisionally those intellectual relations which we assume to hold between the manifold things which exist. Whatever the natures of things may be, and whatever the general kind of relationship between them, the things will not be insusceptible of comparison, and the closeness of the relationship will be capable of unlimited gradation; hence everything by its nature and the totality of its relations to all other things is not only distinguished from all other things and thereby isolated, but also —like a note which has its own immutable place in the scale, or like a truth which has its own definite place in the system, coming between those upon which it depends, and those which depend upon it—everything has its own definite place, in the fabric of reality, between other things which are related to it with different degrees of nearness or contrast. And, moreover, in correspondence with this intellectual order, everything will appear to a soul in which its influence encounters a capacity for spatial intuition, to have that definite place among the images of other things which seems to be assigned to it by the totality of its intellectual relations to them; and this place which it has will seem to change, and the thing

itself to move through the intuited space, if these relations
which it has to the rest of the world are changed.

The spatial appearance of the world does not altogether
result from the mere *existence* of the intellectual order among
things; it is only complete when this order exerts its *influence*
upon those to whom it is to appear. It cannot therefore be
the same to all by whom it is intuited; for in this intellectual
whole of the universe souls themselves occupy places at
different points of the structure; on these parts, which have
different values, the action of the whole is different, and
accordingly that whole wears for them a different aspect; to
each of them there appears but a section of it, and this with
that specially foreshortened projection which corresponds to
the difference of the position in the world which this being,
as compared with its neighbours, occupies in the intellectual
order of things. So as a whole it is indeed the same world
which we see, but to each it is different in detail; one person
could share exactly the view of another only if he could be
transferred from *his own* relations to the world as a whole,
into those in which the other stands—a change which to him
must seem to be a spatial movement of himself through the
space-world which appears to him. An easy continuation of
these considerations teaches—what to exhibit here in detail
would require a superfluous expenditure of words—that as
the images formed by different souls of the space-world
surrounding them on the one hand are not identical, so on the
other hand they are not without connection. Each appears
to every other to have some definite position in the space-
world intuited by that other, and each attributes at the same
time to his own image in the space-world which he beholds
such a position with regard to the image of the other, that in
order to change places opposite movements in the same line
must seem necessary to each; hence within the space-world,
which to each *seems* to stretch between him and the other,
whilst in truth it exists only in themselves, each will be able
to find the other out, and they will be able by definite movements to meet and enter into reciprocal action. It is neces-

sary to think this out thoroughly for oneself; for philosophic theories have little value if they can only be laboriously demonstrated in the lecture room, and in practical life remain uncredited because it is not easy to find the connection between them and everyday occurrences. Without myself making the attempt here in detail, I venture to hope that a further pursuit of the indications I have given will wholly remove the appearance of paradox which the doctrine of the ideality of space generally has at first sight for the common consciousness. Under the above-explained conditions of merely subjectively intuited space, we have in point of fact exactly what would be afforded us by a real objective existence of space if such were possible; no part of the phænomena with which we are familiar and of their persuasive evidence is inexplicable on our assumption as to the real state of the case; even when our presupposition has been accepted in principle and on the whole, it does not make necessary a violent change in received expressions and ideas that refer to details. As we always speak of the rising and setting of the sun, and shall never substitute for them awkward expressions framed according to the actual and well-known condition of things, we may continue to look at the world, as far as all practical details are concerned, as though space were spread around us and we ourselves were floating in it; it is only when we are concerned to establish ultimate principles according to which all the connection of phænomena is to be judged, that we—just as astronomers are—shall be obliged to recur to the true condition of things as the foundation of all the rules of phænomena.

At this point I would exclude from the circle of subjects we are considering, a department of thought the development of which is indeed important in itself, but would require a diffuseness of treatment which, for the object that we have in view, would not be compensated by any counterbalancing advantage. For the fundamental notions of natural philosophy which we must form concerning the concatenation of physical events will, it is plain, take a very different turn when we consider space as a real stage upon which all occurrences are presented, and when we regard it as a mere phænomenon in the semblance of which real action between things, which was originally of quite a different kind, comes afterwards to be clothed. In the latter case we can no longer regard movement in space as a performance by which we overcome distance, as though that distance were a reality ; we cannot speak of forces having a tendency to move bodies nearer to or further from one another in space, or which at a certain distance would encounter a certain amount of opposition to their action. For us all these simplest intuitions of natural philosophy will need reconstruction upon a new foundation. This reconstruction we do not here attempt, and will only remark by the way that for it many oft-discussed difficulties vanish into nothingness, and in their place others arise at points where, to the hitherto accepted mode of thought, nothing whatever suspicious seemed to lurk. But the problems with which we are at present concerned urge us in quite another direction.

§ 3. Let us grant that the reader was in a certain respect deceived when we compared the intellectual order of things, on which we held the order of their spatial appearance to be dependent, to the relations of sound, or to the articulation of a system of abstract doctrines. What we then needed was a provisional illustration, and to it we sacrificed for the moment the exactness which we must now turn back and seek. The two comparisons are inappropriate because they liken the order of immutable and eternally valid systems to an order that is mutable and merely factual. A system of truths is connected together in but one way, and that a way that never changes ; we, choosing different points of departure, may bring its several parts into special prominence in various combinations, and interpret for ourselves the results that flow therefrom ; but these results do not *arise* from our procedure ; *they* are eternally valid, and it is only in our consciousness of them, that is in the real state of a real being, in our own conscious soul, that something happens which has not existed eternally. The relations of sound too are eternally the same; they may, like those of the body of truths referred to, afford material for one, but only for one, spatial symbolization. Rightly constructed, this would for ever express the immutable organization of the scale, and be capable of being exhibited instead of the scale itself, as an object of consideration, and its whole wealth of inner relations being simultaneously present would as known vary with arbitrary movements of attention. Things on the other hand do not constitute a motionless organization of a manifold, in which every individual element, in virtue of its constant nature and the unchangeableness of its total relations to the rest, occupies an immutable position; they are, on the contrary, subject to movement, and obviously change their places in the intellectual whole of the cosmos no less than their phænomenal images change their places in space. Hence it follows either that their natures cannot be immutable, but must be mutable in order that the change in their reciprocal relations (which corresponds to the change in themselves) may explain the mutability of their spatial appearance—or the relations in which things stand to one another must in themselves be accessible to a mutation which does not at the same time affect the nature of the things.

If one is led to this alternative by the attempt to deduce the spatial places of phænomenal things from the intellectual places of real things, it is hardly doubtful that one will prefer to make an attempt to affirm in the first place the second member of the disjunction. For do not things as appearing in space seem to move without any mutation of their nature ? or if they seem to undergo any such mutation, is it not just the change of place which introduces the mutation and constitutes its cause ?   But if we think of things as being enclosed in a net of mutable intellectual relations, or as being moveable within this network, we encounter an inconceivability which is essentially the same as that which we have so long been trying to refute under the name of extension, which though empty is yet real in itself.   For we regarded objective space as unthinkable, not on account of its special geometrical nature, but because of its presenting a system of empty relations as an independent whole. But—to take up again a consideration previously indicated—all relations as such have existence and reality only in the consciousness of him whose mind exercises a definite relating activity; apart from consciousness they have not themselves an independent existence *between* the things related or relatable, but there is a foundation for them in the nature of things which are so framed that consciousness is constrained and enabled by their influence upon it, to connect and estimate by means of these relations the impressions which those things make upon it. Hence in the intellectual world also there is nothing *between* individual beings, nothing by change in which the beings themselves can be removed from or brought near to one another, or have their reciprocal action roused or hindered; but all these relations are part of the appearance which the intellectual world as a whole assumes for each of its parts which is capable of having anything whatever presented to it; moreover, by them there is interpreted only that being which springs up within individual beings, that multiplicity of inner reciprocal actions which in reality things exercise directly upon one another, being upon being, without the mediation of any such middle terms.

It must necessarily be difficult for us, considering the mode of apprehension to which the consideration of daily experience has accustomed us, to carry out the abstraction which we here demand; and it is worth while to elucidate it by some supplementary observations before we go on to deduce its further consequences. It seems to us all so self-evident that if an effect arises which previously did not exist, there must have been some mediating process by which it was brought about, and, moreover, all our previous considerations have so expressly and repeatedly made it a duty to seek the mechanical links in all action, that the demand which we now make will have a confusing effect not only in a general way but also as regards the coherence of our train of thought itself. But notwithstanding, we have for a long time been leading up to this

demand. We have already repeatedly emphasized the assertion that we cannot go on indefinitely requiring intermediary
machinery for the bringing about of the most simple results
and the elucidation of the most simple effects ; at some point
or other the chain of intermediaries must consist of simple
members connected together *immediately* and not requiring
something else to hold them together ; somewhere or other
there must be simple processes of reciprocal action, which
consist in this, that the inner condition of some being, as soon
as it exists, is the direct producing cause of some fresh inner
condition in a second being ; there must be somewhere that
real sympathetic affinity between existent things which a
widespread superstition unfortunately imagines it sees only
where, according to the unanimous testimony of experience, it
does *not* exist. We have already had often enough to convince ourselves that all attempts to explain still further these
most simple elements of action and occurrence, to elucidate
them by showing the way in which they come to pass, must
invariably fail ; but they fail not on account of the imperfection of our knowledge, but because the very existence of that
which they erroneously seek is impossible.

There is no process of action adapted to bring to pass
events which though all their conditions are present are not
as yet actual, but only a process of the gradual completion
of causes as yet incomplete. If any inner state of a being
is the adequate cause of change in another, the change
happens forthwith and does not need any process of realization ; if that state is not an adequate cause, no process of
action could constrain it to a result that does not spontaneously flow from it ; finally, if that state can, through a
series of intermediate links, pass into a second state, which
would constitute the complete cause of such a result, and if
there is a disturbance of states by which this transition is
accomplished, then previously to the accomplishment each of
these intermediaries must be followed by the event which
corresponds to *it* as result to cause, and only when this series
has been completed will that event occur which flows from

the thus established final state of the acting being as its necessary consequence. Hence the only path by which the primarily given state can attain its final operation, leads through these intermediate events; taken together and in the order of their succession, they constitute what we call the mechanism by which a result is realized. Therefore we can never refer to mechanism to explain how a result arises the complete cause of which already exists in actual states of actual things, and we shall always need a mechanism in order, in any real occurrence, to connect the first member with a final member of which the complete cause had not, under the form of inner conditions of existing reality, been realized by that first. For the significance of mechanism never consists in its being a kind of magic artifice, by which is brought about an, event which though all its conditions are complete, yet in some incomprehensible way delays to happen; in every case it is required only in the interests of the constancy and regularity of the cosmic course, which demand not only that every real occurrence should have an adequate ground, but also require that every intermediate link, by which the inadequate passes into the adequate, should itself be previously realized as an actual state of some real being. For only thus is each of these members an active cause, which not only within the being of which it is itself a state carries on the mutation of the inner states of that being, but also becomes the cause of mutations in other beings.

Our previous remarks have been intended to show that reciprocal action is not rendered less thinkable by our not allowing of anything between beings which can separate them or combine them or connect them with one another. It is not external mediation that is needed in action; not a multifarious transporting from this place to that, and from that place to this; we are relieved from all this apparatus by the knowledge that all things being parts of an Infinite that unites them as in one substance, they need no other bond than this in order that the states of one thing may have a determining significance for those of another. Those mediating links were themselves of an internal and intellectual kind; to things which by the universal metaphysical justice of this Infinite cannot in accordance with their meaning follow directly one from another, they give reality by making actual the intermediaries which render it possible that those things should follow from one another in accordance with this meaning. There *is* therefore nothing else than an eternal universal inner stream of reciprocal action in things; its individual waves are not caused by impulses communicated to things from without, they arise from the native consistency, according to which any previous state of a being that is not separated by any gulf from the inner existence of another, *becomes* directly a subsequent state of this other; we must get rid once for all not only of the thought of a network of spatial relations along which the conditions of action run backwards and forwards between things, but also of all idea of supersensuous intellectual bonds of connection which, lying outside of things and sometimes contracting and sometimes expanding, at one time bring things together so as to produce action, and at another break the contact necessary for reciprocal action.

If I have succeeded in making clear what I mean, I shall certainly be expected to give an answer to one other question. The ordinary view took pains to ward off all mutation from the inner nature of things, and held that change was only to be admitted in external relations. Now how can our present view—which puts all action wholly *in* things, and supposes universal mutability of their states — comport with the assumption of the unity which we ourselves have repeatedly pointed out as essential to the nature of everything? I might fairly pass this question over, if I were less in earnest; for even the views which most strongly emphasize the unity of the nature of things must in the end reconcile with that unity not only a change of inner states but also a simultaneous plurality of such states, as otherwise they would be destitute of a source from which to derive an explanation of the way in which events can occur at all. I forego this way of escaping the difficulty, though at this point I cannot fully answer the question proposed ; and, on the contrary, expressly admit that I only wish to dispose of it provisionally by (for the present) merely referring to a previous exposition (cf. i., pp. 168 *seq.*, 536 *seq.*) concerning the meaning of that unity which we really have reason to require in things. That exposition taught us to seek this meaning only in the consistency with which changing states of anything are so connected together that—having regard to the conditions under which they arise—they appear to be varying and manifold expressions of one and the same thought, in the realization of which the being of the thing consists. But we could never require unity in the nature of a thing in the same sense in which we are accustomed to use this expression to denote the monotony of an absolutely homogeneous quality ; unity of this kind can never be real, but is always a property of something else which is Real, and even this not in the signification which it would have if it could ever *be* even a part of the nature of this something else ; on the contrary, it is everywhere but a partial appearance which that thing wears for some consciousness which intuits it. Every simple quality exists only when it is perceived and only for him who perceives it ; if it could exist anywhere independent of him, it would still certainly not be the nature of anything, for in its simplicity it can only be or not be ; it cannot so change as to remain, in some fresh condition of its existence, the same that it was in a previous condition. But only that which is capable of and can outlast change can be substance, and this capacity things must have in order to be things ; the invariable, which can *only* either (1) *be* while it continues entirely homogeneous, or (2) be annihilated and give way to some other that takes its place— that thus may indeed have its turn of existence with others but cannot change *itself*—is always something unsubstantial that may be a predicate, but can never be a subject of predicates. However, I admit that this consideration does not completely answer the doubt expressed, further discussion of which we reserve for a short time.

§ 4. Having given these somewhat detailed explanations, we can now briefly add our later results to the previous ones, in somewhat changed order.

IV. The nature of everything by which it is distinguished from other things is one, as regards its consistency, but never simple in the sense in which a homogeneous quality is simple. An adequate knowledge of that nature (supposing such to be possible) would understand it in the form of a thought or of an Idea, for the unchanging meaning of which there are innumerable differing expressions, appearances, and verifications under differing conditions. With the limitation of never being or appearing, doing or suffering anything that is not a consistent expression of the fundamental thought which constitutes the being of anything—with this limitation everything is mutable, and can only be a thing or substance if it is mutable after this fashion.

V. The objective relations by which the commensurable natures of individual things are brought together for the *realization* of the result of which the content that is *thought* together is the *basis*, do not consist in spatial movements. The case is not that things are in space, in which they can move, but space is in things as the form of an intuition through which they themselves become conscious of their supersensuous relations to one another. The place occupied by any element at any definite moment on account of the totality of the relations which it then has to all the rest in the intellectual order of the world, determines the place in space at which this element must be intuited by the rest; to the change which the element experiences in the intellectual order there corresponds in spatial intuition the movement which hence has to be regarded as change of place, but not—at least not primarily—as a passage through space.

VI. The supersensuous order upon which we suppose that of the apparent spatial cosmos to depend, cannot be regarded as a mere intellectual counterpart of space in such a fashion that it too, like a web of independent and changing but non-spatial relations, comprehends things in itself and extends between them just in the same way as (according to an earlier view) space was supposed to have an independent existence as an encompassing background and as empty extension. *All* relations, even these intellectual relations, exist as relations only in the relating mind at those times when it exercises its relating activity. Therefore the supersensuous order of the world does not consist in a tissue of complicated relations *between* things, sometimes contracting, sometimes expanding, but only in the totality of the reciprocal action between things taking place in the world at every moment. The actions are not produced, changed, and organized by a multitude of impulses running backwards and forwards between things, but they themselves being comparable in meaning, and hence subject to universal laws, produce in one another impulses that become realized without the help of any mediating mechanism, and arrange themselves, according to their meaning (as constituents of the world's content which stand in need of one another), in that intellectual order which is *valid for* them but does not *exist between* them.

CHAPTER III.

THE REAL AND THE IDEAL.

Contradictions in the Notion of Things and in their Formal Determinations—
Idealistic Denial of Things—All that is Real is Mind—What it is that we
must seek to Construct, and What it is that we have to Recognise as
immediately given—Summary.

§ 1. WHAT we have recorded hitherto as the results of
our reflection has been of essentially formal
significance; we have sought to make clear to ourselves the
conditions under which it seemed that reality (*Wirklichkeit*) of
existence could belong to any being whatever its nature, and
reality of occurrence to any event whatever its content; but we
have not yet sought to determine *what* that may be which is,
or happens, according to these conditions. In doing this we
have perhaps had in some measure the feeling of a rich man
who is not concerned for the moment to reckon up his possessions in detail, but contents himself for the present with
marking in such a way his numerous flocks and herds whatever they may consist in, that in case of need he would be able
to recognise and to find his property. But a certain feeling
of perplexity takes hold upon us now that the time is come for
really (*wirklich*) showing in what our possessions consist, and
giving an account of what actually are the things and the events
which really being or happening satisfy the conditions that
we have sketched out. Wherever we may look there seems
to be nothing that we can specify—all that according to the
ordinary view forms the content of reality, the many-coloured
impressions of sense and the multifarious forms and movements of the extended universe, we have been forced to regard
as phænomena which do indeed reveal changing relations in
that which is truly real, but do not point out what it is that
this true reality consists in.

Now one might hope to get rid of this perplexity by a candid confession of human incapacity—by acknowledging that what things are in themselves and what effects they actually have upon one another in reciprocal action, must remain for ever unknown to us ; that only from the varying relations of that which appears is it possible for us to conclude to formally corresponding variations of this unknown, variations of which, however, we can never cognize the actual content. But the more certain we may be that at some point or other we shall be brought to this confession, the more necessary is it not to reach it prematurely, and by so doing avoid investigations which we ought to undertake even though they may promise no other result than the knowledge that we were mistaken with regard to that which our confession of incapacity assumed to be the highest attainment of cognition possible for us.

We shall do well to distinguish two kinds of ignorance. It may be that of anything which we are seeking in order to the fulfilment of some definite requirement of cognition, the general notion under which it should be thought is clear, while we perhaps only lack grounds for deciding among which of the various species of this universal we should reckon that which we seek. It may, however, also happen that nothing is clear to us except the need which we desire to satisfy by that for which we are seeking, and that of the essential nature of that which would be fitted to afford such satisfaction, we have not even, as it were, a generic image showing the possibility of that which we seek. If with regard to the question which at present occupies us, we found ourselves in the first of these two cases, we should be satisfied. To speak figuratively, we should then know not indeed *what* colour things and events would wear, but that they would have *some* colour, that is that their nature would be determined by some species of a genus familiar to us, the existence of which would be guaranteed to us by the generic image that we have of it.

Certainly in the present day people often think that with

regard to the notions by which it has hitherto been attempted to determine the being of things, we are in this comparatively favoured position, and possess, in those notions, truths which rightly mark out the genus of reality, and only leave undetermined the special colouring, the knowledge of which we can if necessary do without.   But to me it seems as though we were in the second and less favoured position, like a geometer who, having before him the result of an analytic calculation, cannot hit upon any geometrical construction by which that which is abstractly required may be presented in intuition. It seems that, as regards the formal conditions which we require from being and action, we are not only not in a position to point out any essential characteristic of that which is Real (*Real*), by which those conditions may be satisfied, but that those demands themselves require, with reference to reality, much of which either we perceive that it can *only* be thought, and cannot *be* and happen elsewhere than in thought, or at least of which we cannot perceive how it can be anything *more* than thought, how it can hold good of reality or occur in reality.   Taking a brief retrospect, I will illustrate the importance of this consideration by the thoughts which we have gradually developed concerning the τὸ τί of things, their unity, and the mode of their existence.

In the popular view the essence of things seems at first sight to consist in sensible qualities.   But it soon becomes plain that these are only states of our sensation, resulting in the most plausible case from reciprocal action between things and ourselves, but neither capable of existing except in him who feels, nor fitted, even if they could so exist, to constitute the nature of a thing.   We took refuge in supersensuous intellectual qualities.   That this name is not a mere combination of words destitute of an object, that there is something corresponding to it, we believed we could show by reference to mental properties which—as the properties denoted by good, evil, holy—seemed as a matter of fact to present examples of a content supersensuous and at the same time like sensible qualities in their simple intuitable definiteness.

But this was only seeming. Having regard to constancy of action, learnt from past experience or assumed for the future, beings might have these attributes imputed to them, and in contrast to the individual actions manifesting the attributes, the attributes themselves then look like original simple qualities; yet in themselves they only indicate a kind of demeanour of things, not what things *are* in order that they should demean themselves thus—this latter, however, being what we sought. And then one might think for a moment what it would be to look away from all illustrative examples, and to seek the being of things in qualities of quite another kind—a kind of which no one can form the slightest idea. But in doing this one would commit the error, so often blamed, of confusing the expression of a necessity of thought with actual knowledge of the object in question, and of believing that demands have been fulfilled by the mere fixation of them in a verbal expression — whilst either it cannot be shown that those demands are capable of being fulfilled by the reality to which they refer, or it can be shown that they are not capable of being fulfilled by it. For the name *unknown qualities* does indeed express, by the name *unknown*, our incapacity of cognizing those qualities; but in calling them *qualities* it keeps up the erroneous appearance of our having at least the general notion under which this unknown may be correctly thought as one of its species. Now not only have we no idea *what kind* of quality constitutes the being of things, but we err even in thinking that we may subsume this under the *general notion* of quality. For this name *quality,* as long as it has any definite meaning at all, always denotes something that by its nature has reality only as a state of feeling of some sensitive being, but which except in such a being, except as felt, cannot *exist* either independently or in dependence on something else.

It would seem then that nothing remains for us to do but to regard the being of things not as an unknown quality but simply as unknown. But even this complete renunciation of all pretensions to knowledge proves untenable; for as long as we wish to speak of things at all—and it is not apparent how we can comprehend phænomena without supposing things—we must assume that things have a nature capable of producing varying appearances under varying conditions. In this respect too, as we have already pointed out, a simple quality, even if it could *be*, would be incapable of constituting the being of things—that being, it seemed to us, could only consist in the unchanging significance of a thought which, without changing its meaning, manifests itself in different ways under different conditions. Now the word *thought* has a double meaning, signifying on the one hand *the activity of the thinker*, in virtue of which all his thoughts are thoughts, and on the other hand the *content thought*, by which one thought is distinguished from another. We have, of course, intended here to employ only the second meaning; things are not the thoughts of a thinker, but their being is so constituted that if knowledge of their content were possible at all, it could be adequate only in the form of a thought, combining many individual ideas by definite relations into one significant whole ; this nature of things itself, however, remaining an undivided unity, and by no means consisting of the plurality of relations and related points which we require for its representation in cognition. That this mode of thought also has its secret defects was betrayed by the difficulty which we had in rather silencing than refuting objections to it. The question how that which is in us the content of a thought can, independent of us, be a thing, we put off by the remark, just in itself, that this difficulty would recur in any case ; that whatever image we may frame in thought as to the nature of the thing, we are still left asking how that which is in us a thought-image can, without us, be a thing ; that therefore we should not seek to know how reality is produced, and that it is enough to know the content which, when realized after a fashion which must be always incomprehensible, is a Real thing. But all this is not quite convincing ; a thought in order to become a thing needs not merely this affirmation of reality that requires only to take it as it is found and posit it, but the thought itself lacks something in order to be that which when posited would be a thing. The thought, however affirmed, posited, or realized, would remain an existing thought and no more, and that this is not quite what we mean by the name *thing*, we certainly feel, although we may find it hard to point out what is lacking. We shall perhaps most easily get a clear notion of it by recalling a view which, little fettered by such scruples as ours, delights to characterize the being of things with the utmost brevity as an operative Idea. Here we see exactly what we want—the possibility of being operative is lacking to the realized thought, if it is nothing more than that. That identity with itself of the thought-content which we presupposed, as confirmed in the most diverse forms of its expression or manifestation, actually has reality only in as far as we think it, and follow it out in a train of thought which, bringing together its different steps, can become conscious of itself; we, the thinkers, in accepting a definite Idea, as determining the direction of our reflection, or in, as it were, putting at the disposal of that Idea the real living power of our thought—we alone it is who realize its identity with itself, by seeking for and finding that identity; it is we alone who by so doing give to the Idea (which yet certainly was a valid truth without any co-operation of ours) the only kind of reality that could possibly belong to it, namely that of being a thought really thought by some thinker. Our intention and our living effort either theoretically to recognise the meaning of the Idea, in all its instances or consequences, as self-identical, and to remove all apparent exceptions to this consistency, or in practice to carry out the Idea under the most diverse circumstances, to get rid of all opposition to it, and to secure an adequate expression of its essential content under the most varied conditions—all this alone it is, this action of our own, that lends to the Idea the appearance of real active efficacy, power of self-conservation, and impulse to development; these appertain to the Idea only in as far as it is thought by us, while according to our previous view they appertained to it in as far as being an unthought and objec-

tive content—thinkable indeed, but only incidentally so—it
constitutes things.   This requirement is one that cannot be
fulfilled; for the permanent and tangible difference between
thoughts and things will ever consist in this, that the contents
of thought, both when differing and when similar, may be put
in opposition without having any effect upon one another;
things on the other hand are disturbed by one another and
offer resistance; it is true that they do this in accordance
with the content of their nature, which is perhaps susceptible
of being expressed by thoughts, but this capacity of conflict
and this active efficacy do not accrue to them *from* that Idea
of their being which they vindicate *through* them.   This then
is what was wanting; if we express the being of things as
actively efficacious Idea, we do, it is true, express correctly
enough what we need, but as a matter of fact active efficacy
does not on that account accrue to the Idea with the ease and
speed with which we can bestow it on the Idea in speech by
means of an adjective.   On the contrary, it remains doubtful
whether the name of *operative Idea* without addition or
omission denotes anything which exists or can exist; the
presumption is *against* its validity, for it is plain that in it
we transfer to Ideas regarded not as thought but as existent,
a power which demonstrably belongs to an Idea only when it
is thought.

The difficulties to which the idea of the unity of the thing
in the course of its mutations is subject, are not merely
connected with what we have referred to above, they are
intimately related to it.   After having convinced ourselves
that things could no longer be things if they had the absolute
rigidity of complete unchangeableness, we found their permanence to consist only in the logical connection between their
internal states.   What then exactly are the states of a being?
We know what we mean by this expression in two cases—
the first is where we are concerned with the various possible
arrangements of a plurality; there is reason for understanding
these arrangements as being not really different facts but
different states of this plurality, only in as far as one feels

justified in regarding this plurality as a coherent whole, and some primary order as an original law destined for the self-conservation of this whole. The second case is presented by our own inner life; in it our ideas, feelings, and efforts appear to be in their nature the states of a being, of the necessary unity of which, as contrasted with them, we are immediately conscious. The first case has no interest for us; and that which in the second case makes inner states possible, does not seem transferable from the Ego to the non-Ego. For these inner events appear to us as states only through the marvellous nature of mind, which can compare every idea, every feeling, every passion with others, and just because of this relating activity with reference to them all, knows itself as the permanent subject from which, under various conditions, they result.

Now it might be said that though on account of its lack of consciousness it may not be possible for a thing to *know* its states as belonging to it, in the same way that we know our states as being ours, yet in the unity of the thing its states may always *exist*, for even our states do not *become* ours by becoming apparent to us. But such reasoning we cannot admit. If a thing within those limits within which we have admitted that it may change, setting out from the value *a*, gradually acquires the values *b, c, d*, . . ., then our thought, comparing these values, may always recognise in them members of a series which, taken altogether, are connected together in the logical coherence of one identical law of development—but in what way could it be shown that those values are *more* than the realized members of that series, simultaneous or successive, yet independent of one another? that they are to be thought not as separate realities that alternate with one another, but as states of one being that changes in them, and holds them together by the continuity of its presence in them? It is of no use whatever to say, We believe that it is so, and have never held any other opinion; the important point is rather to be certain that in real things those conditions are fulfilled under which that which is thought can be actualized. Now the possibility of regarding our inner experiences *as our states* depends not at all upon the bare general predicate of unity, appertaining to every·substance, not to the Ego alone but also to things; but upon the special nature of consciousness, by which the Ego is distinguished from the non-Ego.   It is only because memory and recollection can range the past beside the present, only because a relating activity of attention can comprehend variety and produce in contrast to it the idea of the permanent Ego —in short, only because we *appear* to ourselves to be unity, that in truth we *are* unity.   Supposing that a mind reacted at every moment to external stimuli, and that these reactions taken together would constitute for a second observer a series as logically coherent as the most scientifically developed melody, but that the mind itself knew nothing of this, but was destitute of memory and at every moment absorbed in the action that at that moment it was carrying out, and at every succeeding moment forgot in the new reaction all remembrance of the preceding one—then this mind would no longer be a *changing* unity, a substance *self-conserving* in the midst of change; it would be a series of real existences succeeding one another according to a definite law—existences of which it would be impossible to say wherein their similarity differed from the similarity of substances that were originally distinct and continued to be distinct.   Hence there would not be the slightest ground for calling the members of this series the states of one being, and that unity which we are thinking of when we speak of the states of a being, cannot therefore be simply transferred from the Ego, in which is the special ground of its reality, to things in general, in which this special ground is lacking.

Let us pass on, finally, to our third difficulty.   It seemed that we must characterize the existence of things as relatedness.   But when we tried to give a name to the relations referred to, it seemed that spatial connections (which really afford us the only intuitable example of that which we mean by relation) are received by us as holding not of existent things but of their appearance.   We substituted for them

supersensuous intellectual relations; that this expression
really signifies something that is actually to be met with, we
believed to be testified by all the graduated relationships,
similarities, and contrasts which we find between non-spatial
sensuous qualities or abstract truths. But when we came to
examine these cases more closely, they all turned out to be
something different from what we wanted. It is true that
they all as causes determine the content of some future event
as their result; but we could not regard them (as we formerly
did spatial relations) as variable conditions, which sometimes
bring together things the natures of which remain unchanged,
so as to cause the realization of consequences which have
their basis in those natures, and sometimes hinder this
realization. And here, again, we might for a moment have
amused ourselves by inserting between things changeable
relations of quite another kind, namely such relations as no
one can frame an idea of, and by making the changeable
action of things dependent on their sometimes increasing and
sometimes diminishing closeness. But then we remembered
how perfectly vain it would be to invent a special and
mysterious kind of connection for this end; the general concept of relation is wholly adverse to every attempt at such
objectifying. *No* kind of relation could be assumed as subsisting between things, acting upon them, conditioning, preparing, favouring, or hindering their reciprocal action; but
reciprocal action itself, the passion and action of things, must
take the place of relation. Just when and in as far as things
act upon one another, are they related to one another; there
are no objective relations other than this living action and
passion, and least of all relations in which things merely *stand*
provisionally, without having any effect upon each other's
natures, only coming to act later, as a result of this relatedness; the mode of expression here reprobated is figurative,
and we now no longer doubt that in a metaphysical point of
view it is wholly meaningless.

But have we now reached the conclusion of the matter?
Hardly—for what more could we understand by the action of

a thing than that a change of its states is followed by a
change of the states of some other being ?   To this succession
it is due that as we reflect and compare, we regard the
second event as emanating from the first, because perception
of it is conditioned by perception of the first ; but there does
not exist between things any authenticated connection of such
a kind as that a state of one is wrought by the activity of
the other.   When we call the active element *active*, properly
speaking we say *of it* nothing whatever ; we simply affirm
that a second being suffers in consequence of its states.   But
is this suffering or passion itself clearer and more significant
than that action ?   What meaning has this expression when
applied with such generality to the changes of state of any
existing thing we choose to consider ?   We fear that it has
not any which can be specified.   For in characterizing the
change of any being not merely as the appearance of a new
condition in place of an earlier one which vanishes away but
as passion, our intention plainly is to indicate that the unity
of the being feels and wards off the imputed change as prejudicial to its own permanent nature.   But what we thus
require can never be performed by a being in the nature of
which we presuppose nothing but a capacity of being changed
and also of being not wholly changed, but of preserving or
restoring from change an abiding part of its essential content
—it is only we who, feeling pain and joy, desire and aversion,
measure by them the value of our inner states for our own
being.   It is only in this *feeling* that actual suffering, to
which we have here tacitly referred, really has a place ; and
every time that we apply this word to unconscious existences
its real meaning vanishes, and with it that for the sake of
which we desired thus to transfer it.   That which does not
feel good and ill suffers as little as it acts ; but that which
cannot suffer is no Real (*reale*) unity, and is not for itself, but
only for the apprehension of some other, a whole that deserves
to be called by one name.

§ 2. If we bring together the results of the foregoing observations—which, dry as they are, we could not well avoid—

we find that concerning that nature of things which has to be assumed in order to make the course of the world intelligible, we are forced to make definite presuppositions; but are not only unable to say how things could set about satisfying these presuppositions, but have also to acknowledge to ourselves that the nature of things, thought as we think it, is adverse to the fulfilment of the demands which we make upon it. Three inferences which seem to exclude one another, and yet finally lead to the same goal, make it possible for us to hold such a conviction. Either we content ourselves with ascribing to our notions of things (as we previously did to the intuition of space) only a subjective validity as forms under which there appears to us the unity of the real world, which in its true shape we are incapable of cognizing; or we give up the thought of things, which we cannot work out to a satisfactory conclusion; or finally, we supplement the notion of things in such a way that it includes the conditions under which those demands upon their nature which we could not retract become capable of fulfilment.

Against choosing the first of these three ways no objection can be made, if it is taken to signify a complete breaking off of all investigation, and an unconditional renunciation of all pretensions to knowledge; but as a proposition containing a permanent addition to knowledge in the form of a positive assertion, the view from which this resignation flows cannot be maintained. For however much one may think that the nature of things is in itself beyond the reach of all knowledge, so that even the most unconditional and certain declarations of knowledge concerning things can only be understood subjectively of the mode in which they appear to the cognizing mind—even in such a case our assertions are not intelligible unless we presuppose the existence of things, and reciprocal action between them and us, for only thus can we give to the notion of their appearance a meaning that is intelligible and capable of being stated. Hence we should always in one breath both deny the cognizability—even in the most general way—of the nature of things and of action and (in order that we may be able to speak of their appearance) presuppose afresh the validity of our most general determinations of both; a familiar circle, from which this doctrine of Subjective Idealism has never been able to escape. Now this circle might itself be put to the account of that imperfection of our knowledge which we are forced to recognise, and it might be admitted that *we* certainly cannot explain how the phænomenal world can originate for us except by supposing that things have some kind of influence upon us, but that this reciprocal action of which we have a notion indicates the ground of that appearance, not as it is in truth and fact, but only in a way that is comprehensible to us. But then the things presupposed by us and the action assumed between them, would be wholly emptied of all special content of their own, altogether incapable of being intuited, indeed wrongly called by the names of *thing* and *action*, and would probably signify nothing more than the wholly unknown cause of our perception of the world, or rather our craving for some such conditioning cause. What is maintained from this standpoint would be as follows: thought, in order to make its own activities intelligible, is obliged to suppose a producing cause of them, and to present to itself in idea the conditioning power of this cause as a varying action of external things upon itself, being yet at the same time forced to recognise this whole mode of presentation in idea as only its own explanation of that cause, or of the action and passion which it attributes to that cause—this explanation being one that is not truly accurate. And in this case the notion of things must be reckoned among the ideas by which we seek to interpret our perception of the cosmos; it does not stand alone, established from the beginning by a special revelation, so that it would only be our further metaphysical thoughts concerning the unity and reciprocal action of things that would be incapable of combining with it as established truth; it too is, on the contrary, a product of our thought, the necessity and validity of which may be matter of question.

And here we—following the example of the historical development of philosophy—turn to the second of the ways above pointed out, namely that of Idealism. That all sensuous impressions which supply the content of our image of the cosmos, and all ideas of relations to which its order is due, are subjective states and activities of our mind, is an observation that at an earlier stage (cf. *supra*, pp. 346 *seq.*) seemed to us an inadequate ground on which to found the conviction that the whole phænomenal world which floats before our consciousness is but the product of a mysteriously ordered play of our imagination. But we here reach a similar view with better reason—not the subjective source of our idea of the world, but the very content of that idea, as we seem forced to think it forbids us to concede to it any other reality than that of an appearance in us. In pursuing the course of this Idealism for a while, we will assume that the lonely thinker may have been tempted, at least for a moment, to regard all physical and mental reality as an ordered dream of his personal individual Ego, the only Real thing which he immediately knows; but then his scientific instinct will, by some easily supplied middle terms, have brought him again so near to the ordinary view as to make the reality of other individual minds with which life brings him into contact, as indubitable to him as his own. It is only the realm of things, an intermediate region, which to the ordinary view seems to be spread out between minds, and by its own changes to initiate, keep up, and guide their inner life, that Idealism declares to be a mere appearance *within* minds. According to Idealism conscious beings interpret the connection of their own direct action and reaction by the image of a world of changeable things inserted between them, and acting upon them, in the same way as (according to our earlier assumption) in spatial intuitions the intellectual order of a world of things in themselves then presupposed by us, became transformed to the image of a space-world embracing those things themselves.

At any rate (so this Idealism maintains) the phænomenal world in which all minds have a common interest, and in which yet different minds participate with differences which have a correspondence among themselves, cannot have its ground in individual minds as such.   But why should we seek this ground nowhere but in the presence without us of a multitude of things, when, on the one hand, what these do towards explaining the microcosmic order can be done without them, and, on the other hand, we always fail to understand how things can do that which they must do in order to be things.   For when it comes to the point, the assumption of things has no other use for us than this, that things mark for us fixed positions in the real world, positions in which we find, grouped together and realized, causes which give rise to results, points of departure for some occurrences which we call their effects, and, as it were, the goal of other occurrences which we call their states, although we cannot make it clear how these things possess an inner nature from which actual effects could proceed, or which could experience actual suffering.   To regard these points of intersection of action—which are in themselves wholly empty and selfless, and seem on the one side to bring together that which on the other side they disperse again—as Real beings, may be a fiction convenient for our survey of the connection of phænomena, but must not be affirmed as an established dogma; on the contrary, this assumption must give place to any and every other which affords an equally intelligible explanation of the course of the world, without requiring the impracticable assumption of the Realness (*Realität*) of that which is destitute of all the inner conditions of Realness.

Now such an assumption offers itself to Idealism in a conviction which we have already reached by another path—the conviction that all individual things are thinkable only as modifications of one single Infinite Being. What might be the positive signification of this word *modification* we left in obscurity; it sufficed for us that it denied the independence of things with reference to the Infinite Being.   We did not mean that the Infinite should be conceived after the analogy of some plastic material from the various parts of which all the multitude of different things should be cut out, and become independent objects; but if we now explain our meaning to be that things are states of the action and passion of the Infinite, we do not imagine that they—though without attaining the independence of self-sufficing substances—have reality as such states of the Infinite, elsewhere than in minds; we regard them rather as acts of the Infinite, wrought within minds alone, or as states which the Infinite experiences nowhere but in minds. Manifesting itself in the individual mind, and being in it and in all its like the efficient source of their life, the Infinite develops a series of activities, as to which *how* they take place remains incomprehensible to finite consciousness, which intuits their product, as they occur, under the form of a multiform and changing world of sense. In this appearance which it presents to the eye of our mind, the Infinite exerts its own unity after a double fashion. For to the observing consciousness it first shows that similar consequences are attached to similar causes, and different consequences to different causes, thus revealing the logical consistency of its action which is governed by general laws; and also among the changing phænomena produced by the varying play of its action, there are brought into prominence the images of Things with their perdurable natures, as witness to certain and constant activities that are always maintained in it, and the rich content and significant reciprocal relatability of which it unfolds in the multiplicity of those changing events. Finally, being actively efficacious in all individual minds, as a power which in the whole spirit-world has assumed innumerable harmonious modes of existence, the Infinite brings to pass the exhibition of those same universal laws, by the totality of the various world-pictures which arise in various individuals; and moreover, the constant activities which appear to every individual mind as the real points of contact and intersection for the events within its world, are exercised by the Infinite with such accord in all that the same things—or at any rate the same world of things—appear to all as a common object of intuition, as an external reality common to all and connecting all.

This explanation of the world given by Idealism with
reference to the relation between individual minds and the
Infinite would still leave outstanding some obscurities which
we do not yet wish to draw attention to; but it would
certainly make superfluous the assumption of Real things
in which are lacking all the inner qualifications of Realness.
But whilst Idealism thus reduces to mere appearance that
which as thought could not be a being at all, we held it
possible to take a third path, which amounts to this, that we
add to our idea of things that which their content seemed to
lack in order to make Realness possible for them.   In fact,
if the doctrine of Idealism reserves to spiritual beings the
Realness which it refuses to selfless things (and this it tacitly
does), what hinders us from finding in this mental nature that
addition which the previously empty notion of things needed
in order to become the complete notion of something Real?
Why should we not transform the assertion that only minds
are Real into the assertion that all that is Real is mind—
that thus things which seemed to our merely external observation as working blindly, suffering unconsciously, and being
self-contradictory through their incomprehensible combination
of selflessness and Realness, are in fact better internally than
they seem on the exterior—that they, too, exist not merely
for others but also for themselves, and by this self-existence
are capable of being after the fashion which we have felt
compelled to require of them, though hitherto without any
hope that our requirement could be fulfilled?

This assumption of a soul in all things would be much
nearer common opinion than the more artistic view of
Idealism; we ourselves have previously been led to it by
other causes, and it has so many roots in the human mind
that from the most varied standpoints we might describe
the satisfying and interesting prospects which it opens to us
concerning the connection of things.   But we would now
turn with indifference from all these inducements, and devote
ourselves to some other questions raised by a comparison of
the two views which we have last developed.   As I have

already noticed at an earlier point, their assertions have much more affinity than at first appears, and I fear lest there should be maintained between them a distinction which would rest upon an inadmissible prejudice. Idealism, it will be said, denies that things have Realness, and regards them as being by their nature incapable of detaching themselves from the Infinite, of which they are states, and attaining complete independence; whereas the last-mentioned view allows Realness to things, in that it regards them as having minds, and minds (in the self-existence (*Fürsichsein*) which constitutes the distinctive peculiarity of their nature) possess that which makes them capable of existing not only within or in dependence upon the Infinite, as states of it, but also detached from it and in self-dependence. This mode of expression would involve the thought that the attribute of mentality is merely the legitimate ground in virtue of which beings which have minds can obtain Realness as a form of existence distinguishable from that self-existence to which we have referred. The influence of this thought is frequently encountered in the region of religious speculation, where it gives rise to the familiar question, whether the world, or things, properly exist in God or not, whether they are or are not immanent in Him—the complete dependence of the nature and existence of the world (or of things) upon God being conceded from the first. The answers to this question, whichever alternative they may assert, plainly betray the opinion that it is not existence in God which would make the complete Realness of things indubitable, but only an existence *external to* God, whether that existence were original or due to some creative act of God. Thus they regard Realness as a definite formal relation to God, which they characterize by spatial images that are certainly wholly inadequate; of this relation they presuppose universally that it gives independent existence to any content to which it applies, and they will only admit partially and in detail that it is not every content which can stand in such a relationship, but that the title and the capacity thus to stand must be the result of some peculiar advantages of natural endowment. That this could not be our view, and why it could not, may most simply be made clear by the consideration to which we now proceed, in which for the sake of brevity we shall retain to some extent the phraseology of those religious investigations which we have mentioned, although our doing so is not perhaps quite justified at this stage of our reflections.

Let us assume that in God the idea of a definite content is thought in such a way as to include all the consequences which it has in the world of the divine thought, these thoughts of God being at the same time the very power which is in finite minds the efficacious cause of their intuition of the world; or, in other words, let us assume that in the Infinite a definite activity is so exercised that at the same time—as must happen in consequence of the unity of this Infinite—there are also consistently exercised all those other activities which, in accordance with the universal orderliness of the action of the Infinite, must flow from that one; and that this activity of the Infinite is again the efficacious power which produces in individual minds the image of an external world:—if we assume this, then according to the view of Idealism, these inner acts of the Infinite really are the Real forces which (being in fact efficacious within the Infinite, each calling out and conditioning the other according to law) produce true action, that is at the same time incidentally perceived by individual minds as a world of external things embracing them all. And now we would ask ourselves, What exactly would be gained by these thoughts of God or these states of the Infinite, both of which have now been thought as immanent in God and in the Infinite as states of the one or of the other—what exactly would be gained (to use the phraseology of the discussions referred to) by their being *external to* God, or what exactly would be gained for them by being dissatisfied with this their immanence in God, and finding out for them in addition to this some transcendental existence? Finally, in what would this existence external to God ultimately consist, and what would be the *real* meaning of that which is figuratively intended by this spatial expression *external to* ?

If one ponders these questions it will be found that nothing whatever is gained for selfless unconscious things, but that they rather lose by having ascribed to them that existence external to God; all the stability and all the energy which they exhibit as active and conditioning forces in the changes of that course of events which is visible to us, they—thought as mere states of the Infinite—possess in all the same fulness as if they existed as things external to it; nay more, it is only through their common immanence in the Infinite that they have in any degree—as we saw earlier—that capacity of reciprocal action that could not belong to them as isolated beings detached from that substantial substratum. Thus by doing away the immanence of things in God, we reap no advantage as regards that which things should be and do for one another and in connection with one another; but it is true that as long as things are only states of the Infinite, they are nothing *for themselves*. It is desired that something should be gained for things themselves; this is plainly what is meant by the insistence upon existence external to God; but the more genuine and true Realness of *being something for oneself*, or more generally of *self-existence*, is not attained by things by their being made external to God, as though this transcendency (of which it would be wholly impossible to give the exact significance) were the precedent formal condition to which self-existence were attached as its consequence; but in that a thing is something for itself, consciously refers to itself, apprehends itself as an Ego—by just this, which is its very essence, it detaches itself from the Infinite. It is not that it thereby *acquires* an existence external to the Infinite, but that by the very fact it *has* such existence; it does not fulfil thereby a condition by which is secured to it complete Realness, as a kind of existence including and bestowing something other than is contained in the condition itself—but self-existence or Selfhood (*Ichheit*) is the only definition which expresses the essential content and worth of that which we, from accidental and ill-chosen standpoints, characterize formally as Realness, or independent existence external to God, in contrast to immanence.   He therefore who, constrained by necessity, regards minds as well as things, as being states, thoughts, or modifications of God or of the Infinite, yet as not serving merely to propagate the logical results of the nature of the Infinite from point to point, being connected amongst themselves as links of a chain, but as also feeling that which they do and suffer as their states, in some form of relation to self (*sich*), as events experienced by their self (*Selbst*)—he who assumes this, and yet believes in addition that for these living minds immanent in God, he needs to prove an existence external to God, in order that they may be Real in the full meaning of the word, does not, it seems to us, know what he is about—he does not know that he already possesses the kernel whole and complete, and that what he painfully seeks is but the shell.

The result of these considerations admits of being differently expressed.   If we continue to use the phraseology in accordance with which we designated Reality as the general affirmation which belongs to action as well as existence, then Realness is the special kind of reality which we attribute to or seek for things, as the points from which action sets out and in which it is consummated.   This Realness has appeared to us as dependent upon the nature of that to which it is to belong; it is the being of that which *exists for self*.   But we want the name *self-existence* in order to characterize in a more general way the nature of mentality, which only reaches its highest stage in the self-consciousness of the being that knows itself as an Ego (*Ich*), and is not, because of this being its highest stage, absent in the being which, though far removed from the clearness of such self-consciousness, yet in some duller form of feeling exists for itself and enjoys its existence. Hence to Realness in this sense we can attribute various degrees of intensity ; we cannot say of everything that it is either altogether Real, or altogether not-Real; but beings, detaching themselves from the Infinite with varying wealth and unequal complexity of self-existence, are Real in different

degrees, while all continue to be immanent in the Infinite.
Hence the distinction between Idealism and the standpoint
which we have just taken up does not consist in this, that we
ascribe to things a transcendental and hence Real existence,
while Idealism ascribes to them only an immanent and hence
merely apparent existence; rather there exists between the
two *this* difference, that the idealistic view, convinced of
the selflessness of things, on this account will not allow
that they are more than states of the Infinite; while we,
agreeing herewith in principle, leave undecided, as something
which we cannot know, the question whether this assumption
of selflessness is appropriate, holding, however, that it is far
more likely to be *in*appropriate, and that all things really
possess in different degrees of perfection that selfhood by
which an immanent product of the Infinite becomes what
we call Real.

§ 3. We seem now to some extent to have struggled
upwards out of the helplessness to which we confessed at the
beginning of this chapter. The nature of that which is Real
is no longer so wholly unknown to us and so wholly incapable
of being showed forth as it then seemed; we are no longer
so completely limited to going round about it at a distance
with purely formal abstract notions of Realness and unity
and inner states of passion and action, without being able to
make clear the living meaning of any of these notions by
pointing to some well-known and pregnant intuition. To the
nature of Mind, of the Ego that apprehends itself, that is
passive in feeling and active in willing, and that is one in
remembrance in which it brings past experiences together,
we can now point as to a similitude of that which is the
nature of beings endowed with Realness; or we may believe
that directly and without any similitude we find the thing
itself, the nature of all Realness in this living self-existence.
I will leave undecided whether we are really free to choose
between these two alternatives; in order to cut short the prolixity in which this consideration would involve us, we shall
be satisfied to have it granted to us that at any rate there is

in mind the nature of a Real being, although the nature of things may not be made properly clear to us by the analogy of mental existence, but only imperfectly and figuratively illustrated by it.

But will even this be granted to us ? Shall we not rather be met with the reproach that we have characterized as the original being of things that which, as a late and mediated result, most of all needed that we should show how it was put together out of more simple and more essential material ? For are not ideation, feeling, volition, self-consciousness, events the possibility of which can only be understood by pre-supposing the nature of a Real unconscious being which in itself neither ideates, nor feels, nor wills, and assuming that this nature is stirred by numerous stimuli, and that from the reactions by which—in accordance with its unknown peculiarities—it responds to those excitations, the familiar phænomena of mental life are produced ? Has not the more enlightened psychology of modern times devoted all its strength to this problem, partly with valuable results, partly, so far, without any results at all ? Must not then this mental nature, this self-existence that we have here inconsiderately characterized as the essential nature of Realness, be rather understood and explained as one of the products arising from conditions which act upon the far more recondite nature of that which is properly Real and which is incapable of being intuited, and can only be held fast in the subtlest ontologic abstractions ?

I may easily seem to be contradicting the greater part of what I have already said when I pronounce the under-taking here indicated to be a decided step into the perverse region of those investigations which seek to know by what machinery reality is manufactured, without considering that there cannot well be any machinery unless there has existed previously some reality, from the constituents of which, and according to the already valid laws of which, that machinery could be put together. We are tempted to take this step wherever our interest in investigation has been first aroused by the varying values of certain fundamental phænomena or fundamental facts, for the alternating occurrence of which there must be different conditions that make now the one and now the other necessary. And if we have moreover had full opportunity to remark that even diverse phænomena, which on account of the difference of their content seem at first to be each something special in itself, are yet dependent on mere changes of magnitude of homogeneous conditions, we are likely to be seized by a sort of constructive passion from which nothing is safe, and which would end by deducing the whole positive content of real things—the place of which in the world we have to explain—from mere modifications of the formal conditions upon which the variations of those places depend. However, if this remark is to be of any use to us, I must try to illustrate it by reference to some examples which are not alien to our subject.

Our eye sees sometimes light and sometimes shade, and sees various colours one after the other. Now when the student has learnt that these changing sensations proceed from mathematical differences in the light waves, he generally becomes inclined to assert that colours *are* nothing whatever but different vibrations of ether; though he may perchance bethink himself at this stage of his scientific knowledge and admit that they do indeed *proceed* from those vibrations, but yet are in themselves something new and different, namely special states of psychical excitation in us. But now perhaps he learns in psychology that we have reason to regard even these qualitatively different impressions—indeed even those sensations of the different senses which differ so as to be incapable of comparison—as mere phænomenal forms, under which the soul becomes aware of a countless multitude of excitations, which qualitatively are quite homogeneous, and are only quantitatively or formally different; that perhaps to a sensation of colour as distinguished from the hearing of a musical note there corresponds only a more intense degree of disturbance, or one that takes place with a different rhythm in the succession of its individual nervous shocks, but that this psychical disturbance or movement is always generically the same in both, and indeed in all cases of sensation.   And having learnt this he easily grows accustomed to look down upon the many-coloured qualitative variety of mental phænomena with a certain feeling of superiority as a sort of juggle of which one has penetrated the secret; and this feeling is appropriately expressed thus:—Internal phænomena are not *actually* different from one another at all, they only appear to us to be different, being in truth mere formal modifications of one process which is everywhere in essentials the same.

I do not think that I present this perverse view in too glaring colours; it is a fact that many act as though they believed at the moment when they come to perceive this similarity of the origin of psychical processes, that their dissimilarity has ceased to exist; they forget altogether that it is the mode in which these supposed modifications of one homogeneous process appear to us, that is the very point with which we are concerned.   If it were certain past a doubt that the sensations of light and those of sound depend upon two psychical disturbances which at most differ from each other only as quantitatively and formally as vibrations of ether from sound waves, yet the disparateness of these sensations, in as far as felt, is not thereby done away with, but lasts on afterwards just as it did before; their worth and their reality are not lessened by the fact that both sensations are but modes in which the processes referred to appear to us; these modes of appearance are, on the contrary, real permanent mental facts, of which those external facts of physical sense-stimulation or the psychical disturbances corresponding to them, are indeed the occasioning causes, but the nature of which is not determined by those causes, and the difference between which is not in the least diminished by the slighter degree of difference that exists between their causes.   Or if we hear that feelings and stirrings of the will are really nothing more than manifold pressures and movements which ideas cause in one another by their reciprocal action, ought we to allow this "nothing more"?   If we have made the discovery that they are nothing more, does pain, on that account, cease to hurt, or can we root out from our consciousness the fact that a motion of our will is and remains, always and for ever, something totally different from a non-voluntary rise and subsidence of ideas ? Such explanations even if correct teach us only the occasioning causes to which it is due that the characteristic content of mental events appears upon the stage of consciousness, they do not inform us as to the producing causes of this content; they teach us to know conditions upon the change of which depend alterations of the consequences attached to them, but the dependence of these alterations is regulated in such a peculiar manner that from a comparison of two values of the condition, no thinking can, without using other data as well, divine how that difference of the two results will appear which corresponds to the given difference of the two values of the condition. Hence as far as changing action depends upon altering conditions, so far (taking the sense we have indicated) has science in general, including psychology, to solve explanatory and constructive problems. It may seek out the occasioning conditions of the various forms of presentation in idea, and feeling and willing, and of the varying course of these changing events and the multiform products of their reciprocal action; but it cannot hope to make out, from any data, how it can happen at all that there can be ideas, feelings, and volitions, and that one inner state can influence another; still less may it believe that in the mere explanation of instrumental machinery it has reached the essential meaning of spiritual events, or apprehended that which actually and in truth they are, as contrasted with that which in direct inner experience they appear to us to be.

§ 4. I feel that my remarks so far have been devoted to blaming admitted errors, and that they have not had sufficient reference to the case before us. It may be unhesitatingly admitted that all explanation can but set forth the inner regularity of a *given* reality in its changing development, and cannot deduce back until it reaches either the simplest elements of action, the combinations of which it investigates, or the original proportions between them, the consequences of which it tries to trace.   But within the boundary lines thus drawn may we not yet find a constructive task ?   For the different fundamental phænomena of mental life are not, it may be said, given to us in experience as unconnected occurrences, each of which changes and develops according to its own law and in dependence on the alteration of conditions that are valid for it alone; on the contrary, they occur in our observation as states of beings, and indeed as all states of one being, or at least it is only when regarded as such that they have meaning and significance for us.   And how it is that in one being the possibility of such various manifestations can exist, and can exist in such a way that some appear under some conditions and others under other conditions, is not self-evident, and we are justified in attempting to investigate the inner structure which this being must have in order that it may be mental; nothing being less admissible than to give out that this mental nature is to be recognised off-hand as being in general the original nature of the Real—as though the nature of mind involved no puzzle.

I still, however, hold to my opinion ; only I understand the difficulty of refuting the prejudices opposed to it, because I am fully conscious of the power of those impulses which continually beguile men into such attempts.   We have an ineradicable inclination to regard the laws which enable us to apprehend the development of any real being, *because* it develops thus and no otherwise, as precedent conditions on account of which it is constrained to develop thus ; we have further an ineradicable inclination to regard the " contingent aspects," the analyses, the auxiliary notions and relations by which *we* succeed in *thinking* the connection between real things when they already exist, as actual machinery by means of which those things come to *exist ;* and finally, we are specially inclined to reverence analogies to which we have become accustomed through intercourse with the world of sense as types of universal validity, to which all reality must

conform.   From the first inclination arises the habit of
speaking of a world of truths preceding the world of realities
as something which by its very notion is earlier, an error of
which I shall soon have occasion to speak more at length;
the third inclination produces those materialistic conceptions
of the world of mind to the refutation of which we decline to
return again; from the second bias that we mentioned arises
the mania for giving to that which is most real and most
original a still more secure foundation constructed from its
own consequences.   How this is to be understood I will try
to make clear with such means as I can here make use of.

We may say—not with exactitude, but as helping towards
comprehension—that in all the notions of things—of their
unity, their states, their passion and action—by which we
introduce order and connection into our perceptions, what the
mind in effect does is to copy the general features of its own
nature, and because it feels that itself and its reality subsist
and are contained in them, it seeks to transfer them to
external reality too, and to work them into it, as the only
characteristics of true existence which it knows.   But in being
thus transferred, these features lose the living content which
they had in the mind's sense of self, and which the non-Ego,
observable only from without, cannot be regarded by mind as
possessing likewise; they are transformed in this transference
to forms empty of content which do no more than preserve
and express the modes of connection which both relate the
manifold content of the mind to it, and relate the constituent
parts of that content to one another.   In self-consciousness
experience of the Ego as the subject of mental life is so
immediate that it brings with it also experience of what is
meant by being such a subject; at present it is the fashion
for knowledge to attenuate the living intuition of the Ego
into the formal notion of a substance which in some way, not
intelligible to us, renders to a manifold of external phænomena
the same office of a subject by which the parts of that manifold are held together; remembrance, by which the soul really
connects into one all-embracing consciousness its temporarily

separated experiences, fades into the formal notion of a unity
with self, which in some way which we certainly cannot take
in appertains even to those unconscious and selfless substances; notions of states and actions arise like empty shadows
of the efficient volition, and painful suffering of living experience, and establish between the shadows of things many and
various shadows of connections.   And then the soul having
in its intercourse with the world of sense become accustomed
to the use of these abstractions, it turns, as it were, suicidally
against itself, and imagines that it can comprehend its own
nature only by help of these ontological notions which from
the very beginning had significance only in as far as they were
reflections—though pale and faint—of the mind's own nature.
And finally, it reaches the point of no longer understanding
its own self, and hits upon the device of enriching its nature
by a core of unconscious substance with which in imagination it endows itself, and in which it tries to induce self-consciousness by an ingeniously devised system of stimulation.

That this must always happen, that there is in the very
nature of the soul a craving which drives it to bring all reality,
including its own life, under these forms, and to make it an
object of reflection, is a fact which we do not deny, and to
the inevitableness of which we have already referred (cf. i.
p. 626); it is just here that we find the difficulties with which
at this point we have to struggle.   But it is possible, nevertheless, to be conscious that all those ontological notions are
but products of thought, not conditions of the possibility of
him who thinks or of that which is thought, are but aspects
which truth wears to finite mind, and not the very form of
truth itself; and this true state of the case forces itself
upon us on different occasions with different degrees of clearness.   Thus we happen perchance to say, *Our Ego possesses self-
consciousness;* then struck by the perversity of making out
that our very being is possessed by us, and that the most
essential feature of our nature is a possession of that which
is thus possessed, we amend our expression, and say, *I am a
soul;* but even so we only veil the still unremoved perplexity;

we know now no more than we did before in what essential
relations the subject, copula, and predicate of this judgment can
stand to one another as long as they are thus distinguished.
And we make up our minds to admit that it is vain to
attempt to separate that which is one by expressing it in the
form of a judgment, and then by recombining the parts to
construct a unity which can only be known in direct intuition.
But things will still go on as before, and the attempt will
ever be renewed. Whenever we are considering some isolated
action of any being, the rest of its nature appears as something constant from which the action proceeds ; and continuing
this process, we come at last to contrast the totality of its
actions and properties with a permanent root from which they
arise, and divide the being into—(1) something that is nothing,
suffers nothing and does nothing ; and (2) a host of qualities and
actions which proceed from this something. Here, after some
consideration, there is a division of views ; some resolve beings
into pure activity without anything that acts, while others in
some incomprehensible way connect activity with something that
is inactive ; if we say to both, The thing which acts is itself
the being, then this and every similar expression involves
the error of regarding the article as indicating the true being
which only participates in action. When upon the application of an external stimulus a sensation arises, it seems as
though it, being a reaction of the soul, must have been preceded
by some passion which calls it forth, and to which it corresponds ; thus we come to imagine unconscious stirrings of the
soul, impressions which are *succeeded* by sensation, as by an
elastic rebound ; on the other hand, we reflect that if the
reaction is to proceed from the passion there must be *one*
moment in which they are both coincident as an indivisible
action ; but if for *one* moment, why not for all ? and why not
admit that the distinction between excitation and reaction is a
fiction of theory, as indispensable for many purposes of comparing and combining cognition, but in fact just as unreal as the
movements along two sides of a parallelogram into which we
arbitrarily analyse some given simple movement ? When some

idea hinders or obscures in consciousness some other idea differing in content, and perhaps not capable of being compared with it, or when in external Nature two substances differing in appearance produce in one another movement or equilibrium, we draw the conclusion that both must, notwithstanding, have a hidden similarity in order that they may be able to act upon one another, and regard them as differing values of a homogeneous process or a homogeneous substance. But why not admit that their homogeneity consists in just that capacity of reciprocal action which belongs to them,—that is, that they are in truth only *equivalent*, and *not* homogeneous in the sense that they are really constituted by, or have arisen from, some one third thing,—and that this reduction of elements, qualitatively different but equivalent in working to different quantities of one identical substratum, is indeed a fiction that is very convenient for our calculations, but one that certainly needs a special proof of its essential truth if we are to accept it as valid?

It would be easy to multiply such examples; cognition everywhere seeks to make clear to itself the inner connection of the living nature of reality, by such analysis or reference to co-ordinates as it may find convenient—which afterwards it easily comes to regard as essential determinations of the being of things. The temptation to this is not equally great in all cases. Often the nature of the truth which applies to all reality admits of our reaching the same goal from different starting-points and by different roads, and then we easily convince ourselves that none of these roads is that taken by the thing itself and that the relation of the thing to the system of co-ordinates by the help of which we seek to determine it is a relation of indifference; in other cases—among which we must reckon those simplest and most general ontological notions of which we have been speaking— we have not such a choice, but are constrained always to return to the same modes of conceiving reality. And then these inevitably appear to us as conditions which not only make our knowledge of the thing possible, but make the thing itself possible; and this is the case to such a degree that doubtless the conclusion of this long exposition of mine will be rewarded by the incredulous question, But how must it come to pass that minds can suffer these states and develop these reactions? This is once more the question that demands to know how reality is created; and we once more answer it by saying that it does not seem to us as though it must *come to pass* that this should *be possible*, but that minds *do* so suffer and react that considered in detail there is a process of the development of events one out of another, from point to point. We shall soon have occasion to return to this question; and we will defer until then the explanation of any obscurities which may yet remain in these considerations, the results of which we shall now try to formulate, in the same way as we have done the results previously reached.

§ 5. VII. The notions by means of which we seek to determine the nature and connection of things, make demands with regard to which on the one hand we cannot understand how things thought as selfless can set about fulfilling them, and of which on the other hand it is clear that the nature of things thought as it has hitherto been thought excludes their fulfilment. For anything that we could imagine as an accomplished and concretely intuitable fulfilment of these postulates —not merely a fulfilment demanded and indicated in abstract formulæ—is only possible in some mind, in virtue of the peculiar nature which distinguishes it from that which is not mind.

VIII. If now that which we must require from things as the subjects of phænomena at the same time cannot be performed by them as long as they are things, then either things cannot exist, or they must exist otherwise than they have hitherto been thought to exist. Either only minds exist, and the whole world of things is a phænomenon in minds, or things which appear to us as permanent yet selfless points of departure, intersection, and termination of action, are beings which share with minds in various degrees the general characteristic of mentality, namely self-existence.

IX. The Realness of things and their self-existence are notions which have precisely the same significance. The meaning of this assertion is twofold. First, that a mind which continues immanent in the Infinite as a state, activity, or modification of it, directly that (notwithstanding this immanence) it exists for self, has in this very self-existence the fullest Realness, and does not obtain Realness by being detached from the Infinite and attaining the independence of an existence out of it; self-existence is the positive content of this independence for which we seek, the meaning of which becomes quite incomprehensible if it is regarded as some different kind of formal relation to the Infinite into which that which possesses self-existence has yet to enter.   But our proposition asserts in the second place (and this second assertion is most intimately connected with the first) that Realness is not to be understood as a consequence attached to self-existence as something to be earned by it, and hence dis-- tinct from it.   Even the expression, Mind is Real in virtue of its self-existence, has not in this reference the exactness which we would desire ; for that *in virtue of* allows of the misinterpretation that Realness may depend upon certain general conditions, which mind may fulfil by its self-existence, but which something else, for instance selfless things, may fulfil in some other way.   But there are no such conditions ; there is no law precedent to all reality, according to the pre-scriptions of which Realness and not-Realness are distributed among all that is conceivable.   It is only the living mind that is, and nothing is before it or external to it ; but it exists in such a way that it can only make its own existence and action objects of reflection by giving to their manifold content a framework of abstractions, connections, and other auxiliary constructions by which that content is divided, combined, and systematized—and these easily come to appear to it as not merely conditions of its thought about itself, but as being also conditions of its reality.

CHAPTER IV.

THE PERSONALITY OF GOD.

Faith and Thought—Evidence of the Existence of God—Impersonal Forms of the Supreme Being—Ego and Non-Ego—Objections to the Possibility of the Personality of the Infinite—Summary.

§ 1. OUR exposition, which is now hastening to a conclusion, must, for brevity's sake, be allowed to omit the mention of middle terms so obvious as to be easily supplied by the reader. Our reflections hitherto have been busy about the nature of finite things and the possible modes of conceiving their reciprocal connection, but we have not spent much pains in attempting to elucidate the notion of that One Being which we have notwithstanding regarded as the indispensable presupposition of all intelligibility in finite things. The course of our investigation would now naturally lead us to this attempt; for however perseveringly we may have had to turn away from every expectation of an explanation as to how reality comes to exist, yet in the assertion of a dependence of the finite many upon the infinite One there is involved the assertion of a permanent relation of real to real; and to determine as far as possible the meaning of this relation is a task which we are bound to recognise as admissible. But it would not be useful here to carry on this investigation to further developments logically resulting from the purely metaphysical motives that have hitherto been its mainspring; we find such development of it already existing in the region of religious thought—a rich and full development, having a form which must attract our attention in a high degree for this very reason, that it seeks to satisfy the needs of the heart and the conscience as well as of speculative knowledge. To this familiar development we will turn and take as the object of our reflection, not the metaphysical postulate of the Infinite, but instead of it the full and complete concept of the God who is to realize this postulate.

Here we must think, at least for a moment, of the doubt which may arise at this point, reminding us of the resultlessness of philosophic investigations concerning those ultimate questions which only the new and special faculty of Faith is competent to answer. Whatever may be thought concerning the origin of religious truths, the view taken will unquestionably leave something to be done by scientific cognition. If religion were a pure product of human reason, philosophy would be the only competent organ of its discovery and interpretation. If reason is not of itself capable of finding the highest truth, but on the contrary stands in need of a revelation which is either contained in some divine act of historic occurrence, or is continually repeated in men's hearts, still reason must be able to understand the revealed truth at least so far as to recognise in it the satisfying and convincing conclusion of those upward-soaring trains of thought which reason itself began, led by its own needs, but was not able to bring to an end. For all religious truth is a moral good not a mere object of curiosity. It may therefore include some mysteries inaccessible to reason, but will only do so in as far as these are indispensable in order to combine satisfactorily other and obvious points of great importance ; the secrecy of any mystery is in itself no reason for venerating it ; a secrecy that was permanent and in its nature eternal would only be a reason for indifference towards anything which should thus refuse to be brought into connection with mental needs; and finally, above all things, to revel in secrets which are destined to remain secrets is necessarily not in accord with the notion of a revelation.

But must that which is a secret for cognition be always really a secret ? Does not the nature of faith consist in this, that it affords a certainty of that which no cognition can grasp, as well of *what* it is, as *that* it is ? And does not all science itself, when it has finished its investigations of particulars, come back to grasp, in a faith of which the certainty is indemonstrable and yet irrefragable, those highest truths on which the evidence of other knowledge depends? There is certainly a germ of truth in this rejoinder; but not the less clear is the essential difference that separates such scientific faith from religious faith. It is only in universal propositions, which in innumerable conceivable cases indicate those modes of relating a manifold which occur under definite conditions, that scientific faith places immediate confidence. When it declares that everything which is thinkable is identical with itself, that similar things under similar conditions produce similar results, and under dissimilar conditions dissimilar results, and that every change is preceded by a cause—all these propositions are *universal* truths which tell us indeed what must necessarily happen or take place if any case in which they are applicable should arise, but tell us nothing whatever about the actual occurrence of something real. The essential truths of religion have all an opposite character; they are assurances of the *reality* of some being, or event, or series of events, assurances of a reality of which the content when it has once been recognised may certainly become indirectly a source of universal laws, but which in itself is not a law but a *fact*. Now those universal truths in which scientific cognition puts absolute faith, are at bottom but the very nature of cognizing reason itself, expressed in the form of principles of its procedure, and it is conceivable that reason, unable to escape from its own nature, may be overpowered by the evidence of these rules of thinking, which to it are inevitable. But not more than its own being can be known to the mind in immediate consciousness; it cannot have innate revelations of facts other than itself, however great and incomparable the value and significance of these facts may be.

Religious faith is comparable not to this immediate evidence of ultimate principles but to another element that co-operates in the construction of knowledge—namely to the *intuition* by which content is given to those principles, and by which those universal laws are supplied with cases to which they

may be applied.   Even in sense perception we receive the
content of sensations just as revelations which can only be
accepted as they are; we have no reason, we have no need,
we have no means to prove the reality of an impression of
colour, nor has knowledge any conceivable task adapted to
show how this colour should appear.   It *is*, and *is as it is*, by
immediate revelation which we can but receive.   The same as
we here experience under the influence of physical stimuli,
we may experience from direct divine operation within our
heart; thus faith would be an intuition of those supersensuous
facts revealed to us by this operation.   There is truth in this
—more truth than in the previous comparison.   But every
sensuous impression regarded in itself is but a way in which
we are affected, some phase of our own condition ; in itself it
gives no knowledge of any matter of fact, taken alone it constitutes no experience.   Here again it is only our thought
which, mastering the manifold revelations of sense, compares
and combines them, or interprets given combinations, thus
arriving through them at the knowledge of some fact.   We
can hardly picture to ourselves the workings of God upon the
heart otherwise than after this pattern ; we cannot imagine
the recognition of any fact as something that can be simply
communicated, something that reaches the mind ready made
and without any activity on its part, we can only imagine
that occasion can be given to the mind to, as it were, produce
such recognition by exercising this activity, and in this it is
that every appropriation of a truth must consist.   As sense
in itself furnishes merely an impression, so also this divine
influence would produce merely a feeling, a mood, a mode of
affection ; what is thus experienced becomes a revelation only
through some work of reflection which analyses its content and
reduces it to coherence by clear notions that are capable of
being combined with our ideas of the real world.

It will not always be possible for this to happen ; much of
this inner life of the believing heart must always remain
purely subjective experience, and these incommunicable states
will by no means contain only that which is of least value in

our faith ; on the contrary, that which is best and fairest and most fruitful in our experience will always be realized in us only in the shape of these living emotions which are superior to the forms of knowledge. It cannot be our business to interpret this wealth of inner experience—to interpret either that in it which transcends knowledge, or that which is too insignificant to become matter of knowledge. The only part that can hold our attention is that which is not only beheld by the individual in his rapture, but which every one can communicate to others, which is capable of becoming common property, and which, by arguments that all human reason must recognise, he can either prove as truth, or justify to faith as a convincing probability, by which formidable objections are refuted, and thus a possible solution furnished of problems that press upon us.

§ 2. Reason at one time tried to solve the essential part of this problem of interpreting and defending the content of faith by proofs of the existence of God. It would be unfair to reproach this form of procedure with the contradiction of trying to exhibit that which is highest and (by its own assumption) unconditioned, as being, notwithstanding, the necessary and conditioned result of truths, the validity of which must— since they are to be accepted as grounds of proof—be earlier and more fundamental than the reality of that which is proved by them. Although this error has not always been avoided, yet these proofs—like all investigations which strive to go back from results to their causes—are only intended to mediate our *knowledge* of the principle by those of its consequences which are given, and with this view they presuppose the absolute validity of a truth which knits all the world together, and which allows of our divining the *notiora naturæ* from the *notiora nobis.* But the way in which the undertaking has been carried out seems to show that human insight has not received in sufficient completeness those data of reality which it needs in order that it may, under the guidance of general principles of reason, reach with exactitude and completeness the end to which it strains, and this even if we do not reckon those chance wanderings by the way due to defective criticism of the desired end to which we were pressing on. We will now only take a brief retrospective view of this region of thought, to which previous reflections have already sufficiently introduced us.

The *Cosmological Proof* concludes from the contingent and conditioned character of everything in the world to the existence of a Necessary and Unconditioned Being, and it seems to it that nothing but an absolutely perfect being can be thus unconditioned. We call that *contingent* which in the realization of some intention occurs as an unintended and accessory result — occurs because the means which we must use, possess, besides the properties by which they serve our purpose, others which for our ends are indifferent or even obstructive—properties which, since they are there, cannot be prevented from having their own effect, as far as general laws permit. If we transfer the application of this word *contingent* to the course of Nature, attributing intentional design to that course as a coherent whole, then *contingent* signifies everything that is not part of Nature's plan, but only some unavoidable consequence of the means and laws by which Nature proceeds at every step. Hence the *contingent* being without end and aim, it has only grounds and causes by which it is produced in the coherent whole of reality ; but as external to this whole, neither being nor action considered in themselves can be either contingent or necessary. For that which in such case would be signified by the name contingent—existence which might be non-existent or might be other than it is—is not a special and more imperfect kind of existence, in contrast to which some other and better kind might be imagined, but any part of reality, considered as detached from the rest, is contingent simply in the sense that its non-existence, or its existence otherwise than as it is, is conceivable. There is nothing which is necessary and of which the non-existence is impossible, except the conditioned, which as consequent is determined by some antecedent, as an effect by some cause, and as a means by its end ; but the notion of a being isolated and conditioned by nothing, and yet possessed of necessary existence, is wholly impossible. If, therefore, contingency is so often rejected as belonging to the ultimate reason of the universe and necessity so eagerly claimed for it, this happens because both expressions, having lost their speculative meaning, have come to be used as determinations of value. Taken thus, *contingent* connotes that which does indeed exist, but has not any significance, for the sake of which it need exist; *necessary* connotes something not that must be but that has such unconditional value that it seems in virtue of this value to deserve also unconditional existence. Only in such a sense can it be required that the Supreme Principle of the universe should be necessary. But to demand that God should be represented not only as really existent, but as being obliged to exist, would be wholly erroneous and involve a confusion of notions. All religious needs would be perfectly satisfied by proof of His reality; to wish to prove His necessity, would not only be to exaggerate our demands in a wholly useless manner, but would in fact also lead to the contradiction of conceiving God as dependent upon some being superior to Himself, and containing the constraining cause of His existence.

The other part of the Cosmological Proof also gives occasion for similar remarks. Perfection is an unequivocal predicate only when it denotes agreement between the nature of an object and some standard to which that nature ought to conform. Hence it is only failure to accomplish that which is due which is imperfection, but a thing is not imperfect because we do not find in it some merely conceivable excellence. That we do yet in such a case speak of imperfection, proceeds from the fact that the word perfection has also lost its speculative meaning of conformity to a standard, and has become an independent designation of that which is directly commendable, and worthy in itself. Now if a thing does not fulfil the obligations of its own nature, we may perhaps have reason to assume that it has been restrained by some foreign power from the attainment of that to which it was destined; but the mere absence of some conceivable beauty or excellence does not show that that which in this sense is imperfect is either dependent or conditioned. For, in fact, unconditioned existence may belong to that which is indifferent and petty as well as to that which is significant and great, and is not the exclusive privilege of that which is most excellent.

Thus then the Cosmological Proof could only conclude from the conditionalness and conditioned necessity of all individual real things in the universe, to an ultimate Real Being which, without being conditioned by anything else, simply is, and simply is what it is, and finally may be regarded as the sufficient reason through which all individual reality is, and is what it is.   And this way of looking at the proof clearly shows that it cannot of itself attain to the religious conception of a God, but only to the metaphysical conception of an Unconditioned.   And it is not even able to establish the unity of this Unconditioned.   It is indeed possible that at a further stage of development the demand for unconditionalness may be found to have connected with it a demand for unity too ; but this connection has not been discovered by the proof which we are considering, and hence it does not refute the assumption of an indefinite plurality of cosmic beginnings, of a plurality of unconditioned Real beings, in which, on the other hand, students of Nature may hope to find an explanation of the multiplicity of phænomena more easily than in the unity of the Supreme Principle.

The *Teleologic Proof* seeks to attain certainty of the reality of God from the purposiveness in the world.   In order to be convincing, it would have strictly to fulfil several requirements with regard to which we have long ago seen that it can satisfy them only with various degrees of probability.   It would *first* have to show that there is in the world a purposive connection which *cannot* result from an undesigned co-operation of forces, but must have been designed by some intelligence.   But we have seen that even conscious design can effect the realization of its purpose only by means of instruments, from certain conjunctions of which that which is desired proceeds as a necessary result; and that even the conjunction of instruments for this result is only possible when the conjoining design works also upon each of them with a blind force, which in accordance with general laws is able to move it in the way necessary to bring it into such conjunction with the rest. Hence though it may be in a high degree improbable it yet remains possible that a course of Nature destitute of design may of itself have taken all the steps, which in order to realize a purpose must have been taken under the guidance of design; and therefore this first requirement cannot be fulfilled.

And we do not succeed better in fulfilling the *second* requirement—in showing that purposiveness does not occur merely here and there but that it pervades the whole world harmoniously and without exception, so that not merely do intelligent actions occur in it, but the whole is embraced in the unity of one supreme design. How little does our actual experience suffice to show this! How much seems to us wholly inexplicable, purposeless, even obstructive to ends of which we had assumed the existence! The few brilliant examples of a harmony that we can at least partly recognise, which are presented principally by the animate creation, may well confirm an already existent faith in God, in the conviction that in that also which we do not yet understand the unity of the same wisdom may work purposively; but empiric knowledge of the purpose in the world does not furnish the means necessary for enabling any one to attain indisputable faith who does not yet possess it. Taken alone it would much more easily produce the polytheistic intuition of a plurality of divine beings, each of which rules over a special department of Nature as its special genius, and the varying governments of which agree so far as to attain a certain general compatibility, but not a harmony that is altogether without exceptions.

Not merely the defectiveness of the scientific knowledge which we have through experience but also internal difficulties hinder the fulfilment of the *third* requirement—that, namely, of showing that creative wisdom in carrying out its designs never experiences opposition, and is never forced to produce

that which is even only indifferent as regards its purposes ;
but only if this were so would wisdom be omnipotent.   Not
merely, however, does observation show us much which at
least our limited knowledge can understand only as an accidental and accessory effect of the struggle between a formative
design and the independent and resisting nature of the
material to be formed; but, moreover, general reflection
cannot get clear the notion of design without contrasting
with it some material independent of it by elaborating which
it attains realization; and thus all our consideration of purpose
leads us only to the notion of a governor of the universe and
not to that of a creator, which was what we sought.

Finally, how little men have succeeded in fulfilling the
fourth requirement, and in proving the unconditional worth
and the sacredness of the designs which we plainly see pursued
in the world, is taught by a glance at the development of the
doctrines which attempt this proof.   For has not philosophy
often pointed out to us as supreme and unconditionally sacred
cosmic ends much in which living feeling can find no worth
at all ?   Have not popular faith and dogmatic theology found
cause in the ills of the world, and the logical consistency with
which evil develops, to divide the dominion of the world
between God and the devil, taking comfort in the thought
that even of this apparent discord there may be some explanation inaccessible to human reason ?   But though that which is
inaccessible to human reason may indeed be an object of faith,
it cannot furnish any proof that such faith is true; and
thus the Teleological Proof is destitute of all demonstrative
force, however great and unmistakeable may be the efficacy
with which it brings together for the strengthening of faith all
that is best in secular knowledge.

Perhaps, if we were to ask less, we should on the whole
obtain more, and the fundamental thoughts which animate
these proofs may be not incapable of being turned to account
in another way.   The Cosmological Proof prematurely pushed
its demand for the full and complete concept of God into an
assertion of the supreme perfection of the unconditioned, not

having as yet established the unity of that unconditioned. This it could have done if it had considered more searchingly what is involved in the thought of the conditioned existence of things, in the thought of any ordered course of the world. Not the purpose in the world, for this is subject to doubt, but the fact that there is a cosmic course in which events are connected according to laws, must have led it to the necessary unity of that which is the substantial basis of the world. But we will not now return to this consideration, which we discussed at an earlier stage, and to which we also devoted the beginning of this last division of our inquiry. We found—so we thought—the impossibility of that pluralistic theory of the world which presupposes a plurality of original Real beings, independent of one another, and then imagines that from the reciprocal actions of these according to general laws, a cosmic order may be produced. If it had really considered deeply what is meant by saying that *one* truth holds of *many things*, and that for the many, of which each at first existed in a world of its own, there is yet the possibility of a community, in which these may act upon one another, it would have found that both these conceptions are unthinkable without an original unity of existence of all that is real—the activity of this reality, after it works and whilst it works, being capable of appearing as action which in an orderly fashion is bound together by one universal truth, and produced by connections between the separate elements. This unity of that which while unconditioned conditions all finite things, having been established, it became permissible to try and determine the notion thus obtained—the notion of an Infinite Substance—by those more significant predicates by which it was transformed into the notion of a living God.

What the Teleologic Proof has attempted to contribute here, seems to me to have been more impressively stated in the despised form of the *Ontologic Proof*, though it is true that in the scholastic form given to this proof not much of what I have referred to is to be recognised. To conclude that because the notion of a most perfect Being includes reality as one of its perfections, therefore a most perfect Being necessarily *exists*, is so obviously to conclude falsely, that after Kant's incisive refutation any attempt to defend such reasoning would be useless.   Anselm, in his more free and spontaneous reflection, has here and there touched the thought that the greatest which we can think, if we think it as *only* thought, is less than the same greatest if we think it as existent.   It is not possible that from this reflection either any one should develop a logically cogent proof, but the way in which it is put seems to reveal another fundamental thought which is seeking for expression.   For what would it matter if that which is thought as most perfect were, as thought, less than the least reality ?   Why should this thought disturb us ?   Plainly for this reason, that it is an immediate certainty that what is greatest, most beautiful, most worthy is not a mere thought, but must be a reality, because it would be intolerable to believe of our ideal that it is an idea produced by the action of thought but having no existence, no power, and no validity in the world of reality.   We do not from the perfection of that which is perfect immediately deduce its reality as a logical consequence ; but without the circumlocution of a deduction we directly feel the impossibility of its non-existence, and all semblance of syllogistic proof only serves to make more clear the directness of this certainty.   If what is greatest did not *exist*, then what is *greatest* would not be, and it is impossible that that which is the greatest of all conceivable things should *not* be.

Many other attempts may be made to exhibit the internal necessity of this conviction as logically demonstrable ; but all of them must fail.   We cannot prove by thought, we can only know by experience, that anything endowed with beauty is beautiful, or that any disposition of mind has the approval of conscience—except in those easily intelligible cases in which, taking something compound, derived, or as yet obscure, to which those determinations of worth had already been attached by immediate feeling, we bring it under some universal by a brief logical process of analysis.   And just as little can we prove from any general logical truth our right to ascribe to that which has such worth its claim to reality; on the contrary, the certainty of this claim belongs to those inner experiences, to which, as to the given object of its labour, the mediating, inferring, and limiting activity of cognition refers. As such an immediate certainty, this conviction lies at the foundation of the Ontological Proof; and it too it is which carries the Teleologic Proof far beyond the inferences which could be reached by means of its own impracticable assumptions. For when once the dominion of significant moral forces that operate purposively has been confirmed by experience, though over but a small portion of the world, the silent enlargement of this experience into an assertion that there is a wisdom, a beauty, a goodness, and a perfection that pervade the whole world without exception, rests in this case not merely on the common logical mistake of a generalization of some truth proved to be valid in a particular case, but is supported by the living feeling that to this, which is greatest and most perfect, there belongs a perfect and all-embracing reality.

Lively as this conviction may be, and sufficient as its certainty may be for us, yet it shares the formal indeterminateness which attaches to all the inner experiences of faith. For it leaves us in doubt as to what the reality is which that which is highest and most worthy must possess; it believes only that it knows that this highest and best must be one with the Infinite which speculative philosophy found itself bound to recognise as the true reality. The reasons which justify this attempt to blend the Existent and the Worthy into the notion of the living God belong to those intermediate links of the course of thought which we may fairly skip, and this all the more because the following consideration of that to which the attempt has led will include our opinion as to the right and wrong of that attempt.

§ 3. Two distinct series of attributes through which man tries to comprehend the being of God recall to us the two impulses from which arose the notion of God and belief in Him. Metaphysical attributes of Unity, Eternity, Omnipresence, and Omnipotence determine Him as the ground of all finite reality; ethical attributes of Wisdom, Justice, and Holiness satisfy our longing to find in that which has supreme reality, supreme worth also. We have no need to give a complete account of these attributes or to touch doubtful questions as to their reciprocal limits; the only really important point for us is to reach a conviction as to the mode of existence that is to give a definite form to this essence of all perfection, determining also at the same time the special significance of several of the attributes referred to. If these reflections, which are now struggling to a conclusion, were allowed once more to run into the prolixity of systematic completeness, it would be easy to develop from the preceding investigations as to the nature of existence the answer which we should have to give to this last question as to the nature of that Infinite which we have there discovered. But just because it is easy for the reader to supply this transition we will regard the goal to which it would lead, the notion of a Personal God, as being already reached, and endeavour to defend this against doubts as to its possibility, as being the only logical conclusion to which our considerations could come.

The longing of the soul to apprehend as reality the Highest Good which it is able to feel, cannot be satisfied by or even consider any form of the existence of that Good except Personality. So strong is its conviction that some living Ego, possessing and enjoying Self, is the inevitable presupposition and the only possible source and abode of all goodness and all good things, so filled is it with unspoken contempt for all existence that is apparently lifeless, that we always find the myth-constructing beginnings of religion busied in transforming natural to spiritual reality; but never find them actuated by any desire to trace back living spiritual activity to unintelligent Realness as to a firmer foundation. From this right path the progressive development of reflection turned off for a time. With increasing cosmic knowledge, it grew more clear what must be required in the notion of God, if He were not only to contain in Himself all that is greatest and most worthy, but also to contain it after such a fashion as to appear at the same time as the creative and formative ground of all reality; and on the other hand, in more refined observation of spiritual life, the conditions became clear to which in us finite beings the development of personality is attached; both trains of thought seemed to combine in showing that the form of spiritual life is incompatible with the notion of the Supreme Being, or that the form of personal existence is incompatible with the notion of the Infinite Spirit. And there arose attempts to find more satisfying forms of existence for the Highest Good in ideas of an Eternal World-Order, of an Infinite Substance, of a Self-developing Idea, and to depreciate the form of personal existence which had previously seemed to the unsophisticated mind to be the only one that was worthy. Among the infinitely manifold variations which these views have experienced we will content ourselves with briefly showing, of the three we have mentioned, the grounds of their untenableness.

What noble motives and what moral earnestness may lead to the dissolving of the notion of the Divine Being in that of a Moral World-Order, as contrasted with crude anthropomorphism, must be still fresh in men's remembrance. And yet Fichte was not right when, with inspired words, he opposed his own sublime conception to the common narrow-minded idea of a Personal God; because he sought that which was most sublime, he thought that he had found it in the conception which he reached; if he had followed out to the end the path which he took, he would have recognised that by it that which he sought could not be reached. The question, *How* is it that a World-Order can be conceived as the Supreme Principle? cannot he put off by appealing to the fact that we cannot demand a history of the origin of the Principle itself; he who, regarding Personality as an impossible conception of the Godhead, prefers some other to it, will at least have to show that the one which he brings forward is not contradictory; for nothing will be gained by substituting for an impossibility some other assumption of which the possibility is not proven. Now the fact is that the one sufficient reason which will always forbid that some World-Order should be put in the

place of God, is to be found in the simple fact that no order
is separable from the ordered material in which it is realized,
still less can precede such material as a conditioning or
creative force; the order must ever be a relation of something
which exists, after or during its existence.   Hence if it is
nothing but *Order*, as its name says, it is never *that which
orders*, which is what we seek, and which the ordinary notion
of God (however inadequate in other respects) determined
rightly at any rate in this, that it regarded it as a Real being,
not as a relation.

But in considerations concerning these highest things, which
often make us feel the defectiveness of human language,
names seldom mean exactly what they connote, but generally
more or less; only it mostly happens that what we have to
add or to omit cannot without contradiction be combined
with or subtracted from that part of the signification which is
retained.   For this reason all the manifold views which we
here group together will complain of our interpretation of
their proposition, *God Himself is the Order of the world*, as
being a misinterpretation.—In the first place, the World-Order
can *not* take up that position with regard to the world, which,
according to the common view, is occupied by the extra-
mundane God; this position must remain empty, seeing that
it is an impossible place, which nothing could occupy.   Again,
it will be said, to understand Order merely as a relation established by some ordering being, would only betray an incapacity
to understand the true reality, which, through and through,
without any residuum of dead substance is living activity,
movement, and growth, not indeed indeterminate, but determining itself in unvarying consistency to the coherence of
one thought.   But yet if we more clearly analyse these
enthusiastic ideas, must they not, if they are to mean what
they are intended to mean, return again to that which they
avoid ?   We have already had occasion to argue how little
possible it is by the notion of a law of Nature regulating
mere phænomena to avoid the assumption of reciprocal action
between things, or to explain their apparent effects: even if

what is meant by saying that a law *commands* were clear, it
would still be incomprehensible how things or phænomena
should *obey* it; only an essential unity of all existent things
could cause the states of one thing to be efficient conditions
of the changes of another.   On the universal World-Order
which, claiming to govern the moral world also, takes the
place of that law, we must pass a similar verdict.   To us, too,
it is not doubtful " but most certain, and indeed the ground
of all other certainty, that there is this Moral Order of the
world ; that for every intelligent creature there is an appointed
place and a work which he is expected to perform, and that
every circumstance of his lot is part of a plan, in independence
of which not a hair of his head can be harmed, nor (in another
sphere of action) a sparrow fall from the house-top ; that
every good action will succeed and every evil action certainly
fail, and that to those who do but truly love that which is
good all things shall work together for good." (Fichte,
Sämmtliche Werke, v. p. 188.)   But now how can all this be
thought ? Or more accurately, When we think this, what is it
that we think ?   Could that World-Order ever bring together
any plurality to the unity of any definite relation or maintain
such a unity, if it were not at the same time present in each
individual of the plurality and sensitive to every state
occurring in all the other individuals, and capable also of
bringing the reciprocal relations of all into the intended form,
by an alteration of position determined by reference to their
remoteness from the point aimed at ?   This is no sophistical
construction by which we would attempt to show how this
Order comes to exist, but it is an analysis of that which we
must think, if we would think that which is ascribed to this
Order.   And now, after all our detailed discussions on this
point, we cannot say exactly how this notion of an Order which
is affected by facts, and by reaction correspondent to its
nature and affection alters facts, is to be distinguished from
the true notion of a being.   But on this account to call it
simply *Order* is the mistake of an opposition which, shunning
erroneous conceptions of being, obstinately tries to attach

those juster conceptions of which it is itself possessed, to a notion with which they are wholly incapable of being combined.

Now if the notion of any active order necessarily and inevitably leads back to that of an Ordering Being, the notion of a Moral Order leads further. Is it possible to imagine a Being which, stimulated by the influence of every existing condition of the cosmic course, should, with purposeless and blindly working activity, impart to that course the ameliorating impulses by which the thoroughgoing dominion of what is good is established,—a Being which cannot consciously indicate the place of each individual and appoint his work, or distinguish what is good in a good action from what is bad in a bad action, or will and realize the good with its own living love, but yet acts *as though* it could do all this? It is not open to speculation to decline answering this question, for every view must take account of the necessary points of connection, without which its own meaning would be incomplete; but whoever should seek to answer it by imagining an unconscious, blind, impersonal mechanism, of which yet goodness should be the moving spring, would entangle himself profoundly in those impracticable subtleties among which the great mind whose error we here deplore, thought he must reckon the conviction that Personality is the only conceivable form of the Supreme Cause of the universe. Whether the answering of this question is equally necessary for practical life may seem doubtful; but I believe that it is so. The conviction that there is a World-Order may suffice to guide our conduct and to comfort us concerning its apparent resultlessness; but the religious mind is led to apprehend the Supreme Good under the form of a Personal God both by humility and by the longing to be able to reverence and love, motives which the religion of a mere strict fulfilment of duty has too little regarded.

We cannot consider the remaining views in even as much detail as we have those above referred to. The common admission of substantial unity in the World's-Cause, connects us only apparently with the reverence of Pantheism for the one Infinite Substance; and, moreover, the conceptions which we have formed concerning the meaning of the Real have removed us so far from the circles of thought in which Pantheism moves, that it is not possible to give a brief explanation of our relation to it. It regards as existent being what we can conceive only as phænomenal—the spatial world, with its extension, the figures which it contains and its unceasing movements; it regards it as conceivable that an inexhaustible vital force of the Unconditioned and the One should find relief in manifesting itself in these extended figures and their changes as though in so doing it really accomplished something; but for us all this was but the shadow of true and supersensuous being and action; hence Pantheism might think it possible to understand the spiritual world as an isolated blossom, growing from the strong stem of material Realness that works unconsciously, but to us it seems inconceivable that spirit should arise from that which is not spirit, and inevitable that all unconscious existence and action should be regarded as an appearance, the form and content of which springs from the nature of spiritual life. From a metaphysical point of view, we could only agree with Pantheism as a possible conception of the world if it renounced all inclination to apprehend the Infinite Real under any other than a spiritual form; from a religious point of view, we cannot share the disposition which commonly governs the pantheistic imagination—the suppression of all that is finite in favour of the Infinite, the inclination to regard all that is of value to the living soul as transitory, empty, and frail in comparison of the majesty of the One, upon whose formal properties of immensity, unity, eternity, and inexhaustible fulness it concentrates all its reverence. But this as well as the reason which holds us back from seeing that which is highest in the universe in an infinite and self-conscious Idea, we shall notice later—as far as it is possible to give a mere passing consideration to subjects that have been so endlessly discussed.

§ 4. An Ego (or Self, *Ich*) is not thinkable without the contrast of a Non-Ego or Not-Self; hence personal existence cannot be asserted of God without bringing even Him down to that state of limitation, of being conditioned by something not Himself, which is repugnant to Him.—The objections that speculative knowledge makes to the personality of God fall back upon this thought; in order to estimate their importance, we shall have to test the apparently clear content of the proposition which they take as their point of departure. For unambiguous it is not; it may be intended to assert that what the term Ego denotes can be comprehended in reflective analysis only by reference to the Non-Ego; it may also mean that it is not conceivable that this content of the Ego should be experienced without that contrasted Non-Ego being experienced at the same time; finally, it may point to the existence and active influence of a Non-Ego as the condition without which the being upon which this influence works could not be an Ego.

The relations which we need in ideation for making clear the object ideated, are not in a general way decisive as to its nature; they are not conditions of the possibility of the thing as they are for us conditions of the possibility of its presentation in idea. But the special nature of the case before us seems to involve something which is not generally included—for it is just in the act of ideation that Selfhood (*Ichheit*) consists, and hence what is necessary for carrying out such an act is at the same time a condition of the thing. Hence the first two interpretations which we gave of the proposition referred to seem to run together into the assertion that the Ego has significance only as contrasted with the Non-Ego, and can be experienced only in such contrast. Whether we agree with this assertion will depend in part upon the significance attached to the words used. We see in the first place that at any rate Ego and Non-Ego cannot be two notions of which each owes its whole content only to its contrast with the other; if this were so they would both remain without content, and if neither of them apart from the contrast had a

fixed meaning of its own, not only would there be no ground
for giving an answer one way or the other to the question
which of the two members of the contrast should take the
place of the Ego and which that of the Non-Ego, but the very
question would cease to have any meaning. Language has
given to the Ego alone its own independent name, to the
Non-Ego only the negative determination which excludes the
Ego without indicating any positive content of its own.
Hence every being which is destined to take the part of the
Ego when the contrast has arisen, must have the ground of its
determination in that nature which it had *previous to* the
contrast, although before the existence of the contrast it is
not yet entitled to the predicate which in that contrast comes
to belong to it. Now if this is to remain the meaning of the
term, if the being is to be Ego only at the moment when it
is distinguished from the Non-Ego, then we have no objection
to make to this mode of expression, but we shall alter our own.
For it is our opponents' opinion and not ours that personality
is to be found exclusively where, in ideation (or presentation),
Self-consciousness sets itself as Ego in opposition to the Non-
Ego ; in order to establish the selfhood (*Selbstheit*) which we
primarily seek, that nature is sufficient in virtue of which,
when the contrast does arise, the being becomes an Ego, and it
is sufficient even before the appearance of the contrast. Every
feeling of pleasure or of dislike, every kind of self-enjoyment
(*Selbstgenuss*), does in our view contain the primary basis of
personality, that immediate self - existence which all later
developments of self-consciousness may indeed make plainer
to thought by contrasts and comparisons, thus also intensifying
its value, but which is not in the first place produced by them.
It may be that only the being who in thought contrasts with
himself a Non-Ego from which he also distinguishes himself, can
say *I* (*Ich*) to himself, but yet in order that in thus distinguishing
he should not mistake and confound himself with the Non-
Ego, this discriminating thought of his must be guided by a
certainty of self which is immediately experienced, by a self-
existence which is earlier than the discriminative relation by

which it becomes Ego as opposed to the Non-Ego.   A
different consideration has already (cf. i., pp. 241 *seq.*) led us
by an easier path to the same result, and we may refer the
reader to this passage for explanation and completion of what
is said here.   The discussion referred to showed us that all
self-consciousness rests upon the foundation of direct sense of
self which can by no means arise from becoming aware of a
contrast with the external world, but is itself the reason that
this contrast can be felt as unique, as not comparable to any
other distinction between two objects.   Self-consciousness is
only the subsequent endeavour to analyse with the resources
of cognition this experienced fact—to frame in thought a
picture of the Ego that in cognition apprehends itself with the
most vivid feeling, and in this manner to place it artificially
among the objects of our consideration, to which it does not
really belong.   So we take up our position with regard to the
first two interpretations of the proposition of which we are
speaking, thus :—We admit that the Ego *is thinkable* only in
relation to the Non-Ego, but we add that it *may be experienced*
previous to and out of every such relation, and that to this is
due the possibility of its subsequently becoming thinkable in
that relation.

But it is not these two interpretations but the third that
is most obstructive to that faith in the Personality of God
which we are seeking to establish.   In one form indeed in
which it sometimes occurs we need not make it an object
of renewed investigation ; for we may now consider it as, in
our view, established that no being in the nature of which
self-existence was not given as primary and underived, could
be endowed with selfhood by any mechanism of favouring
circumstances however wonderful.   Hence we may pass over
in complete silence all those attempts which think to show by
ill-chosen analogies from the world of sense how in a being as
yet selfless an activity originally directed entirely outwards is,
by the resistance opposed to it by the Non-Ego (comparable
to that which a ray of light encounters in a plane surface),
thrown back upon itself and thereby transformed into the

self-comprehending light of self-consciousness. In such ideas everything is arbitrary, and not a single feature of the image employed is applicable to the actual case which it is intended to make clear; that outgoing activity is an unmeaning imagination, the resistance which it is to meet with is something that cannot be proved, the inference that that activity is by that resistance turned back along the path by which it came is unfounded, and it is wholly incomprehensible how this reflection could transform its nature, so that from blind activity it should turn into the selfhood of *self-existence.*

Setting aside these follies which have influenced philosophic thought to an unreasonable extent, we find a more respectable form of the view which we are combating occupied in proving that though that self-existence cannot be produced by any external condition in a being to which it does not belong by nature, yet it could never be developed even in one whose nature is capable of it, without the co-operation and educative influences of an external world. For that from the impressions which we must receive from the external world, there comes to us not only all the content of our ideas, but also the occasion of all those feelings in which the Ego, existing for self, can enjoy self without as yet being conscious of a relation of contrast to the Non-Ego. That all feeling must be conceived as (in some definite form of pleasure or displeasure) interested in some definite situation of the being to which it belongs, some particular phase of its action and its passion ; but that neither is passion possible without some foreign impression which calls it forth, nor activity possible without an external point of attraction which guides it and at which it aims. That in any single feeling the being which is self-existent is only partially self-possessing ; that whether it has self-existence truly and completely depends upon the variety of the external impulses which stimulate by degrees the whole wealth of its nature, making this wealth matter of self-enjoyment—that thus the development of all personality is bound up with the existence and influence of an external world and the variety and succession of those influences ; and that such development would be possible even for God only under similar conditions.

It is not sufficient to lessen the weight of this objection by the assertion that this educative stimulation is necessary only for finite and changing beings, and not for the nature of God, which, as a self-cognisant Idea, eternally unchangeable, always possesses its whole content simultaneously. Though this assertion grazes the truth, yet in this form it would be injurious in another respect to our idea of God, for it would make the being of God similar to that of an eternal truth—a truth indeed not merely valid but also conscious of itself. But we have a direct feeling of the wide difference there is between this personification of a thought and living personality ; not only do we find art tedious when it expects us to admire allegorical statues of Justice or of Love, but even speculation rouses our opposition forthwith, when it offers to us some self-cognisant Principle of Identity, or some self-conscious Idea of Good, as completely expressing personality. Either of these are obviously lacking in an essential condition of all true reality in the capacity of *suffering*. Every Idea by which in reproductive cognition we seek to exhaust the nature of some being, is and remains nothing more than the statement of a thought-formula by which we fix, as an aid to reflection, the inner connection between the living activities of the Real ; the real thing itself is that which applies this Idea to itself, which feels contradiction to it as disturbance of itself, and wills and attempts as its own endeavour the realization of the Idea. The only living subject of personality is this inner core, which cannot be resolved into thoughts, the meaning and significance of which we know in the immediate experience of our mental life, and which we always misunderstand when we seek to construe it— hence personality can never belong to any unchangeably valid truth, but only to something which changes, suffers, and reacts. We will only briefly point out in passing the insurmountable difficulties which the attempt to personify Ideas thus would encounter if there were any question of determining the relation between the Ideas so personified and the changing course of the world; it would immediately appear that these could as little do without the additions necessary to transform them into suffering and acting beings as the World-Order to which we have before referred.

Yet the transference of the conditions of finite personality to the personality of the Infinite is not justified. For we must guard ourselves against seeking in the alien nature of the external world, in the fact that it is *Non*-Ego, the source of the strength with which it calls out the development of the Ego; it operates only by bringing to the finite mind stimuli which occasion the activity, which that mind cannot produce from its own nature. It is involved in the notion of a finite being that it has its definite place in the whole, and thus that it is not what any other is, and yet that at the same time it must as a member of the whole in its whole development be related to and must harmonize with that other. Even for the finite being the forms of its activity flow from its own inner nature, and neither the content of its sensations nor its feelings, nor the peculiarity of any other of its manifestations, is given to it from without; but the incitements of its action certainly all come to it from that external world, to which, in consequence of the finiteness of its nature, it is related as a part, having the place, time, and character of its development marked out by the determining whole. The same consideration does not hold of the Infinite Being that comprehends in itself all that is finite and is the cause of its nature and reality; this Infinite Being does not need—as we sometimes, with a strange perversion of the right point of view, think—that its life should be called forth by external stimuli, but from the beginning its concept is without that deficiency which seems to us to make such stimuli necessary for the finite being, and its active efficacy thinkable. The Infinite Being, not bound by any obligation to agree in any way with something not itself, will, with perfect self-sufficing-ness, possess in its own nature the causes of every step forward in the development of its life. An analogy which though weak yet holds in some important points and is to some extent an example of the thing itself, is furnished to us by the course of memory in the finite mind. The world of our ideas, though certainly called into existence at first by external impressions, spreads out into a stream which, without any fresh stimulation from the external world, produces plenty that is new by the continuous action and reaction of its own movements, and carries out in works of imagination, in the results reached by reflection, and in the conflicts of passion, a great amount of living development—as much, that is, as can be reached by the nature of a finite being without incessantly renewed orientation, by action and reaction with the whole in which it is comprehended; hence the removal of these limits of finiteness does not involve the removal of any producing condition of personality which is not compensated for by the self-sufficingness of the Infinite, but that which is only approximately possible for the finite mind, the conditioning of its life by itself, takes place without limit in God, and no contrast of an external world is necessary for Him.

Of course there remains the question what it is that in God corresponds to the primary impulse which the train of ideas in a finite mind receives from the external world? But the very question involves the answer. For when through the impulse received from without there is imparted to the inner life of the mind an initiatory movement which it subsequently carries on by its own strength, whence comes the movement in the external world which makes it capable of giving that impulse? A brief consideration will suffice to convince us that our theory of the cosmos, whatever it may be, must somehow and somewhere recognise the actual movement itself as an originally given reality, and can never succeed in extracting it from rest. And this indication may suffice for the present, since we wish here to avoid increasing our present difficulties by entering upon the question as to the nature of time. When we characterize the inner life of the Personal God, the current of His thoughts, His feelings, and His will, as everlasting and without beginning, as having never known rest, and having never been roused to movement from some state of quiescence, we call upon imagination to perform a task no other and no greater than that which is required from it by every materialistic or pantheistic view. Without an eternal uncaused movement of the World-Substance, or the assumption of definite initial movements of the countless world-atoms, movements which have to be simply recognised and accepted, neither materialistic nor pantheistic views could attain to any explanation of the existing cosmic course, and all parties will be at last driven to the conviction that the splitting up of reality into a quiescent being and a movement which subsequently takes hold of it, is one of those fictions which, while they are of some use in the ordinary business of reflection, betray their total inadmissibility as soon as we attempt to rise above the reciprocal connection of cosmic particulars to our first notions of the cosmos as a whole.

The ordinary doubts as to the possibility of the personal existence of the Infinite have not made us waver in our conviction. But in seeking to refute them, we have had the feeling that we were occupying a standpoint which could only be regarded as resulting from the strangest perversion of all natural relations. The course of development of philosophic thought has put us who live in this age in the position of being obliged to show that the conditions of personality which we meet with in finite things, are not lacking to the Infinite; whereas the natural concatenation of the matter under discussion would lead us to show that of the full personality which is possible only for the Infinite a feeble reflection is given also to the finite; for the characteristics peculiar to the finite are not producing conditions of self-existence, but obstacles to its unconditioned development, although we are accustomed, unjustifiably, to deduce from these characteristics its capacity of personal existence. The finite being always works with powers with which it did not endow itself, and according to laws which it did not establish,—that is, it works by means of a mental organization which is realized not only in it but also in innumerable similar beings. Hence in reflecting on self, it may easily seem to it as though there were in itself some obscure and unknown substance—something which is in the Ego though it is not the Ego itself, and to which, as to its subject, the whole personal development is attached.   And hence there arise the questions—never to be quite silenced —What are we ourselves ?   What is our soul ?   What is our self—that obscure being, incomprehensible to ourselves, that stirs in our feelings and our passions, and never rises into complete self-consciousness ?   The fact that these questions can arise shows how far personality is from being developed in us to the extent which its notion admits and requires.   It can be perfect only in the Infinite Being which, in surveying all its conditions or actions, never finds any content of that which it suffers or any law of its working, the meaning and origin of which are not transparently plain to it, and capable of being explained by reference to its own nature.   Further, the position of the finite mind, which attaches it as a con-stituent of the whole to some definite place in the cosmic order, requires that its inner life should be awakened by successive stimuli from without, and that its course should proceed according to the laws of a psychical mechanism, in obedience to which individual ideas, feelings, and efforts press upon and supplant one another.   Hence the whole self can never be brought together at one moment, our self-conscious-ness never presents to us a complete and perfect picture of our Ego—not even of its whole nature at any moment, and much less of the unity of its development in time.   We always appear to ourselves from a one-sided point of view, due to those mental events which happen to be taking place within us at the time—a point of view which only admits of our surveying a small part of our being ; we always react upon the stimuli which reach us, in accordance with the one-sided impulses of this accidental and partial self-consciousness; it is only to a limited extent that we can say with truth that *we* act ; for the most part action is carried on in us by the individual feelings or groups of ideas to which at any moment the psychical mechanism gives the upper hand.   Still less do we exist wholly *for ourselves* in a temporal point of view. There is much that disappears from memory, but most of all individual moods, that escape it by degrees. There are many regions of thought in which while young we were quite at home, which in age we can only bring before our mind as alien phænomena; feelings in which we once revelled with enthusiasm we can now hardly recover at all, we can now hardly realize even a pale reflection of the power which they once exercised over us; endeavours which once seemed to constitute the most inalienable essence of our Ego seem, when we reach the path along which later life conducts us, to be unintelligible aberrations, the incentives to which we can no longer understand. In point of fact we have little ground for speaking of the personality of finite beings; it is an ideal, which, like all that is ideal, belongs unconditionally only to the Infinite, but like all that is good appertains to us only conditionally and hence imperfectly.

§ 5. The more simple content of this section hardly needs the brief synoptical repetition in which we now proceed to gather up its results and to add them to those already reached.

X. Selfhood, the essence of all personality, does not depend upon any opposition that either has happened or is happening of the Ego to a Non-Ego, but it consists in an immediate self-existence which constitutes the basis of the possibility of that contrast wherever it appears. Self-consciousness is the elucidation of this self-existence which is brought about by means of knowledge, and even this is by no means necessarily bound up with the distinction of the Ego from a Non-Ego which is substantially opposed to it.

XI. In the nature of the finite mind as such is to be found the reason why the development of its personal consciousness can take place only through the influences of that cosmic whole which the finite being itself is not, that is through stimulation coming from the Non-Ego, not because it needs the contrast with something *alien* in order to have self-existence, but because in this respect, as in every other, it does not contain in itself the conditions of its existence. We do not find this limitation in the being of the Infinite; hence for it alone is there possible a self-existence, which needs neither to be initiated nor to be continuously developed by something not itself, but which maintains itself within itself with spontaneous action that is eternal and had no beginning.

XII. Perfect Personality is in God only, to all finite minds there is allotted but a pale copy thereof; the finiteness of the finite is not a producing condition of this Personality but a limit and a hindrance of its development.

CHAPTER V.

GOD AND THE WORLD.

Difficulties in this Chapter—The Source of the Eternal Truths and their Relation to God—The Creation as Will, as Act, as Emanation—Its Preservation and Government; and the Ideality of Time—The Origin of Real Things—Evil and Sin—Good, Good Things, and Love—The Unity of the Three Principles in Love—Conclusion.

§ 1. WE traced back the manifoldness of reality to *one* unconditioned primary Cause; and this One, which can give coherence to finite multiplicity and the possibility of reciprocal action to individual things, we found not in a law, not in an Idea, not in any cosmic order, but only in a *Being* capable of acting and suffering; in Mind alone, self-possessing and having self-existence, and not in a substance developing with blind impulse, did we find in truth and reality the substantiality which we felt constrained to require in this Supreme Being. The rapidity with which we hurried towards this goal of our thoughts carried us past difficulties to which we now return.

Our ideas concerning even God and divine things can satisfy us only when they are in harmony with those general laws of thought and those truths which reason sets before us as having binding force with regard to every object of which we can judge. Hence even that Supreme Being whom we reverence as the unconditioned and creative Cause of all reality, as soon as He becomes an object of our investigation may easily seem to be conditioned by general truths and laws possessing a validity independent of and prior to Him. When we speak of the wisdom of God we seem obliged to think of it as applied to truth, the independently valid content of which is recognised by God, and hence prior to Him; we seem obliged to think of His justice or any other of His ethical

perfections as expressing nothing more than the immutable
and thoroughgoing conformity of His being to an ideal of all
good, the eternal worth of which is independently established;
even creative activity, as it produces real things, is hardly
intelligible to us except as a deliberative choice that summons
into reality whichever it will from the abundance of the conceivable and possible forms of future existence, spread out
before it as a store from which to choose.   All this is incompatible with that unconditionedness which must belong to the
Supreme Reality, not only as regards its existence, but also in
such a way that it determines through itself alone the form
and object of its activity.   We will divide the discussion of
these difficulties, and unite in one inquiry concerning the
origin of eternal truths, an explanation of the relation to the
being of God, of the laws (1) of cognition and (2) of the
course of events, and (3) of the determination of moral worth;
and later we shall turn to consider in what way we must
conceive the forms of reality to have their foundation in the
same divine nature.

§ 2. The philosophy of common sense generally seems to
take it for granted as self-evident that even the divine activity
moves within the limits which the general laws of all being
and action set to any conceivable activity.   When expressly
questioned on this point religious faith may occasionally
hesitate somewhat; but for the most part it admits this tacit
presupposition and recognises eternal truth as primary and
unconditioned, as being an absolutely valid necessity, to which
even the living reality of God is subject.   If we ignore the
contradiction with reference to the unconditionedness of God
which is plainly involved in this view, we yet find that it
involves another contradiction which equally invalidates it—
namely one that concerns the nature of truth.   It is only as regards an individual and finite thing that an individual law before
it is realized in it can appear as a power existing external to it;
for in such a case this law is realized in other things in the
states of which it is embodied, and by the coherent action of
which it becomes possible for it to subject to itself things which

had as yet escaped its dominion. But the whole body of truth cannot precede the whole of reality, or that One Supreme Being from which it flows, as though it were a power existing independently *in vacuo ;* for of truths we can only say that they *are valid,* not that they *exist.* They do not hover among or external to or above existing things ; as forms of connection between multifarious states, they are present only in the thought of some thinker whilst he thinks, or in the action of some existing being at the moment of his action. If they rule not only the present but also the future, they can do this not because they are enthroned in eternal splendour beyond and above all reality and all time, but because, really being *in* that which is real, they are continually produced afresh by its action. Existing things receive through their own action in unbroken continuity, and as it were transmit to themselves from moment to moment, the unchanging forms of their being and their states and the connection between these, and thus they every moment reproduce the conditions of the influence which truth exercises upon them. If it were thinkable that the course of the world should suddenly cease to contain the efficient causes of that which truth commands, then this truth would no longer *be* in the world, and certainly he who should then think of it as existing external to the world in its inactive validity, would not be able to say how it could happen that reality should come to be again subject to it. Hence it is impossible that a realm of eternal truths should in any way exist *external to* God as an object of His recognition, or *before* Him as a rule of His working, and this impossibility does not disappear if we avoid the spatial and temporal expressions, the figurative use of which we have just indulged in. It would only be a useless change of terms if we were to call such truths not external to and before God but in Him and with Him ; thought as universal necessities, to which the Divine Being like all else is subject, they would still continue to lay claim to this impossible validity, preceding and transcending all reality—a validity which we must deny to them, and through which, if they had it, they would be alien and limiting conditions of that by which the being of God is distinguished from all other being in which their influence is shown.

The course of our thought is so thoroughly accustomed to the perverse idea of an independent truth, giving laws to reality, that we do not take offence at the contradictions which it involves. But so much the more is natural feeling hurt by our detraction from that unconditionedness of God which cannot be surrendered, if we regard it as subject to a truth which is independent of it. So a second form of the view we are considering resolves to regard the eternal truths as creations of God, which He might have left uncreated, or have created other than they are. But this opinion too speedily leads to contradiction and is incompatible with the notion of truth. For truths can no more be made than they can exist independent of reality, and no thought which is of questionable validity can by the will, or the recognition, or the command of any one, be made true if it were not so before. Statutes may be enacted ; but statutes are only commands which choose some one thinkable order of relations from among a number all equally thinkable, and that which is chosen they do indeed endow with actual validity, but never with that intrinsic necessity which its nature lacks. But in order that the statute itself should be enacted, there must pre-exist some truth intrinsically and independently valid which enables men to distinguish what is possible from what is impossible, and the cases to which the order that is to be established applies from those to which it does not apply. Now if it is unthinkable that *any* truth should arise by creation, it is still more impossible to imagine creative activity directed to such an impossible aim as the original production of *all* truth. For in whatever way we may picture it to ourselves, as long as we imagine that through this activity something arises which but for it would not exist, we must imagine that the activity takes effect in a certain sequence of events in which as a producing condition it brings its results to pass. But in a world in which as yet there is no truth (supposing such a world to be thinkable) what could be called a condition and what could be called the result of a condition? Where should we find any guarantee of the connection of the one with the other, of any act having any result, or of its having the one at which it aims, and not another at which it does not aim?

The ill-success of these two extreme views is sought to be avoided by a third view which takes a middle path, declaring these eternal truths to be neither objects recognised by God, nor creations of His arbitrary will, but the necessary consequences of His own being. But it is a mistake to think that the difficulty can be avoided in this way. If there is to be any meaning in saying that something proceeds as a logical consequence from the nature of God, we must, in thought at least, oppose to this another something, proceeding as an illogical consequence from the same nature. In order to distinguish the two we need some universal intrinsically valid standard, measured by or compared with which the one something may be recognised as deducible from a definite source, and the other something as not deducible from the same source. Thus we find ourselves led back by a very short road to the necessity of assuming some unconditioned primary truth as having binding force even upon the being of God, in order that by it we may be able to comprehend as logical results of the divine nature those eternal truths which can be deduced from it. However sensible we may be that this attempt follows a true impulse, still this formulation of its results is a failure, and other considerations are needed in order that we may turn to advantage the good which it does contain.

The resultlessness of all these views, which we have presented in their most unmodified and therefore most intelligible forms, is due to the concealed ambiguity with which they apply the name of God. When we doubt whether God recognises truth or establishes it, whether He wills that which is good, or whether if He wills anything it is thereby good, we must first of all get clear the question, Is the God to whom these propositions refer regarded as the God whom our religious consciousness seeks and acknowledges, in His

fulness and completeness; and are the activities or properties which form the predicates of those propositions
already included in the concept of Him in such a way that
the propositional form only serves to set them out afresh for
the sake of explanation after the manner of analytic judgments ?   Or is the name of God here merely a provisional
anticipatory designation of a being to whom the content of
these predicates does not as yet belong, so that the propositions referred to express, after the fashion of synthetic judgments, some process, some activity, or some event, which is
intended to endow that being with these predicates for the
first time ?   That the second of these assumptions is in a
religious sense unmeaning, and is in itself unthinkable, we
will try to show by taking for illustration two familiar questions,
around which the strife of opinions has been concentrated, as
representatives of the metaphysical and ethical difficulties of
the subject.

The first of these questions—whether God recognises or
brings to pass the truth of the proposition $2 + 2 = 4$, and
whether He could make true the proposition $2 + 2 = 5$, does
not very happily express the point which is here in question.
It gives an impression that the point in dispute is whether
God could replace the one proposition which is now true, by
the other which is now false, arbitrarily raising the latter to
the rank of a truth.   In doing this, however, He would not
create truth at all, but presuppose it.   For in order that it
may be possible to express any proposition in the form of an
equation, in order that the correct proposition $A = A$, the
questionable proposition $A = B$, or the erroneous proposition
$A = Non\text{-}A$, should have any imaginable meaning, the truth or
untruth of which we could discuss, it is indispensable that
each of these letters should indicate a content which is in
itself something stable, self-identical, and distinct from everything else, and can hence be called by a name which belongs
to it alone.   Hence from every individual thing that is to be
thought as having any relation, true or false, to some other,
the Law of Identity must be of prior validity as the simplest

truth, without which there can be neither other truths nor any untruth at all. Hence, more generally expressed the question would run thus — Can the will of God establish the Law of Identity so as by means of it to make true some individual relation which contradicts it ? But the answer to this question is devoid of interest; there is no natural and unavoidable motive for raising it. He who should believe that he must answer it affirmatively for the sake of unconditioned divine omnipotence, would be obliged both in the question and in the answer to treat the notions of the contradictory and of the non-contradictory as having an already established definite significance, *before* it could be decided what attitude the divine omnipotence would take with reference to those notions. Now if it should decide to establish as truth the contradictory—that is what was contradictory *before* its decision—it would not create *all* truth, it would not first establish the notion of truth, it would be but an arbitrary will, struggling to upset, as far as possible, truth which it found already binding. No religious need drives us to seek in God omnipotence thus devoid of intelligence. Hence the second clause of this much-debated question must be dropped, and limiting ourselves to the first clause, we ask only whether the truth which is *not yet* can be established by God ? Now an omnipotence which could only accomplish whatever was possible would indeed be merely the greatest among all finite powers, but such as could accomplish the impossible would be none the less finite; for it would presuppose something impossible in itself, that is impossible without the help of omnipotence—something that omnipotence would be able to make possible; but the only true omnipotence must be that which first produces the whole unnameable region within which there is a distinction not previously existent between the true and the untrue, the possible and the impossible.

Now if this is the real meaning of creating truth, who is the God to whom we ascribe this creation ? Is He not the perfect and complete God in whose being we imagine that all truth already is—but if for him who is to create truth it is *not yet* valid, in what does his being and his omnipotence consist, except in a general capacity of doing, that is without content and without direction, and that certainly appears wholly unlimited, but only for this reason, that it neither finds objects with which it could enter into relation, nor rules by which it might regulate its procedure ? This, however, is an idea that signifies nothing which could possibly exist. If from examples of various performances we frame the general notion of capacity or power, we obtain an abstraction logically allowable, and applicable in thought, the content of which, however, does not denote anything that can exist until we supply that which we had previously abstracted. As there is no motion without velocity and direction, and none which could be endowed with velocity and direction after it had come into existence, so we cannot conceive of any power that has not some mode of procedure, nor of any empty capacity that in its emptiness hits upon definite modes of activity. Hence even the divine power cannot be thought as without content and without direction; and the definite mode of action in which it thus consists, and which when we reflect seems to exclude every other conceivable mode of action, is by no means to be regarded as a limitation of its uncon-ditionedness. It would indeed be so for a finite being; for such a being finds the modes of activity from which its nature excludes it existing beside it as regions really subject to the power of other beings, regions which are closed to it, and hence form impassable boundaries of its own activity. There is nothing of this kind in the case of the Infinite; being itself the ground of all reality, it is also the source of the various possibilities of manifold activity which reality contains ; no mode of action beyond its own can be opposed to it as independent of it, or at least as a reality inaccessible and forbidden to it. If it should be asserted finally that the unconditionedness of the Divine Omnipotence is detracted from not only by the reality but by the very conceivability of other action than its own, we deny this also, and the denial will serve to make perfectly clear the meaning of our own view.

For we use this necessity of associating with the notion of any power the thought of some definite mode and kind of action, in order to maintain that just that which we know as the sum of the eternal truths *is* the mode in which Omnipotence acts, but is not *created* by Omnipotence—in other words, this sum of eternal truths is the *mode of action* of Omnipotence, but not its *product*. This signifies, in the first place, that Omnipotence remains an imperfect notion, signifying nothing real, if eternal truth is not associated with it in thought, showing the direction and kind of its action; it signifies further that truth is real not merely in itself, but only as the nature and eternal habitude of the highest activity; and finally, it signifies that truth regarded as truth, that is as a whole of thoughts connected together and conditioning one another, has but a derived and secondary existence in the mind of the thinker by whom it is thought. An intelligence which being itself a part of reality, is itself under the dominion of these eternal habitudes of all action, in comparing the various examples of being and action discovers truths as the general ideas which make comprehensible to it the connection of the details of reality *posterior to* its existence as a whole. And then for such an intelligence there arises for the first time the delusive appearance that this universal, which the individual may think as the precedent and conditioning principle of his thoughts, has also preceded all reality as a destiny existing and ruling in the shadowy emptiness of unreality; it appears to this intelligence that before the existence of the world *and of God,* there existed an ordered realm of possibilities and necessities—that real things which only subsequently come to exist, by assuming some one of these ready-made forms, realizing some one of these possibilities, become thereby finite and limited, and, by the fiat of that already existing necessity, excluded from being some other possible thing which goes on possessing an indescribable existence, in some indescribable locality beyond the world and reality, itself bounding and limiting all reality.

We have here touched an absolutely decisive point in our philosophic theory; but since it has already so often, and

in so many forms, been the topic of our discussion, it is
sufficient here to give it the expression which we have just
done, which gives an answer to the second as well as the first
of the questions raised.   For impossible as it appears to
imagine truth as the creation of Omnipotence for which it is
as yet not valid, equally impossible is it to understand it as
an object of recognition for any being that does not by its own
nature participate in it.   Only he for whom truth is true can
recognise it as truth.   An intelligence which, being destitute
of any innate rule of its procedure, should serve only as a
mirror to bring into view everything existing external to it,
would, if it were possible to imagine it at all, reflect truth
and error with equal impartiality, and without observing the
distinction between them.   The understanding can find truth
only where it sees the content of its thought agreeing with a
standard which it carries within itself, agreeing that is with
the laws of its own procedure in the combination of given
material.   Hence it only recognises truth in as far as it
belongs to its own nature from all eternity; truth that was
originally unconnected with it, it would neither comprehend
as such, nor, as a matter of fact, recognise in such a way that
this could subsequently become a rule of its procedure.   Thus
it appears to be in every way impossible to set up in opposition to truth, a God for whom truth has as yet no validity,
whether we regard Him as its creator or as accommodating
Himself to it; truth cannot be created by His *act*, but it is
only through His *existence* that it subsists; it cannot be
external to him who is to recognise it, on the contrary its
recognition is only thinkable as cognition of one's own being
in it.

It would be superfluous to analyse at equal length the
second example to which we referred.   The Good cannot be
established by any divine will, nor be to it an object of
recognition, unless that will already contains that Good in the
same way as we have said that truth must be contained by
the mind which apprehends it.   If God, without being determinable by ethical predicates, were merely a power developing

in some living form or other, or a will working from the beginning in some one direction, harmony with those forms of development, or movement in that direction of working, would certainly be a condition of subsistence and wellbeing for any finite thing dependent upon Him ; and if there subsisted in this finite thing a consciousness of its existence and position, those conditions would appear to it as commands, the neglect of which it would be dissuaded from by fear, and punished for by remorse. But in such a case the notion of Good as of an ideal having binding force in virtue of its own majesty, could arise only through a somewhat incomprehensible error of limited finite insight; the binding force could not be deduced from such a will, and faith in its unconditioned supremacy would have to be explained as an illusion But for the same reason Good could not be an object of recognition to God. Supposing the unconditionedness of the Divine Being not to be lessened by the fact of its being decided external to and independent of Him, what is good and what is not good, yet even then His will could only recognise the value of the Good thus given, if He Himself in virtue of His own nature had already attached equal value to it, just in the same way as the understanding comprehends given truth as truth only because it is true for that understanding itself. So that in the case of goodness as well as in that of truth it appears inadmissible to separate from God those essential perfections by which only the notion of Him is made complete, and then to assume as an already existing Being an unintelligible Divine Nature to which these perfections are subsequently added by a deed or a series of events which might possibly never have come to pass. Every such attempt mistakes the arbitrary circuits made by our thought in the consideration of its object for a movement of the object itself, which, being eternally the same, is simultaneously all that which our thought can comprehend only in succession.

§ 3. Religious reflection analyzes the relation of God to reality into Creation, Conservation, and Government, and we will now make these three notions of divine working the subject of a question as to the formal conditions of the relation between God and the world which they indicate; we do not as yet touch upon the origin of the inventive thought by which God has given content to that which He created, order to that which exists, plan and direction to that which happens.

Creation cannot be an object of investigation in the sense of our seeking to find the process by which it was brought about; such processes can take place only within a world that already subsists, the constituents of which are capable of action and of being combined in an orderly fashion so as to produce results. But creation, regarded as having taken place, establishes a permanent relation between creator and creature, the meaning and religious worth of which it is the more necessary for us to consider because it is not similarly understood by all. Is reality a production of the divine *will* only? Or is it an *act* of God? Or, finally, is it a non-voluntary *emanation* of His nature? In giving an affirmative answer to the first of these questions, we receive only partial approbation from religious feeling, which, especially in the present day, seems more inclined to regard as an act of God that which it intends to indicate by the notion of creation. For the soul feels that it possesses the living God after which it longs only when it is allowed to speak of a work of creation, in which God, pervading every smallest part of existing reality with His living nature, would in truth *produce* that which, according to our view, would on occasion of His will *arise* as it were spontaneously.

If a movement of our limbs seemed only to follow our volition, we should almost cease to regard it as ours; it would be as foreign to our own being as now those further results appear to be which our action brings forth in the external world—*they* come from us, it is true, but *we* are no longer present in them. But this is not the case; on the contrary, at the moment of movement we think that we directly feel the transmission of active will into our limbs; we think that we directly feel even the smallest remission or increase of tension which the will from moment to moment calls forth in the living members of the body; and all this happens, not as though at a distance from us, so as to be indirectly experienced by us, but we believe that we are ourselves present at every point at which these processes take place; nay more, it seems to us as though we plainly felt how our active force is efficiently transmitted even to the foreign body which we handle, and as it were pervades and restrains the non-Ego in its own domain. It is this self-enjoyment of our own living energy which the view that regards creation as an act refuses to omit from the notion of God, and the worth of this religious need may be recognised, although we must hold that this mode of satisfying it is erroneous.

For a well-known psychological illusion has here misled men into looking for the distinction between our action and that which merely has its origin in us, in a place where it cannot be. The feeling which accompanies our movements is not a sense of volition in the full swing of an activity by which it compels results, but is a perception of the *effects* of volition *after* they have been produced in a fashion wholly imperceptible to us. Our will does not really *produce* the movement in the sense in which this view always holds that it does; but to every volition that arises, in as far as it is a definite state of the soul, there is attached as an inevitable consequence some definite movement in accordance with an ordered connection of natural effects which is equally withdrawn from our insight and our control. Whilst this movement is taking place, or after it has taken place, we receive from the changed condition of the limbs which it brings about, or in which it consists, sensations of which this changed condition is the cause, and which do indeed reveal to us that which has taken place in us as a consequence of volition, but not the slightest hint of the mode and fashion in which this result has been brought about. That by which our act is made our *act* and distinguished from that of which we are merely the *cause*, does not consist in such an outgoing of the active being beyond the limits of its self that it still remains itself in that foreign object of its energy into which it flows with active efficacy ; *all* acts are *consequences* of volition, inevitable consequences, and not requiring any special impulse to realization, provided the volition itself is once definitely present, and the way in which these consequences arise is precisely similar to that in which arise the consequences of other and non-voluntary mental conditions, or the incidental consequences of volition directed to some other end. The essential characteristic of an act is that it is the consequence of a volition which willed it and nothing else, that it is not the consequence of a feeling, or of an idea, or of any other mental state except volition. The will may be prevented from actually realizing its result; but no one can contribute more towards making the result of that volition his own *act* than a steadfast and undistracted volition; it belongs *to us* only because we will it, and do not by divided willing put hindrances in the way of the mechanism by means of which it follows our volition as a necessary result; but nowhere is there any work of ours through which, by fresh activity on our part, it is either necessary or possible for us to bring about the result of our volition.

For a finite being work is the sum of all those intermediate operations which it has to set in action because its will cannot influence directly the foreign objects which it intends and strives to modify ; but the finite being *feels* itself working to the degree and extent to which the connection of natural processes furnishes it with direct sensations of the consequences of its action; hence the movements of our own body are the only part of the result which seem to us to be our own work—those changes which we aim to produce in the external world do not seem so, because we perceive them only mediately as facts that have taken place, and are not made aware of them as our act by an immediate feeling of effort. But in this meaning of work there can be no work for God, for His will does not find in the alien nature of the objects with which He deals the same barrier as ours does; but for the same reason the self-enjoyment of His own vitality and energy belongs to the Divine Being in boundless measure; for standing in an ungraduated and equally intimate relation to all parts of reality that either already exist or are coming into existence, He will be directly conscious of *every* consequence of His will as being what it is, and it is not conceivable that any event proceeding from God's will should be for Him such an alien development of something external, as the last ramifications of a series of events which we initiate must certainly be for us. Hence we may affirm in conclusion that we do not attribute to God any greater vitality by characterizing His creation as work, for all work, in as far as it is indirect action, belongs only to the finite; the divine will does not work out its result, but *is* that result; we do not impute to Him any greater vitality by describing creation as His act and not as a simple consequence of His will, for such a distinction does not exist, every act being but a consequence of volition; but if we drop all notion of mediating activity or of work, or of action that goes out of itself, and regard as equivalent divine volition and its consummation, then we can imagine the living pervasion of the creature by the Creator and boundless enjoyment by the Deity of His own activity—a self-enjoyment which we finite beings can attain to only by the roundabout path of that obliging psychological illusion to which we referred.

If then we do not regard creation as an act, what is the attitude which we take up towards the view which considers it as an efflux of the divine nature, or in the more definite form which alone can interest us, as an emanation of the divine intelligence? Has it been our intention to agree with the view which regards the imagination of God as having indeed designed and planned the possible content of the universe, but as awaiting the realization of the same from the will which is to summon into existence but one, and that the best, of many possible worlds hovering in the realm of potentiality? On the contrary, we must characterize this splitting up of the divine activity as also erroneous.

And above all things it would be not the will but the insight of God which among many possible worlds should

discern the best; not the choice but the realization of that
which was chosen, would be the work of the will.   But I fear
that for this work he only could specify a special content who
should seek reality in a wholly incomprehensible separation
of the world from God, whether as proceeding out from Him
or being established external to Him.   If we drop this impossible spatial image, how shall we distinguish those divine
thoughts which have been realized from those which hover
unrealized in the divine imagination?   How but after the
same fashion as that in which we distinguish our own ideas
of empty possibilities from perceptions of reality, and unfulfilled projects from efficient motives of our action?   All these
empty possibilities too *are* real—as real as their nature (*i.e.*
the nature of their content) permits; they subsist as our
thoughts, as movements of our soul, and have all the influence
upon us of which their content, and the form of their existence
as our states, makes them capable.   But it appears later that
regarded as motives of our action they would *not* be adequate
causes of **a** desirable result, and hence they do *not* become
efficient motives of our action; or it appears that regarded as
perceptions they are not causes of those results in the phænomenal world which we attributed to them, and hence we come
to regard them as illusions, not because they are nothing
whatever or are non-existent, but because they are without
effect in the system of things external to us.   And it is in the
same way that we distinguish the unrealized from the realized
thoughts of God; not by supposing that many possible worlds
hovered before Him and that His will realized one of them by
an act the content of which must remain altogether incapable
of being specified.   For in being all equally possible they
all possessed reality already, and we could conceive nothing
else by which, as by a reality now starting up for the first
time, the elective will might be induced to prefer any one of
them to the rest.   If we may speak of the subject after the
manner of men, then we would say that what remained
unrealized was clearly seen by God from the beginning in its
resultlessness, in its lack of such consistency as would have

made it possible for it to become the basis of progressive
cosmic order, and in its incapacity of combination with that
which God's will had determined as the content of creation.
In us finite beings there may be permanent illusions and
projects incapable of being carried into effect, to which we yet
continue to cling; for the ends at which our action aims are
presented to us by the course of external circumstances so
that we have only an imperfect view of their advantages;
our knowledge of reality is gained not by direct and penetrating insight into things but by interpretation of subjective
excitation. But it is not so with God; and hence our
thoughts concerning His creative action must set out not from
the equal possibility of that which was uncreated, but from its
impossibility which was originally recognised by Him.

But this expression needs some correction and explanation. Above all we cannot mean that the images of different
worlds were present to and known by God as being *in themselves* possible or impossible in the same way as many combinations of our ideas, which we, being conscious of the laws
of a real world independent of us, regard as being in
themselves impossible, or incapable of being carried out in
that world of reality. For God there was no reality *within
which* He had to realize His creation, nor laws which, prior to
Him, *of themselves* determined what was possible and what was
impossible. But when God thought and willed the thought
of *His* world, He created also in it that logical order in virtue
of which it became possible that there should arise empty
images of other realities as incompatible with that world; the
cause and ground upon which is founded a distinction of the
possible from the impossible and from the real, is subsequent
to the reality of the first real existences. And further, we do
not believe God to have drawn such a distinction between
these two realms of thought—that which was willed and that
which is alien thereto—as to induce Him to realize the
content of the first, and by withholding His realizing activity
to consign the second to the eternal nothingness of empty
thought—of thought which is *mere* thought; it is, we repeat,

simply impossible to say in what the distinction between the two could consist, if we consider this distinction to be established by a divine act, and do not seek its significance in the difference between that which has been and that which has not been realized. Both are thoughts of God; but the thoughts of the non-existent are thoughts which on account of their content—of their own resultlessness, their incoherence and the incapacity of development of their constituents—could neither form worlds, nor enter into connection with those thoughts of existing things which are connected and logically consistent. Thus to the consciousness of God they appear as unconnected with the world which He wills, of active interference with which their own content makes them incapable, and to finite beings they appear as non-existent. For the thought of such beings can indeed produce the empty images of them, but it nowhere discovers a trace of their efficient connection with that order of things which from the standpoint of finite beings, is regarded as reality because it is the thought of God in which they themselves have their place and which influences them with all the fulness of its logical consistency. And thus there arises for finite minds the illusion that this reality (that is the active efficiency of real things that results from their content) is due to an act by which that is realized which is in itself merely possible—an act that must always remain insusceptible of definition.

And now at last we need no longer fear that any one will misunderstand us to such an extent as to suppose that we have wished to represent the world as an emanation of the divine intelligence and not as proceeding from His will. We do not indeed use the expression *product of His will* because we do not wish to call up afresh the already rejected thought of a special act of realization. But yet we say that the world was willed by God, and this expression we have already frequently used provisionally. It is only for the finite being that will is principally an impulse towards change, towards the establishment of something which did not exist; but the real nature of will is only the approval by which the being

that wills attributes to himself that which he wills, whether
it is something that is to be realized in the future, or something that exists in eternal reality. The objects upon which
a finite mind is occupied are brought to it in succession by a
cosmic order which is independent of it; and all the more
on this account does it seek its will in the mobility which
produces what was non-existent, changes or abolishes what
was existent, and demeans itself as independent towards those
occasions of its exercise which it cannot with equal independence bring about. And yet at last even for the human mind
that which is most important in will is to be found, not in this
mobility of the change-producing impulse, but in thé approval
or disapproval with which the whole man wills or does not
will, accepts or rejects, himself. It is such an uniform and
unchanging will that we have regarded as connected with or
eternally based upon the divine thought of the world; we
could not understand it as the mere conclusion of deliberations
carried on by unvolitional divine insight without unduly
assimilating the divine being to the image of a finite mind.
And it would not be impossible to show that intelligence
without will is as inconceivable as will without insight; we
are withheld from setting about the proof here by remember-
ing the extent to which we have already penetrated into a
region where countless misunderstandings may attach to each
of the imperfect expressions which we are obliged to employ
in order to indicate in some way those extreme limits of
human ideation of which we are forced to take account.

§ 4. Conservation and governance in as far as they concern
Nature and the course of Nature have already frequently been
objects of our consideration (cf. *supra*, p. 130, and i. p. 446).
And it is only in so far that they belong to the task which
we have set before ourselves, and we never considered it part
of that task to exhaust the relation of God to the spiritual
universe, to the meaning, end, and destiny of all things. But
a single point in this wide world of thought induces us to
make an addition which is called for by what has gone before.
If the world were but a chain of mutually conditioning

events, if all the future were but a logical development of the past, conservation and governance, without being beset by any peculiar difficulties, would be but various expressions of the divine creative activity. But religious faith finds such a mechanism of the cosmic course neither correspondent to its own need nor worthy of being the divine creation; it assumes that the freedom of finite beings introduces into the cosmic course new beginnings of action, which, having once come into being, proceed according to the universal laws of that course, but have not in the past any compelling cause of their appearance. Thus it is that conservation and governance come to have a work to do. But how does this assumption agree with the unconditionedness and perfection of God, how with His omniscience which that perfection cannot lack, and which could not subsist without foreknowledge of the future? To attain by inference to a knowledge of the future which has its causes in the present is a prescience possible for us in a limited degree and belonging to God to an unlimited extent; but what can be the meaning of saying, as people do, that God foreknows that which is to happen through freewill in the future, not as something that *must* come, but as something that *will* come? If the future does *not exist*, how could this non-existent (unless represented in the present by its causes and thus *not* free) stand in any other relation to cognition than that which never will be, and how therefore could it be distinguished from the latter?

It is certainly a somewhat strange proceeding when we finite beings who are so often reminded of the limits of our knowledge, ask questions concerning the possibility and conditions of omniscience, and expect an answer to our questions. We can foresee that we shall end with a postulate, of which we cannot describe the fulfilment, satisfied if the reflections from which we can start do not make that unknown fulfilment appear as a dream that is altogether ridiculous.

It has been attempted to make the unconstrained freedom of fresh beginnings compatible with omniscience by the assumption that time is but a form of intuition under which the world

appears to us, but in which it does not exist. This inversion
of the ordinary view, however, cannot be so easily carried out
as the similar inversion with regard to space; space could be
given up because without it there still remained to us a com-
plexly organized world of intellectual reality, clearly exemplified in our own inner life; but in order that any event should
appear in time, must we not presuppose that there is an actual
succession of its phases, or at least an actual temporal succession of ideas *in us* by which the merely apparent succession
of these phases would then be determined?

Much may be said in answer to this natural objection
without invalidating it. It is true that empty time in which
events take place, or a current of empty time flowing on of
itself, could be neither a producing nor a determining condition of the course of events. The passing moments could not
bring reality with them, they could not choose what should
last or what should pass away, they could not determine the
place at which each event should enter into their current. It
is only through that position with regard to the whole which
every individual occupies in virtue of its significance, of its
being conditioned by one and having itself conditioning force
with reference to another, that the point of its entrance
into time and the length of its duration in time are determined. Now if this one essential of action—the direction
which it takes and the order into which it falls—lies only in
the conditioning bond of the content itself, as it is taking
place, empty time is just as little capable of producing from
this timeless connection the movement and succession of
action. Any given extent of empty time is exactly the same
at the end as at the beginning of its course; however great or
however small we may imagine it, nothing occurs in consequence of its lapse through which there could be produced a
condition of or necessity for the appearance of any event of
which the cause already subsists, more adequate or more
constraining than that cause and that condition respectively
were at the beginning of this vain expenditure of time. If
the causes which then subsisted were not capable of effecting

the realization of the event in question, then no lapse of time of whatever length would be sufficient to supply this lacking motive force. Such reflections favour the attempt to seek true reality only in the conditioning force which every event exercises upon its own result, regarding time, which appears to our imagination as the unending form in which this order is embraced, as a mere form of conception in which, for us only, is spread out the timeless connection of the cosmic content. And it is possible, up to a certain point, to give clearness even to this unusual mode of thought. In themselves, past, present, and future are not different as far as time is concerned, but simultaneous—if we can allow that this phrase, incorrect in itself, is intelligible; in this whole of reality nothing passes away; but the whole is a whole of members which condition one another, and is comparable to a system of truths of which the simplest condition all the rest, and (to make use here of a natural figure) precede them not in time but in importance; not only does the series of consequences proceed in a straight course from them, but also all the propositions which depend in equal degree upon those principles, appear as co-ordinate, simultaneous, and of equal value. Reality, as we know, is no system of truths, and we must allow for the inadequacy of the comparison; but it is so organized by means of relations of reciprocal conditioning, that each of its parts presupposes immeasurable series of causes, draws after it an equally immeasurable series of results, and finds itself at the same distance as countless other members, from the first causes, or from any given member of the whole. It is this organization which is intuited by cognition in temporal succession; the condition precedes that which it conditions, the latter follows, the causes and results which are most closely connected are in immediate juxtaposition, the more distant results are divided from their immediate cause by a space of time which is filled up by the successive intermediate links which connect it with that cause. And it is not to a cognizing mind, standing without and regarding it as some alien mechanism, that this organization appears thus; all finite beings are themselves members of this series, and to each, in its due place in the series, the assumptions which it involves, as far as it knows them, appear to be past, its consequences, as far as it is certain of them, to be future; and for the rest, the whole of its unknown causes and of its incalculable effects appear as an endless past and an endless future.

This view admits of development so far, and is right as far as regards the order and connection of the events as they take place; but if it denies with cogent arguments the existence of unending empty time, which even according to our natural way of thinking is never held to exist thus, but is regarded as unceasingly passing away and then again coming into being, it cannot by any ingenious torture of thought really avoid that unceasing ebb and flow, the temporal *succession* of events. It is indeed true that it does not fail because the idea of that which we think as earlier must precede in time our idea of that which we think as later; on the contrary, it is only a consciousness that comprehends both in one wholly indivisible act, that is in a position to compare them and to assign them their different places in the apparent extension of time; but even these indivisible acts are repeated and follow one another. Any finite being placed at some particular spot in a timeless system would always necessarily see as its future some one special content whether clear or obscure, and some other as its past; life which makes the former more and more clear and the latter gradually more and more faint, is not conceivable without a real stream of occurrence which carries consciousness past the content of the world, or the content of the world past consciousness, or lets both change together.

But this necessary recognition of the course of time is connected in us with a strong feeling that the recognition cannot contain any final utterance on the subject. We are very ready to declare that what is gone is gone for ever— but are we fully conscious of all that this declaration implies? Is all the wealth of the past wholly non-existent? Is it entirely broken off from all connection with the world, and not in any way whatever preserved as part of it ? And is cosmic history nothing but the infinitely narrow and incessantly changing streak of light which we call *the present,* glimmering between the obscurity of a past which is done with and is no longer anything, and the obscurity of a future which is also nothing ? In expressing these questions thus, I follow that turn of thought which seeks to modify the monstrousness of their content. For these two abysses of obscurity, however empty and formless, are yet supposed to *exist,* and to constitute an environment of which the unknown interior offers a kind of dwelling-place for the non-existent—a place into which it has disappeared or whence it comes. But if one tries to do without even these images and not even to imagine the emptiness which bounds existence in both directions, one will find how impossible it is to do with the naked contrast of existence and non-existence, and how ineradicable is men's desire to be able to regard even the non-existent as being in some wonderful way a constituent of reality. Hence we speak of the distant future and the distant past, this spatial image satisfying the need we feel of not letting aught of that which does not belong to the present escape from the greater whole of reality.

Unable as we are to specify how the lapse of time comes about, and how the condition of any given moment passes from existence into non-existence, in order to make room for the condition of the succeeding moment, we are equally unable to say how on the other hand there comes to pass this comprehension in a contemporary or supratemporal reality, of that which is ever flowing on. But accustomed to find the world more wide and rich than thought which tries to follow its marvellous structure, I entertain no doubt as to the fulfilment of this postulate, of which indeed we can only speak in a limited human fashion. There does not exist for God the condition which binds us to one definite spot in the universe, making it possible to refer to this region of our immediate experience—our present—as past or future every-

thing that is or happens external to us ; God Himself being
not a member of this whole but its all-embracing essence is
as near to any one part of this reality as to any other, and
although there lie open to His all-penetrating knowledge those
inner relations by which this whole would be systematized
into temporal order, yet for Him no particular point has
exclusively the specific worth of *the present ;* for God, this
belongs to the infinite whole.

And finally—to return to the point which gave rise to
these reflections—free actions also find their place in this
timeless reality ; not as non-existent and future, but as
existent.   For although not conditioned by the past, they
would be unmeaning unless they had reference to present
occasions which furnish the ends at which they aim, and
unless they attained reality by producing results.   Hence
their place in this timeless existence is determined not by
members which preceded them as conditions, but by members
which succeed them as conditioned, or are co-ordinate with
them ; hence omniscience need not foresee free action as
something that will be, but can observe it as something real,
which, regarded as a temporal phænomenon, has its place at
some definite point in the future.

§ 5. We have already (cf. i., p. 384) so unreservedly
acknowledged that it is impossible to derive from anything else
the inventive thought from which spring the forms of natural
reality and also (as we may now add) those of the historical
course of the world, that we need not now venture on any
fresh attempt in this direction.   But as far as we are concerned, one of the motives which generally urge men to such
efforts has become inefficacious.   We no longer hold that a
realm of eternal truths, of formal necessities, of abstract
outlines of all later reality, is absolutely prior to all else in
the Divine Being in such a way that the rich and varied
forms of reality when compared to it must appear as something wholly new, as some spontaneous action which, showing
itself under forms that we cannot calculate beforehand, submits
to this alien Being.   The eternal truths are for us only the

modes in which creation itself proceeds; they subsist not before it but after it as laws to which the products of creative activity appear subject. And at this point we must go back to a more exact determination which in what precedes we let drop for the sake of clearness. Properly speaking, neither a law nor the sum of eternal truths can be accepted as the direct mode of procedure of any power; for truths and laws determine only the reciprocal behaviour of the various manifestations of any force, but they do not give the very content that can be broken up into these various manifestations. Hence if we cannot derive from universal necessary truths the reason why this particular reality and no other subsists, we have also to remark that it is no longer any part of our task to make such an attempt—that direction of the eternal power which led to the existing world of forms *is* the original first and only reality, and whilst it acts or when it has acted it appears to thought (which itself is included in it as its product) from the double point of view of living creation in a definite direction, and of an activity which in its procedure follows universal laws; and it is then that for the first time occasion is given to thought to dream of other directions of that creative activity which do *not* exist, and the possibility of merely thinking which depends upon the reality of the direction which *does* exist and of the inner order of the creative force that works in it.

But this consideration does not furnish us with a complete conclusion. Even upon the assumption that we are only concerned with a natural cosmic order, and are not called upon to give any account of worth and goodness, it would still only be satisfactory if it could be shown how from the content which the creative force strives to realize, the sum of the eternal truths results, as an abstraction which separates that which is universal in the self-evident procedure of the force that produces all the parts of this very content itself. There is no hope of any such achievement. Stress may with justice be laid upon a difficulty which would make it impossible for *us,* even if the connection to be pointed out did actually subsist—it is only a very small part of reality, only Nature as terrestrial that we know ; we do not know the forms of existence and action which subsist elsewhere, or the connection between these and our own sphere of experience; hence we are quite unable to comprehend the tendency of creative force, the inventive and formative thought which it obeys, in *one* notion which should characterize it completely, exhaustively, and impartially ; hence it is impossible for us, from the fragmentary view of the connection and meaning of Nature, which is all that has been accorded to us, to deduce the universal laws of its procedure as they might be gathered from the complete content of the creative Idea as an abstract expression of its action, by any one who knew that content. I do not doubt that such an all-embracing knowledge of Nature as a whole would oblige us to give up a number of our ordinary points of view, would cause many perplexities to disappear, and would wholly transform many difficulties; but I need not undertake the perplexing inquiry whether it really would do what is expected of it, and make possible the achievement referred to, for I have a conviction (which I trust the reader shares) that just this boundless insight into Nature would show the invalidity of the assumption with which we set out ; it would appear that there is not working in the world any bare formative force, but that the inventive thought which determines cosmic forms is indissolubly connected with the realm of Worth and Good. The lesser question, How are the universal laws connected with the formative thought ? is absorbed into the more important one, In what connection do both stand to that which has eternal worth ?

Religious faith is accustomed to consider some supreme good as the guiding end, free creative divine imagination as the means by which the end is realized, eternal truth as the law according to which this imagination and its products work. Now if we beheld in the world unequivocal and thoroughgoing harmony between these three principles, the attempt to combine them might be regarded as practicable.

Creative imagination indeed could never properly be derived from the Supreme Good ; for no end, regarded abstractly and in isolation, determines more than certain general requirements which seem capable of being fulfilled by various means; and just as little could laws be derived from the direction taken by that imagination. But it might perhaps be shown that just as the power is not conceivable in itself but only as acting in some definite direction, so also Good, thought in its universality, is but an abstraction from some definite existing good, which would not be opposed to the coming reality as a formless end, the mode of carrying it out being as yet undetermined ; but would be directly identical with that which we called the direction of the creative imagination. And then there would be only *one* thing : only the one real power appearing to us under a threefold image of an end to be realized—namely, first some definite and desired Good, then on account of the definiteness of this, a formed and developing Reality, and finally in this activity an unvarying reign of Law.

Before giving to this view, which is a confession of my philosophic faith, the furthest elucidation which I am able to give, I would lay stress upon the decisive and altogether insurmountable difficulty which stands in the way of its being carried out scientifically—that is, upon the existence of *evil* and of *sin* in Nature and in History. It would be quite useless to analyse the various attempts that have been made to solve this problem. No one has here found the thought which would save us from our difficulty, and I too know it not. It may be said that evil appears only in particulars, and that when we take a comprehensive view of the great whole it disappears ; but of what use is a consolation the power of which depends upon the arrangement of clauses in a sentence ? For what becomes of our consolation if we convert the sentence which contains it thus—The world is indeed harmonious as a whole, but if we look nearer it is full of misery ?—He who justifies evil as a means of divine education, ignores the suffering of the inferior animals and

all the incomprehensible stunting of the life of Mind which
we see in history, and limits the omnipotence of God; for
evil is only used as a means of education because there is no
other means. And finally, we are not satisfied even if this
limitation is admitted not secretly but openly as by Leibnitz,
who in every case of irreconcilable difference between the
omnipotence of God and His goodness, believed himself bound
to decide for the latter, and to explain evil by reference to the
limits imposed by the primeval necessity of the eternal truths
even upon the free creative activity of God. For of all imaginable assertions the most indemonstrable is that the evil
of the world is due to the validity of eternal truth ; on the
contrary, to any unprejudiced view of Nature it appears to
depend upon the definite arrangements of reality, beside which
other arrangements are thinkable, also based upon the same
eternal truth. If there were retained the separation (which,
however, we do not admit) between necessary laws and the
creative activity of God, in our view evil would undoubtedly
belong not to that which must be, but to that which is freely
created. Let us therefore alter a little the canon of Leibnitz,
and say that where there appears to be an irreconcilable
contradiction between the omnipotence and the goodness of
God, there our finite wisdom has come to the end of its tether,
and that we do not understand the solution which yet we
believe in.

§ 6. I was no doubt wrong when I first offered these
reflections to the courtesy of my readers, in passing over this
gap in our philosophic theory, which cannot be filled, with
words which though they seemed to me emphatic enough were
yet but brief. What moves me to the following remarks
is not the hope of now filling up the gap, but the wish that
no doubt should remain as to the meaning and end of all
the reflections in which the reader has hitherto been so
obliging as to accompany me. I have never cherished an
assurance that speculation possesses secret means of going
back to the beginning of all reality, of looking on at its
genesis and growth, and of determining beforehand the

necessary direction of its movement; it seems to me that philosophy is the endeavour of the human mind, after this wonderful world has come into existence and we in it, to work its way back in thought and bring the facts of outer and of inner experience into connection, as far as our present position in the world allows. I acknowledge my steadfast adherence to the old-fashioned conviction, that not only is our scientific knowledge but fragmentary, but that also there are ways to lead us to fuller light which are as yet hidden from us ; the task of our philosophy is not vast and cosmic but modest and terrestrial—it has to construct the image of the world as projected on the plane surface of our mundane existence. I might work out this simile, and appeal to the fuller dimensions of true reality in which may be reconciled supreme goodness and the existence of evil, which in our view must always conflict ; but all that I should accomplish with such a juggle of words would be to veil the admission which we must frankly make, that we cannot even imagine the direction in which the unknown conciliation of the difference is to be sought.

If I still hold fast my confidence in the existence of a solution which we do not know, what I wish to give expression to is not a didactic affirmation to be bolstered up by some kind of speculative support, but only the watchword of a struggle in which I desire that my readers should participate —a struggle against the confidence of views which impoverish faith without enriching knowledge. But to regard the course of the world as the development of some blind force which works on according to universal laws, devoid of insight and freedom, devoid of interest in good and evil—are we to consider this unjustifiable generalization of a conception valid in its own sphere, as the higher truth ? Is it not rather the unsatisfying conclusion to which weary thought may come back at any moment, if it gives up its unattainable but not the less certain goal ? But as to all that is good and beautiful and holy—will the arising of this light out of the darkness of blind development be really more intelligible than is, for us, the shadow of evil in the world which we believe to be cast by that light ?

§ 7. If we go back to the facts which cause us to form the notions of *Good* and *good things*, we find that our conscience approves and enjoins definite kinds of disposition and volition — what are thus approved we call good ; we find further that certain objects and their impressions upon us are felt by us to be helpful and agreeable—as being thus helpful they are called good when they correspond to some permanent and general need of our nature, but useful in as far as they are conducive to some isolated end, the importance of which as regards our whole destiny is left undetermined. Conscience and feeling by their indemonstrable but irreversible declarations directly assign these values to very various objects ; but on the other hand, the similarity of the declarations urges us to seek in these various objects similarity of the grounds upon which those declarations have been made concerning them. This path of abstraction leads us to find by comparison of individuals the invariable condition in virtue of which any content is good, useful, or beautiful. But the end of this path may be conceived in two different ways. Either there appears as such a condition only a universal formal relation which has reality not in this universality, but only in any one of the individual forms from which it was abstracted ; or a hope is entertained of reaching some universal which actually exists in such universality, and in fact *is* that which it indicates as a quality in the individual real thing.

With regard to what is useful and what is agreeable, we all think that we are in the first case, and the scientific instinct of our time does not, like that of antiquity, seek what is useful *in itself* or what is agreeable *in itself*. We are content if we can find general notions of both, which are not in themselves that which they denote in other things. For as in general no notion is that which it connotes—as the notion of red is not red, and the notion of sweet is not sweet —so also the contents of the general notions of what is useful and of what is agreeable are not themselves agreeable and useful, but they are conditions under which agreeableness or usefulness appear as predicates *of something else*, that is of the individual real thing which fulfils these conditions. On the other hand, Beauty-in-itself and Good-in-itself are still goals and even starting-points of manifold speculations. In these two cases men seek to find in the universal not only conditions under which something other than the universal itself, something fulfilling the conditions, is beautiful or good, but also seek to find something which is *in itself* the goodness or beauty which we originally know only as a quality in the individual. I leave beauty to the reflection of the reader, and only pursue the question whether and how the peculiar nature of what is good makes it possible to carry out in its case the task which is not, in all cases, practicable.

Actions are not good simply as events that occur, nor their results simply as facts that have been established—it is only the will from which the actions proceed that is good. And the will itself is regarded as good not as a mere impulse to execution, but as the outflow of a frame of mind which is not simply knowledge of a command but also agreement with it, and this agreement is not—like the obedience of any natural force to the law which it follows—a mere factual agreement, but is a case of compliance where non-compliance was possible. And it must be not simply a possibility of disobedience which is perceived, but the disobedience, by its own worth, which it opposes to the worth of the command, must withstand the tendency of the will to compliance. But worth can exist only for a sensitive subject; whatever may proceed from an intelligence that feels neither pain nor pleasure and from a will guided thereby, no moral judgment could be passed upon it. And finally, we should not even call good the frame of mind of him who, by a choice involving no sacrifice, should simply prefer the worth which is greater, both objectively and to him, to the worth which is lesser; on the contrary, that which for the feeling mind is the nearest and most urgent worth, must be sacrificed to some other worth, which to it, as feeling, is not greater—the welfare of self must be sacrificed to the content of a recognised command.

From this point our path diverges from that of the popular view with which hitherto in this hasty recapitulation of familiar points of view it has coincided. For as we long since acknowledged, we do not agree with those who seek this higher worth in an Idea of the Good which requires men to strive after some formal relation of wills to one another, or the realization of some particular condition of things as a directly binding duty or as the Supreme Good. No relation however profound between conditions and events which merely occur without their harmony being enjoyed by any one, is a good in itself, and no will is good because, being conscious of the complete unfruitfulness of such relations, it yet devotes itself to establishing them. If any heart postpones its own good to some other good, this other can only be found in the happiness of some one else, and the sacrifice is good only because it is made on this account. *Good* and *good things* do not exist as such independent of the feeling, willing, and knowing mind; they have reality only as living movements of such a mind. What is good in itself is some felt bliss; what we call good things are means to this good but are not themselves this good until they have been transformed into enjoyment; the only thing that is really good is that Living Love that wills the blessedness of others. And it is just *this* that is the *Good-in-itself* for which we are seeking; this, having reality as a movement of the whole living mind which feels, wills, and knows itself, is just on that account not merely a formal general condition the fulfilment of which by any other thing would entitle that other to the appellation of good, without the condition itself being good; but this it is which alone in the true sense has or *is* this worth, and all else—resolves, sentiments, actions, and special directions of the will—all these share with it only derivatively the one name of good. We finite beings, included in a world the plan of which is not revealed to us, cannot allow benevolent love to act unregulated in the hope that however it may be directed by our defective foresight, it will lead to the good at which it aims; our conscience holds up before us in a number of moral commands the general laws under the guidance of which our action, however variously caused, is sure of taking the right path—but there is not set before the Divine Being in like manner a Good-in-itself that takes the form of a command valid even for Him. No kind of unsubstantial unrealized and yet eternally valid necessity, neither a realm of truth nor a realm of worth is prior as the initial reality; but that reality which is Living Love unfolds itself in one movement, which for finite cognition appears in the three aspects of the good which is its end, the constructive impulse by which this is realized, and the conformity to law with which this impulse keeps in the path that leads towards its end.

In returning for the last time to this thought, which, from the beginning of these concluding considerations, has been hovering before us, I would recall the confession of its scientific impracticability made at the commencement of this Book (supra, p. 572 seq.). This limitation of our capacity has in a general way been confirmed by numerous attempts, which we cannot but respect, and which in individual cases have borne much fair fruit, in clearing up and establishing our vague convictions. Christian ethics would be likely to succeed best in exhibiting particular moral Ideas as the various forms which active Love must prescribe to itself. It would be able to show that all the sterner and apparently more exalted forms of morality which distinguished the heathen heroism that "scorned delights," are yet nothing compared to the gentleness of Love, and nothing unless they have their root in it; that all the commands which, in a scientific point of view, particularly attract our attention, by the definiteness of their content and the ease with which they may be drawn out into a series of sharply defined maxims, are nothing more than a mechanism devised for its own development by the principle of Love, which seems comparatively formless and, as it were, merely potential. On the other hand, the attempts

to explain existing reality from the same principle will always
be far less convincing. In the first place, not one of them, in
trying to express its meaning, has in its description of the
tasks and the needs of that Everlasting Love which is regarded
as the source of the universe, been able to avoid such an
extended use of analogies drawn from the life of the human
soul, as must necessarily displease scientific instinct. We cannot
otherwise than unwillingly see the core of our conviction, of
which in its simplicity we are sure, developed into a system, if
this has to illustrate the origin of things by ideas the meaning
of which only becomes clear through references to connections occurring much later in the course of the world which
we are explaining, and to reduce the figurative expression of
which to its real significance (which in this case is admissible)
would be an almost interminable task. This general insecurity
is intensified by the frequent endeavour to immediately derive
particular forms of reality from particular impulses which are
supposed to be discovered in the nature of the Supreme Principle. Whatever the world may be in which Creative Love
manifests itself, that world is undoubtedly devised as a whole
by that Love; from the whole of the ideal picture which
Creative Love sets before itself, Nature and History as wholes
have their task as a whole assigned to them, and carry it out
by means of a connected system adapted to its realization.
The labour of deduction would have to be directed in the first
place to developing the existence of an universal mechanism
in the procedure of all things from the notion of Supreme Love,
and then to developing from the total content of that which
this Love designs, that definite form of the mechanism which
is adequate to the production of all reality, with steady order
and unvarying fidelity. The fulfilment of this task, as we
have already noticed, can hardly be carried out in the form
of an unbroken deduction, starting from the principle itself—
it will be possible only in more modest measure, as an explanation, by reference to the principle, of actually existing facts.
For we do not possess either of Nature or of History such
complete knowledge as would enable us to guess the whole of

the divine plan of the universe ; the attempts that have been made to determine this from meagre earthly experience betray only too plainly the unfavourable nature of our standpoint, which, with all the one-sidedness of its limited outlook, wishes to be taken for that topmost summit, from which the whole world may plainly be seen spread out below.   This lack of a commanding view is the reason why those attempts so often err in estimating the reality of their particular objects ; they present as the immediate ends of the creative Idea that which even an empirical knowledge of things regards as only a very incidental consequence of general laws, and thus they fall into permanent disagreement with physical science, which in its own less lofty region, rules with an incomparably superior exercise of exact knowledge.

But it is not only the different moral Ideas and the forms of reality that would have to be explained from the same source of Eternal Love—the eternal truths also, the sum of that which, as it seems to us, we must necessarily think, and which could not be otherwise, must be similarly explained. If the scientific solution of this task appeared to me possible, I would employ all my powers in trying to carry it out; for only thus could I furnish a complete justification of my belief that the sphere of mechanism is unbounded, but its significance everywhere subordinate.   I should have to show that the fact that truth exists at all cannot be understood by itself, and is only comprehensible in a world of which the whole nature depends upon the principle of Good that we learnt to know in Living Love itself ; and no less should I have to point out specially how it is but of the nature of this Love, and, as it were, its primary work, to establish an universal order and regularity, within which various individuals, comparable in kind, could be brought into a connection of reciprocal action.   If this eternal sacredness and supreme worth of Love were not at the foundation of the world, and if in such a case there could be a world of which we could think and speak, this world, it seems to me, would, whatever it were, be left without truth and order.   I should further have to call to remembrance that the strongest pillar of all truth, the Law of Identity, of which we were conscious as a sheet-anchor amid the complications of those contradictory phænomena of reality which we have just been considering, might easily appear to us as a truth that prevails of its own power and uncaused; but that even its own content is but the formal reflection of that significant trueness to itself, the immediate connection of which with the supreme worth of Goodness we are again strongly conscious of, when we assume the eternal identity of God with Himself not merely as a logical perfection of the notion of God, but also as an ethical perfection of His nature. I should then have to show what is meant by saying that there is something which we call adequate cause, and causal connection; however impossible it may seem to us that either of these should have been other than absolutely primary, we are yet just as directly conscious that a world would be unmeaning in which one thing should be established or produced by another merely in order that things should be or should happen after this fashion. If the natures of things are such that two can join in any way so as to become the adequate cause of some third, this marvel is to me intelligible only in a world in which what is aimed at is not mere occurrence of some kind, but deeds that are to have results, and the freedom of which presupposes an universal reign of law as well as fruitfulness in the production of new results in the world of things, results which furnish this freedom with aims and objects of its endeavour. From this consideration of the metaphysical principles of all our cognition, we should have to go on to mathematical truths and their validity in the world of reality. We would not indeed commit the solecism of trying to deduce mathematical propositions from other principles than the fundamental ideas of mathematics itself; but of those fundamental ideas — the ideas of magnitude, recurrence, equality, unity, plurality, addibility, divisibility— we should have to show that the fact of their thinkableness is not a bare and uncaused fact, but an essential presupposition of that order which the Good as Supreme Principle imposes on the world, and which another principle (to express the empty thought for clearness' sake) would not have imposed upon it in a similar fashion. We should have to refer to the dominion of mathematical truth over reality, and to show as regards it that it is only in a period of as yet imperfect elaboration of mechanical science that the regularity of Nature seems to be of an unique kind, recognisable only by means of the magical rules of an arithmetic abounding in formulæ, and not capable of being reduced to simple ideas; the further mechanics progresses, the more do we see its most general results revert to the form of propositions, the easily understood sense of which (pointing out everywhere what is most simple and rational as the law of action) may be expressed in notions, and needs a mathematical dress only in order that the signification of these notions may be made susceptible of those precise determinations of magnitude which they require in application to the concrete. And so the time may come in which these simplified propositions of all mechanics will become more directly connected with the Supreme Principle, and will admit of being interpreted as the last formal offshoots of that Good which is the beginning and the end of the whole universe.

Much might yet be said upon this subject; but I will not part from the reader with a profession of holding back some important knowledge concerning these questions. On the contrary, any further development that we might seek to give to these thoughts would not satisfy us, but in its inevitable incompleteness would be open to the reproach of being mere sentimental trifling. I participate fully in the scientific instinct whence this reproach would spring; and since everywhere in these discussions I have contented myself with an explanation of those intelligible principles which may be of use in the examination of our doubts, and on the other hand have never entered upon those vast regions which hitherto have been filled only by the vague imaginings of poetic fancy, it may here be sufficient to express once more my faith in a goal from attaining which we are held back by a chasm which it seems impossible to fill up.

§ 8. It is but seldom that after a long journey we have the satisfaction of being able to say to ourselves that we have not passed by any eminence which promised a good outlook, and have examined all the best points of view, and that we have never, through lingering in any one spot longer than was fitting, on account of some insignificant attraction, neglected to seek out any more important prospect obtainable from a neighbouring point. And still less shall we succeed in grouping together the manifold moods and thoughts which arose in us by the way, into one simple memory-picture without giving up much which in the brightness of its living individuality attracted and enchained us. Such self-reproaches and such difficulty do I feel in parting from a work of which I desire to express yet once more the essential meaning, unburdened by the special explanations which I have undertaken in it. It would be vain to attempt this in any other way than by emphasizing once again the scientific attitude which has guided and been at the foundation of the whole— on the one hand a struggle against veneration of mere empty forms, and over-estimation of what is but presupposition or result, means or mode of manifestation, of that which is truly worthy and living and real; and connected with this the struggle against all fanaticism which would like to see the Supreme Good active in some other way than that which it has itself chosen, or which believes that Good to be attainable by some shorter path than the roundabout way of formal orderliness which it has itself entered upon.

From this attitude arose our respect for the scientific worth of mechanical investigation in Nature and History, and from it likewise our obstinate refusal to see in all mechanism anything more than that form of procedure—susceptible of isolation in thought—which is given by the Highest Reality to the living development of its content, which content can never be exhaustively expressed by this form alone. And this struggle has been not only against materialistic views, but also and equally against that Idealism which imagines itself to be fighting against them for the right. It seemed to us wholly indifferent whether the most essential core of reality from which all else is to proceed as a matter-of-course accessory should be sought in soulless atoms, blind forces, and mathematical laws of action, or in necessary notions of any kind, in relative or absolute Ideas, and the jugglery of their dialectic movements. All these views uniformly degrade Nature and History by making them representations of something absolutely indifferent and worthless, the presence of which in the world of thought is only comprehensible when it is thought as the final formal reflection of the living mind and its living activity.

And as in knowledge so it seemed to us in life also to be the sum and substance of wisdom neither to neglect what is small nor to give it out as great; to be enthusiastic only for that which is great, but to be faithful even in the least. We agreed neither with endeavours to arrange human relations in accordance with ingenious suggestions, without regard for the universal mental mechanism by which Right is realized, nor with schemes which having stiffened into rigidity in the service of this mechanism can further nothing but the establishment of orderly conditions. It seemed to us that everywhere the universal was inferior as compared with the particular, the class as compared with the individual, any state of things insignificant as compared with the good arising from its enjoyment. For the universal, the class, and the state of things, belong to the mechanism into which the Supreme articulates itself; the true reality that is and ought to be, is not matter and is still less Idea, but is the living personal Spirit of God and the world of personal spirits which He has created. They only are the place in which Good and good things exist; to them alone does there appear an extended material world, by the forms and movements of which the thought of the cosmic whole makes itself intelligible through intuition to every finite mind.

It may be thought that our conclusion is fanatically enthusiastic; still we would repeat here an avowal that we have made before—the avowal that when we view the world as a whole we see everywhere wonders and poetry, that it is only limited and one-sided apprehension of particular departments of the finite that are prose. But to this we would add that it is the business of men not to take the name of these wonders and this poetry in vain, and to revel in continual contemplation of them, but above all things to cultivate that more modest realm of scientific knowledge which is able not indeed to lead us into the promised land, but to keep us from wandering too far out of the road that leads to it.

INDEX.

THE END.